VOLCANIC HAZARDS

To the memory of my father, 1910–1984

I have thought it worth while, in view of the exaggerated, absurd, and hysterical statements that have appeared in many papers, to make a sober estimate, as close as the facts and figures at my disposal permit, of the actual damage caused by the eruption. A certain amount of excuse must be made for those who have lived through what must have seemed a cataclysm, with its dangers of unknown extent, the darkness, the contagious terror and natural discomfort, but viewed in the cold light of day the area affected is really small, the permanent damage [limited]…, and the number requiring organized help no more than such a rich neighbourhood, assisted by the generous contributions already poured in from all directions, can easily afford to a class of people so frugal and easily satisfied in the matter of the necessities of life.

DELMÉ-RADCLIFFE, 1906: 5.

VOLCANIC HAZARDS

A Sourcebook on the Effects of Eruptions

R. J. BLONG
MACQUARIE UNIVERSITY,
SYDNEY

ACADEMIC PRESS

A Subsidiary of Harcourt Brace Jovanovich, Publishers

Sydney Orlando San Diego San Francisco New York
London Toronto Montreal Tokyo São Paulo

ACADEMIC PRESS AUSTRALIA
Centrecourt, 25–27 Paul Street North
North Ryde, N.S.W. 2113

United States Edition published by
ACADEMIC PRESS INC.
Orlando, Florida 32887

United Kingdom Edition published by
ACADEMIC PRESS, INC. (LONDON) LTD.
24/28 Oval Road, London NW1 7DX

Printed in Australia

National Library of Australia Cataloguing-in-Publication Data

---

Blong, R. J. (Russell J.).
Volcanic Hazards.

Includes index.
ISBN 0 12 107180 4.

1. Volcanoes. 2. Disasters — Physiological aspects. 3. Disasters — Sociological aspects. 4. Disasters — Environmental aspects. I. Title.

363.3'495

Library of Congress Catalog Card Number: 83-73405

CONTENTS

PREFACE

On September 12, 1977, tremors heralded the beginning of an eruption in the East Rift Zone of Kilauea on the Big Island of Hawaii. Lava began to flow on September 13. Because Kilauean rift eruptions are nonexplosive and lava flows are localised down narrow valleys, such eruptions are easy to monitor and available experts can be concentrated in quite small sectors. The eruption, terrain and land use characteristics allow time, for consensus decision-making in the Emergency Operations Centre (EOC), and for sequential meetings and orderly mass evacuation. Thus, residents of Kalapana were warned on September 13 and an evacuation center was opened on September 14.

This 1977 eruption in the Kilauea East Rift Zone was the sixteenth in 20 years. Because eruptions are frequent on the Big Island, a highly developed monitoring, warning, decision-making and evacuation system has been set up with experienced scientists and public officials working as a team in a command network with clearly defined roles. Within the EOC, only one or two officials from the Hawaiian Volcano Observatory are permitted to make press statements. This formalised and restricted contact limits media pressure and allows the EOC to function with little disturbance.

Because there are few roads in this Puna district of Hawaii, and because the terrain and lush tropical vegetation inhibit access to the rift area, it is relatively easy for officials to police roadblocks and to control evacuation. Furthermore, in 1977 a card system for evacuation existed detailing the number of occupants of each threatened dwelling, the volume of belongings requiring transport, the number of vehicles per dwelling and the presence of any medical problems amongst the potential evacuees.

The population of the Puna district is largely semi-rural, and relatively small, with a tightly knit social structure and numerous community links. The residents are aware that the lava flows move rather slowly and that warning times are adequate. The residents know about Kilauea eruptions because they have experienced them before. They know that past damage by lava flows has generally been relatively minor and that the Emergency Operations Center is in command of the situation (Sorenson and Gersmehl, 1980).

As it happens the 1977 Kilauea eruption stopped and started several times. The 24 hour alert was relaxed on September 22, but a new alert was issued on September 28. On September 29, it was predicted that the lava flow would reach Kalapana and 97 people were evacuated. Four days later the lava flow stopped and the evacuees returned to their homes. On October 5 human activities were back to normal. Two days later the eruption officially ended (Sorenson and Gersmehl, 1980).

The 1980 eruption of Mount St Helens in Washington State, U.S.A., provides a number of interesting contrasts with the 1977 Kilauea eruption.

Seismic activity around Mount St Helens was first noted in late March, 1980. Minor eruptions of steam and ash occurred frequently during April and early May accompanied by the ominous deformation of part of the north face of the volcano. Volcanologists from Hawaii and elsewhere hurried to the scene. Despite their expertise and equipment the exact nature of the threat was not determined. Mount St Helens had last erupted in a very minor way in 1921. Although there was no local experience to rely on, the character and distribution of deposits from earlier eruptions had been mapped and provided a basis for delimiting hazardous areas (Crandell and Mullineaux, 1978).

On May 18, 1980, Mount St Helens erupted sending a massive avalanche down the north face and a cloud of volcanic ash high into the atmosphere. There were no short-term warning signs, the eruption had a very rapid onset and it was highly explosive. An avalanche buried the adjacent landscape by as much as 200 m and an accompanying lateral blast destroyed an area of more than 400 km^2 within a few minutes. In the next 24 hours ash fell across an area of more than 80 000 km^2.

Although this eruption was not particularly large when viewed on a world scale, it was unprecedented in the 150 years of European occupation of the area around the Cascade Range. The monitoring system used was hastily assembled and the volcanologists, for the most part, had only limited local knowledge; they had little baseline data to compare with the early 1980 activity.

Furthermore, there was no formal command network in the Emergency Operations Center set up in Vancouver, Washington. The Washington State Emergency Services director had been appointed only a few months previously. There were limited funds, and knowledge of the hazard was sketchy. Numerous "experts" made statements to the media. The Emergency Operations Center was a makeshift affair. In fact, it was four days after the May 18 eruption before United States Forest Service personnel were seen to be in command of rescue operations. The death toll in the blast area north and west of the volcano was the cause of strong media interest and, before long, media criticism.

The extensive network of forest roads in the area made policing of roadblocks almost impossible, especially when maps became available before May 18 specifically indicating routes around roadblocks. Errors in the delimitation of the potential hazard zone, the minimal warning time once the major eruption began, the absence of precedents, and the total inexperience of residents and visitors with volcanic eruptions all contributed to the death toll.

To the north-east of the volcano, tephra (volcanic ash) was dumped over a wide area. In the aftermath numerous problems emerged such as would breathing tephra cause silicosis, and what would be the short and long term effects on crops, livestock, business activity, and revegetation? In the absence of prepared information from the Emergency Operations Center, the State Emergency Services, or the Federal Emergency Management Agency, media comments were provided by officials and others, with and without expertise.

In the period following the May 18 eruption several smaller eruptions occurred showering Portland and other localities with thin layers of tephra. Dome growth, sealing the vent, occurred on several occasions to be destroyed each time by renewed activity. At the time of writing (December, 1983) it seems unlikely that eruptions will end soon.

From the foregoing, it is clear that only part of the contrasts in the effects of these two eruptions can be attributed to differences in eruption magnitudes and styles. Certainly, the effusive spread of lava at Kilauea owes much to its origin in hot spot volcanism as the Pacific plate moves northwestward across a mantle plume. This style of volcanism is characterised by the relatively frequent, nonexplosive eruption of nonviscous basaltic lavas which advance slowly and fairly predictably down topographic lows. On the other hand, the violent and explosive eruption of Mount St Helens with the generation of large pyroclastic flows and widespread tephra fall is typical of the andesitic (and dacitic) volcanism of areas of convergent plate tectonics such as the Cascade Range. Clearly, the differences in eruption style result in large part from differences in location relative to the shifting surface of the earth.

However, the contrasts in effects also stemmed from differences in the preparedness and organisation of the public officials. On the one hand, the EOC was tightly run; on the other, efficiency came more slowly. On Hawaii, each official performed specific rehearsed functions, the high level of co-ordination resulting from experience. In Washington State, roles had to be learned once they had been belatedly defined.

Similarly, contrasts in the behaviour of the citizens stemmed from contrasts in their experience of volcanic eruptions and, to a lesser extent, from their views on the value of official pronouncements and actions.

To the geologist, seismologist or volcanologist the study of volcanic hazards is mainly concerned with volcano monitoring and hazard delineation. Through the use of a variety of instruments the scientist takes the pulse of the volcano and attempts to predict future behavior. Such studies are also based on the volcano's past behavior; by mapping and dating the variety of deposits around a volcano the geologist can elucidate volcanic history and delineate with some certainty the areas of greatest hazard. Attempts can then be made to mitigate the hazard, usually by prediction, warning and evacuation of the populace and/or by relocating or engineering structures so that they are resistant to hazard impact.

The social scientist generally takes a broader view of volcanic hazard. Volcanic predictions and hazard delineation are but elements of the geophysical subsystem which interact with both citizen and public official subsystems. Heathcote (1979: 3), for example, defines natural hazards as "extreme geophysical events greatly exceeding normal human expectations in terms of their magnitude or frequency, and causing significant material damage to man and his works with possible loss of life. As such they [are] recognised to occur as an interaction between systems of human resource management and systems of geophysical events".

While we can recognise enormous contrasts between the Kilauea and Mount St Helens, eruptions much greater contrasts are possible. Certainly, the elements comprising the two geophysical subsystems are towards opposite ends of the spectrum but the pubic officials and their regulations, the citizens and their beliefs, attainments and aspirations are drawn from the same cultural milieu. Contrasts might have been greater if we had chosen for one example a volcano in a developing country, with low per capita income and a low level of technological skill in the community, and in an area remote from a poorly organised and corrupt or apathetic central bureaucracy isolated by poor communications from a population inexperienced in volcano lore.

Even with consideration of just three interacting subsystems an almost infinite number of hazard scenarios can be constructed. Where third world examples are chosen, international disaster relief – another subsystem – must often be grafted into the analysis of the hazard and its aftermath.

In these terms volcanic hazards influence diverse aspects of human existence with ramifications far beyond the immediate hazard impact zone. Religion, literature, politics, public safety, and science are but a few of the aspects of human existence affected by erupting volcanoes.

For example: 12th century Cistercian monks in Iceland spread rumours that the volcano Hekla was the gateway to hell, or even Hell itself, in order to threaten sinners. During the 1960 Kapoho eruption on the island of Hawaii some inhabitants read Hawaiian language chants to the goddess Pele and made traditional offerings of breadfruit, bananas, pork and tobacco, the gifts wrapped in Christmas paper.

Lord Lytton's book, *The last days of Pompeii* (1834), was closely modelled on and inspired by excavations at Pompeii. The opening lines of Alfred Lord Tennyson's *St Telemachus* were inspired by the brilliant sunsets over Europe resulting from the 1883 eruption of Krakatau.

The coats of arms and other insignia of several Central American republics include volcanoes. The argument that Nicaragua's erupting volcanoes would adversely affect the canal helped sway the United States Senate in favour of the Panama route. The continued eruption of Irazú, in 1964-65 caused Costa Rica to appoint a Cabinet Minister of the Volcano.

Signs of impending eruption of the Soufrière, St Vincent, in 1979 led to the evacuation of thousands of people from the northern end of the island. Lifestyles were disrupted for up to two months, yet the deposits produced by the eruption were reworked and removed by natural processes so that they all but vanished within a few days leaving no permanent stratigraphic record, except on the upper slopes.

Such diverse effects are part of the fabric of volcanic hazards. Thus, this book is not just about volcanic eruptions. It is concerned with the effects of eruptions on people, their possessions, their lifestyles and their environments. This volume, therefore: (i) provides background data on volcanic activity (Chapter 1) with attention directed specifically at those types of activity and those characteristics which are hazardous (Chapter 2); (ii) establishes the direct effects of volcanic eruptions on humans in terms of death and injuries (Chapter 3) and social aspects such as perception of eruption hazards, evacuation, panic, looting, and religious beliefs Chapter 4); (iii) establishes the indirect consequences of volcanic eruptions for humans by illustrating the effects on buildings (Chapter 5), utilities, communication networks and machinery (Chapter 6), agriculture (Chapter 7) and commercial activity (Chapter 8).

There are three specific aims involved. The first is to present a sourcebook on the effects of volcanic eruptions. For the most part, this book is based on actual examples of what happened (or, what was reported as happening!) before, during and after various eruptions. While some sections of the book provide an inventory of all known examples of particular effects, limitations of space have elsewhere prevented such an all-inclusive approach. In each case, an effort has been made to illustrate the diversity of effects produced by eruptions.

The second aim is to draw conclusions from the available data, in so far as it is possible to interpret or analyse them. This has not proved an easy task as the data are of variable quality and the effects are the result of many influences, as the Kilauea and Mount St Helens examples suggest. Where appropriate, the effects of volcanic eruptions are compared with the effects of other natural hazard impacts.

The third aim is to offer pragmatic suggestions, generally by the use of examples, that may reduce vulnerability to volcanic hazard impacts and/or the aftermath of such impacts.

This volume should be of interest to planners, engineers, city administrators, agriculturalists, and emergency services personnel who have to do something about the effects of volcanic hazards; to volcanologists and geologists who did not know eruptions affected so many things; to geographers, environmentalists, and natural hazard scientists who are interested in the interrelatedness of phenomena; and to citizens who have experienced, or might yet experience, some of these effects.

ACKNOWLEDGEMENTS

In gathering information for and writing this volume I have become indebted to a very large number of people.

My primary debts are to Professor George Walker of the University of Hawaii and Dr Chris Newhall, United States Geological Survey, Vancouver, Washington State, for their careful comments on much of the manuscript, and to Dr Bob Bernstein, United States Public Health Service, Centers for Disease Control, Atlanta, for his detailed dissection of Chapter 3. I also wish to thank Dr Tom Simkin, Dr Richard Fiske, and many others at the Smithsonian Institution National Museum of Natural History for their help in numerous ways; Drs Bob Schuster, Bob Decker, and Bob Tilling of the United States Geological Survey for their hospitality and assistance with information; Dr Jeremy Anderson, Eastern Washington University, and Professor Marion Marts, University of Washington, for aid with unpublished material; Dr Wally Johnson, Australian Bureau of Mineral Resources for his prompt and enthusiastic support in a number of matters; and Dr Bob Attwood, Yakima Health District, Washington State, for his interest, hospitality and contacts.

Much of this volume was prepared while I was a visiting professor at the University of Hawaii; my thanks to Professors Roland Fuchs, Brian Murton, George Walker, and to numerous others for making me so welcome and to the University for financial assistance. I would also like to acknowledge the financial support I have received over the years from Macquarie University Research Grants, Australian Research Grants Scheme and the Smithsonian Institution National Museum of Natural History toward fieldwork and library research in many lands.

I also owe a considerable debt to librarians and libraries for the assistance they have given me with a variety of cataloging systems and hard-to-find papers. In particular I wish to thank staff at the Mitchell Library (Sydney), the Papua New

Guinea Volcano Observatory, the Smithsonian Institution National Museum of Natural History, the British Museum, Cambridge University, the University of Washington, the University of Hawaii, and the United States Geological Survey at Reston, Virginia, Vancouver, Washington State, and Washington, D.C., and the Hawaiian Volcano Observatory. At my own institution I thank Jocelyn Gardener, Val Barnes and Corinne Unger for their help with matters bibliographic.

My debt to numerous photographers will be evident throughout the book, but I would particularly like to thank Barbara Shattuck, *National Geographic;* Wil Dooley and Dr Bob Schuster, United States Geological Survey; William Cox, Smithsonian Institution Archives; Dr Wally Johnson, Australian Bureau of Mineral Resources; Dr Grant Heiken, Los Alamos National Laboratory; Dr Peter Francis, The Open University; Professor Kazuo Saito, Hokkaido University; and Pat Gilbo, American Red Cross.

My special thanks are also due to John Cleasby and Rod Bashford, Macquarie University, and to Kathie Gibbs, University of Hawaii Cartographic Laboratory, for their skilled drafting of the diagrams. I wish also to acknowledge the assistance in numerous practical matters of Katie Takeshita, Monica Okido, Julie Chun and El'se Maeda, all of the University of Hawaii. To Gwen Keena, Macquarie University, my sincere gratitude for innumerable but skilled hours at the word processor, her many kindnesses, and her unfailing humor even when faced with yet more manuscript pages.

Finally, I would like to thank Jeremy Fisher and Grant Walker, Academic Press Australia, for their friendly and professional assistance.

1

VOLCANIC ACTIVITY AND VOLCANIC HAZARDS

1.1 VOLCANIC SETTINGS

The world's active volcanoes are not evenly distributed across the earth's surface. More than 80% of the volcanoes shown on Figure 1.1 lie north of 10°S and two-thirds are located around the so-called Pacific "ring of fire".

A comparison of a map of plate tectonic margins with Figure 1.1 would indicate that active volcanoes occur in three general settings: (i) near convergent plate margins; (ii) at divergent plate margins; and (iii) at mantle plumes or hot spots.

Broadly speaking, both hot spot volcanoes such as those of the Hawaiian chain and those at divergent plate margins (for example, Iceland and the Galapagos Islands) erupt oceanic basaltic magmas. In these eruptions most of the energy is released as thermal energy contained in the erupted lavas with only a small fraction released as kinetic energy in lava fountaining and minor explosions.

Volcanoes at convergent plate margins include those on island arcs such as Japan, at areas of collision between continental and oceanic crust (the Andes, for example), and at areas of continent–continent collision (for example, Italy). Eruptions at convergent plate margins are generally regarded as andesitic (53%–63% SiO_2) in composition but in fact they range from basaltic (less than 53% SiO_2) to rhyolitic and dacitic (greater than 63% SiO_2). In these eruptions a larger fraction of the total energy is released as explosive kinetic energy.

In general, increased silica content means increased viscosity of the magma which in turn implies increased explosivity; thus, there is a general relationship between location of volcanoes in plate tectonic terms and eruption style. However,

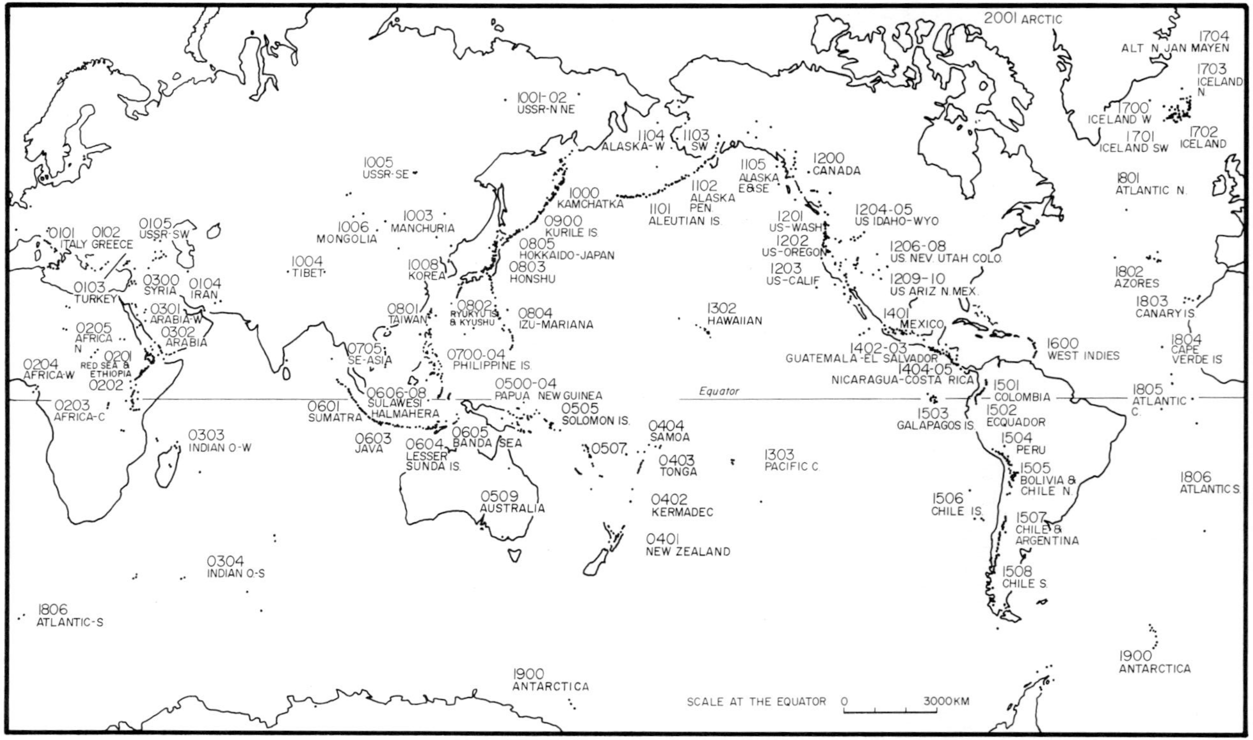

FIG. 1.1 Volcanoes of the world (after Simkin *et al.*, 1981). The numbers refer to the system used in the *Catalogue of Active Volcanoes of the World* (22 vols) and the Smithsonian Institution Volcano Reference File.

this relationship cannot be regarded as too firm. In Hawaii where effusive volcanism is characteristic the 1924 eruption of Kilauea was dramatically explosive because of the access of ground water to the magma. Similar phreatomagmatic eruptions occurred in the Azores (Capelinhos, 1957-58)* and in Iceland (Surtsey, 1963) along distensive plate margins where non-explosive effusive volcanism is more normal. Despite these not uncommon exceptions, we can recognise a general relationship between plate tectonic location and both the type and degree of volcanic hazard.

1.2 TYPES OF ERUPTION

Where volcanic activity is from a vent or pipe cone building usually occurs (Fig. 1.2) but where activity is centered on a fissure sheets of lava are normally produced. However, even in these situations a series of cones may be built along the fissure (Fig. 1.3). Data on thousands of eruptions collected by Simkin *et al.* (1981) indicate that nearly half of the eruptions recorded were central crater (summit) eruptions, 9% were from parasitic craters, and 8% were fissure eruptions.

Problems are inherent in most classifications; certainly classifications of eruption style, such as that presented in Table 1.1, are no exception. In this classification only 10 eruption types are recognised. They are organised in order of increasing hazardousness. Thus eruption types near the top of the table are either non-explosive, involve slow-moving eruption products, or directly effect rather limited areas. Those types in the lower half of the table are fast-moving explosive and/or affect large areas.

With this classification several problems need to be kept in mind:

1. Most eruption types are named after volcanoes that exemplify that style of behavior, but Rittman (1962: 14) points out that while a Strombolian eruption is one characterised by the persistent ejection of lava fragments and scoriae, Stromboli also undergoes periods of rhythmic vapor emission, violent eruptions and solfataric activity. Similarly, although a Surtseyan eruption is characterised by violently explosive fragmentation of hot magma as a result of contact with groundwater or seawater (a phreatomagmatic eruption), the 1963–1967 eruption of Surtsey included phases characterised by the relatively quiet, Hawaiian-style emission of magma and the production of lava flows.
2. Frequently, the character of an eruption changes as it progresses. For example, the 1883 eruption of Krakatau began with Vulcanian activity and then changed to the cataclysmic phase in which extremely powerful Plinian and Peléean

*In general, volcano names in this volume follow the usage and spelling provided in the Smithsonian Institution Volcano Reference File. The first mention usually includes the country or region in which the volcano is located. The index also provides locations. Data on coordinates, elevation and eruptive history can be found in Simkin *et al.* (1981).

FIG. 1.2 Parícutin in eruption in May, 1943, photographed from the northeast. The cone has been built by repeated eruptions of tephra and scoria. A lava flow is seen in the left middleground. Tephra mantles the foreground (Smithsonian Institution Archives, Record Unit 7281, William F. Foshag Papers, F17).

FIG. 1.3 Lava fountaining at Kilauea in February, 1983. A series of small spatter cones has formed along a fissure in the East Rift Zone. Pahoehoe lava in the foreground.

Table 1.1 Types of eruption

Eruption type	Nature of activity and character of deposits	Examples
Icelandic (Basaltic flood)	very widespread flows of very fluid lava; minor amounts of tephra	Laki, Iceland, 1783
Hawaiian	lava fountaining; bombs; thin flows of fluid lava, some tephra	Matavanu, W. Samoa, 1905-1911; Kilauea, Hawaii, 1977, 1983
Strombolian	ejection of fluid lava blebs; scoriae, bombs and glassy tephra; thicker flows of moderately fluid lava; cinder cones	Heimaey, Iceland, 1973
Vulcanian	moderate to violent ejection of viscous blocks, sometimes short, thick lava flows; tephra and block cones	Vulcano, 1888-1889; Ngauruhoe, 1975
Phreatic	explosive eruption of non-magmatic material	Kilauea, 1924: Waimangu (Tarawera, New Zealand), 1917
Surtseyan (Phreatomagmatic)	violently explosive, highly fragmented material; glassy tephra	Capelinhos (Fayal), Azores, 1957; Surtsey, Iceland, 1963; Taal, Philippines, 1965
Plinian	paroxysmal ejection and wide dispersal of tephra (including pumice); sometimes associated with caldera collapse	Soufrière, St Vincent, 1902; Vesuvio, AD 79
Peléean	collapse of a Plinian or Vulcanian eruption column or dome or cone collapse to produce *nuée ardente* or glowing avalanches (pyroclastic flows)	Mt Pelée, Martinique, 1902
Bandaian	collapse of part of volcanic edifice producing massive landslide or avalanche	Bandai, Japan, 1888; Mount St Helens, 1980
Katmaian	voluminous production of ignimbrites	Katmai-Novarupta, Alaska, 1912

activity occurred in quick succession. In this case and many others (Sparks and Wilson, 1976), Peléean activity in which *nuée ardentes* flow rapidly down the sides of the volcano, results from the collapse of parts of the rapidly expanding and rising Plinian or Vulcanian cloud.

3. The character of the one eruption can also vary from place to place on a volcano. During the main Plinian phase of the 1906 eruption of Vesuvio (April 6 to 8), with ejection of massive quantities of pumice and tephra from the main vent effusive emission of lava occurred from a number of vents on the south east flank of the volcano. Similarly, it is quite common during eruptions of Etna for explosive liberation of gas to occur from the summit while relatively quiet effusion of lava occurs from lateral vents.

Simkin *et al.* (1981) summarise data providing a wealth of information on eruptive characteristics. Of the 5564 dated eruptions recorded, 24.1% included lava flows, 8.8% were of phreatomagmatic type, 5.1% produced pyroclastic flows (*nuées ardentes* or ignimbrites), 5.6% generated mudflows or *lahars* and 5.9% were submarine eruptions. An overwhelming 72.8% of the dated eruptions involved "normal" explosive activity (Simkin *et al.*, 1981: 71). This generally means that activity was of Strombolian, Vulcanian or Plinian type.

1.3 VOLCANIC ACTIVITY

Figure 1.1 does not locate all of the world's volcanoes. It shows only those volcanoes that are known to have erupted during the last 10 000 years. Furthermore, it locates only a few of the thousands of active submarine volcanoes present along divergent plate margins. This omission is not serious from our point of view as the terrestrial volcanoes are far more hazardous.

Many texts describe volcanoes as "active", "dormant" or "extinct", but Walker (1974) has pointed out that this categorisation is unsatisfactory for two reasons. First, it has a variable meaning as the historic period covers more than 2000 years in the Mediterranean but only the last hundred years or less in Antarctica and Papua New Guinea (see also Baker, 1979: 341). Simkin *et al.* (1981: 23-25) have demonstrated that the apparent increase in the reported frequency of volcanic eruptions toward the present day results not from an actual increase in volcanism but from factors such as the invention of the printing press, the "discoveries" made during early Spanish–Portuguese explorations, and war. It should also be noted that even recent catalogs of early Mediterranean eruptions are frequently in error concerning eruption dates and eruption histories (Stothers and Rampino, 1983).

Second, some volcanoes only erupt once — they are monogenetic. Parícutin, Mexico, erupted from 1943 to 1952 but it is unlikely to erupt again. Similarly, Auckland (New Zealand) is built on more than 50 monogenetic volcanoes (Fig. 1.4). None of these volcanoes has been active in the historic period but the Auckland volcanic field must be considered as dormant not extinct. In fact, only three — Mt Smart, Mt Wellington and Rangitoto — are *known* to have erupted in the last 10 000 years. Although there is a high probability that another eruption will occur in the Auckland area it is unlikely that the eruption will be from one of the known volcanoes (Searle, 1964).

The categorisation — active, dormant, extinct — is also unsatisfactory because some apparently extinct volcanoes become active. Walker (1974) examined the records of 180 volcanoes listed in the *Catalogue of Active Volcanoes of the World* (22 volumes) for which historic records spanned more than 500 years. He calculated the average eruption rate per century, shown as a logarithmic probability plot in Figure 1.5 (Walker, 1974: 36). Although there are many uncertainties in this diagram, it shows that the median volcano considered had an eruption rate of 0.45 times per century — that is one eruption per 220 years and that 50% of the volcanoes erupted less frequently than this. If the straight line fit is extrapolated it appears that about 20% of volcanoes erupt less than once every 1000 years and about 2% less than once every 10 000 years. Although this might seem fanciful the eruption on Heimaey, Ice-

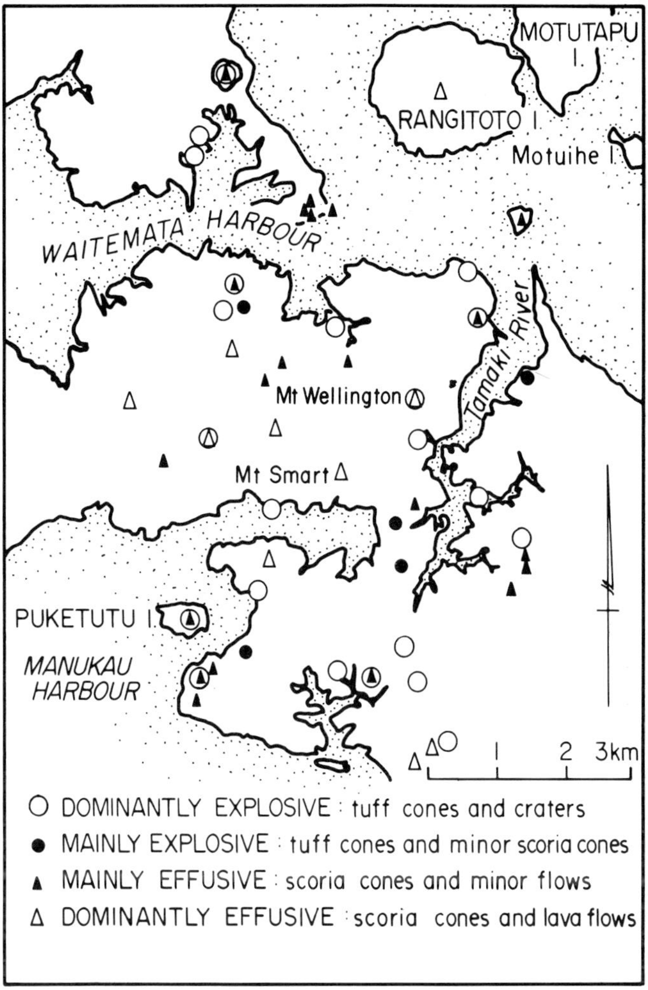

FIG. 1.4 Volcanoes of the Auckland, New Zealand area (after Searle and Mayhill, 1981: 47).

land, in 1973 was the first for 5000 years and the devastating eruption of Lamington, Papua New Guinea, in 1951 was probably the first for at least several thousand years. Similarly, the 1982 eruption of El Chichon in Mexico was the first recorded eruption although solfataric activity had occurred previously (Simkin *et al.*, 1981: 90). A UNESCO (1979) report suggests that, on average, one "extinct" volcano erupts every five years. Thus it is certain that some volcanoes have intervals between eruptions that are much longer than the historic record. Simkin and Siebert (1981: 342) show that while the median repose interval for 4875 eruptions with known prior eruptions is only five years (Figure 1.6a) much longer intervals commonly precede

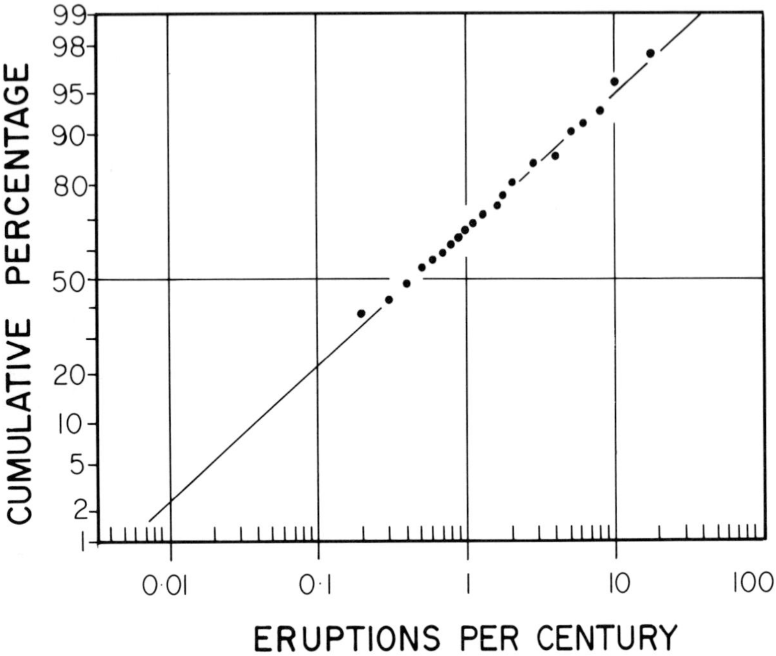

FIG. 1.5 A logarithmic probability plot of the cumulative proportion of a sample of 180 volcanoes which have eruption rates less than the value given (Walker, 1974: 36).

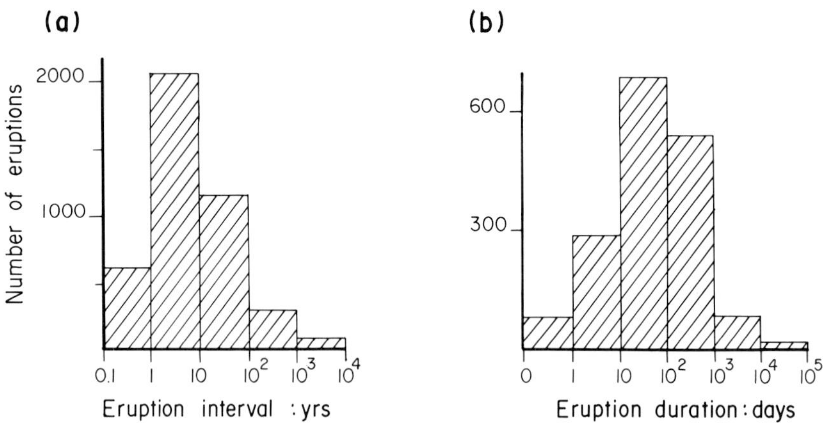

FIG. 1.6 Eruption frequencies for volcanoes active in the Holocene: (a) the interval between eruptions; (b) the durations of eruptions. See text (modified after Simkin and Siebert, 1981: 343).

very explosive eruptions. In fact, for the 25 most violent eruptions for which data are available the median repose period is 865 years.

The investigation by Simkin *et al.* (1981) of volcanic activity during the Holocene indicates that a total of 5564 dated eruptions are known to have taken place from 1343 volcanoes. Of these, 529 volcanoes erupted in the historic period. These data suggest that more of the world's approximately 10 000 volcanoes are likely to become "active" in the future.

Figure 1.6b presents a histogram of eruption duration for 2139 eruptions. The median duration is about 60 days. However, we should recognise that "an eruption" can cover an extended period and include a multiplicity of events. For example, Sakurajima (Kyushu) has been in more or less continuous eruption since 1955 with explosions occurring on most days, while the current eruption of Stromboli began before 450 BC! Similarly, the eruptions of Mount St Helens, Washington State, on May 18, May 25 and June 12, 1980, are regarded by the citizens of Castle Rock, Yakima, Portland and other affected places as separate events, but in a catalog of eruptions the activity beginning on March 17, 1980, and continuing through 1983 is but one eruption. The most violent phase of the 1980 eruption began on May 18 and lasted about nine hours. The duration of the climactic phases of other well-known eruptions is presented in Table 1.2. From a hazards point of view these phases are of greatest significance.

Figures 1.5 and 1.6 emphasise that volcanoes neither erupt with equal frequencies nor produce eruptions of equal duration. Although volcanoes on convergent plate margins produce only 10%–13% of the magma reaching the earth's surface annually, these volcanoes provide 84% of the known eruptions and, even more importantly from a hazard point of view, 88% of the eruptions with fatalities (Simkin and Siebert, 1981: 342).

In an attempt to determine eruption frequency–duration for each convergent margin volcano belt Simkin and Siebert (1981) calculated the number of years each volcano had been active since 1880, summed these for each belt and divided by belt length (Fig. 1.7). Their results indicate that the Halmahera (Indonesia), Calabria (Italy) and New Zealand belts have been the most active. However, this does not necessarily imply that these areas present the greatest volcanic hazard.

Table 1.2 Durations of climactic eruptive phases

Volcano	Year	Duration	Source
Vesuvio	79	~24 h	Lirer *et al.*, 1973: 768
Fogo	1563	~ 2 days	Lirer *et al.*, 1973: 769
Asama	1783	4 days	Lirer *et al.*, 1973: 769
Tambora	1815	~ 2 days	
Cosiguina	1835	~ 3 days	Rittman, 1962: 42
Krakatau	1883	~45 h	Docters van Leeuwen, 1936: 9
		~36 h	Ball, 1906: 324
Santa Maria	1902	~18 h	Rittman, 1962: 41
Katmai-Novarupta	1912	<20 h	Curtis, 1968: 182
Hekla	1947	1 h	Thorarinsson, 1967
Mount St Helens	1980	9 h	Christiansen and Petersen, 1981: 25

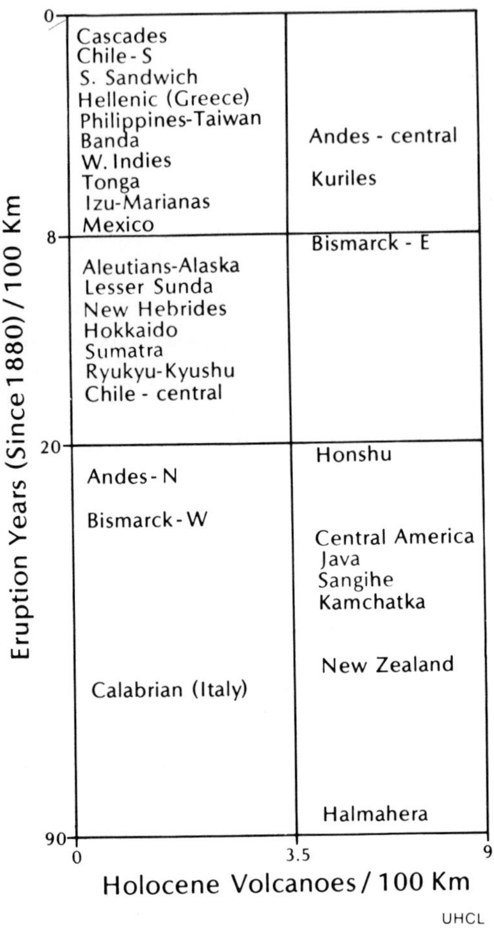

FIG. 1.7 Volcanic activity since 1880 per 100 km of arc length for a number of volcanic belts (Simkin and Siebert, 1981: 343).

1.4 ERUPTION SIZE

The "size" of a volcanic eruption can be estimated in a number of ways. For example, by the total volume of erupted products (magnitude), by the volumetric or mass rate at which these products left the vent (intensity), by the area over which the products are distributed (dispersive power), by the explosive violence or by the destructive potential of the eruption (Walker, 1980; Wilson and Walker, 1981). A commonly used measure of the size of an eruption involves an estimate of the total thermal energy released, derived from Hédervári's (1971) formula based on the volume of erupted material. For a moderately large eruption the total energy released lies in the range 10^{15}–10^{18} joules. By comparison, a one kilotonne atomic explosion liberates about 4×10^{12} J (Bolt, 1976: 39-41). Table 1.3 sets out estimates for a number of well-known eruptions. However, such a measure of the energy liberated in a volcanic eruption is not always directly correlated with the degree of volcanic hazard.

While it is generally thought that violently explosive eruptions such as that of Krakatau in 1883 are the "biggest" eruptions this is not necessarily the case. For example, Macdonald (1972: 60-61) indicates that the nonexplosive eruptions of Mauna Loa, Hawaii, in 1859 and 1950 were as big, in terms of total energy released, as the Krakatau eruption (Table 1.3).

Newhall and Self's (1982) volcanic explosivity index (VEI) provides a simple descriptive measure appropriate to a discussion of volcanic hazards. This index combines the total volume of eruptive products, the height of the eruption cloud, duration of the main eruptive phase and several descriptive terms into a simple 0–8 scale of increasing explosivity. Estimates for a number of eruptions are presented in Table 1.3. It is noteworthy that of all historic eruptions only the 1815 eruption of Tambora (Indonesia) has been assigned a VEI of 7 while none have a VEI of 8. More than half of the volcanoes for which data exist have had "large eruptions"— that is, eruptions with a VEI of 3 or greater (Simkin *et al.*, 1981: 21).

Table 1.3 Examples of eruption magnitudes

Volcano	Year	Eruption type	Total energy* (joules)	Volcanic Explosivity Index†
Santorini, Aegean Sea	*c.* 1500 BC	Plinian–Katmaian	1.0×10^{20}	6
Laki, Iceland	1783	Icelandic	8.6×10^{19}	4
Tambora, Indonesia	1815	Plinian–Katmaian	8.4×10^{19}	7
Krakatau, Indonesia	1883	Plinian–Katmaian	$\sim 1.0 \times 10^{19}$	6
Vesuvio, Italy	1906	Plinian	1.7×10^{17}	3
Katmai-Novarupta, Alaska	1912	Katmaian	2.0×10^{19}	6
Sakura-jima, Japan	1914	Plinian–Peléean	4.6×10^{18}	4
Mauna Loa, Hawaii	1950	Hawaiian	1.4×10^{18}	0
Oshima, Japan	1950-51	Strombolian	9.4×10^{16}	2
Kilauea, Hawaii	1952	Hawaiian	1.8×10^{17}	0
Bezymianny, Kamchatka	1955-56	Bandaian	2.2×10^{18}	5
Capelinhos (Fayal), Azores	1957	Surtseyan–Strombolian	4.0×10^{17}	2
Agung, Bali	1963	Vulcanian	4.5×10^{17}	4
Surtsey, Iceland	1963	Surtseyan–Strombolian	1.9×10^{17}	3
Taal, Philippines	1965	Surtseyan	1.0×10^{16}	4
Arenal, Costa Rica	1968	Vulcanian	1.0×10^{15}	3
Mount St Helens, USA	1980	Bandaian–Plinian	$\sim 5 \times 10^{16}$	5

*Mainly from Bullard, 1976.
† From Simkin *et al.*, 1981.

Table 1.4 Volcanic hazard summary

Volcanic hazard	Frequency	Post-1600 Hazard level* (1-10)	Post-1900 Hazard level* (1-10)	Frequency of Adverse Effect/Damage/Death — Distance (km): Less than 10	10-30	30-100	100-500	500-1000	Greater than 1000
Lava flows	24%	<1	<1	F	C	VR			
Ballistic projectiles	~60%	<<1	<<1	C					
Tephra falls	~60%	4	1	VF	F	F	C	R	
Pyroclastic flows and debris avalanches	5%	10	10	A	F	R	VR		
Lahars and jökulhlaups	6%	4+	1.5	F	F	R	VR		
Seismic activity and ground deformation	~50%	<<1	<<1	C	C	VR			
Tsunami	<1%	10	<<1	A	F	C	R	VR	
Atmospheric effects	~60%	<<1	<<1	C	C	R	VR	VR	
Acid rains and gases	~40%	<<1	<<1	F	F	R	R	VR	VR

*Hazard level is based on the relative frequency of deaths given that the specific type of activity occurs.
A = Always; VF = Very Frequent; F = Frequent; C = Common; R = Rare; VR = Very Rare.

1.5 VOLCANIC HAZARDS

Although volcanoes at convergent plate margins do not erupt more frequently or produce eruptions that last longer than those of other volcanoes (Fig. 1.7) they are, in general, characterised by more explosive eruptions. This suggests that the median convergent margin eruption lies nearer the bottom of Table 1.1 than does the median divergent margin or hot spot eruption. In turn, this relation implies a difference in degree and type of volcanic hazard.

Table 1.4 indicates the relationship between the type of hazard, the relative frequency of damage, the range at which damage occurs and the type of eruption. A similar summary, based on Icelandic experience, was presented by Thorarinsson (1979: 126). In Table 1.4 nine types of volcanic hazard are recognised (each is discussed in detail in Chapter 2).

The frequency column is based partly on data presented by Simkin *et al.* (1981), and partly on subjective assessment. Tephra falls, atmospheric effects, earthquakes and acid rains and gases are clearly the most frequently occurring volcanic hazards. This assessment is in general accord with Thorarinsson's (1979: 126) estimates.

However, the frequency column (like Thorarinsson's data) only indicates the relative frequency of occurrence of the hazard, not the severity of the hazard. As will be seen in later chapters data on damage caused by volcanic hazards are frequently subjective and difficult to compare even in similar situations. However, a considerable body of material indicates the numbers of people killed in eruptions and the hazards responsible. The post-1600 and post-1900 hazardousness scales are based on the relative proportions (Scale 1–10) of people killed by volcanic hazards. Details of deaths are presented in Chapter 3. For the present it is sufficient to note that if human deaths are accepted as an adequate measure of hazardousness, pyroclastic flows and tsunami have been by far the most hazardous volcanic agents. However, in the twentieth century volcanically-generated tsunami have been unimportant causes of human death whereas pyroclastic flows have maintained their leading role. It will become clear in Chapter 3 that both sets of data are biased by a few eruptions causing large numbers of deaths.

In summary we can note that pyroclastic flows and tsunami have been the greatest, and that lahars and tephra fall are not unimportant, destroyers of human life. However, we should also note that this solidly-based conclusion does not necessarily imply that these four volcanic hazards are the most injurious or those most likely to cause damage. As plants, insects, animals, humans, buildings, commercial activities and communication links differ dramatically in their mobility (and in their resistance to heat, impact shear stresses, and noxious gases *inter alia*), generalisations about hazardousness are clearly difficult. However, in order to clarify the relationship between volcanic hazards and their effects it is necessary to have a clear idea of the characteristics of those volcanic hazards listed in Table 1.4.

2

HAZARDS PRODUCED BY VOLCANIC ERUPTIONS

As suggested in Table 1.4, the hazards produced by volcano eruptions can be divided into nine groups. This chapter outlines the physical character of each type of hazard and examines those aspects which create the hazard.

2.1 LAVA FLOWS

Three types of lava flow are generally recognised: pahoehoe; aa; and block lava.

Pahoehoe flows are characterised by ropy, smooth or billowing crusts (Fig. 1.3). Numerous thin flows, less than one meter thick, are frequently superimposed one on another, with tough skins forming quickly on exposure. Pahoehoe lavas may develop lava tubes which can be up to several kilometers in length and 30 m in diameter. Excellent examples occur on Rangitoto, New Zealand (Fig. 1.4), in Hawaii Volcanoes National Park (Fig. 1.3) and around Newberry volcano in Oregon, U.S.A. As lava tubes allow still fluid lava to drain away beneath a chilled crust, flows with tubes can travel great distances, creating an additional hazard.

Aa flows are generally thicker than pahoehoe flows and are characterised by a surface of rough clinkery and spinose fragments (Fig. 2.1). The interior of the flow is usually massive lava. Aa flows frequently form linear channels and gutters can be up to 30 m deep. Good examples of these flows are found on Kameni (Santorini), Etna and on the Modoc Plateau in northern California.

Block lava flows are like aa flows except that the surface is composed of large polyhedral blocks. These flows are best developed in intermediate and siliceous lavas that were glassy close to the vent (for example, the Newberry obsidian flows). They are less common than the other types.

FIG. 2.1 Aa lava flow advancing toward Parícutin village, June 1944 (Smithsonian Institution Archives, Record Unit 7281, William F. Foshag Papers, F2116).

While this three fold division seems straightforward, many flows are more complex. Pahoehoe flows commonly change into aa flows with distance. The early 1983 flows into the Royal Gardens subdivision on Hawaii provide good examples. Such an alteration is accompanied by an increase in thickness and a change from fluid flow to a bulldozing movement. A change from an aa type to a block type, as in the Oni-oshidashi flow produced during the 1783 Asama eruption, further increases flow thickness (Aramaki, 1956: 224-228).

About 90% of lava flows involve basaltic magmas. About 8% are andesitic and the remainder are siliceous magmas (Williams and McBirney, 1978: 97). The rhyolitic and dacitic lava flows are shorter and stubbier than basaltic flows. Basaltic flows from fissures tend to spread farthest. Factors which influence this variation in flow morphology with silica content include temperature, viscosity, yield strength and effusion rate.

Measurements during a number of eruptions indicate that temperatures recorded in basaltic lavas reach 1050–1130°C (for example, in flows and fountains at Kilauea — Fig. 2.2) while temperatures in basaltic andesites at Etna (1966) and Sakura-jima (1914) range from 880°C to 1050°C (Macdonald, 1972: 57–58). As the ignition temperatures for cloth, paper and wood are less than 250°C these differences may be rather academic from a hazards point of view!

FIG. 2.2 Dallas Jackson, a Hawaii Volcano Observatory scientist, determining the temperature of a pahoehoe flow with a thermocouple during the February, 1983, Kilauea eruption. The temperature recorded in this flow was about 1118°C.

Crystals nucleate and grow as soon as an erupted magma begins to cool — 1200°C is close to the maximum temperature for a lava flow although higher values have been measured with optical pyrometers where ignition of hydrogen and gases from burning vegetation have enhanced laval surface temperatures. At the other extreme, in partly crystalline lavas 750°C is close to the limiting conditions of flow; a temperature of 750°C was recorded in a still mobile flow at Parícutin while a barely moving flow from Mauna Loa in 1949 had a temperature of 745°C.

Lava flows lose heat by both radiation and conduction but the former is by far the more important. Once the surface temperature falls below around 800°C heat losses result in a lowering of temperature of only a few degrees per hour as the thin plastic skin that forms rapidly on a pahoehoe cooling unit, for example, drastically cuts heat loss. Furthermore, losses by conduction to both air and ground are only one or two degrees per hour (Williams and McBirney, 1979: 122). Rainfall, or water applied artificially to the surface of a flow, hastens cooling.

The crust on a pahoehoe flow thickens over time. It is possible, but hazardous, to walk on a crust a few centimeters thick over still moving lava (Macdonald, 1972: 73). On the other hand, thick flows can still have quite a high internal temperature five to six years after emplacement. Hot water from the 1783 Oni-oshidashi flow from Asama was conducted along a gutter to Ôzasa village 6 km distant and utilised as a hot bathing spring for 23 years (Aramaki, 1956: 229).

The viscosity of lavas varies widely with temperature, the size and proportions of crystals in the magma, the amount of volatiles (including H_2O) in solution and several other factors. The effects are interrelated. However, because viscosity is strongly inversely dependent on temperature and because basaltic melts have higher temperatures than more acidic magmas it can be estimated that the viscosity of the latter is generally more than 1000 times greater than that of the basaltic lavas.

Because viscosity varies inversely with temperature, viscosity commonly increases with distance from the vent. Measurements at Oshima (1951) show an increase in viscosity from 5.6×10^3 to 2.3×10^5 poises in a distance of 500 m (Williams and McBirney, 1979: 123). Similarly Walker (1967) estimated an increase in viscosity from 0.4×10^5 to 1.5×10^7 poises in a similar distance in a small Etna flow. On fairly steep Etna slopes, lavas stop flowing at viscosities of from 10^9–10^{11} poises.

The yield strengths of volcanic rocks increase dramatically when temperatures fall below liquidus values. Laboratory measurements by Murase and McBirney (1973), for example, show an increase in yield strength from 0 at about 1200°C to about 6 kN m^{-2} at 1150°C. A further 30°C fall in temperature resulted in only a small increase in yield strength. Etna (1975) lava flows with temperatures of 1080°C-1090°C had measured yield strengths of 280-290 N m^{-2} (Sparks *et al.*, 1976: 270.) In physical terms an increase in yield strength increases not only the stress required to sustain flow but also the apparent viscosity of the fluid. For example, silicic magmas form fewer, thicker cooling units than do basaltic lavas, largely because the former have higher yield strengths. Similarly, pahoehoe flows with low yield strengths form more cooling units than aa flows which have higher yield strengths (Wadge, 1978: 503). Thus, yield strength has an important influence on flow size and shape.

Flow morphology is also influenced by effusion rate. The most detailed studies have been undertaken on Etna (Walker, 1974; Wadge, 1978). Figure 2.3 indicates that the lengths of Etna lava flows depend on effusion rate averaged over the duration of the eruption. This plot applies only to aa flows and not to tube-fed pahoehoe flows which may achieve much greater lengths at equivalent effusion rates (Malin, 1980). Wadge (1978) has also shown that the effusion rate of Etna lavas is quite variable, that the maximum rate is usually achieved when the flow front is about 3 km from the vent and that on a typical Etna slope of 10° the maximum length of the lava flow is reached after about 10 days. The most common effusion rate is in the range 15-45 m^3 s^{-1}.

Rates are usually highest at the beginning of an eruption with the subsequent decline usually more pronounced for basaltic eruptions. In the 1947 eruption of Hekla the effusion rate for lava flows was about 1000–1250 m^3 s^{-1} per kilometer of fissure at first but fell on the second day to about 100 m^3 s^{-1}. The effusion rate subsequently fell to 4–5 m^3 s^{-1} before increasing again. At the other extreme, effusion rates during the 1783 Laki fissure eruption probably reached more than 5000 m^3 s^{-1}. These and other examples of effusion rates indicate that effusion rates may vary by up to three or four orders of magnitude even during the one eruption.

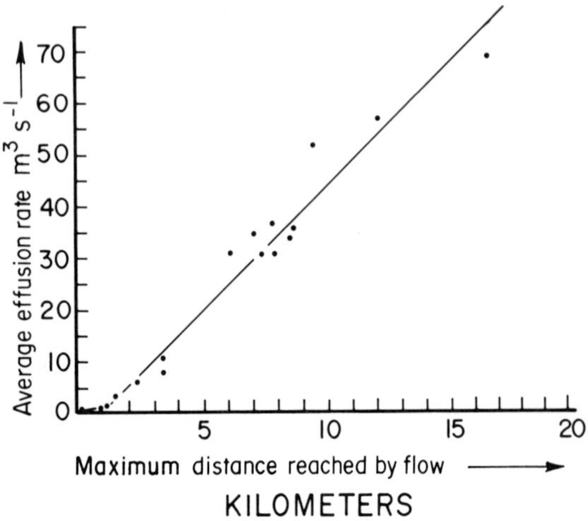

FIG. 2.3 Plot of average effusion rate during eruption against maximum distance from vent achieved by aa flows from various historical flank and summit eruptions of Etna (after Wadge, 1978: 504).

Lava flow velocity is clearly affected by temperature, viscosity, yield strength, effusion rate, gradient and topography. Other factors being constant, flow velocity diminishes away from the source. Where the surface is not crusted the lava flow moves fastest at the surface and in the middle. At Parícutin, Krauskopf (1948) estimated the surface velocities of lavas in channels near the vent as twice the mean velocity of the total flow. Where crusting occurs the hottest and fastest moving parts are beneath the flow surface (Williams and McBirney, 1979: 100).

In general, silicic lava moves more slowly than basaltic lava while pahoehoe flows have higher velocities than aa flows. However, lava flows often pulse, surge and pond so that the rate of forward movement may be quite variable. Some examples of flow front velocities, the most significant velocity from a hazards viewpoint, are given in Table 2.1. Where flows are confined within lava channels or tubes, velocities may be several orders of magnitude higher. Tazieff (1977: 189-200) reported an average flow velocity of 30 km h^{-1} and an initial velocity in excess of 100 km h^{-1} when the Nyiragongo lava lake drained suddenly on January 10, 1977. Velocities of 64 km h^{-1} were reported on slopes with an incline of from 10°–25° during the 1855 Mauna Loa eruption; 1940 flows from the same volcano reached 30–40 km h^{-1}. Despite the high velocities in 1855 the flow front advanced at only 9 km h^{-1} (Williams and McBirney, 1979: 100).

Although prehistoric lava flows such as the Roza flow of the Columbia River basalts had a volume of around 1200 km^3 and one Yakima flow 30 m thick covered 18 000 km^2, historic flows have attained only a fraction of these dimensions. The

Table 2.1 Lava flow front velocities

Volcano	Year	Velocity (m h^{-1})	Remarks	Reference
Matavanu, Samoa	1906	3-960	~6° slopes	Anderson, 1910: 623-625
Sakura-jima	1914	245	later decreased	Aramaki, 1956: 228
Kilauea	1963	375	pahoehoe flow	Moore & Koyanagi: C6
	1983	<250	aa flow	
Parícutin	1943	5		Williams & McBirney, 1979: 72
Heimaey	1973	3600	narrow aa flow into harbor	Clapperton, 1973: 483, 486

biggest, the flows from the Lakagigar (Iceland) fissure eruptions of 1783-84, covered more than 550 km^2 with a maximum length from source of 88 km and a total volume of 12.3 km^3 (Thorarinsson, 1969: 913) (Fig. 2.4).

Flows from fissure eruptions tend to be more voluminous than flows from central vents. The largest historic flow from Mauna Loa (1859) reached a length of 50 km; the average of all flows on that volcano since 1843 is about 25 km with volumes less than 0.5 km^3. Kilauea flows rarely exceed 11 km in length with volumes less than or equal to 0.25 km^3. The 1983 Kilauea eruptions produced, by the end of October, lava flows up to 8.5 km long covering an area of 23 km^2 (see Fig. 4.1), but with a total volume of only about 0.12 km^3 (*SEAN Bull.*, 8(10): 9). Most historic lava flows are measured in fractions of cubic kilometers and flows exceeding 4 km^3 are rare (Williams and McBirney, 1979: 97-99). The largest historic flow on Etna (1669) has a volume of approximately 1 km^3. Nine years of activity at Parícutin produced flows with a total volume of approximately 1.4 km^3.

Walker's (1973) study of the dimensions of more than a thousand lava flows recognised significant differences in the median sizes of high and low viscosity flows. Low viscosity flows (averaging around 10^4 poises) had a median length of 4.1 km and an average thickness of 10 meters. High viscosity flows (in the order of 10^9 poises) had a median length of 1.3 km and an average thickness of 100 meters. More importantly perhaps, Walker's analysis suggests there is a 1% probability that a high viscosity lava flow will exceed 11 km in length and a low viscosity flow 45 km.

The evidence presented above suggests that some general conclusions can be reached concerning volcanic hazards from lava flows of different types. However, it is clear that the available data are biased in favor of lava flows from the Hawaiian volcanoes, Vesuvio, and Etna — nearly all basaltic lavas. This bias in the data perhaps tells us as much about relative hazards as does reasoning from the data!

The fastest moving, hottest, most mobile, least viscous flows stem from Hawaiian and Icelandic-style eruptions. *A priori*, this should make these flows the most dangerous, but they are relatively thin and composed of numerous flow units when of the pahoehoe type, thus cooling comparatively rapidly. The low viscosity and low yield strength of pahoehoe flows and the thinness of individual flow units

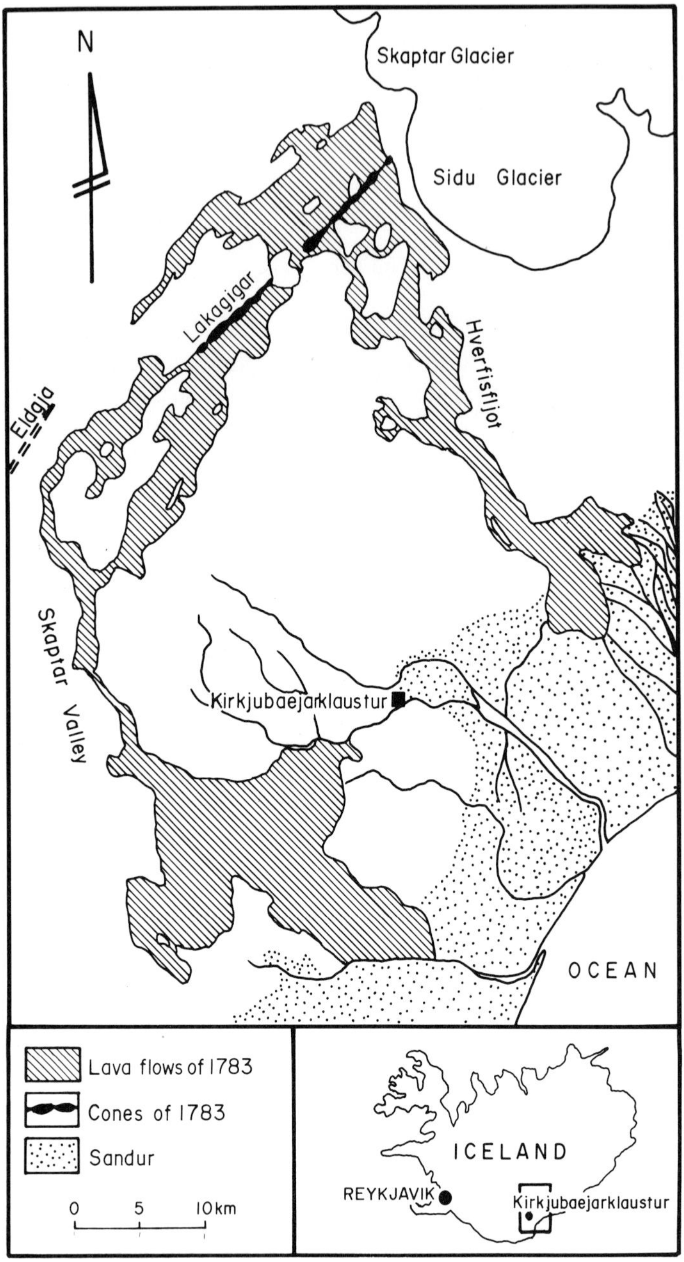

FIG. 2.4 Map of southeast Iceland showing extent of 1783–1784 Laki lava flows (modified after Thorarinsson, 1970, and Macdonald, 1972).

suggests that such flows may be readily diverted by surface irregularities and man-made objects such as stone walls and houses though the high temperatures would increase the danger of ignition of such objects. Clearly the hazard from these lava flows extends over relatively large areas.

Although aa and block flows tend to move more slowly than pahoehoe flows and have lower temperatures, they probably represent no less a hazard. Temperatures are still well above ignition points for many manmade materials and vegetation. Also, flow front velocities may be high where effusion rates are high. Furthermore, flow fronts are thicker and have higher yield strengths so that they are less easily diverted. Avalanches of lava blocks from steep, high, hot flow fronts represent an additional hazard. On the other hand, such lava flows cover more limited areas.

2.2 BALLISTIC PROJECTILES AND TEPHRA FALLS

The term *tephra* includes all airborne products of a volcanic eruption except gas. It thus includes tephra (or volcanic ash) falls and pyroclastic flows (Thorarinsson, 1974).

In eruptions of the Strombolian, Vulcanian, Surtseyan and Plinian type (Table 1.1) a vertical or near-vertical eruption column forms. It is composed of a jet of dispersed solids and gas. The dispersion is initially driven upwards by decompression and gas expansion (the so-called gas thrust phase). This column can continue to rise as a convective plume to heights of 30 km or more providing surrounding air is entrained and heated and the density of the dispersion is further lowered as large clasts fall out.

Volcanic bombs, blocks and lapilli (Fig. 2.5) which leave the vent with ballistic trajectories are referred to here as *projectiles*. In general, a volcanic bomb is ejected as a clot of magma, greater than 64 mm in diameter. Bomb shape depends on factors such as the fluidity of the magma, length and velocity of flight, rate of cooling, rate of gas escape, and deformation on impact (Williams and McBirney, 1979: 128). Although quite elaborate classifications of volcanic bomb types have been constructed (see for example Macdonald, 1972: 125–128), they are of limited relevance to a study of volcanic hazards. Volcanic blocks are projectiles composed of old volcanic rock derived from the walls of the vent. They tend to be angular and roughly equidimensional. Particles which fall from the eruption column or plume are here called *tephra*. In most cases airfall tephra consists of ash and lapilli (Fig. 2.5), particularly pumiceous lapilli with low specific gravity which can be transported large distances downwind. The term lapilli refers to fresh magma clots, cinders, and old rock fragments in the size range 2–64 mm. Such fragments are frequently highly vesicular so that scoriaceous and pumiceous lapilli are common. Lapilli composed of lithic fragments tend to become projectiles while the pumice lapilli tend to be part of the tephra fall component of the eruption plume. Both ballistic projectiles and tephra falls are significant hazards.

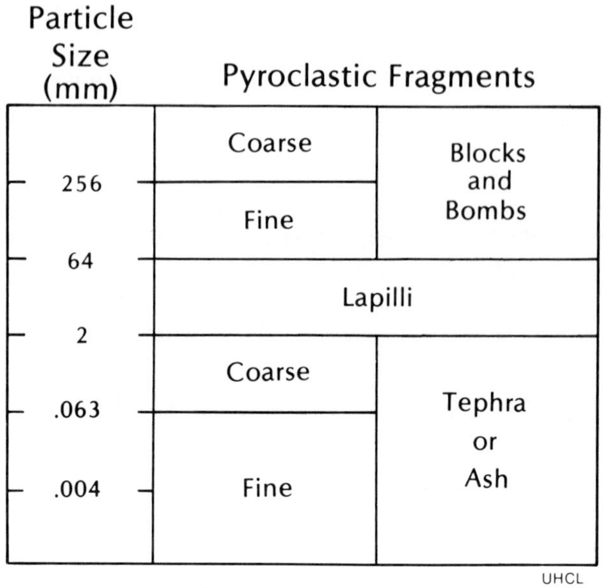

FIG. 2.5 A modified version of Fisher's (1961) widely used classification of pyroclastic fragments. In practice, the relationship between various terms and particle size is not as precise as the diagram suggests.

2.2.1 Projectiles

> Most of the fragmental material ejected from the volcano [Parícutin, 1943 — Fig. 2.6] in this stage of its history consisted of bombs rather than of ash or cinders. With each explosion the bombs were blown 2000 to 3000 feet into the air. Most of them landed on the cone; the greatest distance at which I found a bomb was 3500 feet from the center of the volcano. The bombs went so high that it took from 10 to 15 seconds for most of them to fall, after they had reached their greatest height. They were roughly spherical and ranged in size from a walnut to a big house. Most of them were between three and five feet in diameter. The largest I saw was a block 50 feet in diameter, which was blown 300 feet above the top of the crater; that is, 850 feet above the ground. Nearly all the bombs when they landed were so thoroughly solidified as not to change in shape, though many when they struck the sides of the cone, broke into pieces. A few were still liquid when they landed and splattered out like pancakes on the ground. Bombs of this type did not penetrate the earth for more than three inches. Others rotated slowly in the air, gradually thinning in the middle, and before they fell separated into two tear-shaped bodies. Some after coming to rest on the cone, smoked for a considerable time, certainly for as long as 15 minutes. Most of the bombs consisted of highly vesicular basalt.
>
> [Trask, 1943: 503].

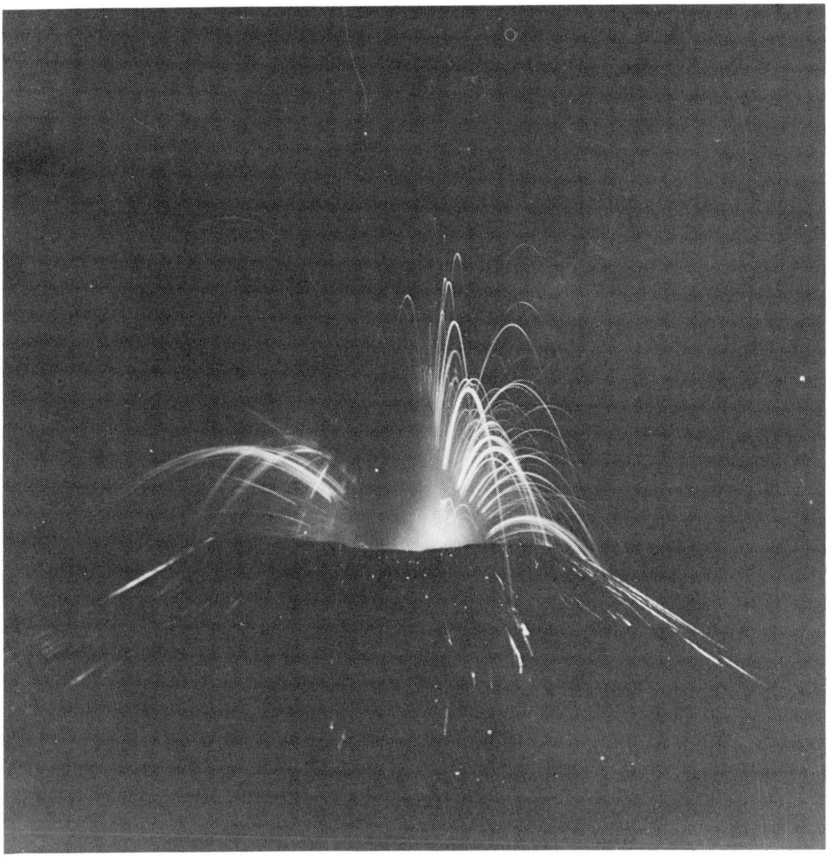

FIG. 2.6 A time exposure of incandescent volcanic bombs ejected from the crater of Parícutin during early June, 1943 (Smithsonian Institution Archives, Record Unit 7281, William F. Foshag Papers, F31).

Although Wilson (1972) calculated the maximum theoretical ejection velocity to be equivalent to 600 m s^{-1}, measured velocities are much lower. At Heimaey, 1973, for example, they were in the range 75-110 m s^{-1} (Self *et al.*, 1974: 547); at Asama and Arenal (1968), 180–200 m s^{-1} (Minakami *et al.*, 1969: 788), at Karkar in 1979, 250–320 m s^{-1} (McKee *et al.*, 1981: 84). Similarly, maximum distances will be achieved with initial trajectory angles of 45° (Fudali and Melson, 1972: 886), but observations at Stromboli found that about half the initial trajectory angles were greater than 80°. Except for projectiles ejected from very shallow depths in the vent, or in the case of directed blasts, ejection angles are likely to be much greater than 45°. However, as at Heimaey in 1973, ejection angles may vary from

A

B

FIG. 2.7 (a) Bomb crater — unknown location, but probably Arenal, 1968. US Department of Interior, Geological Survey Photo File No. P19-68-57a. (b) A bomb crater about 1.5 km from the Karkar, March, 8, 1979 vent. Craters in this area were as much as 7 m across and up to 2 m deep. Bombs were probably ejected during the phreatic eruption at angles of 30°-60°. Prior to the eruption this area was covered with tropical rainforest.

vertical to less than 45° during the one eruption (Self *et al.*, 1974: 542). Around Ngauruhoe, New Zealand, in 1975 final trajectories of andesitic blocks appear to have been vertical as the projectiles remained in their impact craters. This was also the case at Arenal (Fig. 2.7a). Although ejection angles at Karkar were believed to be 30° – 60°, the craters produced (Fig. 2.7b) are not elongated (McKee *et al.*, 1981: 71). Minakami's (1942) data for eruptions of Asama during the 1930's indicate final trajectories of 60° on 10° slopes. However, many projectiles landing on steep slopes will roll downhill.

Table 2.2 provides some examples of maximum projectile sizes and ejection distances for specific eruptions. Theoretical considerations and the examples given suggest that ballistic projectiles will rarely reach more than 5 km from the vent.

The kinetic energy of impact of a volcanic projectile is dependent on both mass and terminal velocity, both being a function of projectile diameter and projectile density. Walker *et al.* (1971) found that terminal velocities determined experimentally were well matched by velocities computed for smooth cylinders. As a projectile has only to fall some tens of meters to attain terminal velocity the kinetic energy of impact (= ½ mv^2) can easily be determined for a range of projectile sizes, densities and masses. As volcanic projectiles range from lithic fragments derived from the vent wall to highly vesiculated pumice, the densities of projectiles also range widely, from about 0.15 to about 3.6 g cm^{-3}. Figure 2.8 indicates that even relatively small projectiles cover several orders of magnitude of impact energy. As projectile density spans more than an order of magnitude this diagram also demonstrates that it is impact energy rather than size or density alone that determines projectile hazardousness.

Sparks and Wilson (1976: 443) have shown that magmatic particles larger than about 5 mm will fail to maintain thermal equilibrium with the surrounding gas dispersion; that is, they remain hotter than the convecting cloud, the rate of cooling being dependent on particle radius.

Data on projectile temperatures are sparse. Initial temperatures are likely to range up to about 1100°C for projectiles generated from new magma but there is little evidence that lithic fragments torn from the vent wall need reach thermal equilibrium with the gas dispersion.

Table 2.2 Ballistic ejection distances

Volcano	Year	Projectile size	Distance	Reference
Vesuvio	1929	2-3 kg	3 km	Rittman, 1962: 33
Stromboli	1930	30 t	3 km	Rittman, 1962: 37
Ngauruhoe	1975	0.8 m diam.	2.8 km	
Ruapehu	1945	0.9 m diam.	1.2 km	Beck, 1950
Asama	1937	>1.0 m diam.	3.5 km	Minakami, 1942: 77
Asama	1783	0.5 m diam.	11 km	Aramaki, 1956: 197
Usu	1977	0.3 m diam.	2 km	Katsui *et al.*, 1978
Kilauea	1924	10 t	1.1 km	Stearns, 1925: 4
		8 t	1.1 km	Williams & McBirney, 1979: 19
Arenal	1938	~340 t	0.3 km	Minakami, 1942: 86
	1968	>0.5 m diam.	>4 km	Minakami *et al.*, 1969: 787
Soufrière	1902	~21 kg	6 km	*Nature*, 1903, Jan 15: 257

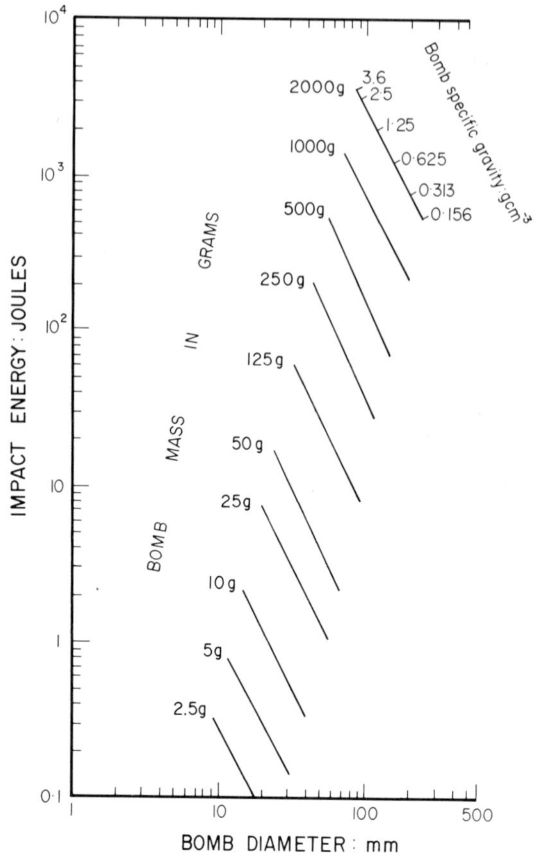

FIG. 2.8 Projectile impact energy as a function of bomb density, mass and diameter (modified after Blong, 1981).

However, temperatures on impact are of more significance from a hazard point of view. In 1783 at Asama, Japan, pumice bombs 50 cm in diameter landed red hot at a distance of 11 km from the crater (Aramaki, 1956: 197). At Sakura-jima in 1914, forest undergrowth was set on fire by falling projectiles at a distance of 4.7 km (Omori, 1916b: 210). These two examples indicate that projectile impact temperatures may well be above the ignition point for vegetation and a variety of man-made objects.

Although the hazard from volcanic projectiles may stem from either high impact energies or temperatures above ignition points, it seems likely that only relatively large projectiles with high kinetic energies will arrive at high temperatures.

2.2.2 Tephra fall

Tephra fall occurs from the rising eruption column and from the plume as it drifts downwind. The rate of rise of the eruption cloud can be spectacular: the May 22, 1924, Kilauea cloud rose to 4.5 km in nine minutes while a Surtsey (1963) cloud reached 8 km in 11 minutes and the Soufrière, St Vincent (1979) column climbed 9 km in five minutes (Stearns, 1925: 4; Anderson *et al.*, 1965: 1179; R.S.J. Sparks, 1980, personal communication). These and other data suggest that rates of rise of 8–30 m s^{-1} may be common for small to moderate eruptions.

As the eruption cloud rises it expands laterally as the surrounding air is entrained and passes into zones of varying wind speed, direction and rates of lateral shear. These effects are illustrated by the expansion of the 1947 Hekla cloud (Fig. 2.9). Data presented in Table 2.3 suggest that rates of downwind drift of tephra clouds are usually in the range 20–100 km h^{-1} (6 – 30 m s^{-1}).

As the tephra cloud rises, its density declines so that the larger and heavier particles fall from the cloud. Thus, particle sizes decrease downwind; Figure 2.10 indicates the general relationship between distance from the vent and median particle size.

As the tephra cloud disperses downwind the concentration of particles declines so that the thickness deposited also declines. Figure 2.11 shows the general decrease in deposit thickness with increasing distance along the main axis of the depositional ellipse for eight eruptions. Figure 2.12 illustrates the simple depositional ellipse for Pompeii pumice produced during the AD 79 eruption of Vesuvio and the deposit at Pompeii. This fallout pattern is typical of those produced where the

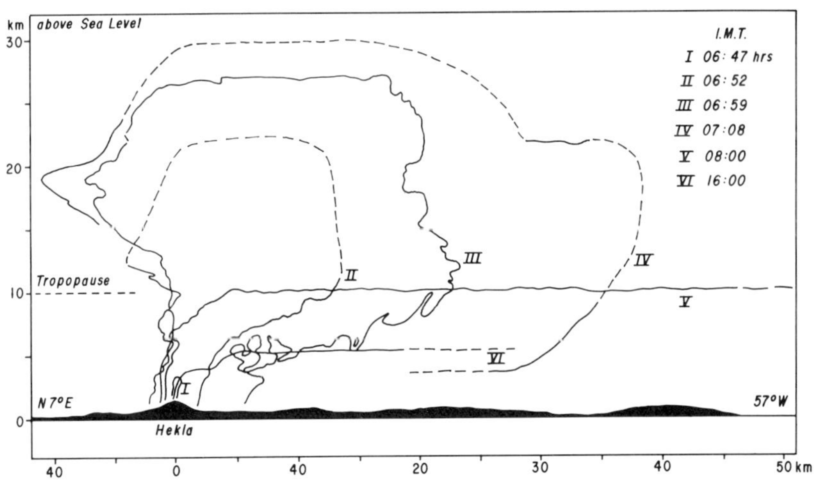

FIG. 2.9 Growth of the 1947 Hekla eruption cloud (after Thorarinsson, 1954).

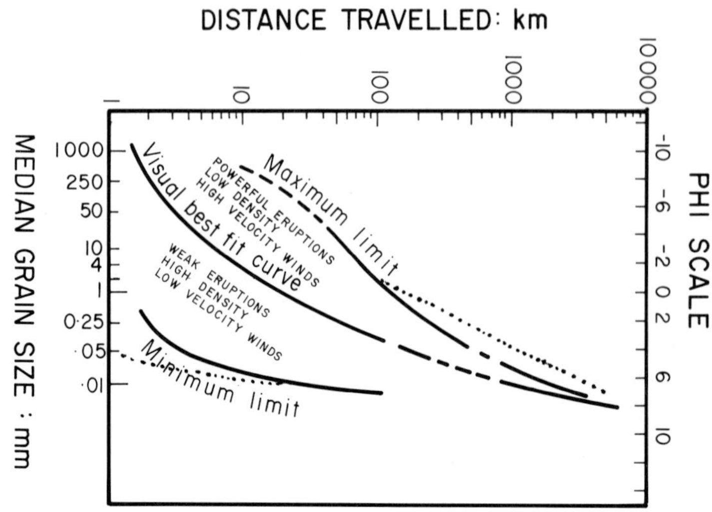

FIG. 2.10 Airborne tephra particle size versus distance travelled. Dotted line represents the maximum and minimum limits (after Fisher and Schmincke, 1984).

Table 2.3 Downwind velocities of tephra clouds

Volcano and date of eruption	Distance (km)	Time interval (h)	Velocity (km h^{-1})	Reference
Taal, Philippines, 1911	20	0.2	100	Martin, 1911: 88
Sakura-jima, Japan, 1914	560	24.0	29	Jaggar, 1924: 467
	960	48.0	20	Jaggar, 1924: 467
	9-10	1.2	9	Jaggar, 1924: 447
Asama, Japan, 1958-1961	49	1.9	27	Murai & Hosoya, 1964: 211
	224	2.3	100	Murai & Hosoya, 1964: 211
Bezymianny, USSR, 1955	45	0.5	94	Gorshkov, 1959: 86
	80	1.6	49	Gorshkov, 1959: 87
Shiveluch, USSR, 1964	80	1.45	55	Gorshkov and Dubik, 1970
	400	6.00	68	Gorshkov and Dubik, 1970
Katmai-Novarupta, Alaska, 1912	160	2.0	80	Wilcox, 1959: 417, 427
	88	3.5	25	Erskine, 1962: 27
Spurr, Alaska, 1953	135	6.0-7.0	21	Wilcox, 1959: 420-422
Mount St Helens, USA May 18, 1980	220	2.3	98	Sarna-Wojcicki *et al.,* 1981: 588
	460	3.8	120	
Cosiguina, Nicaragua, 1835	1120	24.0?	47	Cladcleugh, 1836: 30
Cerro Azul (Quizapu), Chile, 1932	2960	120.0	25-60	Francis, 1976: 176
			65	Wilcox, 1959: 433
Soufrière, St Vincent, 1812	160	4.5	35	Anderson and Fleet, 1903: 475
1902	160	3.5	45	
	1320		100	Bullard, 1976: 144
Hekla, Iceland, 1947			80	Thorarinsson, 1954
1970			74	Thorarinsson and Sigvaldason, 1972: 273
Askja, Iceland, 1875	1800	38.0	47	Thorarinsson, 1971: 225

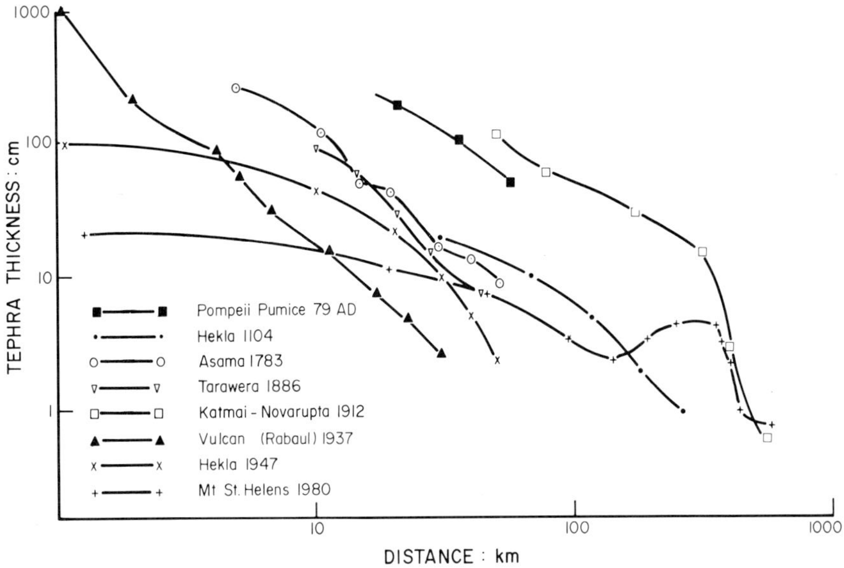

FIG. 2.11 Airborne tephra thickness versus distance from vent. Note the unusual pattern presented by Mount St Helens tephra (modified after Blong, 1982; Mount St Helens data from Sarna-Wojcicki *et al.*, 1981: 583; Asama data from Aramaki, 1956; Rabaul data from McKee *et al.*, in press).

winds at all elevations blow in the same direction during the eruption (see also Fig. 2.9). The pattern shown in Figure 2.12c is produced when winds blow in more than one direction, or when there is little wind and ejecta are distributed in all directions around the vent.

Clearly, the areas experiencing tephra fall in a major eruption can be very large. For example the area covered by pumice and fine ash around Krakatau in 1883 was about 827 000 km^2 (*CAVW*, 1: 55). However, much of this area received only a very thin deposit. Data in Table 2.4 for the 1970 Hekla tephra fall illustrates the point; 0.01% of the area experiencing tephra fall was covered to a depth of more than 20 cm and only 1% to a depth of more than 2 cm. Similarly, although the Cerro Negro, Nicaragua eruption of 1971 deposited ash over an area of about 5760 km^2, only 760 km^2 experienced falls greater than 2.5 mm (*SICSLP*, 1971).

While the hazards produced by tephra fall result principally from the accumulation of the deposit (Fig. 2.13), dangers also exist where large pumice lapilli have not reached thermal equilibrium before impact. During the 1883 Krakatau eruption large blocks of pumice were still hot enough to burn holes in cloth at a distance of 70–80 km from the vent (Docteurs van Leeuwen, 1936: 27). Similarly, birch trees were charred 10 km from Hekla by rhyolitic pumice during the 800 BC eruption

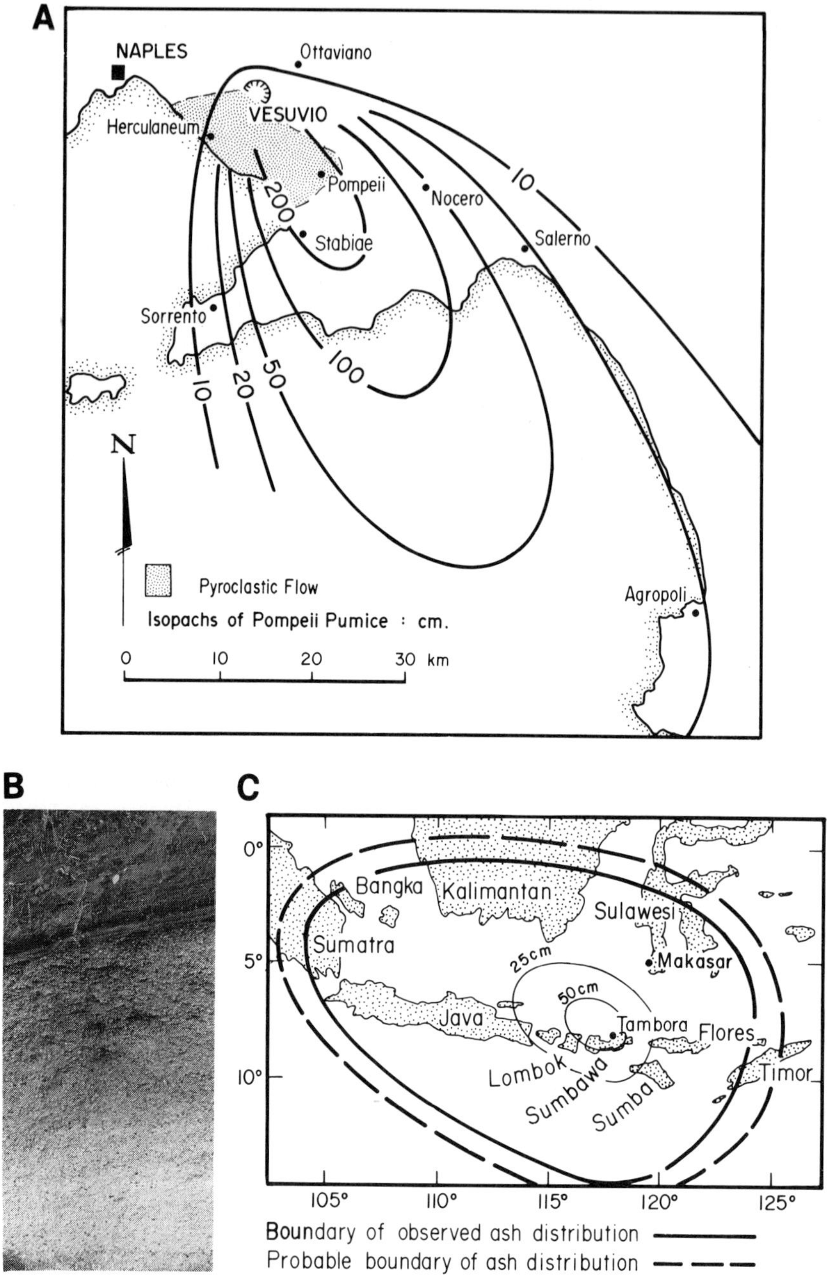

FIG. 2.12 (top) Tephra isopach map of Pompeii pumice, produced by the AD 79 eruption of Vesuvio (after Lirer *et al.*, 1973). (bottom, left) The Pompeii pumice deposit at Pompeii. The lapilli layer is 2.6 m thick. (bottom, right) Distribution of the 1815 tephra from Tambora (modified after Petroeschevsky, 1949).

FIG. 2.13 (top) Tephra mantle on the western side of Sakura-jima island in 1914 (© National Geographic Society); (bottom) Tephra-mantled landscape and house near Parícutin, July, 1943 (Smithsonian Institution Archives, Record Unit 7281, William F. Foshag Papers, F97h).

Table 2.4 Thickness of tephra and area covered by 1970 Hekla eruption

Thickness (cm)	Area (km^2)
>20	9
10-20	33
5-10	95
2-5	500
1-2	2080
0.5-1.0	3300
0.1-0.5	6950

From Thorarinsson and Sigvaldason, 1972: 175

(Thorarinsson, 1971: 229). Thermally oxidised pink pumice blocks in a number of Plinian deposits, including those of Taupo Pumice (*c.* AD 180) 30 km from the vent (G.P.L. Walker, 1983, personal communication), indicate that the impact temperatures of lapilli may be high enough to ignite a variety of materials at considerable distances.

Extreme darkness during tephra falls may also create dangerous situations. An account from Makasar, nearly 400 km from Tambora (Fig. 2.12c), indicates the nature of the problem:

> It was now evident that an eruption had taken place from some volcano, and that the air was filled with ashes or volcanic dust, which already began to fall on the decks. By 11 the whole of the heavens was obscured except a small space near the horizon to the eastward;... The ashes now began to fall in showers, and the appearance altogether was truly awful and alarming... The darkness was so profound through the remainder of the day, that I never saw any thing equal to it in the darkest night; it was impossible to see your hand when held up close to the eye:– the ashes continued to fall without intermission through the night... by half past 9 the shore was distinguishable, the ashes falling in considerable quantities, though not so heavily as before. The appearance of the ship, when day light returned, was most extraordinary; the masts, rigging, deck, and every part being covered with the falling matter.
>
> [Anon, 1816: 252-253]

Considerations presented in this section indicate that the hazard presented by tephra fall should decrease exponentially with distance from the volcanic source. This suggestion is a reasonable first approximation providing the elliptical pattern of fallout and dominant wind directions are kept in mind.

However, the hazard presented by a tephra fall is also influenced by a number of other factors. Consideration of the Hjulström curve and Bagnold's fluid threshold velocity (see Blong, 1981) suggest that tephra deposits with an exposed grain size of about 0.1–0.4 mm will be more easily eroded by wind and water than either coarser or fine-grained deposits. Fine-grained moist tephra deposits are highly cohesive. Furthermore, airfall deposits with surface grain sizes less than about 0.1 mm have a tendency to form surface crusts which encourage runoff and inhibit plant penetration. In chapter 2, Section 8 it will also become clear that toxic fluorine compounds tend to be concentrated on fine-grained particles. These examples suggest that thin and/or fine-grained deposits of tephra may, in some circumstances, represent a greater hazard than thick and/or coarse-grained tephras.

2.3 PYROCLASTIC FLOWS AND DEBRIS AVALANCHES

In this volume the term *pyroclastic flows* is used to refer to three classes of volcanic phenomena:

1. Pyroclastic flows (*sensu stricto*), also called *nueés ardentes* or glowing avalanches. These involve flow over the ground surface of fragmental material (pyroclasts) as hot, gravity-controlled, concentrated gas–solid dispersions which accumulate in topographic lows (Sparks, 1976). These flows are generated when the convective thrust phase of a rising tephra column fails to take over from the initial gas thrust phase. If entrainment of air is not sufficient to maintain the column density below that of the surrounding atmosphere, the convective thrust fails and gravitational column collapse occurs with the lateral movement of the dispersion down the flanks of the volcano as a pyroclastic flow. At this point a Vulcanican or Plinian eruption becomes a Peléean eruption (Sparks and Wilson, 1976; Wilson, 1976). Pyroclastic flows may also result from the collapse of an actively growing lava extrusion or dome (Wright *et al.*, 1981: 459-460.
2. Pyroclastic surges involving turbulent lateral movements of expanded low-concentration gas–solid dispersions. Surges accumulate in depressions but also mantle the topography. They may be hot or cold (Walker, 1981: 393).
3. Directed blasts resulting from collapse of part of the volcano. This category includes the eruption styles of Bezymianny 1956 and Mount St Helens, 1980 (Gorshkov, 1963; Hoblitt *et al.*, 1981).

Although the use of *pyroclastic flow* to cover these three broad types of phenomena is a loose use of the term (see, for example, Wright *et al.*, 1980) it is justified in the sense that the three phenomena are parts of a continuum (Sheridan and Wohletz, 1981: 1388). Furthermore, terminology in the literature is very confused. For example, the type *nuée ardente* at Mount Pelée (May 8, 1902) was originally interpreted by Lacroix (1904) to be a directed blast. Furthermore various authors in the United States Geological Survey volume (Christiansen and Peterson, 1981: 378) on Mount St Helens, 1980, use directed blast, pyroclastic surge and pyroclastic density flow to refer to the one set of deposits.

A fourth category of flow phenomena, here called *debris avalanches,* are also considered in this section. The term is reserved here for the large-scale collapse (the rockslide avalanche of Voight *et al.*, 1981) such as occurred at Mount St Helens before the directed blast and at Bandai in 1888. Again, the literature is confused (Neall, 1976, for example classifies the Bandai event as a lahar); furthermore, it is difficult, from a hazard point of view, to differentiate many of the effects of large debris avalanches from those of pyroclastic flows.

Two descriptions indicate some of the similarities between these phenomena. The first provides an eye witness account of a *nuée ardente* from Mt Pelée, Martinique, on the evening of July 9, 1902 (that is, after the main eruptions of May 8 and May 20, 1902):

> As the darkness deepened, a dull-red reflection was seen in the trade-wind cloud which covered the mountain summit. This became brighter and brighter and soon we saw red-hot stones projected from the crater, bowling down the mountain slopes and giving off glowing sparks. Suddenly the cloud was brightly illluminated... In an incredibly short space of time a red-hot avalanche swept down to the sea... It was dull red with a billowy surface, reminding one of a snow avalanche. In it there were large stones, which stood out as streaks of bright red, tumbling down and emitting showers of sparks. In a few minutes it was over.
>
> [Anderson and Flett, in Russell, 1902: 431]

The second description is of a gigantic rockfall (or debris avalanche) during the 1888 eruption of Bandai:

> This slab fell from a place about 300 metres high. The great masses of earth and rock were shattered as they fell, and broken up into pieces, ever growing smaller as they descended. The behaviour of this pulverized mass resembled the rush of a headlong torrent. Although boulders measuring 10 metres of more in diameter were mixed up with finer matter, as a whole the movement approximated to that of a fluid. No words can describe the fierceness and force of that impetuous downpour - its mad surgings this way and that, and the bold leaps with which it would now and then bound over low ridges that hindered its progress, and shot onward down the neighbouring depression.
>
> [Sekiya and Kikuchi, 1889: 106-107]

Velocity, temperature and flow dimensions are the three aspects of interest in an evaluation of pyroclastic flow hazards.

The velocities of a number of pyroclastic flows are recorded in Table 2.5. In most cases these measurements have been made from photographs taken at a considerable distance from the flow and the measurements refer to the more distal parts of the flows. Furthermore, nearly all the measurements refer to small-scale pyroclastic flows rather than the grand scale radial collapse of Plinian eruption columns.

Table 2.5 Velocities of pyroclastic flows

Eruption	Velocity ($m\ s^{-1}$)	Reference
Lamington, Papua New Guinea, 1951	27-33 (and much slower)	Taylor, 1958: 43-44
Anak Krakatau, Indonesia, 1927	14	Moore, 1967: 25
Merapi, Indonesia, 1930	16	Stehn, 1935, in Taylor 1958: 43
Mayon, Philippines, 1968	9-63	Moore & Melson, 1969: 617-618
	av. 31	Moore & Melson, 1969: 617-618
Asama, 1958	25-30	Murai & Hosoya, 1964: 209
Bandai, 1888	21	Sekiya & Kikuchi, 1889: 108
Bayonnaise Rocks, Japan, 1952	>65	Sekiya & Kikuchi, 1889: 20
Tokachi, Japan, 1926	50	Murai, 1960: 57
Mount St Helens, 1980	170 slowing to ~28	Moore and Sisson, 1981: 437
Mt Pelée, May, 1902	≤55	Fisher *et al.*, 1980: 473
Mt Pelée, 1902, 1929-32	20-33	Perret, 1937: 96
1929-32 (small nuées)	10-20	Perret, 1937: 96
1902	32	Anderson, 1908: 493-497
Soufrière, St Vincent, 1902	>11-13	Anderson & Flett, 1903: 393

Thus they represent the low velocity end of the pyroclastic flow spectrum compared with those predicted in the theoretical models of Sparks *et al.* (1978) and with those inferred for prehistoric ignimbrites by Wilson and Walker (1981: 442). The latter infer velocities of from 70–200 m s^{-1}. Velocities of 20–120 m s^{-1} have also been inferred for large debris/rock avalanches (Scheidegger, 1973: 234).

High velocity pyroclastic flows and debris avalanches thus have the momentum to climb substantial topographic barriers (Francis and Baker, 1977); for example, the 22 000 or so year old Ito Ignimbrite in the Aira (S. Kyushu) caldera climbed over a topographic barrier of more than 600 m, 55 km from source, the 1800 year old Taupo (New Zealand) Ignimbrite overrode hills more than 1000 m above the vent, the 9000 year old pyroclastic flows around Fisher caldera (Alaska) flowed up slopes at least 500 m high, and the May 18, 1980, debris avalanche from Mount St Helens surmounted a ridge 360 m high just north of the volcano (Yokoyama, 1974: 26; Walker *et al.,* 1980; Voight *et al.,* 1981: 351; Miller and Smith, 1977: 175). These and numerous similar examples indicate that lee-side locations may also be at risk.

High velocities also indicate that erosion by the flow (or surge) may be substantial. Significant erosion occurred close to the vents in the Mayon, 1968, and Ruapehu, 1975, flows even though velocities were only moderate (Table 2.5).

Pyroclastic flow velocity may also be quite variable. Moore and Melson (1969: 618) analysed 16 mm movie film shot at 24 frames per second to show that the forward velocity in one *nuée* varied from 9–63 m s^{-1} in a 66 second period. Perret (1937) suggested that variations in velocity within a pyroclastic flow resulted from additional gas release and expansion produced by mechanical shock when the flow met obstacles or dropped over steeper portions of its path. Taylor (1958) suggested that continued production of the pyroclastic flow caused a series of surges in the source area which produced waves in the moving mass. These alternatives seem equally valid. It is interesting to note that the description of the Bandai, 1888, debris avalanche presented earlier also suggested considerable variation in forward velocity.

Although Perret (1937) believed that many of the variations in destruction left by the early pyroclastic flows were removed by subsequent ones, marked variations in the effects of pyroclastic flows can occur across quite small areas. Perret (1932) records that branches were left on trees in St Pierre while nearby monuments were destroyed. Taylor (1958: 40) also noted the extremely variable destruction in some small areas around Lamington, Papua New Guinea, in 1951. Deiss (1966: 18) makes this interesting comment on the effects of the pyroclastic flow from Vesuvius that buried Herculaneum in AD 79:

> The force of the stream of mud [sic] also varied greatly. At some points its pressure felled walls and columns, smashed furniture, and carried away statuary... At other points it rose so gently that it oozed over eggs without breaking them and filled rooms without disturbing the positions of pots on a kitchen stove or a cradle in a nursery.

More complex flow movements also occur. Perret (1937: 22-33) believed that pyroclastic flows swept around (with a rolling rather than a sliding motion), and not over some areas that were only slightly elevated. A more recent analysis indicates that the May, 1902, ash clouds at Mt. Pelée expanded seawards, developing a lateral component at the coast strong enough to destroy St Pierre. These flows also had the ability to refract (Fig. 2.14) around obstacles as shown by the total destruction of Ste Philomene northwest of the path of the main flows (Fisher *et al.*, 1980: 474-5). Similarly, at Bagana, Bougainville, in 1966, Bultitude (1976: 326) recorded that vortices produced by the movement of pyroclastic flows over irregular terrain felled trees in the direction from which the flows descended.

The primary evidence for high emplacement temperatures in pyroclastic flows stems from evidence of welding (the sintering together of vesicular fragments and glass shards) and the complete carbonisation of tree trunks. Temperatures measured or inferred for pyroclastic flows are presented in Table 2.6. It should be noted that some of these measurements were made long after emplacement of the flows. However, heat is retained in thick flows; charred buried tree stumps have been reported to spontaneously ignite on exposure even months after burial around both Lamington, 1951, and the Soufrière, St Vincent, 1902 (Taylor, 1958; Anderson and Fleet, 1903: 432).

Two features are of importance. First, data from Lamington indicates that at Higaturu in the zone of total devastation (Figure 2.15) temperatures of around 200° were sustained for only a minute or two (Taylor, 1958). Similarly, Moore and Sisson

Table 2.6 Temperatures of pyroclastic flows (measured after emplacement)

Eruption	Temperature and comment	Reference
Karkar, Papua New Guinea, 1979	~300°C, scorching of trees	McKee *et al.*, 1981: 69
Lamington, 1951	~200°C — softening of plastics for 1-1½ minutes	Taylor, 1958: 46-47
Ngauruhoe, 1975	90°C, 90 minutes after eruption	Nairn *et al.*, 1976: 390.
Mayon, 1968	98°C in lower part at 60 cm 5 days after emplacement; temp. in large block on surface ~240°C in 3 cm deep crack	Moore and Melson, 1969: 612
Taal, 1965	<100°C	Moore *et al.*, 1966: 6
Asama, 1783	~900°C — strongly welded	Aramaki, 1956: 223
Augustine, Alaska, 1975	604°C at 5 m — others >400°C several days after emplacement	*BVE*, 1978, 16: 55
Mount St Helens, 1980	~350°C — charred trees	Moore & Sisson, 1981: 437
Mt Pelée, May, 1902	1075°C — gold coins distorted	Trechmann, 1938: 4367
	>700°C — bottle glass deformed	
	>300°C — clothing ignited	Will in *HMSO*, 1903: 10
Soufrière, St Vincent, 1902	<300°C — clothing not ignited on E coast	Will in *HMSO*, 1903: 14
Soufrière, St Vincent, April-May 1979	300-350°C infra-red spectrometry of carbonised wood	Bardintzeff & Boudan, 1979: 892

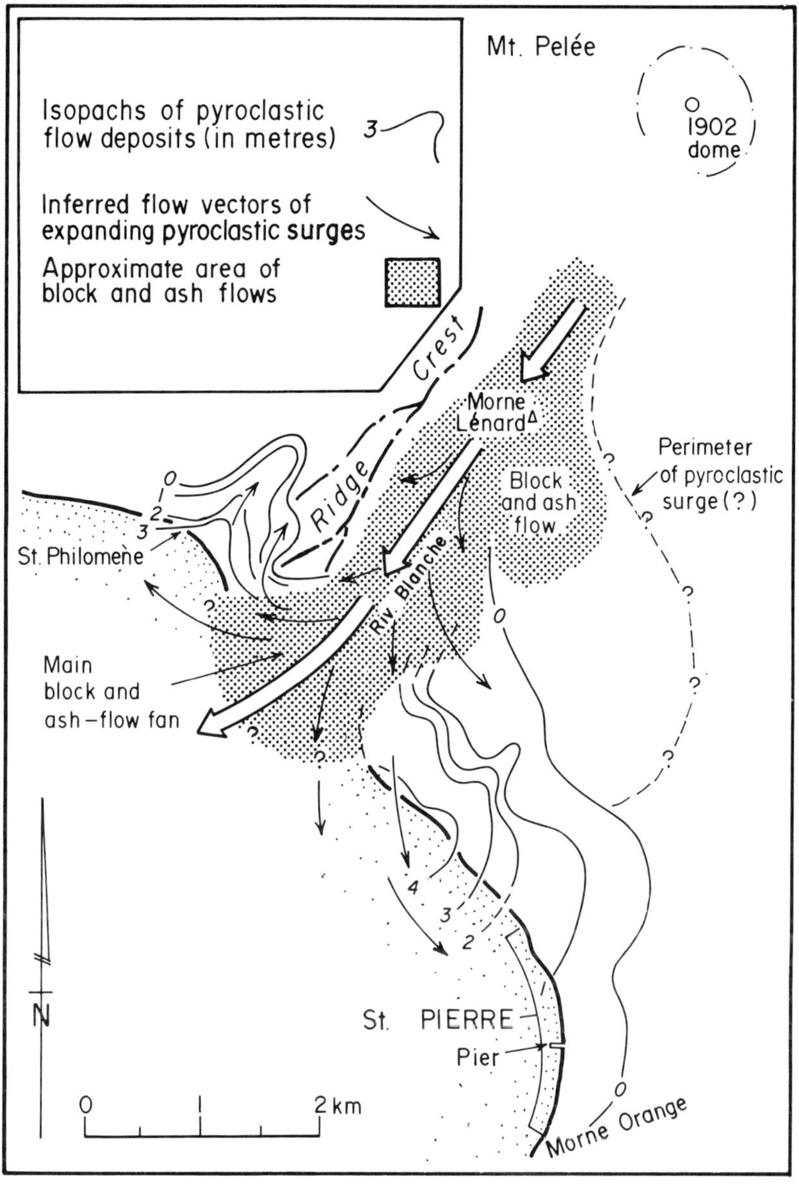

FIG. 2.14 Deposits left by pyroclastic flows produced by Mount Pelée on May 8 and May 20, 1902. The main flows moved down the River Blanche. Note the reversed flow direction in the St Philomene area and the thinness of the deposit in St Pierre (after Fisher *et al.,* 1980: 474).

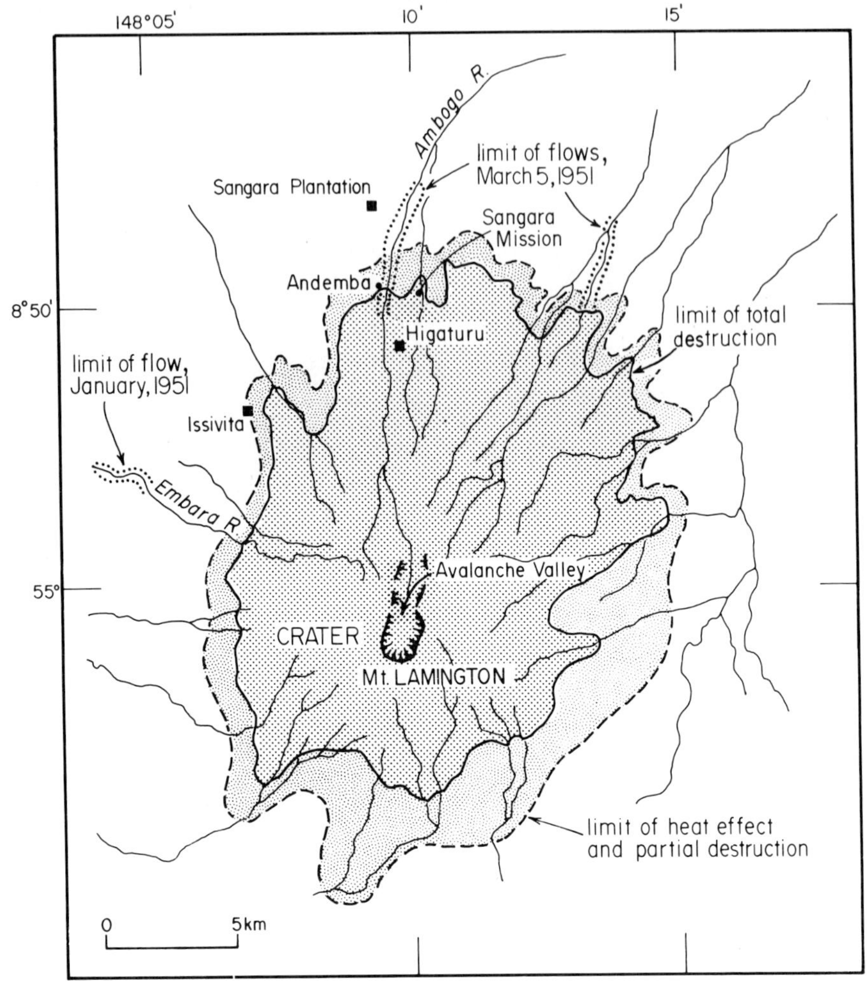

FIG. 2.15 The areas of total and partial devastation by pyroclastic flows around Lamington in 1951 (after Taylor, 1958; Bullard, 1976).

(1981: 424) suggest that the Mount St Helens directed blast produced temperatures of 350 ± 50°C near source and from 50°–200°C near the pyroclastic flow margin. These temperatures were maintained for less than two minutes.

Second, data from several sources emphasises the extreme variability of temperature effects over short distances. Deiss (1966: 18) again comments on the effects at Herculaneum: "In some places it was hot enough to carbonize foodstuffs and wood [Fig. 2.16]; in others, hot enough merely to scorch cloth and papyrus; and in still others wood, rope, eggs and even fishnets were left totally unburned. In some

FIG. 2.16 A wall opening at Herculaneum filled with flow deposits and showing the carbonised wooden frame.

houses wax tablets remained unmelted". Considerable variation was also reported in St Pierre, 1902. Although Perret (1937: 101-103) suggested that some of the anomalies in temperature and pressure/velocity effects were removed by subsequent pyroclastic flows it seems equally possible that it is the passage of successive flows at different velocities and temperatures which created large variations across small areas. Certainly, further variations in the temperature effects noted stem from lightning strikes which are frequent during the passage of the cloud and from the ignition of fires which consume small areas. Such fires may also be extinguished by the passage of subsequent flows.

Pyroclastic flow deposits are highly variable in thickness but thin laterally. In areas of partial destruction (Figs. 2.15, 2.16) the basal concentration is very thin — little deposit is left but ash clouds or ash surges can still be very hot with high velocities. Even in these marginal areas almost total destruction of human life and property can occur even though little material is deposited. This is also true near to the source where erosion, rather than deposition, is likely to take place (Fig. 2.17).

Isopachs of thickness for the May 8 and May 20, 1902, pyroclastic flows from Mt Pelée are shown on Figure 2.14. Boulders in these flows ranged up to 9 x 7 x 7 m, 5 km from the crater (*Nature,* 1903, Jan 15: 258), even though the combined thickness of deposit from a number of successive flows on these two days was only 0–3 m in St Pierre. Over much of the area around Soufrière, St Vincent the May 7,

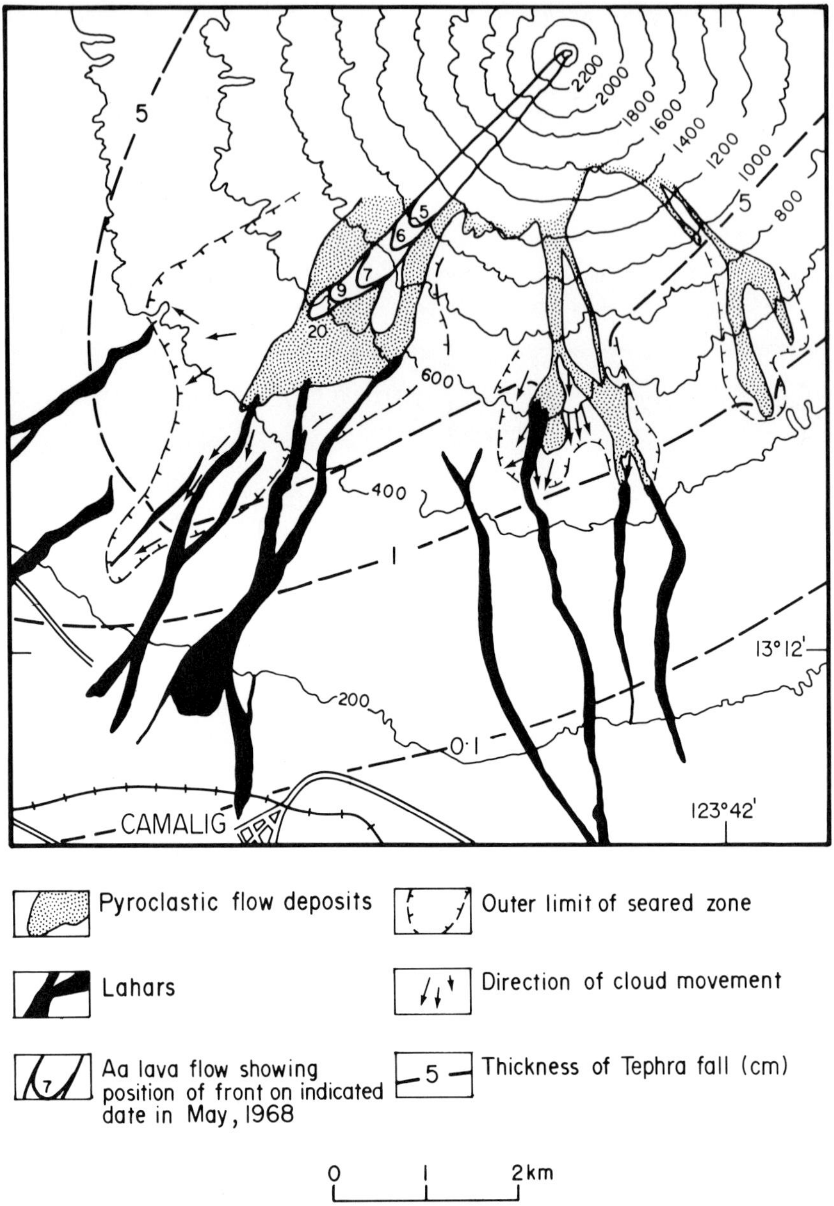

FIG. 2.17 Map of the deposits produced during the 1968 eruption of Mayon (after Moore and Melson, 1969: 602).

1902, eruption left hot pyroclastic flow deposits less than one meter thick. However, in some areas, presumably river valleys, deposits were more than 12 m thick (Anderson and Fleet, 1903: 393).

The areas affected by some pyroclastic flows are shown in Table 2.7 and in Figures 2.14, 2.15 and 2.17. While the areas affected seem impressive, especially when compared with areas covered by historic lava flows, it should be remembered that the examples represent pyroclastic flows produced by large Vulcanian and small Plinian eruptions. Although very large-scale Plinian eruptions have not yet occurred in the historic period, prehistoric pyroclastic flow units covering thousands of square kilometers are not uncommon; for example, the 1400 year old Rabaul pyroclastic flow covered 1200 km^2 to an average depth of 2 m and the Taupo Ignimibrite emplaced 1800 years ago buried more than 15 000 km^2 under a veneer deposit averaging 1 m in thickness. Such low-aspect ratio ignimibrites (Walker *et al.*, 1980) present the most severe volcanic hazard.

It is important to recognise that pyroclastic flows cover a broad spectrum of magnitude and violence (see Chapter 1, Section 4). Relatively small flows are strongly controlled by topography, affecting but a sector of the area around the source, and have insufficient velocity to climb anything but the smallest topographic obstructions. On the other hand, large flows travel radially outwards to roughly equal distances in all directions, frequently flowing transversely across quite deeply incised valleys and climbing over obstructions tens of meters high even tens of kilometers from the source. Wilson and Walker's (1981: 446-447) selection of the Valley of Ten Thousand Smokes ignimbrite (Katmai-Novarupta, 1912), one of the largest pyroclastic flows produced so far this century, as an example of a relatively small flow, emphasises that extreme hazards may exist!

As indicated in Table 1.4 pyroclastic flows cause a substantial proportion of deaths. Deaths result not only from the extreme mobility of pyroclastic flows but also from the (sometimes) high temperatures. Although high temperatures may

Table 2.7 Areas affected by pyroclastic flows

Eruption	Area affected (km^2)	Reference
Bandai, 1888	~70	Sekiya and Kikuchi, 1888: 106
Soufrière, St Vincent, 1902	77	measured from *HMSO* (1903) map
Mt Pelée, May 1902	41	Measured from Heilprin's (1903) map
August 1902	58	
Taal, 1911 (practically total devastation)	230	Pratt, 1911: 63
Katmai-Novarupta, 1912 (Valley of Ten Thousand Smokes only)	126	Bullard, 1976: 166
Lamington, 1951 (area of complete and partial devastation)	230	Taylor, 1958: 39
Mount St Helens, 1980	~600	Hoblitt *et al.*, 1981: 404

damage or destroy buildings and crops, more damage is likely to result from the very high kinetic energies. As kinetic energy is proportional to the square of impact velocities, the impact energy of a pyroclastic flow is likely to be one to four orders of magnitude higher than that of an equal mass of tephra fall. Furthermore, the rate at which energy is expended is very much higher in the case of pyroclastic flows and debris avalanches.

2.4 LAHARS AND JÖKULHLAUPS

The term *lahar* is now used fairly loosely for "a mudflow containing debris and angular blocks of predominantly volcanic origin" and for deposits resulting "from rapid mass flowage of rock debris mobilized by water and that have originated on the slopes of a volcano" (Crandell, 1971: 3).

There is little point in attempting to define *lahar* rigidly. Mass flow phenomena on volcanic slopes form a continuum from hot extremely mobile low concentration pyroclastic flows through mudflows to turbid streams. Because pyroclastic flows follow topographic lows (that is, stream lines) water is easily admixed as flows progress down valley. Thus a pyroclastic flow can become a lahar as cooling, outgassing and fluid uptake occur progressively with distance from the source. Further down valley, as large clasts and other debris are deposited and as water uptake from the stream bed continues, the diluted high concentration flow becomes a muddy stream.

Although some authorities (Neall, 1976, for example) regard *jökulhlaups* or "glacier bursts" as a type of lahar the term applies particularly to water floods created by the bursting of subglacial lakes. The periodic jökulhlaups from Katla and Grimsvotn in Iceland are believed to result from glacial melting by volcanic activity. Enormous volumes of water may flood rivers. For example, Thorarinsson (1957) reports that the 1934 Grimsvotn eruption created a jökulhlaup with a volume of 7 km^3 while the 1918 Katla jökulhlaup reached an estimated peak discharge of 300 000–400 000 $m^3 s^{-1}$ (for comparison, the average discharge of the Mississippi River at its mouth is about 18 000 $m^3 s^{-1}$).

The following brief descriptions provide some idea of lahar characteristics and indicate some of the difficulties of differentiating between lahars and jökulhlaups:

1. Irazú, Costa Rica, 1964–1965, (Fig. 2.18): "Less than 30% of water is enough to transform dry volcanic ash into liquid mud. As a mudflow builds up and surges to the bottom of a valley, its initial volume is swollen with debris of all kinds and sizes that the mud sweeps along with it; trees are uprooted, rocks and stones are torn away from the sides and bottom of the valley. It has been found that a lahar may increase its volume sixfold in this way" (Tazieff, 1965: 23).
2. Kautz Creek, Mt Rainier National Park, Washington State, October 1947: "During the downpour [149 mm fell at the nearby ranger station], a cloudburst

FIG. 2.18 (top) Hydrographs of lahars in Rio Reventado resulting from removal of fresh tephra from slopes of Irazú 1964 (after Waldron, 1967). (bottom) An andesite boulder with an estimated weight of 200 t transported down the Rio Reventado by a lahar (Photo: H.H. Waldron, US Geological Survey).

evidently occurred in the upper part of the Kautz Creek drainage basin, and an additional amount of water may have been contributed by a 'glacier outburst flood' from Kautz Glacier (Richardson, 1968: D83). The discharge of Kautz Creek increased greatly during the daylight hours of October 2, and between 7 p.m. and 8 p.m., high water washed out the [park] entrance roadbridge over the creek. The first lahar occurred between 10 and 11 p.m. and was heralded by a dull roar and vibration of the ground as it advanced down the valley. A succession of lahars flowed downvalley during the night, and the last occurred at 8 a.m. the next morning. The lahars were described by eyewitnesses as having the consistency of wet concrete, carrying along vegetation and boulders as large as 13 feet [4 m] in diameter" (Crandell, 1971: 46). The total volume of the lahars and flood deposits was estimated to be at least 39 million cubic meters.

3. Lassen Peak, California, 1914. "About three o'clock [a.m.] we tried to get back to the house which had moved 53 feet [16 m] and lodged against a tree and the yard fence, but could not reach it at that time [the lahar struck at 11-12 p.m.] . . . On leaving the house we left the door open and the flood entered about two and a half feet [0.75 m] deep, then on the ebb flow part of the mud and water ran out, but when the mud dried it was about sixteen inches [40 cm] deep in the house. A green log also had floated in, a rough old thing, which was about all four of us could do to get it out of the house. The mudflow looked more like mortar than water, and when it ran over the ground it left it slick and smooth like a pavement" (Hill, 1970: 220-221).

The Mt Rainier (Washington State) lahar was initiated by heavy rains and (possibly) a glacier burst. The Lassen Peak (California) example stemmed from melting ice and snow during a lava eruption and the Irazú (Costa Rica) lahars were caused by heavy rains on slopes with freshly-fallen tephra. While there is little point in attempting to rigidly categorise lahars we can note that the Lassen Peak example is a primary lahar in that it stems directly from the eruption. The other two are examples of secondary lahars, or rain lahars (Legowo, 1981: 73), in that their origin is at least one step removed from the eruption. Both primary and secondary lahars can be initiated in a number of ways (Table 2.8). Some lahars may be initiated by several factors; for example, the Toutle River lahars (Fig. 2.19) from Mount St Helens (Washington State) on May 18, 1980 resulted from the descent of pyroclastic flows into the river and, to a lesser extent, from the melting of snow and ice (Lombard *et al.*, 1981).

From the available data it is clear that heavy rains and high runoff volumes, discharge from crater lakes, and rapid melting of snow and ice are the most common causes of lahar formation. Neumann van Padang (1951) and Zen and Hadikusumo (1965) report that 45 out of 56 lahars on 16 Indonesian volcanoes were associated with volcanoes having a crater lake (Neall, 1976: 234). Simkin *et al.* (1981: 18-20) report that lahars have been recorded for 5.6% of Holocene eruptions from a total

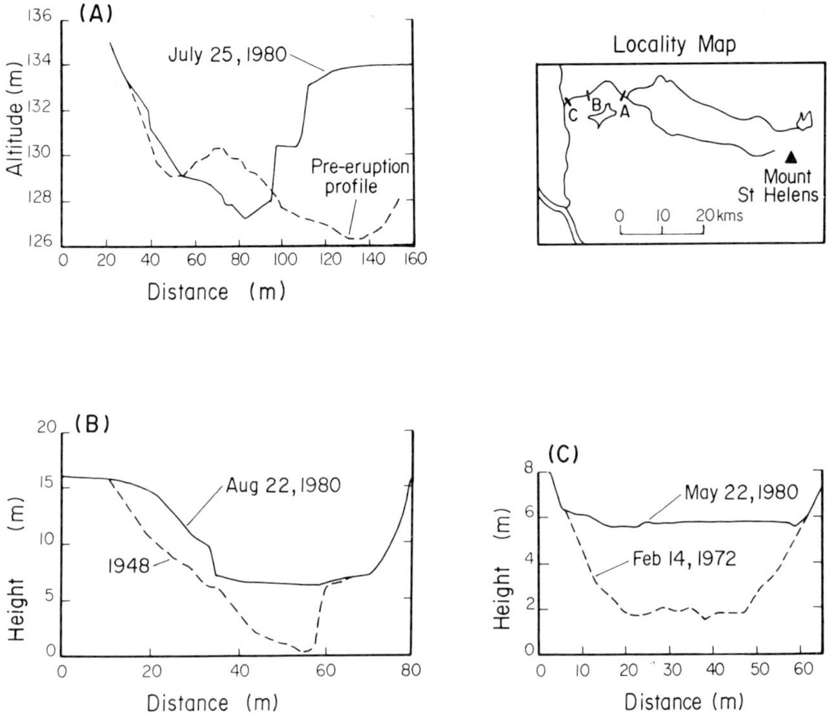

FIG. 2.19 Three cross sections of the Toutle River showing alterations to the valley floor elevation resulting from lahars produced during the May 18, 1980, eruption of Mount St Helens (after Janda *et al.*, 1981: 475).

Table 2.8 Causes of lahars

Causes	Examples
A. Primary lahars — associated with eruptions	
1. Eruption through a crater lake	Kelut, 1919 Ruapehu, 1969
2. Melting of snow and ice by eruption products	Cotopaxi, 1877 Tokachi, 1926 (Fig. 2.20)
3. Descent of pyroclastic flows into streams	Asama, 1783 Bezymianny, 1956
B. Secondary lahars — not directly associated with eruptions	
1. Heavy rains on tephra, perhaps associated with surface crusting and an increase in the runoff coefficient	Santa Maria, 1902 Mayon, 1968 Irazú, 1963-64
2. Rapid melting of snow or ice	Mt Shasta, 1926, 1931 Kautz Creek, Mt Rainier, 1947 Usu, 1977
3. Collapse of crater lake impoundment	Ruapehu, 1953 (Fig. 2.21)
4. Initiated by earthquakes	Mauna Loa, 1868

of 110 volcanoes. Most jökulhlaups have been produced on Icelandic and Andean volcanoes. Kelut, Indonesia, with 28 and Cotopaxi, Ecuador, with 25 lahar-producing eruptions, have been the most frequent sources of primary lahars. However, most secondary lahars are not included in the eruption catalog of Simkin *et al.* Notwithstanding the probability that volcanoes such as Mayon, Philippines, located in the west Pacific typhoon belt, have produced many more (secondary) lahars than either Cotopaxi or Kelut, Indonesian experience indicates that primary lahars are more hazardous than secondary lahars (Legowo, 1981: 73; Neall, 1976: 234).

Because most lahars flow down valleys, areas likely to be at risk can be predicted fairly readily. However, lahars can travel great distances down valley with great velocity, burying large areas under thick deposits. Examples of the dimensions and velocities of some historic lahars are presented in Table 2.9. Clearly, lahars span a range of velocities, thicknesses and travel distances. Because fluid densities and viscosities vary not only from lahar to lahar but also downstream in the one lahar, velocities and shear stresses will cover quite a range. Despite such variations those lahars which reach furthest downstream, those which are thickest (that is, flow overbank), and/or which move most rapidly should prove to be the most dangerous.

Unfortunately little research has been undertaken to define the critical states which produce hazardous lahars. In the case of lahars formed from eruptions through crater lakes or where crater lake walls have been breached the degree of hazard is proportional to the volume of water in the lake. For secondary lahars resulting from heavy rains, lahar initiation will depend on tephra thickness and particle size distribution, slope angle, rainfall amount and intensity, and runoff/infiltration amounts. Critical conditions can thus arise in a number of ways. Experimental work in the Nishiyamagawa, Japan, watershed where 500 mm of relatively coarse tephra fell during the Usu, 1977, eruption, indicated that the critical rainfall for lahar occurrence on 17°–18° slopes of 300 m length was greater than or equal to 20 mm with an intensity of more than 10 mm h^{-1} (Yamaoka *et al.*, 1978). Cold lahars occurred in some valleys on the slopes of Usu in October, 1978, when rainfalls of 15–20 mm occurred in 10 minutes (Kadomura *et al.*, 1979). For the 86 lahars which occurred before the end of October, 1978, the minimum amounts of rainfall which triggered the mudflows were 20 mm for continuous rain, 9 mm h^{-1} for one hour or 12–18 mm h^{-1} for 10 minutes (Kadomura *et al.*, 1980: 176). It was also demonstrated that saturated pumiceous tephra 400–640 mm thick would fail on 35° slopes when overlain by about 300 mm of snow (Yamada *et al.*, 1978). Around Taal, Philippines, in 1965 extensive movement of material off slopes occurred where tephra was more than one meter thick (Moore *et al.*, 1966: 6). Following the Galunggung, Indonesia, eruptions in 1982 lahars did not present a serious threat until the onset of the wet season late in the year.

Data summarised in Table 1.4 suggest that lahars are far less hazardous than pyroclastic flows. Certainly, lahars move more slowly and follow more predictable courses. While lahars may reach much further downstream, their delayed arrival,

Table 2.9 Lahar velocities and dimensions

Volcano	Lahar information
Asama, 1783 (A3)*	Lahar formed from pyroclastic flow debris dam collapse travelled 90 km down Agatsuma River at average 22 km h^{-1} (Aramaki, 1956: 198).
Galunggung, 1822 (B3)	Lahar travelled 24 km; deposit 100 x 10^6 m^3 (Neall, 1976: 235).
Cotopaxi, 1877 (A2)	Traveled 300 km at average speed of 27 km h^{-1} (Francis, 1976: 86).
Santa Maria, 1902 (A3 or B1)	Lahars travelled more than 100 km (Neall, 1976: 237).
Lassen, 1914 (A2)	200 t boulder carried 8-10 km (Hill, 1970: 216).
Kelut, 1919 (A1)	Lahar travelled 38 km at up to 64 km h^{-1} (Neall, 1976: 235)
Tokachi, 1926 (A2)	29 km^2 inundated by lahar about 3 m deep on average reaching speeds of 180 km h^{-1}; estimated volume 20 x 10^6 m^3. (Murai, 1960: 57-58; see Fig. 2.20).
Ruapehu, 1953 (B3)	Traveled >24 km; speed on gentler slopes ~16 km h^{-1}; peak discharge at Tangiwai ~850 m^3 s^{-1}; volume 1.9 x 10^6 m^3 (Neall, 1976: 234-235; Weir, 1983: 202; see Fig. 2.21).
Calbuco, 1961 (A2)	Peak discharge of one flow >3000 m^3 s^{-1}; velocity 20 km h^{-1} (Klohn, 1963)
Irazú, 1963-64 (B1)	Flow 12 m thick in places; peak discharge 497 m^3 s^{-1}; velocities up to 36 km h^{-1}; 200 t boulders moved (Fig. 2.18); sediment concentration 200 000 to 788 000 ppm (Waldron, 1967).
Kelut, 1966 (A1)	Lahars reached up to 25 m thick; covered 45 km^2 (Neall, 1976: 235).
Usu, 1978 (B2)	Maximum velocity 10 m s^{-1} covering 0.266 km^2 with 120 000 m^3 of debris from Nishiyama catchment; maximum thickness of flow 2 m; peak discharge 115 m^3 s^{-1} (Kadomura *et al.*, 1980: 169).
Mount St Helens, 1980 (A3)	Velocities of various lahars ranged from 140 km h^{-1} to 5 km h^{-1} near downstream limits (Janda *et al.*, 1981).

*(A3) refers to lahar type; see Table 2.8

particularly in the case of secondary lahars, often allows adequate warning. However, the hazard may continue for months after the eruption has ceased with even the danger areas migrating as the locus of deposition and flooding moves down valley or as lahar-filled rivers change courses (for example, around Kelut, Indonesia, in 1966 — Neall, 1976: 235), as stream bank erosion by lahars reactivates old landslides (near Irazú, Costa Rica, 1964 — Waldron, 1967), as debris dams rivers to produce lakes (near Spurr, Alaska, 1953 — Wilcox, 1959; Toutle River, Mount St Helens — Youd and Wilson, 1980; Steward, 1982), or as coastal areas are inundated by lahar deposits (west of Santa Maria, Guatemala, 1902 — Kuenzi *et al.*, 1979; Williams and Self, 1983: 38). Similar secondary hazards may be produced by lava and pyroclastic flows but they appear to be more commonly associated with lahars.

2.5 EARTHQUAKES AND GROUND DEFORMATION

From a hazards point of view, earthquakes related to volcanic eruptions are important for two reasons: (i) they can cause significant damage and injury; and (ii)

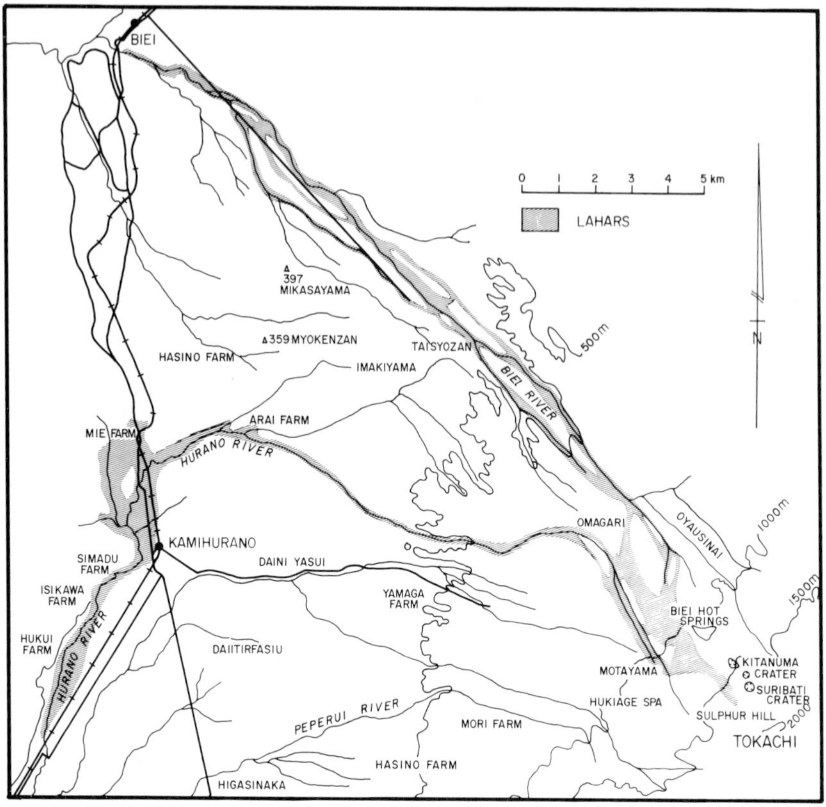

FIG. 2.20 Areas along Hurano and Biei Rivers devastated by lahars from Tokachi in 1926 (modified after Tanakadate, 1927).

they frequently occur prior to eruptive activity. Where the portent is realised, measures to ameliorate the hazard impact can be invoked.

Most volcanogenic earthquakes result from the movement of magma, the formation of cracks through which it can move, and gas explosions within the conduit. Such earthquakes originate within a few kilometers of the surface. Other earthquakes result from redistribution of the load of the volcano and from caldera subsidence. Still others, tectonic earthquakes in the vicinity of a volcano, may allow magma to rise. Thus, the connection, in both spatial and temporal terms, between seismicity and volcanic activity is sometimes debatable.

There is also some debate among volcanologists about the occurrence of earthquakes prior to an eruption. Gorshkov (1972: 9) wrote: "Nearly every eruption or increase in volcanic activity is preceded by a swarm of earthquakes". He believed that no eruption should come as a surprise where there is constant seismological surveillance, but Minakami (1972: 17) has suggested that precursor earthquakes

FIG. 2.21 On December 24, 1953 the ice impoundment of the crater lake on Ruapehu (background) collapsed. The lahar which formed in the Whangaehu River travelled about 24 km destroying road and railway bridges. At the Tangiwai bridge shown here, peak discharge reached about 850 $m^3\ s^{-1}$ (Photo: *Auckland Star*).

are not as common as Gorshkov indicated (see also Tazieff, 1979: 94). Certainly, no earthquakes were felt before the 1783 eruption of Asama, Japan (Aramaki, 1956: 195) or the 1914 eruption of Lassen Peak, California (Day and Allen, 1925: 5-6). Similarly, Shimozuru (1972: 35) notes that there were no instrumentally detected earthquakes before the 1950 Strombolian-type eruption of Oshima, Japan, and Minakami (1974: 315-16) reports that there was no remarkable seismic activity prior to the eruptions of Miyake-jima, Japan, in 1940 and 1962.

Nonetheless, it seems that earthquakes occur before most eruptions. Whether or not they are felt by observers depends not only on earthquake magnitude and distance from the epicenter but also on substrate conditions and the activity or preoccupation of the observer. For example, Frank Perret reported at the Vesuvio Observatory in 1906 that by lying on his bed and gripping the brass bedstead in his teeth he could feel tremors and monitor their frequency. In most circumstances instrumental monitoring is likely to prove more cost-effective!

Details of the seismic activity prior to and during the Soufrière, St Vincent, eruption of 1979 are of interest as they clearly related human and instrumental recordings of earthquakes. The first volcanic earthquake was instrumentally recorded on June 12, 1978. By April 12, 1979, continuous tremor became recognisable above the background microseismic level. The level of tremor increased on June 13 with 70 seismic events in 2.5 hours. Then, increasing tremor saturated the instruments. The government warned of a highly unusual situation but local residents 6 km east of the crater reported that they had neither seen nor felt anything unusual up to this time. Four and a half hours later tephra began to fall (Shepherd *et al.*, 1979: 25).

The period of premonitory earthquake activity can vary from a few hours to more than a year even allowing for differences in the quality of observations. All earthquakes were confined to the day before the 1835 eruption of Cosiguina, Nicaragua (Williams, 1952: 32). In Katmai village, Alaska, 32 km from Katmai-Novarupta, earthquakes began five days before the 1912 eruption (*Alaska's Volcanoes,* 1976: 32). At Nushugat, 200 km northwest of the volcanoes, earthquakes were felt two days prior to the eruption (Martin, 1913: 143). At Matavanu, Samoa, in 1905 a series of earthquakes began on July 15, the eruption starting on August 1. Similarly, the 1955 eruption of Sakura-jima, Japan, was preceded by five months of premonitory earthquakes as magma moved upwards toward the summit crater (Shimozuru, 1972: 34). At Soufrière de Guadeloupe local earthquakes increased in November, 1975, reached anomalously high levels in March 1976 (Fig. 2.22) with the first small phreatic eruption occurring on July 8 (Simkin and Fiske, 1977: 42). As a final example, a 14-month period of felt earthquakes preceded the May 7, 1902, eruption of Soufrière, St Vincent (Shepherd *et al.*, 1979: 24).

While the patterns of premonitory earthquake activity illustrated in Figure 2.22 are typical of those preceding many eruptions other patterns also occur (Fig. 2.23). For example, at Raoul Island in the Kermadec Group in 1964 seismic activity decreased dramatically over a period of 10 days (Fig. 2.23b) before the eruption be-

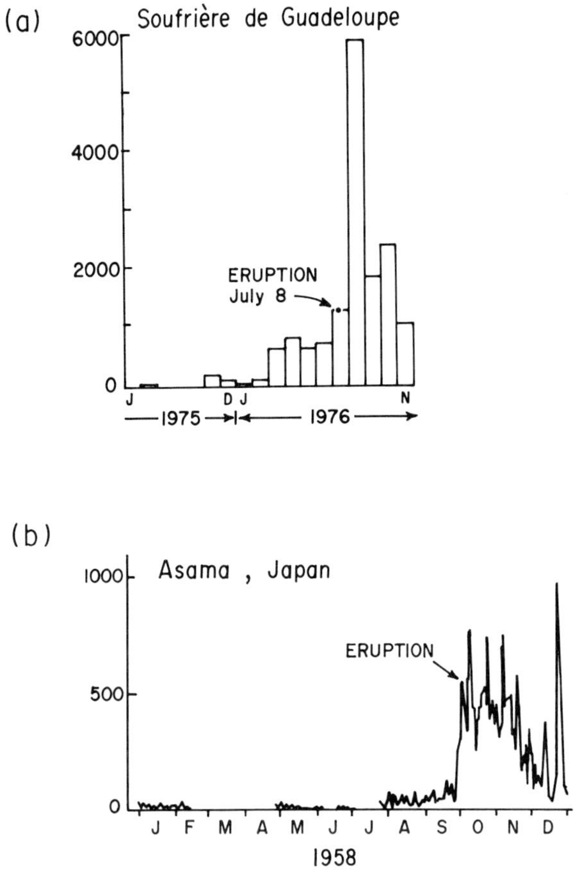

FIG. 2.22 (a) Earthquakes/month at Soufrière de Guadeloupe in 1975–1976 (after Simkin and Fiske 1977: 42). (b) Daily frequencies of earthquakes at Asama, 1958 (after Shimozuru, 1971).

gan. A rather similar pattern occurred at Mount St Helens between mid-March and the end of May 1980 (Endo *et al.*, 1981: 101).

It is clear that precursor earthquakes can provide useful warnings of approaching volcanic hazard. However, the significance of the warning must be perceived before evacuation or any other appropriate avoidance action can be undertaken. For example, precursor earthquakes occurred for several days with increasing magnitude and frequency before the 1951 Lamington eruption (Taylor, 1958: 18-19). However, the significance of this seismic activity was not understood (in fact, Lamington was not even known to be a volcano!) and evacuation was not contemplated. However, where some significance is attached to the earthquakes, evacuation often occurs. Indians at Katmai village were frightened by earthquakes

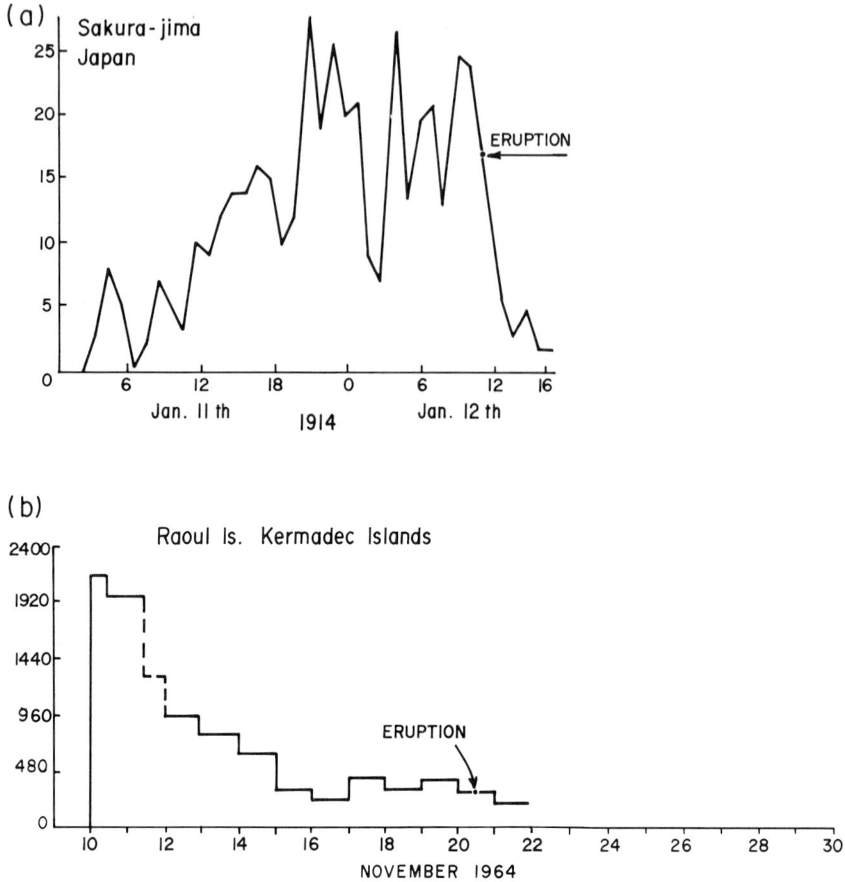

FIG. 2.23 (a) Daily earthquakes at Sakura-jima before the 1914 eruption (after Shimozuru, 1971). (b) Local earthquakes per day before the November 1964 eruption of Raoul Island in the Kermadec Group (after Adams and Dibble, 1967).

and left two days before the 1912 eruption of Katmai-Novarupta (Griggs, 1922: 17). At Tristan de Cunha in 1962, premonitory earthquakes (and rockfalls, ground deformation, and the destruction of the water supply) persuaded the population to abandon the island before the eruption began.

As the foci of volcanic earthquakes are generally at less than 10 km depth, damage from volcanic earthquakes is usually confined to a relatively small area (Rittman, 1962: 159; Shimozuru, 1972: 19). Three points requiring consideration stem from this statement.

First, earthquakes of supposed volcanic origin have been felt at great distances from eruptive centres. Raffles (1830) records that houses shook perceptibly in East

Java, 780 km from Tambora, during the 1815 eruption. Earthquakes during the 1886 eruption of Tarawera, New Zealand, were felt at Wanganui 250 km away (Grayland and Grayland, 1971: 135), during the 1888 eruption of Bandai at 80 km distance (Sekiya and Kikuchi, 1889: 170), and during the 1902 Soufrière, St Vincent eruption at 130 km from the volcano (Anderson and Flett, 1903: 406). Ninety kilometers from Cosiguina in 1835 at San Miguel "some who were walking in a penetential parade were thrown to the ground by a long and violent quake" (Williams, 1952: 32). It is likely, however, that some of these reports result not from earthquake waves but from air shock waves (see Section 2.7.1 below). On the other hand, it is certain that seismic instruments in Petropavlovsk recorded the 1964 eruption of Shiveluch at a distance of 430 km (Gorshkov and Dubik, 1970: 281).

Second, many volcanoes lie in active tectonic areas where earthquakes, not connected with any eruption, can be expected at any time. Relatively strong seismic events have been experienced near volcanoes prior to eruptions. For example, an M = 8.3 earthquake five months before the violent 1902 eruption of Santa Maria, with an epicentre close to the volcano (Williams and Self, 1983: 35), destroyed the town of Quezeltenango 15 km away (Rose, 1972: 31). Similarly, the town of Nicolosi, 15 km from the summit of Etna, was destroyed by earthquakes prior to the 1669 eruption (Rodwell, 1878: 90). Widely felt earthquakes preceded the eruption of Taal in 1911; 39 km from Taal, cracks appeared in the ground (Pratt, 1911: 69).

One of the most quoted examples concerns the AD 79 eruption of Vesuvio. In AD 62 or 63:

> Pompeii suffered an earthquake in winter, a time when we used to be told there was no danger of one. This was on February 5, and caused great destruction in Campania. Part of Herculaneum collapsed and the remainder is damaged; and the colony of Nuceria, although not destroyed has much to complain of. In Naples there was no danger to public buildings, but there was to town houses, and some country houses around were destroyed. We are also told that a flock of 600 sheep died, that statues were split in two, and that some people went out of their mind and wandered around out of control.
> [Seneca, *Naturales Quaestons,* 6:1, quoted in Dilke and Dilke, 1979: 743]

Further earthquakes around Vesuvio occurred in AD 64. Nero was singing in the Naples theatre when an earthquake was felt. Nonchalantly he finished his song — after the performance was finished and the audience had left, the theatre collapsed (Dilke and Dilke, 1979: 743). In the month or so before the eruption, house walls were cracked and columns were felled at Pompeii by further tremors.

It remains a moot point whether these are examples of the damage that can be caused by volcanic earthquakes or examples of tectonic earthquakes in volcanic regions. Certainly Sapper believed the Santa Maria earthquakes in 1902 to be tectonic rather than volcanic in origin (Anderson, 1908: 475-6). Although some authorities (for example, Vitaliano and Vitaliano, 1971; Bullard, 1976; Prinz, 1979: 43) have regarded the earthquakes in the middle of the first century of the common era as signs of renewal of activity on Vesuvio (strictly Somma Vesuvio), Francis

(1979: 750) has pointed out that earthquakes were never rare around the Mediterranean and that there is little geologic evidence to link the earthquakes with the eruption 15 to 17 years later (see also Rittman, 1962: 159).

The third point concerns the amount of damage that can be caused by volcanic earthquakes. Some experienced volcanologists doubt that volcanic earthquakes cause any damage: "Volcanic earthquakes I have known in Indonesia were always weak and never caused damage. They were observed near the volcano only . . . Only shocks which accompany collapse resulting in a caldera may be strong enough to cause damage" (Neumann van Padang in discussion, Vitaliano and Vitaliano, 1971: 100). A.G. Galanopoulos agreed with Neumann van Padang.

However, a few well documented examples suggest that volcanic shocks can and do cause considerable damage. During the seismic crisis of May 12–13, 1957, at Capelhinos, Fayal, in the Azores a large number of houses were destroyed by earthquakes (rather than by ground fracturing). Many of the 450 felt earthquakes reached a Modified Mercalli intensity of X (Parsons and Mulford, 1958: 15; Machado *et al.*, 1962: 3523). One of the greatest known earthquakes of volcanic origin occurred at Sakura-jima in 1914. This earthquake had a focus at 13 km depth and a Richter magnitude M = 6.7 (Shimozuru, 1972: 25). The earthquake damaged masonry in Kagoshima, caused landslides from the cliffs next to the city and killed a number of people (Jaggar, 1956: 105). Similarly, the M = 7.2 earthquake on November 29, 1975, at Kilauea, closely associated with an eruption and a tsunami, caused considerable damage in several areas on Hawaii (Tilling *et al.*, 1976). The extent to which these damaging earthquakes can be regarded as volcanogenic remains unresolved.

Ground deformation frequently occurs before and after eruptions. At Kilauea and other volcanoes deformations amounting to a few microradians of tilt are measured before and after eruptive activity. Such minor deformation is of no consequence from a hazard viewpoint. At the other extreme, areas north and east of Usu were uplifted by as much as 50 m in the six months before the 1944 eruption (Minakami *et al.*, 1952: 264-265). Uplift of several meters also occurred before the 1538 Campi Flegrei eruption (Barberi and Gasparini, 1976: 222). Following eruptions, subsidence may occur. After the 1914 Sakura-jima activity, Omori (1920) measured subsidence of about 800 mm on the island and 300–400 mm in Kagoshima City. In all an area of more than 1000 km^2 subsided by more than 150 mm. According to Huggins (1902: 16) the area around Wallibou, southwest of Soufrière, St Vincent, subsided by as much as 18 m after the 1902 eruption.

Other areas and times of ground deformation have been interpreted as precursors to future eruptions. At Pozzuoli in the Campi Flegrei field, ground deformation of nearly 800 mm in 1970 presented a significant hazard to buildings. In mid-1983 continuing ground deformation in the Rabaul and Long Valley calderas and at Pozzuoli may or may not constitute a future hazard.

2.6 TSUNAMI

Tsunami, or seismic sea waves, are generated by disturbances of the ocean floor. Most tsunami are caused by earthquake faulting but a significant number are induced by volcanic activity. Data summarised in Table 1.4 indicate that tsunami have caused a significant number of the deaths attributed to volcanic activity.

A substantial displacement of water is required to produce a tsunami; ocean floor displacement may act as a paddle shoving a great volume of water. In the open ocean, elevation of the seismic sea wave is generally less than one meter, with a period (the time between successive waves) of five to 60 minutes. The velocity (v_1) of a wave can be calculated from $v_1 = \sqrt{gd}$ where g = acceleration due to gravity and d = depth of water. In deep water such as in the mid-Pacific, tsunami reach speeds of 800 km h^{-1}.

Tsunami are propagated normal to the source area. Thus where tsunami are produced by fault displacement they have a strong directional trend. On the other hand, those resulting from caldera collapse spread radially. In the latter case wave energy dissipates more rapidly with distance from the source. The decline in wave amplitude results from spreading out of the wave and frictional dissipation of tsunami energy (Yokoyama, 1978: 277). The amplitude from a point source decreases in inverse proportion to the square root of the distance travelled (Vitaliano and Vitaliano, 1971: 93).

If the ocean floor was all of one depth the wave front would continue normal to the area of initiation; however, shallower water slows forward movement so that the wave is refracted. Thus as a wave approaches land, tsunami wave height and direction are modified by the topography of the sea bottom, by the continental shelf, the shoreline shape, and even the state of the tide. The hazard offered by a tsunami, then, is a function of conditions at landfall as well as conditions at source. The mean water velocity (v_2) in m s^{-1} on land can be approximated by $v_2 = (h^{0.7}i^{0.5})/n$ where h is the mean inundation height above ground level in meters, i the inclination of the water surface and n is Manning's n, the surface roughness coefficient. Hatori (1964) used $n = 0.023$ for a town area, $n = 0.012$ for an open field, and i ranges from 0.001–0.0025 except in the vicinity of large obstacles such as sea walls and hillslopes. Inundation velocities estimated for several tsunami on Japanese coasts ranged from 1.3–5.6 m s^{-1}.

Latter (1981: 470) records 92 cases of tsunami of volcanic origin. For 23 of the 92 cases no clear causes could be determined; of the remaining 69 cases, 22% stemmed from earthquakes (excluding tectonic earthquakes) accompanying eruptions (for example, Sakura-jima, 1914). A further 27% were ascribed to the impact of pyroclastic flows or surges on water (for example, Krakatau, 1883; Augustine, Alaska, 1885; Mt Pelée 1902; Taal, 1911; 1965), while submarine explosions accounted for 19% (for example, Rabual, 1937). Although caldera subsidence is generally considered in the literature to create most volcanogenic tsunami, Latter (1981) presented this cause for only 9% of cases including the collapse of Ritter Island in 1888. Latter determined six further causes of tsunami from volcanic sources: landslides or avalanches of cold rock 7% (for example, Unzen, 1792); avalanches of hot material 6% (for example, Stromboli, 1944); lahars flowing into water 4.5% (for example Mt Pelée, May 5, 1902); atmospheric shock waves 4.5% (for example, Bezymianny, 1956); and lava avalanching into the sea 1% (for example, Matavanu, 1906). This listing probably substantially under-represents tsunami resulting from submarine eruptions.

By far the best studied volcanogenic tsunami are those produced by the Krakatau eruption of 1883. At Tjabang 11 waves were experienced in 15 hours on August 27 and 28 while at Batavia (now Jakarta) 14 waves of steady period arrived in 36 hours. Table 2.10 indicates the maximum wave heights, distances from Krakatau and travel times calculated by Wharton (1888). The town locations and the area inundated are shown on Figure 2.24. As the figure indicates, run-up distances of more than 5 km occurred where hinterlands were low-lying. The 10 a.m. tsunami on August 27 seems to have been the only one to have travelled any distance beyond Sunda Strait, arriving at tide gauges around the world as a train of long low waves. Latter (1981: 481, 484, 486) suggests that the tsunami around Krakatau were caused by pyroclastic flows impacting on the sea, by rockfalls, by atmospheric shockwaves and as a result of caldera subsidence.

The Minoan (*circa* 1500 BC) eruption of Santorini in the eastern Mediterranean probably also produced significant tsunami (Marinos and Melidonis, 1971). Using rather slender evidence Yokoyama (1978) deduced that the tsunami height at source was 63 m and that the wave travelled to Anàphi in 10 minutes, to Heraklion (Crete) in 25 minutes and to Jaffa–Tel Aviv in 105 minutes.

Both the Santorini example and the Krakatau–Sunda Strait evidence presented in Table 2.10 indicate that tsunami can be hazardous even at distances of more than 100 km from the volcano. Because tsunami may be generated by pyroclastic flows and debris avalanches which can travel tens of kilometers, a significant tsunami hazard can be generated by volcanoes many kilometers from a shoreline (Siebert and Kienle, 1983). Tsunami travel so rapidly that arrival times in areas close to the volcanic source allow inadequate warning and evacuation. Thus, while in Hawaii,

Table 2.10 Tsunami heights and travel times in Sunda Strait, 1883

Town	Tsunami height (m)	Distance (km)	Elapsed time (minutes)
Java			
Merak	30–40	52	45
Anyer	10+	41	37
Tyringin	15	38	30
Princes Island	15	40	?
Sumatra			
Kalimbang	24	30	?
Telukbetung	22	70	61

After Wharton, 1888

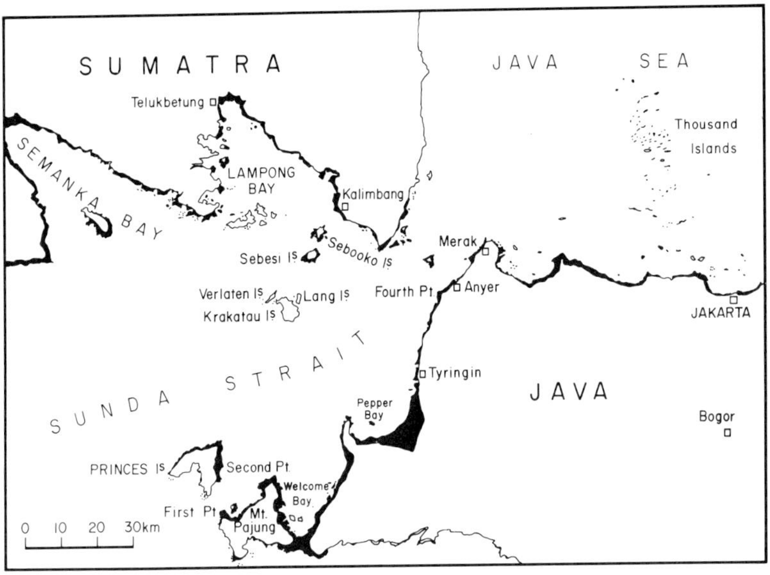

FIG. 2.24 Krakatau and the Sunda Strait coasts of Sumatra and Java. The areas shaded black are those affected by the August 26-27, 1883, tsunami (after Judd, in Symons, 1888).

for example, warnings issued by the Seismic Sea Wave Warning System allow adequate evacuation periods for tsunami originating around the Pacific margin, tsunami generated locally do not allow official warnings to be made. Thus, any earthquake strong enough to cause difficulty in walking or standing (say Modified Mercalli intensity greater than V) should be regarded as a tsunami warning by coastal residents.

2.7 ATMOSPHERIC PHENOMENA

A number of interesting atmospheric phenomena can be associated with volcanic eruptions (see, for example, Burdecki, 1964; Corliss, 1977). They range from harmless spectacular sunsets and Bishops rings through the noisy and widespread sounds of explosions to damaging pressure waves and electrical discharges. Attention here is concentrated on the more hazardous aspects.

2.7.1 Atmospheric sound and shock waves

Erupting volcanoes produce a wide variety of sounds and pressure waves. Sound waves may be heard over quite stupendous distances (Table 2.11). The classic example was the 1883 eruption of Krakatau. Figure 2.25 indicates the locations

Table 2.11 Examples of distances at which volcanic explosions have been heard

Volcano and eruption	Sound heard at	Distance from source (km)	Reference
Tarawera, 1886	North Cape	500	Grayland and Grayland, 1917: 135
Taal, 1911	—	240	Pratt, 1911: 66-67
Sakura-jima, 1914	—	500	Omori, 1916a: 98
Asama, 1911	—	400	Murai and Hosoya, 1964: 211
1958	Sandai	150	
	Gifu	400	Murai and Hosoya, 1964: 211
Bandai, 1888	Takai–Kori	164	Sekiya and Kikuchi, 1889: 120
Katmai, 1912	Juneau	1200	Martin, 1913: 131
	Ketchikan	1400	Erskine, 1962: 52
Santa Maria	Costa Rica	850	Rose, 1972: 31
Cosiguina, 1835	Bogota	1740	Coleman, 1946: 72
	Guatemala City	430	Williams, 1952: 33
Mt Pelée, 1902			
May 8	—	1280	Perret, 1937: 87
August 30	St Kitts	430	Heilprin, 1903: 239

where the explosions were heard and the approximate area over which the sounds of August 26 and 27 reverberated. The Indian Ocean islands of Rodriguez and Diego Garcia, respectively 4700 km and 3580 km from Krakatau, are generally accepted as the furthest points at which noises heard have been attributed to the eruption.

Most nineteenth and early twentieth century accounts of the sounds of volcanic explosions liken the noise to that of gunfire, particularly naval gunfire. Of the 95 verbatim comments concerning the noise of the 1883 Krakatau eruption listed by Strachey (in Symons, 1888: 78-88), 36 quite specifically compare the sounds of the explosions with the firing of cannons or guns. One of the most graphic descriptions of the sounds produced during an eruption comes from the Vesuvio eruption of 1717. From the top of Vesuvio itself the sound was described as "a sort of Murmuring, Sighing, Throbbing, Churning, Dashing (as it were) of Waves, and between whiles a Noise like that of Thunder or Cannon, which was constantly attended with a clattering like that of Tiles falling from the tops of Houses on the Streets" (Anon., 1717: 798).

The sounds of volcanic explosions are frequently not heard uniformly around the source. Omori's observations around Asama in 1909–1911 illustrate this phenomenon. Detonations were heard more than 280 km from Asama. In some cases the direction of the greatest extent of the area in which detonations were heard did not differ much from that of prevailing surface winds. In other cases it was quite contrary. In two cases the sound area consisted of two detached portions, one including the volcano, the other 80–100 km to the west and there was no relation between the distortion of the sound area and the direction of the surface wind (Davison, 1912: 488).

In 1947 the eruption of Hekla was heard at a distance of as much as 280 km but there was a fairly broad belt about 70 km distant in which little noise was heard

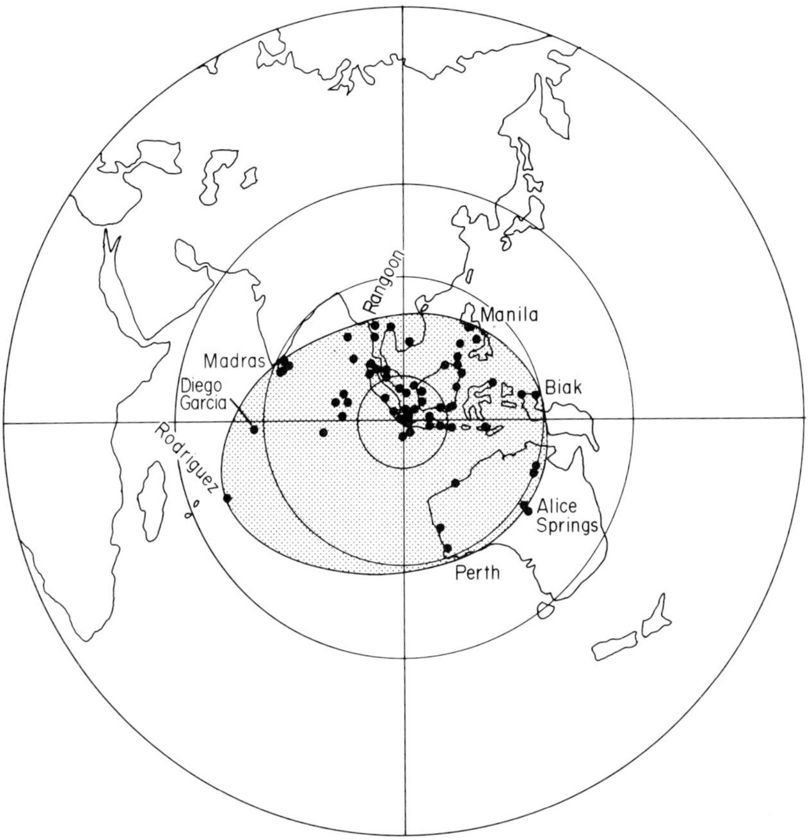

FIG. 2.25 The shaded portion indicates (approximately) the area over which the sounds of the August 26-27, 1883 eruption of Krakatau were heard (after Symons, 1888).

(Thorarinsson, 1970: 25). Similarly, the 1966 eruption of Kelut was heard at distances of 230 km and 310 km but not at 35 km and 40 km (Jennings, 1969: 67). The deadening of sounds during tephra falls has also frequently been noted. Many accounts indicate that close to the volcano the progress of the eruption is recorded as a continuous loud noise with no distinctive explosions. At greater distances individual explosions are clearly heard. This "not uncommon anomaly" as Taylor (1958: 25) called it has been noted during the eruptions of Tambora (1815), Krakatau (1883), Cotopaxi (1877), Tarawera (1886), Soufrière (1902), Vesuvio (1906), Katmai (1912), Lamington (1951), Asama (1958), Fernandina (1968), and Mount St Helens (1980). Bolt (1976: 15), explains that the velocity of sound in the atmosphere first decreases and then increases with altitude so that at some distances the sound may focus, like rays of light through a lens. In describing the sounds produced by nuclear explosions Bolt writes: "At short range, the zone of silence is disturbed only by weak sound waves diffracted along the ground . . . At ranges of 50 kilometers or more, jet streams may duct the sound waves considerably. In U.S. tests at the Nevada nuclear proving grounds, as much as fifteen-fold amplification in sound pressure compared with normal has been observed 65 kilometers from the explosion" (Bolt, 1976: 15).

Although the sounds produced by volcanic explosions have obviously created a great deal of interest they generally only constitute a minor hazard to humans and (presumably) animals. However, atmospheric shock waves are of more significance as they may also cause physical damage to buildings.

Air waves or shock waves originate in the sudden and localised increase in pressure caused by outbursts of large quantities of steam and other gases. Visible shock waves, called "flashing arcs", occur when the velocity of the gas jet exceeds the speed of sound (which ranges from 332 m s^{-1} in air to 830 m s^{-1} in steam at 1200°K). Figure 2.26 presents Perret's diagram of the flashing arcs observed during the 1906 eruption of Vesuvio. Perret (1912) reported that at a time when the frequency of explosions varied from one every three to four seconds to at least three per second, at the instant of each explosion, before being sensed by the ear, a thin luminous arc flashed upward and outward from the crater. After this the sound of the explosion was heard and gas and debris was projected above the lip of the crater. Similar visible shock waves were observed during the 1974 Ngauruhoe eruption.

Atmospheric shock waves have also been recorded on barographs and other instruments at large distances from eruptions such as Krakatau (1883), Mt Pelée (May and August, 1902), and Shiveluch (1964). In these cases shock waves circled the earth. Barographs produced at the time of the Krakatau eruption showed first a sudden increase in pressure and then a fall. At Mauritius, for example, seven well-marked barometric disturbances were recorded (*Nature,* 1884, 30: 279). In Manila (60 km from Taal), shock waves in 1911 caused barometers to rise suddenly and then fall by 1 mm (Pratt, 1911: 66-67).

A loud deep boom originating in the caldera collapse at Fernandina in the Galapagos Islands on June 11, 1968, was recorded as far away as Alaska. In Colorado, the shock waves had a period of 9 minutes and an amplitude of 2–4 Pa. More than two hundred kilometers from the volcano a barograph showed a positive deflection of 120 kPa. A drop in pressure, such as that following the initial rise at Krakatau, was not recorded (Simkin and Howard, 1970: 431).

The Fernandina collapse is particularly interesting because witnesses throughout the Galapagos Archipelago reported earthquakes with Modified Mercalli intensities of III–IV at about the time of the explosion but Simkin and Howard (1970: 431) indicate that a search of the records reveals no such earthquakes. It seems certain that effects attributed to earthquakes were caused by the air shock wave or that there was substantial air to ground coupling.

FIG. 2.26 Flashing arcs observed during the 1906 eruption of Vesuvio (after Perret, 1912).

Such confusion of earthquake and atmospheric waves is probably not uncommon. As reported in Chapter 2, Section 5 volcanogenic earthquakes usually have shallow hypocentres and can thus only be felt in a relatively restricted area. It is likely then, as the Galapagos example demonstrates, that many of the reports cited earlier refer to airshock waves and not earthquake waves. Table 2.12 refers to distances at which airshock waves have been reported. In many of these cases, damage to buildings has been recorded at these distances. During the 1958 eruption of Asama, damage to houses from airshocks occurred across an area of 430 km^2 (Murai and Hosoya, 1964: 211). Figure 2.27 indicates the much larger area within which air shock waves were strong enough to shake houses during the 1914 eruption of Sakura-jima. The areas affected are clearly much larger than the areas experiencing authenticated volcanogenic earthquake damage.

Table 2.12 Distances at which airshock waves have been reported

Volcano	Year	Distance (km)	Reference
Krakatau	1883	160	Judd in Symons, 1888: 27
	1883	830	Verbeek, 1886: 363-364 (in Vitaliano and Vitaliano, 1971: 95-96)
Tambora	1815	370	Anon., 1816
Sakura-jima	1914	585	Omori, 1916a: 98
Asama	1783	>300	Aramaki, 1956: 198
Asama	1973	70	*SICSLP*, 1973: 131
Katmai-Novarupta	1912	240	*Alaska's Volcanoes*, 1976: 35

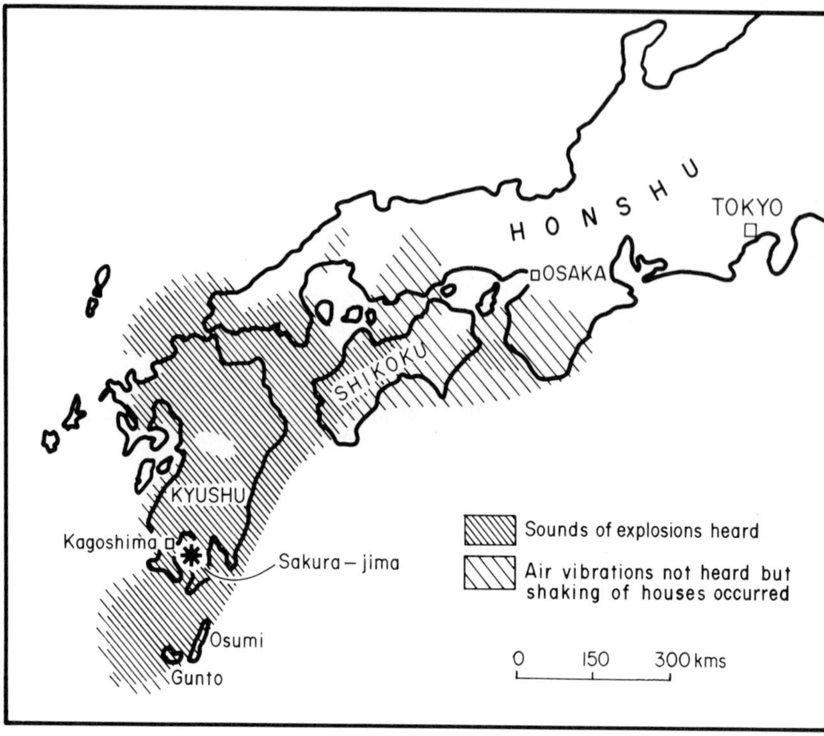

FIG. 2.27 Map showing the propagation of air waves from Sakura-jima, January 12-13, 1914 (after Omori, 1916a, Plate 20).

2.7.2 Electrical discharges

> The electric phenomena accompanying the [Tarawera, 1886] outburst must have been on the grandest scale. The vast cloud appears to have been highly charged with lightning, which was flashing and darting across and through it; sometimes shooting upwards in long curved streamers, at others following horizontal or downward directions, the flashes frequently ending in balls of fire, which as often burst into thousands of rocket-like stars. Fireballs fell at the Wairoa and other places...
>
> [Pond and Smith, 1886: 352]

Powerful, sometimes spectacular, lightning displays occur during volcanic eruptions particularly where an abundance of fine debris is produced together with strong gaseous expansion (Fig. 2.28). The potential differences which create lightning discharges result from friction between tephra particles and steam and other gases (Kikuchi *et al.*, 1978). Lightning may strike upwards or downwards or obliquely as at Ruapehu, New Zealand, in 1945 and Taal, Philippines, in 1911 (Pratt,

FIG. 2.28 Lightning display photographed on May 29 during early stages of Rabaul, 1937, eruption. Viewed from the east (Photo: Roger Davies, Rapindik Native Hospital; Bureau of Mineral Resources, Australia, G. 2008).

1911: 67). It may have a "rik-a-tak" spark-like quality as at Vesuvio in 1906 (Jaggar, 1906: 107), form broad bolts as at Fernandina (Simkin and Howard, 1970: 431), or be forked, branched or displayed as chain lightning as at Taal, 1965 (Carroll and Parco, 1966: 6). Lightning may or may not be accompanied by thunder, and the number of flashes may vary from almost incessant to a few per hour.

Where lightning discharges take place from cloud to cloud or from eruption column to the eruptive center they create a hazard only to communication systems. However, discharges to earth around the volcano are not uncommon, posing a threat to life and property. During the Soufrière, St Vincent, eruption of 1979, lightning strikes were recorded 9 km from the volcano (Shepherd *et al.*, 1979: 26).

The electrical discharge phenomena known as St Elmo's fire refers to a more or less continuous, luminous electrical discharge of weak to moderate intensity and emanating from elevated objects when in the neighbourhood of a strong electrical field. The "fire" appears as a bluish or greenish colour but sometimes appears white or violet. It is generally accompanied by a crackling sound (McIntosh, 1963:

219). A ship 65 km from Krakatau in 1883 reported "balls of fire rested on the mastheads and extremities of the yard-arms" (Judd, in Symons, 1888: 20). Similarly, Perret (1906: 8098) noted from the Vesuvio Observatory during the 1906 eruption: "St Elmo's fire was everywhere – on the railway poles, on a stick held point upwards and even on one's hat". Similar dancing sparks were observed by climbers on Mt Adams 55 km from Mount St Helens on May 18, 1980 (*Volcano*, 1980: 78). In most circumstances, the presence of St Elmo's fire is likely to be frightening but not dangerous.

2.7.3 Volcanic activity and climatic variation

Benjamin Franklin noted in 1784 that much of the northern hemisphere in the summer of 1783 was enveloped in a dry summer fog, not easily dissipated by the sun's rays:

> Hence the surface was early frozen.
> Hence the first snows remained on it unmelted, and received continual additions.
> Hence the air was more chilled, and the winds more severely cold.
> Hence perhaps the winter of 1783-4 was more severe than any that had happed for many years.
>
> [Franklin, 1784]

In Philadelphia winter temperatures were nearly 5°C below the long-term winter average (Sigurdsson, 1982: 601). Franklin speculated *inter alia* that volcanic activity in Iceland (the Laki eruption) was responsible for the "smoke".

Following the 1815 eruption of Tambora, 1816 was known in the northern hemisphere as "the year without a summer" (Stommel and Stommel, 1979). After an extensive analysis Self *et al.* (1981) concluded that large Plinian eruptions (for example, Krakatau, 1883; Santa Maria, 1902; Katmai, 1912; and Cerro Azul, 1932) possibly produce a temperature decrease of 0.2°–0.5°C for periods of up to five years. Sigurdsson (1982) suggested that the critical factors are eruption magnitude and the sulfur content of the magma. Kellogg (1983) indicated that the height of injection of fine particulate matter and aerosols is also important; low-lying particles absorb sunlight and warm the lower atmosphere ("the greenhouse effect") whereas particles high in the stratosphere scatter sunlight and induce cooling at the surface.

Not all the variables in the equations are fully understood but the eruptions of Mount St Helens, 1980, and El Chichon, 1982, revitalised research on the effects of volcanic activity on global climates. That topic is beyond the scope of the present volume but some perspectives can be offered: (i) the purported temperature changes are similar in magnitude to background temperature variations (Self *et al.*, 1981); (ii) sulfur dioxide (SO_2) emissions to the atmosphere from the 1783 Laki eruption equal only about 40% of the current annual emissions from industrial and other anthropogenic sources (Sigurdsson, 1982); and (iii) a world-wide fall in mean annual surface temperatures of 1°C would have a most profound effect on world food production (see, for example, Stommel and Stommel, 1979). Additional perspectives are provided by Lamb (1970; 1971), Bryson and Murray (1977), Hammer *et al.* (1980), Hirschboek (1980), Axelrod (1981), Robock (1981), and Porter (1981).

2.8 ACID RAINS AND GASES

The volatile component of the magma form volcanic gases during the eruption spreading in three main ways from the vent: (i) as acid aerosols; (ii) as compounds adsorbed on tephra particles; and (iii) as salt particles (Thorarinsson, 1979: 142). Óskarsson (1980) suggests that the more explosive the eruption, the higher the proportion transported as adsorbed compounds. Rose's (1977) study of the moderately explosive Fuego, Guatemala, 1974, eruption indicated that 33% of the sulfur and 17% of the chlorine were scavenged from the eruption column by tephra (that is, adsorbed onto tephra surfaces).

Water vapor and sulfur in various forms are major constituents of eruption clouds. Samples from fume clouds from Halemaumau, Hawaii, during 1966 (when no eruption was occurring) and 1977 indicated that particles collected from the fumes were more than 95% dilute sulfuric acid droplets containing some cations in addition to H^+ such as NH_4^{++} and Ca^{++} (Cadle and Frank, 1968). As SO_2 dissolves to form sulfurous acid which slowly oxidises in the presence of air to form sulfuric acid, the concentration of sulfuric acid in an eruption cloud will increase, at least up to some point, both with time and distance downwind.

Analyses of eruption clouds from explosive eruptions are less common but analyses of cloud samples from Pacaya, Fuego and Santiguito in 1978 indicate that hydrochloric acid (HCl) is nearly as important as SO_2 although the amounts vary considerably (Cadle *et al.*, 1979). However, similar variations in Kilauea, Hawaii, fumarolic gases collected in 1959–1960 were reported by Murata (1966). One sample when condensed formed from the water-rich gas a solution that was essentially 2N HCl. Most of the gas samples analysed earlier, however, contained more sulfur compounds such as S_2, SO_2 and SO_3, than chlorine. Murata assumed the composition was temperature dependent.

Thus, although numerous popular accounts of tephra falls describe strong smells of sulfur, analyses have shown other gases such as hydrochloric acid, hydrofluoric acid, carbonic acid and ammonia to have been present in equal or even greater quantities (Wilcox, 1959: 442).

From a volcanic hazards viewpoint, the distribution and amount of rainfall relative to the formation of the fume cloud or eruption plume is important. Kratky *et al.* (1974) determined that the pH of rain water from a single heavy rainfall passing through the Kilauea 1970–1973 eruption cloud was less acidic than those from cumulative samples of lighter rainfalls. Their results indicate, as expected, that the first portion of a rainfall absorbs most of the pollutants from the air.

Although water vapor, sulfuric acid and hydrochloric acid are likely to be the major components of volcanic gases (at least distant from the vent) other minor components may be of significance. Fluorine reached a concentration of about 1700 ppm by weight in the particulate material of the 1970 Hekla eruption (Cadle and Blifford, 1971). Jaggar (*Volcano Letter,* March 4, 1926) noted that 1919 analyses of Katmai fumaroles showed 99% water vapor but hydrofluoric acid occurred in amounts comparable with hydrogen sulfide and nitrogen, "a quantity unheard of elsewhere". Certainly, the concentration of fluorine in the Hekla tephra (1970) is sufficient to cause a hazard.

Figure 2.29 illustrates the fallout ellipse of soluble fluorine from the Hekla 1970 eruption. Óskarsson (1980) also found that absorption of soluble fluorine to form calcium fluorsilicate on the glassy surfaces of tephra grains occurred at temperatures below 600°C, that is, at mid-levels in the eruption column. Because the fluorine is adsorbed onto the tephra surfaces and because smaller grains have higher surface area/unit volume ratios, fluorine is concentrated on finer grains. The fluorine deposition pattern shown in Figure 2.29 is thus a function of both particle size distribution and tephra thickness. Eruptions of Hekla before 1970 were also rich in fluorine, producing a variety of adverse consequences, particularly for grazing animals. Similarly, the Eldfell (Heimaey) eruption of 1973 contained 700 ppm fluorine, Askja (1961) 400 ppm and Surtsey (1963-64) 370 ppm. With the possible exception of the Katmai-Novarupta (1912) eruption cited above, fluorine is not known to have been a hazard produced by eruptions outside Iceland.

In summary, dispersion of gases and tephra clouds with time and distance from the vent suggests that the hazard offered by acid rains and gases will also decrease downwind. Whilst this is probably generally true, scavenging of eruption clouds by rainfall may occur at a variety of times and distances from the vent, thus on occasions presenting localised hazards at distances of hundreds of kilometers. Similarly, the concentration of fluorine and other potentially toxic substances on smaller tephra grains raises the possibility of adverse hazards at distances quite remote from the erupting volcano.

2.9 SOME CONCLUSIONS

The evidence presented in previous sections indicates that while a number of factors may influence the degree of hazard presented by various aspects of volcanic activity, the most significant are: (1) distance travelled from the volcano and the area covered; (2) the velocity; (3) the temperature; (4) the length of warning; and (5) the frequency of occurrence. Each of these physical characteristics influences the capacity of the phenomenon to inflict damage to buildings and other structures and the likelihood that death or serious injury will result. These summary tables supplement Table 1.4.

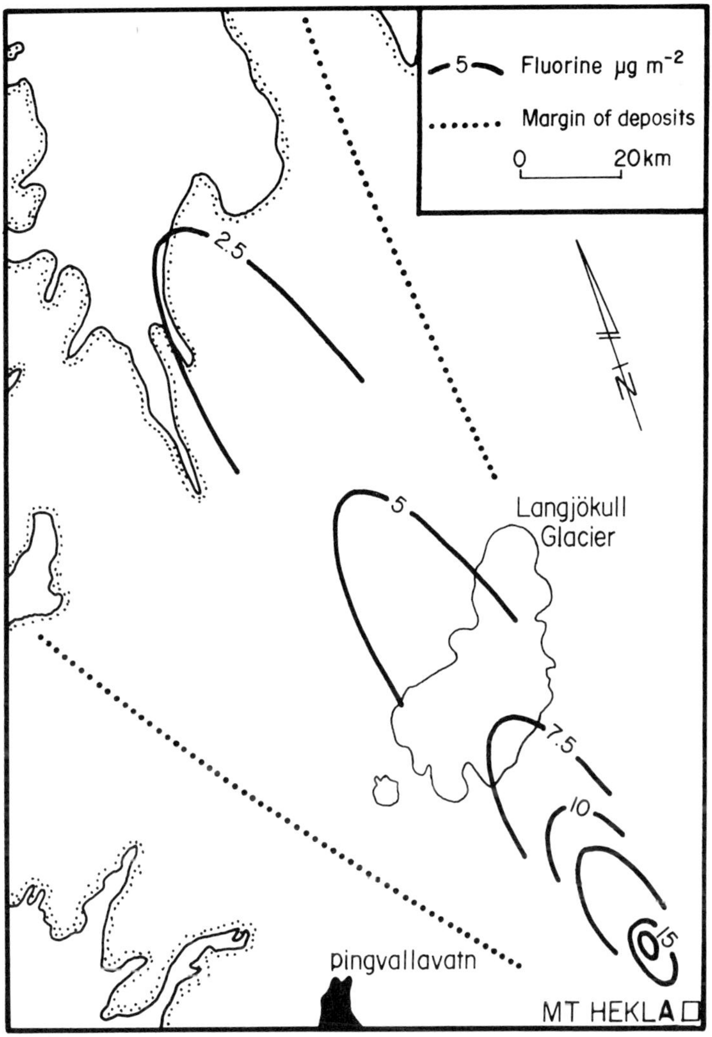

FIG. 2.29 Soluble fluorine distribution from the Hekla 1970 eruption. Numbers are in μg m^{-2}. Dotted lines show the margin of the tephra deposition, 5μg m^{-2} of soluble fluorine fell in a deposit of about 3000 g m^{-2} mass (modified after Óskarsson, 1980: 253).

Table 2.13 Summary estimates of the physical properties of volcanic phenomena

Phenomena	Distance at which effects experienced		Area effected		Velocity		Temperature (°C)
	Average (km)	Maximum (km)	Average (km^2)	Maximum km^2)	Average ($m\ s^{-1}$)	Maximum ($m\ s^{-1}$)	
Lava flows	3-4	100+	<2	1000+	<5	<30	750-1150
Ballistic projectiles	~2	5+	~10	~80	50-100	100+	<1000
Tephra falls	20-30	800+	>100	100 000+	<15	~30	Usually ambient air
Pyroclastic flows and debris avalanches	<10	100+	5-20	10 000+	20-30	<100	<600-700
Lahars and jökulhlaups	~10	300+	5-20	200-300	3-10	30+	<100
Seismic activity	<20	50+	>1000	7 000+	<5500	<5500	—
Ground deformation	<10	<20	~10	100	$<10^{-5}$	$<10^{-5}$	—
Tsunami	<50	500-600+	<10 000	1 000 000+	200	200+	ambient water
Airshock waves	10-15	800+	<1000	100 000+	>300	>500	ambient air
Lightning	<10	100+	<300	3 000	12×10^5	12×10^5	above ignition point
Acid rains and gases	20-30	2000+	<100	20 000+	<15	~30	ambient air

In many cases the summary figures are no more than estimates, as too few data exist to calculate meaningful averages. Estimates of maximum distances reached and areas affected are included as a reminder that the historic period has been too short to include lava flows and pyroclastic flows of the magnitude of those in the geologic record. While such large-scale events as pyroclastic flows covering many thousands of square kilometers are clearly events of a low probability of occurrence, equally clearly, such events will occur again.

Tables 2.13 and 2.14 also indicate that there are various ways of measuring the magnitude of volcanic hazards (see also Chapter 1, Section 4) in terms of mobility, temperature, or the brevity of the warning period. Tephra falls, tsunami, airshock waves and acid rains and gases affect the largest areas but the summary death data in Table 1.4 suggest that, of the four hazards, only tsunami present a significant threat. Furthermore, while tsunami may cover large areas of ocean they are only hazardous in shallow waters and along coastlines.

While lava flows have the highest temperatures, they are relatively slow-moving and confined largely to topographic lows. Thus, the hazard to people is low but very severe to buildings and other structures in their path. To a less extent, because of greater velocities and shorter warning periods, the same is true of lahars and pyroclastic flows — people may escape to high ground. However, while small pyroclastic flows and debris avalanches follow valleys, surges and large pyroclastic flows and avalanches mantle the whole ground surface; in these cases, human safety and damage to structures is measured not in elevation above adjacent river valleys but in distance from the volcano.

These complexities emphasise the difficulties of assessing the degree of hazard based on the physical properties of volcanic phenomena. Hazardousness is, in fact, more easily assessed by judging effects.

Table 2.14 Summary of warning periods and likely effects of volcanic hazards

Hazard	Warning period	Capacity to cause severe damage	Likelihood of severe injury or death
Lava flows	Usually hours or days	Extreme	Very high
Ballistic projectiles	Seconds	Extreme	Very high
Tephra falls	Minutes to hours	Minor-moderate	Low-moderate
Pyroclastic flows and debris avalanches	Seconds	Extreme	Extreme
Lahars and jökulhlaups	Minutes to hours	Very high	Very high
Seismic activity	None	High	High
Ground deformation	Hours to weeks	Moderate	Very low
Tsunami	Seconds to hours	Very high	Very high
Airshock waves	Seconds to minutes	Minor	Very low
Lightning	None	Moderate	Very high
Acid rains and gases	Minutes to hours	Very low	Usually very low

3

EFFECTS ON HUMANS: DEATHS AND INJURIES

3.1 INTRODUCTION

Planning for future volcanic eruption impacts requires specific data about past eruptions — information about the agents of death, the number of people at risk, whether evacuation was ordered before the eruption began, the proportion of bodies recovered, the proportion of those who died during the eruption versus those who died later, the clinical cause of death or injury, and the extent and nature of injuries sustained.

Those familiar with natural hazard data will know that such information is rarely collected (Tubessing, 1979) and that the data that are available are frequently unreliable with contemporaneous accounts giving widely varying totals of death and injury. Estimates also change rapidly through time after impact, the problem of accurate statistics being compounded where rescue efforts are poorly co-ordinated and/or where only a proportion of the bodies are recovered.

Despite the dearth of accurate and verifiable data on deaths and injuries, snippets of information from a wide variety of sources can be pieced together to provide some general statistics about the most likely agents of death and injury. Furthermore, case histories from around the world can be used to elaborate on a number of aspects including clinical causes of death, death rates (or more optimistically, survival rates) and precautions which may minimise death and injury.

3.2 SOME GENERAL STATISTICS

The probability that volcanic activity will result in death or injury clearly depends on a variety of factors including the distance from the volcano at which peo-

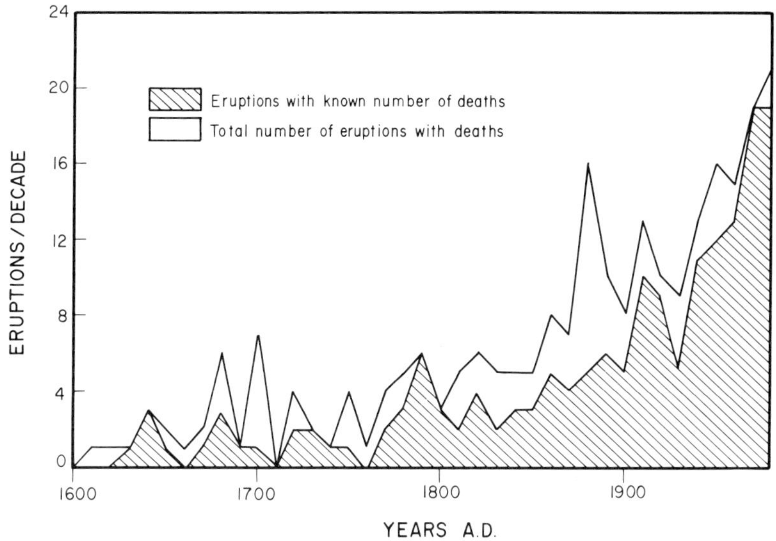

FIG. 3.1 Number of eruptions per decade since AD 1600 where the number of deaths is known (shaded) compared with the total number of eruptions with deaths.

ple live, the size of the population, the types of volcanic hazard occurring (Tables 1.4, 2.13, 2.14), and the magnitude and frequency of activity. On this basis we might expect some regions and some volcanoes to be more hazardous than others (Fig. 1.7).

Although estimates of the numbers killed in eruptions prior to AD 1600 are available in some cases a reasonably complete world-wide coverage can only be assumed for the last four centuries. Figure 3.1 illustrates (i) the number of eruptions per decade with known (that is, reasonable estimates) of the numbers killed, and (ii) the total number of eruptions per decade in which deaths are believed to have occurred. The increasing number of eruptions per decade shown since the mid-nineteenth century almost certainly results from more complete reporting, especially from regions such as Melanesia, rather than from increased eruptive activity (Simkin *et al.*, 1981). Despite this changing base, the post-AD 1600 data enable some interesting conclusions to be drawn. The discussion presented here is based on data abstracted from the Smithsonian Institution Volcano Reference File (VRF). Some examples of problems with the data base are presented in later sections.

Table 3.1 summarises the numbers killed in eight regions (Fig. 1.1) during the last four centuries. It is noteworthy that 67% of the known deaths have occurred in the Indonesian region, 13% in the Carribbean, and 8% in Japan, leaving only 12% for the other 15 regions. It is also noteworthy that there are no recorded deaths in the

Kuriles or Kamchatka, only three in Alaska and none in the continental United States prior to the 1980 Mount St Helens eruption.

Table 3.2 summarises human deaths attributed to each volcanic hazard for the periods 1600–1982 and 1900–1982. Source data and the reasons for attributing deaths to specific hazards are included in the Smithsonian Institution Volcano Reference File. For the 1600–1982 period nearly 40% died from disease and starvation resulting indirectly from eruptions. Twenty-three per cent of deaths were caused by pyroclastic flows and a further 19% by volcanogenic tsunami. Lahars and tephra falls accounted for 6% and 5% respectively. Table 3.2 also indicates that relatively few people have been killed by either lava flows, gases and acid rains, atmospheric effects or seismic action associated with volcanicity. If only the twentieth century data are considered, pyroclastic flows become by far the most important single cause of human deaths in volcanic eruptions.

Numerous problems arise from these data. First, any conclusions reached from these data are influenced significantly by the death tolls from a few eruptions with many casualties (Table 3.3). Of the 238 867 deaths given in Tables 3.1 and 3.2, five eruptions out of more than 170 accounted for nearly 76% of the total. It is also interesting to note that the primary cause of deaths is different for each of the three biggest eruptions.

Table 3.1 Deaths — by regions, 1600–1982

Regions	Total deaths	
Indonesia	160 783	(67.3%)
Caribbean	30 761	(12.9%)
Japan	19 240	(8.1%)
Iceland	9 368	(3.9%)
Central America	5 445	(2.3%)
Mediterranean	3 982	(1.7%)
Philippines	3 700	(1.6%)
Papua New Guinea	3 477	(1.5%)
All others (10 regions)	2 111	(0.9%)
Total	238 867	(100%)

Table 3.2 Deaths — by volcanic hazards

Hazards	1600–1982 Total deaths		1900–1982 Total deaths	
Lava flows	985	(0.4%)	85	(0.2%)
Tephra falls and ballistic projectiles	10 953	(4.6%)	3 019	(5.8%)
Pyroclastic flows and debris avalanches	54 995	(23.0%)	36 787	(70.4%)
Lahars and jökulhlaups	14 746	(6.2%)	6 438	(12.3%)
Seismic activity	89	(0.04%)	32	(0.06%)
Tsunami	44 356	(18.6%)	407	(0.8%)
Atmospheric effects	63	(0.03%)	3	(0.01%)
Gases and acid rains	185	(0.08%)	183	(0.4%)
Disease, starvation, etc.	95 313	(39.9%)	3 163	(6.1%)
Unknown causes	17 182	(7.2%)	2 133	(4.1%)
Total	238 867	(100%)	52 250	(100%)

Second, in the modern world with instantaneous communication systems and massive international disaster relief, deaths due to disease and starvation following an eruption would probably be minimal compared with the enormous tolls resulting from the Laki (1783) and Tambora (1815) eruptions. Most of the twentieth century deaths from disease followed the 1902 Santa Maria eruption. If the 1900–1982 death tolls are recalculated after eliminating deaths by disease, starvation and unknown causes, there are only three significant causes of death in volcanic eruptions: pyroclastic flows account for 78% of deaths, lahars 13.7%, and tephra and ballistic projectiles, 6.4%. These figures are, of course, greatly influenced by the 1902 eruption on Martinique (Table 3.3). Finally, if this eruption is also disregarded, 43% of deaths result from pyroclastic flows, 36% from lahars and other flows and 17% from tephra falls and ballistic missiles.

Three major points arise from this review. First, Indonesia, the Caribbean and Japan emerge as the most hazardous volcanic environments. Thus, as indicated in Section 1.3, the Halmahera, Calabrian and New Zealand volcanic belts may be the most active (Fig. 1.7), but they are not the most hazardous areas. Although a large number of people have been killed in the Caribbean the deaths result largely from two eruptions on consecutive days in 1902. While this does not depreciate the hazardousness of the Caribbean volcanic environment, it does indicate a low frequency of eruptions resulting in human deaths. As indicated earlier, a few volcanic regions have no recorded deaths largely because of low population concentrations.

Second, volcanic hazards such as lava flows, gases and acid rains and atmospheric effects are very unimportant as agents of death,though this does not deny the possibility that they may be important agents of injury.

Third, almost any result can be achieved depending on the mode of analysis of the incomplete data available. Clearly, a record covering only this century is too short a span of volcanic history from which to draw conclusions. On the other hand, the 382-year record might be not only too short but also more inaccurate in that some major causes of death are less applicable now than two to three centuries ago.

Table 3.3 Volcanic eruptions since AD 1600 in which more than 8000 deaths occurred

Volcano	Eruption date	Number killed	Remarks and References
Laki, Iceland	1783	9350	Killed by starvation (Jackson, 1982)
Unzen, Japan	1792	14300	Approximately 70% killed by cone collapse, 30% by tsunami (*CAVW*, 11: 43)
Tambora, Indonesia	1815	92000	About 90% killed by starvation (Neumann van Padang, 1971: 60)
Krakatau, Indonesia	1883	36417	About 90% tsunami, less than 10% pyroclastic flows and tephra falls
Mt Pelée, Martinique	1902	29025	Pyroclastic flows (*CAVW*, 20: 35-36)

Nonetheless, four important agents of death are evident — pyroclastic flows, lahars, tsunami, and tephra fall. The remainder of this chapter, indeed of this book, focuses particularly on the effects of these four agents.

One final point should be mentioned as it places deaths in volcanic eruptions in a broader natural hazard perspective. Although certainly an underestimate for the period, the grand total of 238 867 deaths (625 deaths per year) is but a small fraction of the total number of deaths from natural hazards. For example, an even less complete list of earthquake casualties provided by Bolt (1978) estimates about 2.8 million deaths (7330 deaths per year) for the same period.

3.3 LAVA FLOWS

It is clear from Table 3.2 that lava flows have been responsible for only a tiny fraction of the deaths attributable to volcanic eruptions. Table 3.4 sets out the known cases where lava flows have resulted in deaths; however, some of these reports may be apocryphal.

The 700 deaths attributed to lava flows in the 1631 eruption of Vesuvio provide an example of the scepticism with which some statistics in Table 3.4 (and following tables) must be treated. This eruption was certainly a fairly major eruption (VEI = 4). On December 16, lava invaded the towns of San Giorgio a Gremano, Portici, Pugliano, La Scala, and the western part of Torre del Greco. On December 17, Resina was also entered. Six towns were destroyed by lava flows, nine by lahars and 300 mm of tephra fell in Naples. Bullard (1976: 211-212) reported that 4000 were killed. Alfano and Friedlander (1929: 27) gave the death toll as about 4000, Sapper (1917: 12) as 3000 and an anonymous account (1766: 285) stated the death toll as upwards of 4000. The Smithsonian Institution Volcano Reference File (VRF) in the absence of more decisive information records the death toll conservatively at 3500, notes the uncertainty, records each of the above estimates and rather arbitrarily

Table 3.4 Deaths in lava flows

Volcano and date of eruption		Number killed	Reference
Vesuvio	1630	700	Bullard, 1976: 211-212
	1872	22	Rittman, 1962: 52
	1944	2	*New York Times,* March 22, 1944: 13
Etna	1832	several	Guest *et al.,* 1980
	1843	56	Rodwell, 1878: 104
	1928	5	Duncan *et al.,* 1981: 164
Nyiragongo	1977	70	VRF
Niuafo'ou	1853	25	*Tonga Chronicle,* Sept. 9, 1983: 10
Aoba	1670?	many	Warden, 1970
Api Siau	1976	1	*VRF*
Kilauea	1823	some	Macdonald, 1972: 411
Hekla	1947	1	*Bull Volc.,* 38 Supp.: 56-57
San Jorge	1808	60	Dabney, 1809: 343-344

assigns 20% of the deaths to lava flows. This arbitrary figure seems high but the lava flows certainly travelled quickly, reaching the sea 6 km distant in two hours (Bullard, 1976: 211-212). Palmieri (1873: 100) also noted without providing an estimate of the death toll that the eruption surprised the surrounding population so suddenly "that many perished miserably surrounded or covered with lava".

The Niuafo'ou (Tonga) eruption of 1853 also provides an interesting example of the problems with such data. Coleman (1946: 116-118) provides the most detailed background to this eruption. According to legend the village of Ahau was founded by a man and a woman who rebelled against the strict laws concerning legal marriage. The village headman spoke with heated oratory against the taxes imposed on his people by the high chief on the opposite side of the island and called upon the gods to send a sign from heaven to destroy all his people rather than to permit them to pay this tax. On June 25, 1853, the ground opened beneath the headman's house, lava flowed and two-thirds of the village's population was killed. Coleman (1946: 118) gave the deaths as more than 60 or 70, Wilcoxson (1966: 151) suggested the toll was "more than half the population of Ahua [sic]." and Sapper (1917: 216) noted "many persons killed". On the other hand, Ramsay and Plumb (1938: 258-259) write in their book, *Tin Can Island, a story of Tonga and the swimming mail man of the South Seas:* "One night, without warning, craters opened up among the houses, the earth split, and molten lava destroyed the village. Everyone escaped, with the exception of the chief and seven of his people. As the story goes, one crater exploded directly under the chief's house, and people point out the spot to this day". This account supposedly refers to the village of Selusalema or Jerusalem. More recently, the *Tonga Chronicle* (September 9, 1983: 10) gave the village destroyed as Ahau and the death toll as 25.

Difficulties also occur with data for quite recent eruptions. The January 10, 1977, eruption of Nyiragongo, Zaire, resulted in the very rapid drainage of the lava lake, with the average velocity of lava flow estimated at 30 km h^{-1}, and the initial velocity believed to be in excess of 100 km h^{-1} (Tazieff, 1977: 189-200). The Volcano Reference File suggests that 70 were killed, and the *Bulletin of Volcanic Eruptions* (*BVE,* 1979, 17: 13) estimates 50–100 people killed, but Tazieff estimates 60–300 people were "burnt alive".

Whatever qualifications are made about the value of some of the available data, Table 3.2 makes it clear that relatively few people are killed by lava flows. Where more than a few people have been killed by the one flow it is generally for one of the following reasons:

1. The lava flow advanced very rapidly as in the Vesuvio, 1631, and Nyiragongo, 1977, examples given above. The deaths in a Kilauea, 1823, lava flow occurred for the same reason. According to Macdonald (1972: 411) the lava flow advanced so rapidly on a coastal village that some old people and small children were caught and killed.

2. Advance of the lava flow cut off retreat. Such an occurrence caused the death of 22 young people including eight medical students on Vesuvio in 1872. The party of 22 went to the Valle dell'Inferno to see an outpouring of lava at close range without noticing that their retreat was cut off by two tongues of lava which trapped them between the precipices of Monte Somma and the advancing flow (Palmieri, 1873: 87; Rittman, 1962: 52).
3. Explosions at the lava front as a result of contact with water or ice. One such case concerns the 1808 eruption of San Jorge which inundated the town of Ursalina and the surrounding area with lava. According to Dabney (1809: 343-344), some people endeavoring to save their furniture remained in the vicinity of the lava too long and were scalded by flashes of steam, which, without damaging their clothes, took off not only their skin but also their flesh. About 60 people were scalded, some of whom died. In 1832, a lava front on Etna moved over ice, resulting in steam explosions, and the deaths of several people (Guest *et al.*, 1980). A rather similar event occurred on Etna in 1843 when lava flowed into a cistern. The resulting explosion, rather like a phreatic eruption, killed many people (Sapper, 1917: 40). These deaths have been described in more dramatic terms by Rodwell (1878: 104); near the end of the year, "while a number of the inhabitants of Bronte were watching the progress of the lava, the front of the stream was suddenly blown out as by an explosion of gunpowder; in an instant red-hot masses were hurled in every direction; and a cloud of vapour enveloped everything. Thirty-six persons were killed on the spot and twenty survived but a few hours". A similar steam explosion on March 21, 1944, at San Sebastiano on the slopes of Vesuvio killed two children "when lava reached a water tank, which exploded in a great hiss of steam, throwing them high in the air" (*New York Times*, March 22, 1944: 13).

The other cases recorded in Table 3.4 involve the deaths of only a few people. One spectator was killed and another injured by the collapse of a lava flow flank as it reached the edge of a steep valley during the 1976 eruption of Api Siau (*SEAN Bulletin.*, 1976, 1(13): 1). The only death resulting from the Hekla, 1947, eruption occurred when a scientist was overrun by a glowing lava block which rolled downslope while he was filming the lava flow (*Bull. Volc.*, 38 Supp.: 56-57). A near miss resulting from similar circumstances was reported during the Vesuvio 1822 eruption (Anon, 1824: 56).

Surprisingly few references appear in the literature to injuries resulting from lava flows. There must be numerous cases where burns have resulted from close approaches to lava flows, and where water has flashed to steam. Similarly, injuries must have resulted where onlookers have fallen through partially solidified crust. Certainly, during the January–February 1983 Kilauea eruption a volcanologist was burnt on the hands and legs when he broke through hollow pahoehoe crust and fell into an incandescent lava tube. Severe trauma, mental disturbance, or even death must all have occurred as a result of the flight from the advancing lava front or on seeing one's home or village destroyed. Such occurrences seem to have gone largely unrecorded. Nonetheless, they need to be taken into account in emergency plans.

Deaths and injuries in lava flows form a disproportionately small number of the total deaths in volcanic eruptions given the frequency of occurrence of lava flows (Table 1.4). The low number of deaths presumably results from the generally slow advance of lava flows in well marked courses and the relatively short distances travelled from the vent. However, it is noteworthy that a significant number of the deaths described above can be considered avoidable in that they resulted from close approaches to lava flows by onlookers rather than from danger offered by the hazard itself. This forms a contrast to deaths resulting from most other volcanic hazards. It also forms a warning for emergency service co-ordinators.

3.4 BALLISTIC PROJECTILES

Various lines of evidence presented in Chapter 2, Section 2.1 make it clear that the density and terminal velocity of ballistic fragments ejected from the vent can vary greatly, producing large variations in the kinetic energy of impact (Fig. 2.8). Three examples given by Booth (1979: 331-332) emphasise this variation: (i) at Heimaey in 1973 the impact of 90 mm diameter fragments on fiberglass helmets and padded arctic-type clothing caused no discomfort; (ii) at Etna in 1974 spectators received severe lacerations to unprotected scalps from scoriae fragments less than 80 mm in diameter; and (iii) at Etna in 1971 dense 50 mm projectiles penetrated at least one of several fiberglass helmets left on the mountain overnight.

Other information available in the literature is usually of a more general nature and of only limited quality. Thus, we can note that about 10 000 were killed around Tambora in 1815 as a result of bomb impacts, tephra falls and seismic sea waves (Neumann van Padang, 1971: 60), but we have no certain knowledge of the proportion that should be ascribed to ballistic impacts. Furthermore we have no information on the number injured.

Deaths and serious injuries resulted from ballistic impacts during two eruptions of Soufrière, St Vincent. In the 1812 eruption, 56 people, mostly slaves, were killed. According to Anderson and Flett (1903: 468) most were killed by falling hot stones and in the collapse of huts. "People sought shelter in cellars, under rocks or anywhere; for every where was nearly the same; and the miserable negroes [sic] flying from their huts where [sic] knocked down or wounded, and many killed in the open air . . . Had the stones that fell been proportionately heavy for their size, not a living creature could have escaped without death. These having undergone a thorough fusion, they were divested of their natural gravity, and fell almost as light as pumice, though in some places as large as a man's head" (Anon, 1813: 100). The 1902 eruption was a similar Peléean eruption, but killed between 1565 (Shepherd, *et al.*, 1979: 24) and 1680 people (*HMSO*, 1903: 10-11). Although most of the deaths and injuries resulted from hot pyroclastic flows, three children were admitted to hospital with fractured skulls. The injuries of all three resulted from bomb impacts. One boy aged nine who had a large scalp wound with a fracture and depression of the right parietal bone developed tetanus and died. The other two children survived (Will, in *HMSO*, 1903: 11-12).

Brief accounts of deaths resulting from ballistic projectile impacts include the following:

Vesuvio, March 1944: Three killed at Terzigno more than 5 km from the vent by falling bombs (*New York Times,* March 26, 1944: 25).

Etna, March, 1536: A physician of Lentini named Negro di Piazza, having approached too near the scene of the eruption, was killed by a volley of red hot stones (Rodwell, 1878: 87-88).

Rabaul, circa 1850: Many men killed by a fall of huge blocks of pumice (Fisher, 1939: 14).

Azuma, June 1893: Two scientists killed by blocks, ash and mud (*CAVW*, 11: 153).

Cleveland, June, 1944: A man was believed crushed by falling rocks and later covered with lava during evacuation of an army installation. Four other men were safely rescued (*New York Times*, June 26, 1944: 15).

Sangay, 1976: Two killed and four injured near summit. All were members of the British Vulcan Expedition (*BVE* 16: 65).

Two accounts which contain considerable detail are of more interest. The death of a man during the 1924 phreatic eruption of Kilauea is noteworthy because it is the only death caused *directly* by a Hawaiian volcano in historic time. Truman A. Taylor, determined to photograph the interior of the Halemaumau pit crater between explosions, ventured too close despite a warning from a volcanologist that another explosion was due. An explosion at 11.09 a.m. knocked Taylor down in a barrage of stones. One leg was severed by a falling rock. He suffered slightly from hot mud burns but was able to tell his rescuers that he had not been troubled by poisonous or asphyxiating gases. He died later the same day (Finch, 1924: 218; Stearns, 1925: 4,10; Macdonald and Abbott, 1970: 80).

The account of the injuries suffered by Lance Graham on Sunday June 14, 1914, during the eleventh eruption of Mt Lassen is also of interest. A condensed version of his verbatim report to the *Shasta Courier* while in hospital, has been presented by Hill (1970: 215):

> As we stood on the rim of the crater looking into the irregular great cavity, the three of us became nervous. I climbed a higher point than the others. Perched on a high rock, I was seized with a premonition of the impending danger, and my one impulse was to get away. Somebody suggested that we beat it, I was entirely agreeable to this, but for some unknown reason, though the others started down the mountain at once, I remained behind for about three minutes.
>
> Just as I turned to leave the crater's rim, there was a puff of blue smoke, followed by a tongue of red flame that resembled the discharge of an old fashioned cannon. My impression is that there was a dull rumble accompanying this outbreak, but of this I am not certain. I started to pick my way down the mountain, but in a trice I was enveloped in a

> cloud of smoke, while a perfect hail of small volcanic bombs and cinders beat down upon me. Then I was struck by a stone about twice the size of my fist that felled me to the ground.

The rock cut a gash in Graham's shoulder, pierced the thoracic cavity and broke his collarbone (Holway, 1915: 296).

The most obvious protective measure against projectile impacts is to stay away from erupting volcanoes. The examples presented above indicate that most of the deaths attributed to missiles have occurred close to the vent and have involved curious sightseers, photographers, and scientists. Other defensive measures are obvious. Sigurdur Thorarinsson reports (1980, personal communication), standing on the beach at Surtsey, watching incandescent bombs arcing towards the first landing party, then sidestepping at the last minute. The party kept this up for nearly three hours until activity declined. Tazieff (1952: 15) reports a similar activity on Nyiragongo. Padding of any type obviously protects the head. On Deception Island in 1969 a British party sheltered under rucksacks and then tore sheets of corrugated iron from the hut walls to protect their heads (Baker, 1969: 125). As indicated earlier, fiberglass helmets provide some shelter from at least small, low density projectiles. However, as the specific injuries recounted here were to a shoulder and a leg, more general protection is required than that afforded by a helmet.

3.5 TEPHRA FALLS

3.5.1 Deaths

The most important point about tephra falls is that they kill relatively few people, being responsible for only about 4.6% of recorded volcano-related deaths since AD 1600 (Table 3.2). The chances of surviving even quite thick tephra falls are very high particularly where roofs remain intact and where noxious gases do not accompany the tephra. For example, Walker (1974) reports walking unharmed on Heimaey through scoria falling at a rate of 1.1 kg m^{-2} min^{-1} (that is, 120 mm h^{-1}) even though the median diameter was 6 mm and the maximum size was 80 mm. Three sets of examples are presented below: (i) eruptions where tephra falls were the major cause of deaths; (ii) eruptions where tephra falls were cited as important causes of deaths; and (iii) eruptions where a significant majority of the populace survived a major tephra fall. Clearly, these three categories are not mutually exclusive.

Although Vesuvio is well known for lava flows (see, for example, Table 3.4), several tephra falls around the volcano have produced a significant number of deaths. The most famous of these is the AD 79 eruption which buried Pompeii. Estimates of the number killed at Pompeii vary considerably. Bullard (1976: 200-202) and Sigurdsson *et al.* (1982: 51), report that about 2000 skeletons have been found in Pompeii which has been 70%-75% excavated (Jashemski, 1979). However, Herbet and Bardossi (1968: 12-13) believe that about 16 000 were killed, although it is not clear whether this figure refers only to Pompeii or to the whole of Campagna. A death toll of 2000 for the excavated portion of Pompeii suggests a total death toll for the city of about 2700. Although some of the skeletons found had fractured skulls (Hamilton, 1772: 95; Maiuri, 1961), suggesting roof collapse, many bodies have been found in positions suggesting either ash asphyxiation or the presence of noxious gases (Fig. 3.2). More significantly, it is certain that some citizens survived as much as 2.8 m of tephra fall (Francis, 1979: 750-751) only to be killed by a pyroclastic surge (Sigurdsson *et al.*, 1982: 49). Thus, the total number of people killed by tephra falls at Pompeii is unknown but probably only a few hundred.

Figure 3.3 outlines the area affected by tephra fall during the 1906 eruption of Vesuvio. Both the thickness of tephra and the number of deaths vary enormously from source to source. Ottaviano was undoubtedly among the towns more severely affected but estimates of tephra thickness vary from 60 cm to 7 m (see Blong, 1982: 173). Lacroix (1906: 227; 233) estimated 200 fatalities mainly as a result of house roofs in Ottaviano collapsing under the weight of lapilli. Delmé-Radcliffe (1906), the British military attaché in Rome, put the death toll at 300, with 300 more injured. In Ottaviano 90 people were killed. At San Guiseppe 105 were killed (150

FIG. 3.2 Casts of humans at Pompeii evidently overcome by ash asphyxiation or noxious gases.

according to Rittman, 1962: 38) in the collapse of the roof of the church and 90 more were injured. The roof of the church was known to be unsafe and the priest tried to dissuade the people from crowding in. "The superstitious people, however, insisted on remaining in the church of their patron saint" (Delmé-Radcliffe, 1906: 3). About 20 more were killed in the collapse of house roofs around the church. In San Gennaro, 29 more were killed. In Ottaviano some survivors were dug out alive. One man of 80 years was extricated after having dug in the dark for five days with a piece of broken water pot — without knowing it away from safety, and without food for the whole time (Delmé-Radcliffe, 1906: 4). Survivors reported bad air but no asphyxiating gas (Jaggar, 1906: 113).

A total of 26 people were killed by the March, 1944, eruption of Vesuvio. Of these deaths, 12 at Nocera and nine at Pagani (Fig. 3.3) resulted from roof collapse as a result of tephra accumulation (*New York Times*, March 26, 1944: 25).

Other death tolls resulting from tephra falls include the following:

Tarawera, 1886 (June 10): More than 150 people were killed in the villages of Te Ariki, Moura and Te Wairoa (*CAVW*, 22: 63). Most deaths resulted from the collapse of house roofs under the tephra load.

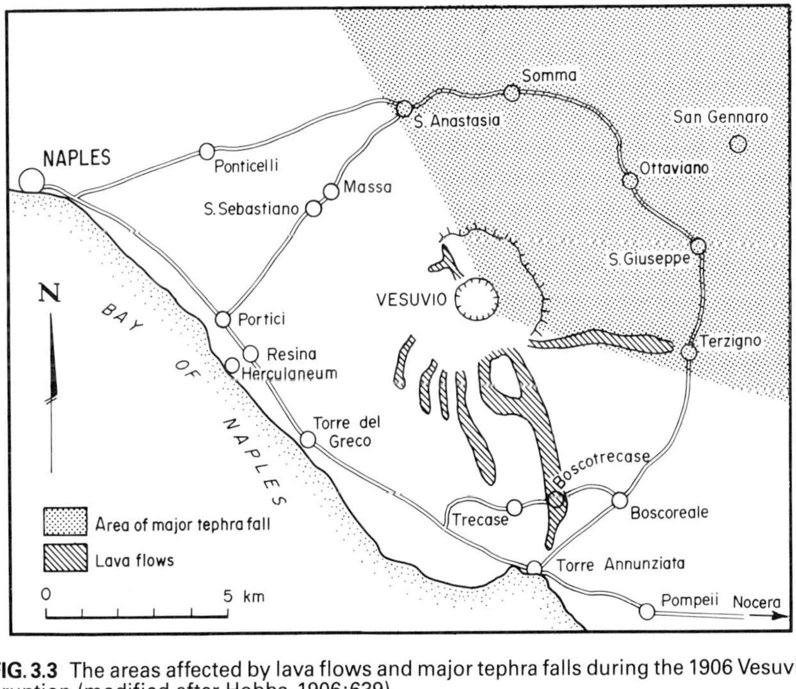

FIG. 3.3 The areas affected by lava flows and major tephra falls during the 1906 Vesuvio eruption (modified after Hobbs, 1906:639).

Rabaul, 1937 (May 29): About 375 of the 441 killed appear to have died as a result of the tephra fall, Cilento (1937: 47) giving the main cause of death as suffocation and/or burial.

Dieng Volcanic Complex, 1944 (December 4): There were 114 people killed or missing and 38 injured by hot mud, coarse ashes and bombs. Evidently all deaths occurred in seven villages within 1 km of the crater (van Bemmelen, 1949: 24).

Agung, 1963 (March 17, May 16): About 163 deaths and 201 injured (out of the official toll of 1148 killed and 296 injured) were the result of tephra falls (Surjo, 1965: 26). Other totals range from 1022 killed and 286 injured (Surjo, 1965: 26) to 1584 killed and 200 injured (*Keesings Contemporary Archives*, July 13, 1963: 19521).

Santa Maria, 1902 (October 24): Probably about 40% of the more than 5000 deaths resulted from collapsing house roofs under the weight of tephra (Anderson, 1908: 482; Sapper, 1917: 253).

Furnas, 1630 (September): A violent Vulcanian explosion ejected pumice over most of the island killing 191 people (*CAVW*, 21: 46).

El Chichon, 1982 (March 29): Most of the 153 deaths on this date resulted from roof collapse and fires ignited by incandescent tephra (*SEAN Bull.*, 7: 3-5).

Many of the estimates of the number or proportion killed by tephra falls are at best "guesstimates" based on careful perusal of available documents. For eruptions where information is available (for example Tarawera, 1886) most deaths seem to have been caused by the collapse of house roofs under a burden of tephra.

Arguments presented in Chapter 3, Section 2 indicate, whatever the basis of calculation, that less than 8% of known deaths are attributable to tephra falls (including ballistic ejection). In many ways it is not the deaths from tephra falls that should be emphasised but the survival rates. Several examples illustrate this view:

1. At Kodiak, Alaska, in 1912 tephra from Katmai-Novarupta fell to a thickness of 300–500 mm. Accounts vary (see Blong, 1982: 173-174), but certainly no more than two or three townsfolk died as a result of the eruption and its aftermath.
2. There are no accurate statistics for the death toll from the Santa Maria 1902 eruption but two small examples illustrate that death rates in tephra falls can be quite low. One building at Suiza collapsed under about 2.3 m of tephra killing 18 people. However, 22 were extracted alive from the wreckage. On this plantation at Suiza, the Helvetia plantation, a total of 50 lives were lost, mainly through roof collapse under heavy loads of tephra. However, about 950 others on the plantation survived (Anderson, 1908: 479,487)!
3. The third example concerns Pompeii where an estimated 2700 people died in AD 79. Although only a few hundred deaths are attributable to tephra fall, the population of the city was probably about 20 000 (Sigurdsson *et al.*, 1982: 39). Furthermore, as mentioned earlier, many citizens survived 2.8 m of pumice lapilli before succumbing to a small pyroclastic flow. According to Leppmann (1966: 36) some survived even this, but, trapped, eventually died of starvation.

3.5.2 Acute medical aspects

Tephra falls evidently cause few injuries other than those resulting from collapse of roofs. However, tephra falls may affect eyes and respiratory systems:

Vesuvio, 1767: "...having rode out to take a nearer view of the volcano, I was obliged to gallop home with my eyes shut, as I could no longer open them from the pain these ashes put me to" (Anon., 1771: 284)

Heimaey, 1973: "Wind whirled up ashes from the ground and the roofs and lashed them into one's face with great ferocity. The finest particles crept under one's collar and seeped down to one's shoes. Worst of all, however, was getting it in the eyes" (Gunnarsson, 1973:73-75).

Augustine, 1976: Alaskan skiers with contact lenses found that their eyes became irritated as tephra filtered behind the lenses (Miller, 1976: 2).

Mt Pelée, 1902: In the period April 29–May 5 (that is before the cataclysmic eruption of May 8) tephra fall increased steadily in St Pierre, breathing became more uncomfortable, eyes smarted and throats were sore (Hill, 1902a: 246).

Katmai-Novarupta, 1912: Babies on Kodiak Island, 160 km downwind, were affected by running noses, severe coughs and eyes full of mucus during the week after the eruption (Erskine, 1962: 206).

Karkar, Papua New Guinea, 1979: Records at the Abison Aid Post southeast of the volcano and in an area where 10–20 mm of tephra fell show a dramatic increase in treatment for upper respiratory tract infections and diarrhea. In general these conditions were exhibited in different patients. Although personnel at the Gaubin Hospital on the southwest of the island and outside the fallout zone kept a watch for possible gastroenteritis there was no evidence of a specific concentration in the tephra-affected area.

However, four more detailed examples — Irazú (1963-65), Usu (1977), Soufrière St Vincent (1979) and Mount St Helens (1980) — indicate *inter alia* that the medical effects reported appear to be proportional to the intensity of scientific study.

Irazú, 1963–1965

Irazú erupted almost continuously from March 1963 to February 1965. On numerous occasions during this period tephra fell on San José, the capital city of Costa Rica and about 30 km west of the volcano. Medical evidence indicates that "acute conjuctivitis was produced by ash particles in the conjunctival sac, leading to redness and burning of the eyes without any marked swelling... Throat irritation, sometimes accompanied by a dry cough, was common; also inflammation and burning of the throat without much swelling of the tissues" (Horton and McCaldin, 1964: 927). Both effects cleared up very quickly after exposure ceased.

Some people also suffered nasal irritation and discharge, which, when combined

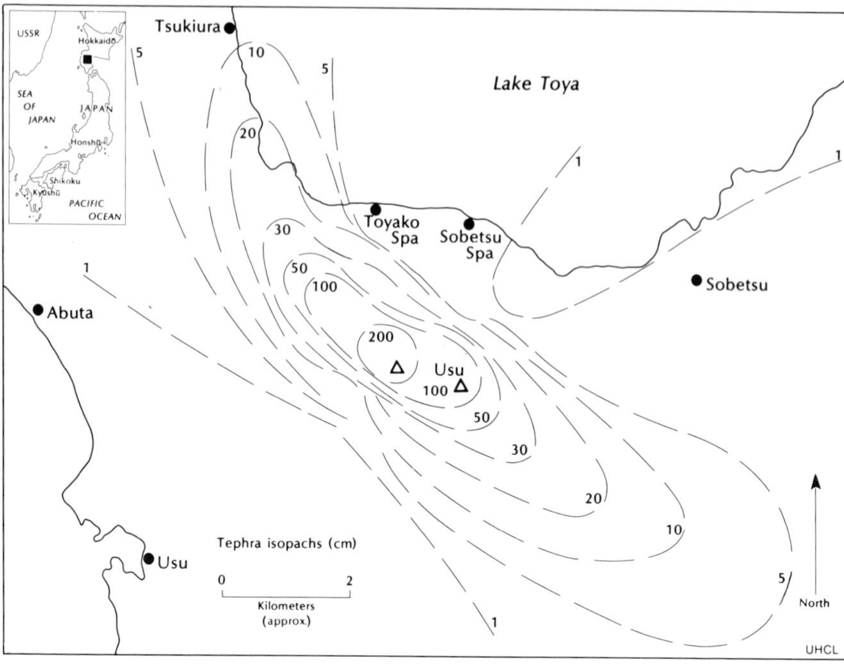

FIG. 3.4 Isopach map showing tephra fall from major Usu, 1977, eruptions. Smaller eruptions in the same year spread minor amounts of tephra to the WNW, SSE, NE and NW (compiled from various figures in *Usu eruption and its impact on environment,* Hokkaido University, 1978).

with upper respiratory infection, produced severe and prolonged stresses. People with pre-existing chest complaints developed severe bronchitic symptoms which lasted some days beyond exposure to the tephra, but local physicians argued that the effects were not severe enough to produce any deaths (Horton and McCaldin, 1964: 927-928).

Usu, 1977

Studies around Usu were much more intensive than those around Irazú. The limits of the various tephra falls, which began on August 7, and the locations of towns referred to in the text are shown in Figure 3.4.

Medical examinations of the nearby populace after the eruptions were placed in the context of pre-existing ailments (Saito *et al.*, 1978). In the Toyako Spa, Abuta Honcho and Tsukiura areas, 25% of the ailments were believed to be eruption-related. These included: (i) eyes felt as though there were foreign particles in them; (ii) eyes were painful, itchy or bloodshot; and (iii) eyes had a sticky discharge or tearing. In the Nagawa, Kannai and Kan areas 32% of the symptoms were considered to be eruption-related. Coughing and sputum were symptoms additional to those listed above.

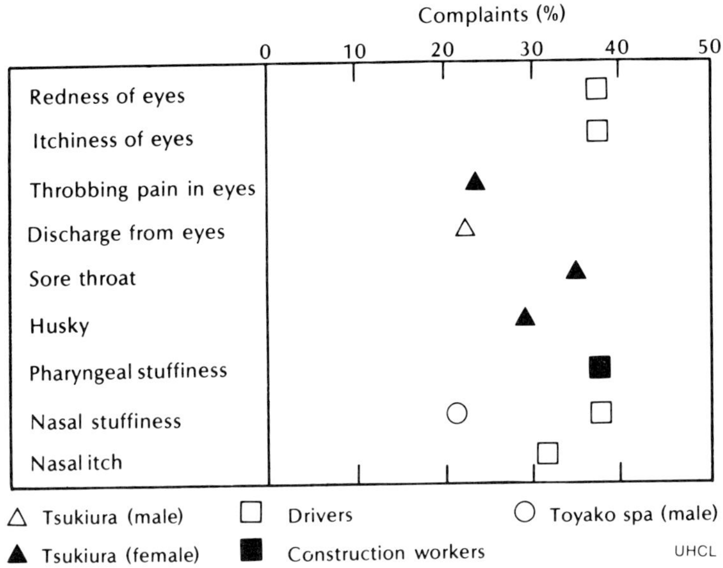

FIG. 3.5 The most prevalent health complaints mentioned in interviews in Toyako Spa, Tsukiura and Abuta areas following tephra fall from Usu in 1977. Locations and tephra thicknesses are given on Fig. 3.4 (modified after Saito *et al.,* 1978: 185).

In addition, 331 people from Toyako Spa and Tsukiura and a number of drivers and those engaged in pipe-laying in those areas were interviewed about respiratory problems and eye, nose and throat conditions. The most significant results, with more than 20% of the sample responding positively, are summarised in Figure 3.5. These results indicate that eye and respiratory complaints were most common among outside workers and drivers and occurred more commonly among the rural folk of Tsukiura than among the urban people of Toyako Spa.

Various eye, nose and throat symptoms were also exhibited by primary school pupils in Abuta and Toyako Spa even a month and more after the eruption (Fig. 3.6a,b,c). The extent to which these results are affected by infectious diseases, psychological reactions and/or re-entrainment of the tephra by wind is not known. However, Figure 3.6d shows the general decline in cases of insomnia reported by the pupils (Saito *et al.*, 1978, Saito and Seki, 1983).

Soufrière, St Vincent, 1979

The Soufrière, St Vincent eruption began on April 13, 1979. Because of evacuation of towns around the volcano, the population in the "collection area" for the Kingstown hospital increased by about 20%. Figure 3.7a illustrates the increase in admission rate to the pediatric ward for the week in which the eruption began. Although the greatest single cause of admission to this ward was gastroenteritis, the increase shown in the figure was largely the result of increased admissions of

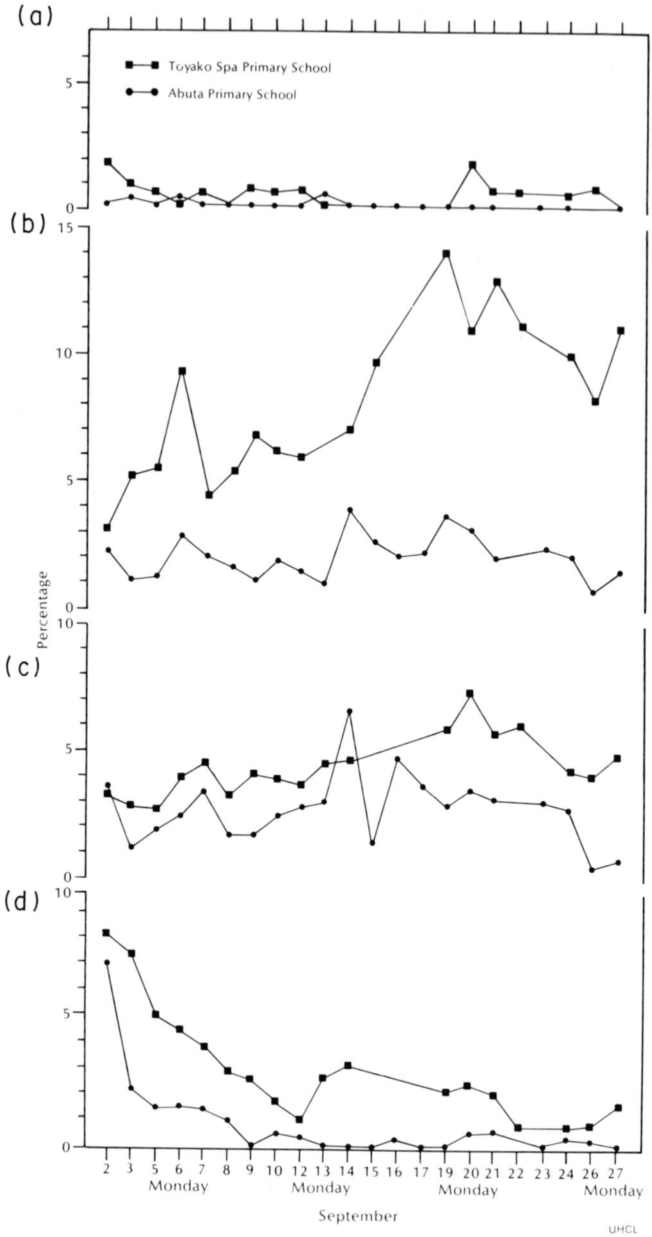

FIG. 3.6 Incidence of complaints of: (a) red eyes, (b) coughs, (c) sore throats, and (d) insomnia amongst primary school pupils in Toyako Spa and Abuta during the 1977 Usu eruptions (modified after Saito *et al.*, 1978: 197).

cases of asthmatic bronchitis (Fig. 3.7b) affecting both sexes equally and all ages except children under the age of one year (Fig. 3.7c).

Leus *et al.* (1981: 68) indicated that interpretation of these data is difficult but narrow the choice to: (i) irritation of the mucus membranes lining air passages as a result of volcanic gases and tephra fall; or (ii) asthmatic crisis induced by the psychological tension caused by the eruption, the evacuation and uncertainty. The authors also noted that there was no change in the incidence of asthmatic bronchitis on Barbados, 180 km downwind where a thin layer of tephra also fell on April 13 and 14, 1979.

However the 1902 tephra fall of 5 mm in Bridgetown, Barbados, did produce some effects on throats and lungs. According to the *Barbados Agricultural Reporter* of Friday May 9, 1902: "Somewhere in the vicinity of 6 o'clock p.m. . . . [May 7] a party of Salvation Army preachers attempted to hold a service in the Upper Green near the Nelson Statue. The usual singing and exhaultation [sic] was begun by the detachment, but the falling dust was too much for mortal throat and lungs to endure in the open air, even though engaged in religious work. Besides this there were few found so foolhardy as to accept the invitation to combat the forces of nature, so the party had to desist and return to cover like the common non-combatants" (Anderson and Flett, 1903:409).

Mount St Helens, 1980

The health concerns reported from the tephra fall area east of Mount St Helens cover a greater range than heretofore reported, probably only because of more intensive investigation. Figure 3.8 shows the areas affected by tephra fallout; few of the areas discussed received more than 10–15 mm and none more than 50 mm of uncompacted tephra.

Figure 3.9 indicates the dramatic increase in visits to the two Yakima Emergency Rooms associated with hospitals. Most of the increase appears to be the result of respiratory or eye problems. A number of these patients were diagnosed as suffering airway irritation with asthma or bronchitis. The asthma patients represented the largest category influencing the increase in Emergency Room visits. The post-eruption increase in the incidence of bronchitis was obscured by a relatively high background of reported bronchitis with an upper respiratory tract infection. In some cases this was diagnosed as acute alveolitis with tightness in the chest and headache; in other cases, it was diagnosed as gastroenteritis. At least some patients suffered a recurrence of these symptoms the day after exposure to tephra (during clean-up operations). Some symptoms might indicate a reaction to stress. While there is no doubt that the increase in Emergency Room visits was partly due to anxiety and apprehension, the majority of patients had objective signs of respiratory problems. Sixteen patients, mostly females of adolescent age or older and without a pre-existing condition, were diagnosed as having hyperventilation syndrome. Most of these patients arrived at the Emergency Rooms a few days after the tephra fall and calmed down fairly quickly (CDC-Mt St Helens Volcano Health

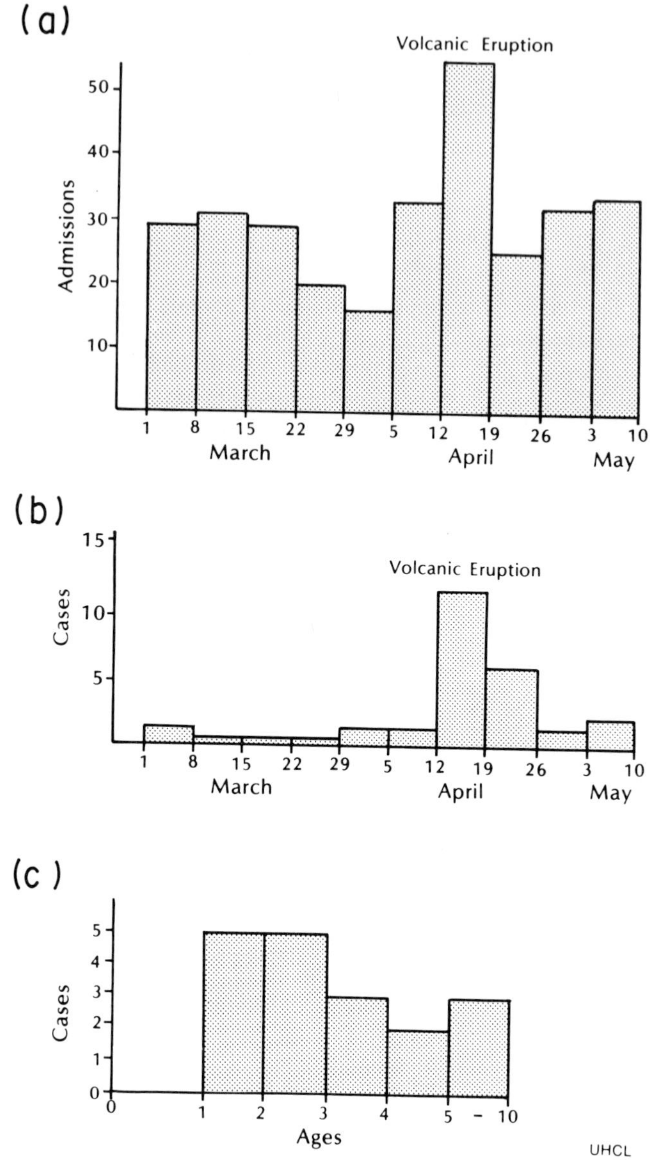

FIG. 3.7 Admissions to the pediatric ward of the Kingstown General Hospital, St Vincent, before and after the April eruption, showing (a) a substantial increase in the number of admissions, particularly (b) of those suffering from the asthmatic bronchitis; children of all ages groups (c), except those under the age of one, were affected (modified after Leus *et al.*, 1981: 67).

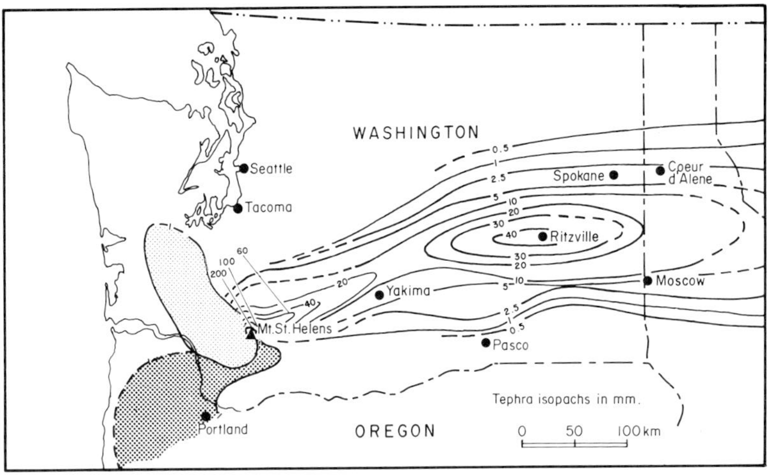

FIG. 3.8 Isopach map of tephra fall from the Mount St Helens eruption of May 18, 1980. Isopachs indicate uncompacted thicknesses in millimeters. Stippled portions show areas receiving more than 1.0 mm of uncompacted tephra from May 25 and June 12, 1980 eruptions (compiled from maps in Sarna-Wojcicki *et al.,* 1981).

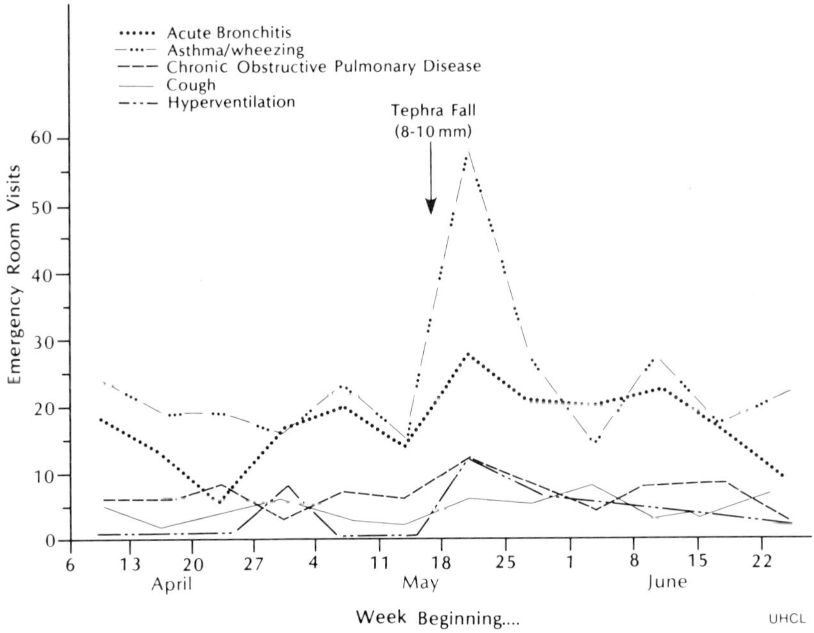

FIG. 3.9 Weekly totals of patients diagnosed as suffering main airways-related problems before and after May 18, 1980 tephra fall. Data based on logbooks of two Yakima emergency rooms (after Baxter *et al,* 1981: 2588).

Report, 17, August 8, 1980; Dr R. Attwood, Yakima, June, 1980, personal communication). A Yakima ear, nose and throat specialist noted no respiratory problems amongst his patients until a period of strong winds began five days after the tephra fall. As this period of the year is normally one of high pollen counts, he suggests that pollen was purged from the air by the tephra fall, returning with the stronger winds (which also resuspended the ash). However, many patients stopped having allergy shots because they had no symptoms, the allergy response of the community as a whole not returning to normal levels until the autumn (Dr M. Murphy, Yakima, May 1981, personal communication).

There were similar increases in the data for Emergency Room visits largely related to respiratory problems in other towns (CDC-Mt St Helens Volcano Health Reports No. 19, August 22, 1980; No. 22, October 21, 1980; Baxter *et al.*, 1981). However, it is important to note that the magnitudes of the increases were influenced by: (i) stress reactions; (ii) problems with pollen and allergies; (iii) weather patterns; (iv) changes in work/activity patterns during the period of tephra clean-up; and (v) exhortations not to drive vehicles. From a volcanic hazards point of view it does not much matter whether the symptoms of respiratory disease are real or imaginary; medical services still have to cope with the influx of patients.

Detailed studies of timber cutters and rigging crews, much exposed to recirculating tephra in the months following May 18, 1980 (Bernstein *et al.*, 1980), suggested that both respirable dust levels and acute respiratory symptoms fell significantly with time. By September, 1980, there was little difference between exposed and non-exposed groups except in the incidence of headaches and eye irritation (Bernstein *et al.,* 1982: 212).

The available data on eye problems associated with the tephra fall and its aftermath in Yakima are also of interest. The records of 129 patients with eye-related problems who visited the two Yakima Emergency Rooms during the period May 18 to June 14, 1980, indicated that 42 (33%) of the visits can be regarded as ash-related (Table 3.5). Ten of the 42 patients were under the age of 15 years. Four of the 10 were playing in or throwing the tephra when the eye problem occurred. Of the adults, 11 of the 32 were working in or cleaning up the ash, and a further seven were riding in or on vehicles. Of the 42 patients, 64% experienced foreign-body sensation, 29% irritation or redness, 14% pain, 5% tearing, and 5% swelling. Patients with conjunctivitis also experienced itching, burning and photosensitivity. These complaints can be related to the findings of physical examinations which indicated the presence of a foreign body (57%), corneal abrasion or scratches (36%), conjunctival injection/edema (33%), hemorrhage (2%), conjunctivitis (17%), swollen eyelid or periorbital edema (2%) and exudate (2%). This total of 149% indicates that many patients suffered more than one condition (CDC-Mt St Helens Volcano Health Report No. 18, August 15, 1980). A telephone survey conducted in several ash-affected communities suggested that 4-8% of the population experienced eye irritation but that only one in ten of these sought medical advice. Of course, many

Table 3.5 Tephra-related eye problems at Yakima Emergency Rooms, May 18 to June 14, 1980

Eye problem	Number of patients
Foreign body	24 (57%)
Corneal abrasion	6 (14%)
Conjunctivitis	12 (29%)
Total	42 (100%)

Source: CDC-Mt St Helens Volcano Health Report No. 18 August 15, 1980: 2.

of those reported as experiencing eye problems may have only suffered from "red eyes" or found contact lenses irritating and reverted to wearing glasses.

Skin irritations may also have occurred in some people exposed to the tephra fall and clean-up. Physicians interviewed in Yakima reported no increase in chronic skin irritations following May 18, but it is unlikely that such relatively minor problems would come to the attention of medical authorities. *US News and World Report* (June 9, 1980: 13), mentioned, however, that many people still itched from "ash rash" caused when the skin absorbed volcanic dust. Axillary (armpit) skin irritations were also reported in Moses Lake (CDC-Mt St Helens Volcano Health Report No. 4, June 6, 1980).

The tephra fall in central Washington State also produced another interesting medical side effect. At Moses Lake (Fig. 3.8) visits to the Emergency Room per week fell from 63 to 43, largely because recreational activities were impossible (56 down to 26). However, motor vehicle accident injuries jumped from four to 11 and falls from house roofs and other fall injuries increased from three to six (*Morbidity and Mortality Weekly Report*). Although the sample size is rather small it appears that most of these variations can be ascribed to the indirect effects of tephra falls.

With the possible exception of the effects of tephra fall on community accident rates, few of the effects from around Mount St Helens are new. It now seems well-documented that increases in respiratory disease and eye irritation will be reported following even minor ash falls. The proportion of the population reporting effects undoubtedly reflects the "sophistication" of the society and the available medical resources and priorities as well as the period of exposure to tephra, meteorological conditions, the proportion of respirable particles in the tephra and (probably) the presence or absence of irritating gases such as SO_2. However, the Mount St Helens studies have provided a great deal of statistically valid data which will allow accurate comparisons of effects resulting from future eruptions.

3.5.3. Chronic medical aspects

However, one important aspect of the reported medical effects resulting from tephra falls does seem to be specific to the May 18, Mount St Helens eruption; the possibility of contracting chronic bronchitis, pneumo(vol)coniosis, or silicosis as a result of breathing an ash-laden atmosphere. Silicosis is a lung disease resulting from the inhalation of fine particles of free crystalline silica which have toxic effects on the lungs causing fibrotic changes. For silicosis to develop the victim must be exposed to crystalline silica (that is quartz, cristobalite or tridymite) of a respirable size range (that is, less than 10 μm). Pneumoconiosis can occur with the inhalation of any kind of toxic respirable dust such as talc, silicates etc. Most inhalable dust is captured by the upper respiratory defence system and/or exhaled – it is only the fine respirable material that lodges deep in the lungs to encourage, after one or more years of exposure, the development of silicosis or, more generally, pneumoconiosis.

The United States Environmental Protection Agency (EPA) standards relevant to industrial particulate emissions (SO_2 – particulate complexes) for the most sensitive persons are given in Table 3.6. In addition the National Institute for Occupational Safety and Health (NIOSH) recommends that exposure to respirable free silica be limited to 50 μg m^{-3}. These criteria are designed to protect the health of workers for a 40 hour workweek over the working lifetime. As the total suspended particulate (TSP) concentration measured in Yakima between noon and 4 p.m. on May 18, 1980, for example, reached 35 621 μg m^{-3}, a fairly high level of concern in the community was not surprising. Even five days after the eruption TSP levels in Yakima still exceeded 1000 μg m^{-3} (Bernstein *et al.*, 1980: 3). Furthermore, analyses of Mount St Helens tephra eventually showed 4%–6% free crystalline silica in less than 10 μm range (CDC-Mt St Helens Volcano Health Report No. 3, June 3, 1980).

Following the initial concern, it appeared that only those workers with long exposures to high concentrations of respirable tephra, such as loggers, were at any risk of developing pneumo(vol)coniosis. Detailed studies must be continued for lengthy periods before proper evaluation of the occupational risks to workers in dusty environments can be ascertained (Green *et al.*, 1982), but studies of rats injected intratracheally with 10 mg of dry respirable tephra from Spokane (7.2% by weight of crystalline silica) showed interesting reactions. Rats killed the following day showed focal acute inflammatory response at the level of terminal bronchioles and alveolar ducts. At periods of one to three months after injection, rats showed a widespread granulomatous reaction, with some showing a central necrotic area and containing foreign body giant type cells. Studies of several loggers who died in the

Table 3.6 US Environmental Protection Agency ambient air quality standards — 24 hour average exposure

Air quality warning	TSP concentration μg/m^3
Primary	260
Alert	375
Warning	625
Emergency	875
Significant harm	1000

Source: CDC-Mt St Helens Volcano Health Report, No. 3, June 3, 1980: 2.

blast area indicated that one logger showed intra-alveolar foci resembling the giant cell granulomatous lesions seen in the rats (Green *et al.*, 1981: 216-7).

While these studies indicate that the tephra is moderately fibrogenic and should be considered a pneumoconiosis risk in heavily exposed individuals, comparison with the dead loggers' lungs are confounded by factors of injury, infection, burn therapy and acute exposure period. Nonetheless, the results clearly indicate the need for heavily exposed workers, including logging crews, road gangs and agricultural workers to take precautions.

3.5.4 Protective measures

Of course, populations everywhere have automatically developed simple methods for protecting themselves from falling tephra. The following examples indicate the universality of some protective measures:

Vesuvio, 1906: During this eruption, which deposited about 30 mm of black fine sand on Naples, citizens wore glasses and held extended umbrellas to protect themselves (Lacroix, 1906: 234); Delmé–Radcliffe (1906: 1) reported this tephra fall to be 150 mm thick in Naples noting that it muffled all street sounds: "the few people in the streets went about in great-coats, with umbrellas up and goggles as protection against the softly falling ash".

Soufrière, St Vincent, 1902: At Bridgetown, Barbados, 180 km to the west of the volcano, 5 mm of tephra fell on May 7, 1902. The *Barbados Agricultural Reporter* of the following day reported: "Various forms of shades were used for protecting the eyes from the dust, some persons using coloured glasses and others using veils. It was a funny sight to see grave city men wearing the sun veils commonly used by ladies. But the object arrived at was the protection of the visual and nasal organs against the inrush of dust, and this the veil did effectually" (Anderson and Flett, 1903: 410).

Spurr, Alaska, 1953: Newsboys in Anchorage, where about 1 mm of tephra fell, donned surplus military masks to combat the ash-laden air (Fig. 3.10).

Irazú, 1963-1965: The tephra fall forced the 230 000 inhabitants of San Jose to wear goggles, bandanas or even gas masks almost every day (Wilcoxson, 1966: 66).

Around Mount St Helens in 1980 there was, of course, no shortage of advice concerning personal protection from tephra fall. The citizens of Ritzville were told to wear wet rags (effective in filtering out the water soluble gases) as makeshift masks; then they were told dry rags were better. Later still, wet rags were better after all (*Volcano*, 1980: 68). The Red Cross suggested drinking lots of clear fluids like water or tea during a tephra fall and avoiding hot chocolate and milk because the mucus build up attracts dust. They also suggested that people with respiratory problems or scratchy throats moisten the air by turning on the shower or putting on a steam kettle (*Seattle Post-Intelligencer,* May 28, 1980: A12).

FIG. 3.10 *Anchorage Daily Times* newsboys used surplus military masks during the 1953 tephra fall from Spurr. The newspaper headline was misleading – only one volcano erupted and the "blanket of ash" in Anchorage was only 3–6 mm thick! (Photo: *Anchorage Daily Times,* July 19, 1953; Alaska Geographic Society).

Federal Emergency Management Agency (FEMA) Mount St Helens Technical Information Network Bulletin No. 3 was produced on May 27, 1980, with the title "Precautions in handling volcanic ash". This three page report indicated that the general public should avoid unnecessary exposure and wear a light facemask out-of-doors. Table 3.7 presents the Appendix to Bulletin No. 3 detailing the masks deemed suitable. Commercial advertisements presenting instructions for wearing respirators also appeared in newspapers (Fig. 3.11). Bulletin No. 14 (June 6, 1980: 137-138) concentrated on "Protecting children from volcanic ash-related health hazards" and stated:

> The most obvious step that can be taken to protect children from inhaling airborne ash is to *keep children indoors when ash is visible in the air.* Children should certainly be advised against strenuous play or running when ash is in the air, since exertion leads to heavier breathing, drawing small particles more deeply into the lungs...
>
> If children must be outdoors when ash is present in the air, they should wear a mask, preferably one marked with "TC-21C" and approved by the National Institute for Occupational Safety and Health (NIOSH). However, commercially available masks are manufactured for industrial use and therefore are designed to fit adults rather than children. One such approved mask, Number 8710 of the 3M Corporation, is stated by the manufacturer to be adaptable to children as young as 5 years.
>
> Fabric masks, including those improvised from handkerchiefs, will filter out the larger dust particles and should be dampened with water to be used when the dust is visible in the air if no approved mask is available. For children old enough to understand the difference, it might be helpful to breathe through the nose, where natural filtration of large particles takes place.
>
> Small children may at times swallow some of the ash. Evidence to date suggests that ingestion of ash is not a hazard to the health of children and adults...

Table 3.7 Approved small particle masks

Masks which provide protection against lung damage from small mineral particles are approved by the National Institute for Occupational Safety and Health (NIOSH). Masks with printing on the manufacturer's container indicating NIOSH approval for protection against "pneumoconiosis" and "fibrogenic dusts" are the masks which will filter out these small particles. The masks will also carry a coded marking starting with "TC-21C" and followed by another two or three digit number.

Single use approved face mask (throw away when they become difficult to breathe through, or structure of the mask is destroyed).

Manufacturer	*Number*
3M	6983
3M	8710
3M	9900
3M	9910
3M	9920
American Optical (AO)	R1050
Binks	40-150
Norton	7170
Willson	1400
Willson	1410

NIOSH approved masks for multiple use (filter elements need be replaced only when they become difficult to breathe through).

Manufacturer	*Number*
American Optical (AO)	R209ON
American Optical (AO)	R4030
American Optical (AO)	R5030
American Optical (AO)	R6030
MSA Dust FO	66
MSA Dust FO	77
MSA Dust FO	88
MSA Comfo II	459440
Willson	560
Willson	1210
Willson	1211
Willson	1212

After Federal Emergency Management Agency Mt St Helens Technical Information Network Bulletin No. 3, May 27, 1980.

3.6 PYROCLASTIC FLOWS AND DEBRIS AVALANCHES

Considerations presented in Chapter 2, Sections 3 and 9, indicate that pyroclastic flows pose threats to human life because of their high velocities and temperatures. Table 3.2 indicates that pyroclastic flows have been the most common direct cause of death in the last 400 years and have accounted for more than 70% of all deaths in volcanic eruptions this century.

Table 3.8 outlines some of the eruptions in which pyroclastic flows have been important agents of death. Most of the estimates of the total number of deaths and the numbers attributed to pyroclastic flows are much more uncertain than the table suggests. Some of the problems with data on deaths and further illustrations of modes of death are elaborated in the following nine examples, presented in chronological order.

Kilauea, 1790: The following account of the journey of Keoua's army along the south west rift zone of Kilauea is taken from I. Dibble's *History of the Sandwich Islands* (1843):

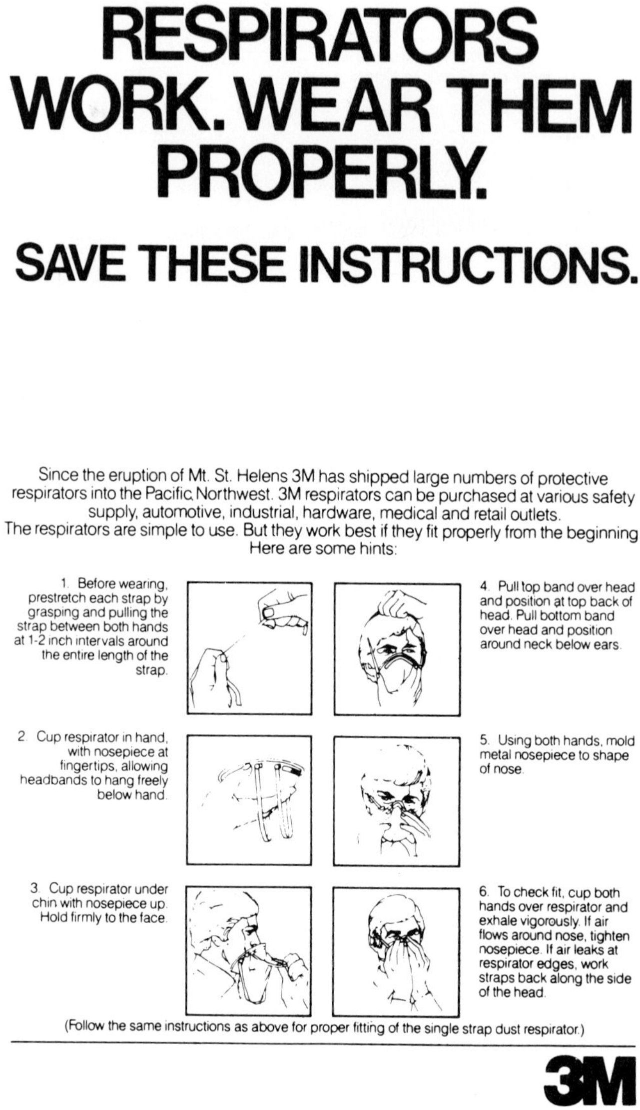

FIG. 3.11 This advertisement from the 3M Company appeared in many Pacific Northwest newspapers in early June, 1980.

> The army of Keoua a Hawaiian chief, being pursued by Kamehameha, were at the time near Kilauea. For two preceding nights there had been eruptions, with ejections of stones and cinders. The army of Keoua set out on their way in three different companies. The company in advance had not proceeded far before the ground began to shake and rock beneath their feet, and it became quite impossible to stand. Soon a dense cloud of darkness was seen to rise out of the crater, and almost at the same instant, the thunder began to roar in the heavens and the lightning to flash. It continued to ascend and spread around till the whole region was enveloped, and the light of day was entirely excluded.
>
> The darkness was the more terrific, being made visible by an awful glare from streams of red and blue light variously combined through the action of fires of the pit and the flashes of lightning above. Soon followed an immense volume of sand and cinders, which were thrown to a great height and came down in a destructive shower for many miles around. A few of the forward company were burned to death by sand, and all of them experienced a suffocating sensation.
>
> The rear company, which was nearest the volcano at the time, suffered little injury; and after the earthquake and shower of sand had passed over, hastened on to greet their comrades ahead on their escape from so imminent peril. But what was their surprise and consternation to find the center company a collection of corpses. Some were lying down, and others were sitting upright, clasping with dying grasp their wives and children, and joining noses (the mode of expressing affection) as in the act of taking leave. So much like life they looked that at first they supposed them merely at rest, and it was not until they had come up to them and handled them that they could detect their mistake. Mr Dibble adds: A blast of sulphurous gas, a shower of heated embers, or a volume of heated steam would sufficiently account for this sudden death. Some of the narrators, who saw the corpses, affirm that though in no place deeply burnt, yet they were thoroughly scorched.
>
> [Stearns, 1925: 11-12].

Despite the evidence of scorching presented here, Macdonald and Abbot (1970: 71) believed that the "bodies were not burned, or apparently otherwise injured". They suggest that the company must have been caught in a cloud of poisonous gas. However, death in a pyroclastic flow of the base surge type seems to be the most likely explanation (see also Swanson and Christiansen, 1973).

Bandai, 1888: A debris avalanche buried 461 people but only 117 bodies were recovered, some battered by falling stones. Seventy people were injured, mostly burnt and scarred by a hurricane of hot ashes and falling stones (Sekiya and Kikuchi, 1889: 137-138). Wilcoxson (1966: 183) cites a *London Daily News* of the time:

> The wounded are receiving treatment in the schoolhouse at Inawishiro, but their condition is terrible. Some have fractured skulls, the majority broken limbs, while others are fearfully burned. Five villages have been totally buried. The dead appear as the victims of a huge boiler explosion. Many are cut to pieces, and others parboiled, so that it is difficult to distinguish sex. But the most ghastly sights which met the eyes of the helpers were bodies dangling on the branches of trees. Thrown into the air by the awful violence of the eruption, their descent had, in many cases, been arrested by the trees. And there the victims hung, their bodies exposed to the cruel and well nigh ceaseless rain of red hot embers and burning ashes. From appearances death speedily relieved them from their agony, yet, short as the time was, their suffering must have been past belief. In other places flesh hangs from the branches of trees as paper from London telegraph wires.

Soufrière, St Vincent, 1902: An official report (*HMSO,* 1903: 10-11) gives the total number killed as 1680, attributing the deaths to pyroclastic flows and accompanying hot gases. Shepherd *et al.* (1979: 24) put the death toll at only 1565 but

Table 3.8 Deaths in pyroclastic flows

Eruption	Number killed by pyroclastic flows	Total killed	Reference
Vesuvio, AD 79	670	3 360	*VRF*
Stromboli, 1930	4	4	Rittman, 1962: 38
Lamington, 1951	2 942	2 942	Taylor, 1958: 48-49
Papandayan, 1772	2 957	2 957	*CIRV*, 1979: 793
Galunggung, 1882	~400	4 011	Anon., 1832: 415
Agung, 1963	820	1 148	Surjo, 1965: 26; VRF
Hibok-hibok, 1950	68	68	*Time*, Dec 17, 1951: 36
1951	500	500	Macdonald and Alcaraz, 1956: 173
Mayon, 1853	34	34	*VRF*
1897	350	350	Sapper, 1917, 153
Taal, 1911	1 335	1 335	Sapper, 1917: 152
1965	150	150	Tolentino, 1973
Kuchinerabu-jima, 1933	26	26	*CAVW*, 11: 16
Unzen, 1792	10 000	14 300	*CAVW*, 11: 43
Asama, 1783	820	1 377	*VRF*
Bandai, 1883	461	461	Sekiya & Kikuchi, 1889
Tokachi, 1926	144	144	Tanakadate, 1927: 786
Kilauea, 1790	80	80	Stearns, 1925: 11-12; Sapper, 1917: 101
El Chichon, 1982	34	188	*VRF*
Santa Maria, 1929	23	23	*CAVW*, 6: 44
Arenal, 1968	78	78	*Bull. Volc.*, 8, Supp.: 12; 55
Dona Juana, 1899	55	55	*Bull. Volc.*, 8, Supp.
Mt Pelée, 1902			
May 8	2 800	2 800	*CAVW*, 20: 35-36
August 30	1 000	1 000	*CAVW*, 20: 35-36
Soufrière, St Vincent, 1902	1 680	1 680	*HMSO*, 1903: 10-11
Mount St Helens, 1980	~50	57	Eisele *et al.*, 1981

Huggins (1902: 5) reckons the number killed at more than 2000. A report in *Nature* (1903, January 15: 257) notes that no autopsies were performed but attributes deaths to the following causes, in order of importance: (i) asphyxiation by hot dust-laden steam and air; (ii) burns due to hot stones, lapilli and dust; (iii) blows by falling stones; (iv) nervous shock; (v) burning by steam alone; and (vi) probably lightning strikes. Sulfur dioxide (SO_2) and H_2S were probably also present and helped to kill people. The positions in which many bodies were found indicated death by asphyxiation. Burns under clothing where the cloth was not charred were attributed to the action of steam (Hovey, 1902: 457). Huggins notes (1902: 5): "Most of the dead were frightfully burned on the exposed parts of their bodies, many of the survivors being burnt in the same manner. Some had lost their legs and arms whilst others burnt and charred all over..." Charles Alexander, a labourer aged 40, who suffered severe burns about the face and neck and on the wrists and legs, gave the following account which agreed on all main points with the narratives of other survivors:

> I lived at Overland [Fig. 3.12]. On the morning of the eruption I was cutting canes on Tourama Estate [about 5.5 km east of the volcano]. About 9 o'clock there was a drizzle of rain with ashes falling, but I still goes on work, then ashes fall heavier. About 12

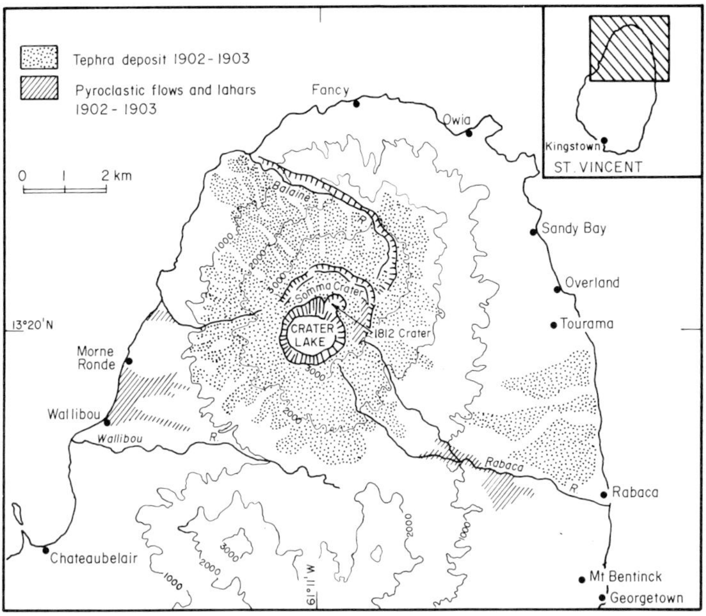

FIG. 3.12 Map of the northern half of St Vincent showing tephra deposits, pyroclastic flows and lahars from the 1902-1903 Soufrière eruption. Contours in feet. (After *CAVW*, Vol. 20, 1966: 45.)

o'clock I start to go home, the ashes still fall and small stones, and when I get home I hear a great noise in the direction of the mountain like a rushing river. All the people then start to run from the village, Tourama way, but when we get half part of the way we turned back, because we meet the Tourama people coming to Overland. As soon as we reach Overland again, large stones begin to drop, and this cause all the people run into the houses. With many others I run into Victor Sutherland's shop. It have a strong galvanised roof, but soon some large stones fall through the roof. About 2 o'clock great darkness come on, and we shut the doors and windows. After this a great heat come with hot ashes through the chinks of the doors and windows, and through the holes in the roof. The hot ashes get into our mouths, and stop our mouths as fast as we try to breathe. We toss backwards and forwards for about two seconds, then everyone fall down. I did not lose my senses, but cannot tell exactly what happened after I fall. I feel choked with the hot stuff going down my belly, and smell plenty sulphur. This did not last too long, only two or three minutes, then I try to get up, but two people both dead lie across me, and after a struggle enough I get on my feet. I then know I was burned, I think I was burned when lying on the floor. I now open a window a little to get air, this revived me a little, but stones still dropped, and more ashes come in, I shut it again, I watched the great black cloud for about two minutes, while the window was open, it was full of fire like lightning, and I see stones drop on the ground break in pieces and glow hot. At this time many people in the shop done dead, and some of the houses outside were burning.

Stones now ceased to drop, and there was a calm — this would be about 4 o'clock, and it was still dark. We now open the windows and doors, and sit up till day clean, then I walked to the Hospital at Georgetown. I know of 6 people being alive in the shop next morning, they were all burned. I think about 80 people die in the shop, most after we fall, but some during the night.

[Will, in *HMSO*, 1903: 13-14]

Some survivors, caught on the edge of a pyroclastic flow, while in a rowing boat off the west coast of St Vincent, reported the gas cloud to be hot, highly sulfurous and suffocating, irritating the throat and nostrils and making people cough. The feeling lasted for only a minute or two (Anderson and Flett, 1903: 392). Will (in *HMSO*, 1903: 14) provides other information based on the accounts of survivors and his own observations:

> Further evidence showed that every person who left shelter during the wave of hot ashes was killed in a few minutes, that inhaling the ashes, &c., caused a feeling as if the windpipe was being compressed, and that this feeling was less acute while the sufferers held in their breath.
>
> The burns seem to have been entirely caused by the hot dust falling on exposed parts at a temperature high enough to cause vesication and destruction of the skin, but not at a sufficiently high enough temperature to ignite the clothing or the thatched roofs of the houses. Many houses were burned, but they seem to have been ignited by the "fire-stones". None of the burns presented the appearances characteristic of those inflicted by lightning and the depth of tissues destroyed was greater than that usually found in burns by steam.
>
> The large number of deaths caused by the eruption appear to have been almost entirely due to asphyxia by the hot ashes and heated air, the latter being probably somewhat deficient of oxygen. The cloud of dust was highly charged with electricity, and it is likely that some of the deaths outside the houses were caused by lightning, also that a few, especially children, may have been killed by the falling stones.

Of the 224 admitted to or attending hospital after the May 7 eruption, 221 suffered burns; 79 of these subsequently died. The burns were so terrible, reported the Governor, that dressing the wounds required nerve and experience.

> The 191 severe cases of burns admitted were characterised more by the extent of the areas implicated than by the depth of destruction of the affected surfaces.
>
> The sites were chiefly the face, ears, neck, forearms, and backs of the hands, the legs and feet, and about 20 per cent had body burns as well, situated in most of these cases on the shoulders and buttocks.
>
> In one case only were there as few as 2 burns, the majority had at least 4, and some as many as 8 or 10.
>
> The degree of burn varied from destruction of the cuticle to that of the true skin and subcutaneous tissues. The latter degree occurred in 5 cases only, implicating in 2 of these the ankles, in 2 the extensor tendons of the fingers, and in 1 the finger tips.
>
> In most of the cases several of the degrees were combined in the surface affected.
>
> All the cases when first seen had an adherent coating of fine dust.
>
> In consequence of the degrees and extent of the injuries, pain was intense.
>
> There was a remarkable absence of shock, which is always an invariable sequel of intense burns in Europeans. Its absence may be accounted for, either by the lower nervous organisation of the coloured race, or by the fact that the injured received their burns between 3 and 4 p.m. on the 7th, and were not removed to hospital till next morning. It is probable that some may have succumbed to shock in the interval. On the other hand many of those who had sustained the most extensive and severest burns started at daylight on the following morning and walked to Georgetown, a distance, in several instances, of 7 miles [14 km]. It may, therefore, be assumed that shock was not a prominent feature, and to its comparative absence must in a great measure be ascribed the large number of recoveries.
>
> The total number of deaths resulting from injuries caused by the eruption was 80, of which 79 were due to burns. The immediate cause of death was in most of the cases

> exhaustion from septicaemia a result it was impossible to obviate owing to the extent of surface involved in each case. A few died from secondary complications, viz., pneumonia and pleurisy, and 4 from tetanus.
>
> [Will, in *HMSO*, 1903: 12]

Mt Pelée, 1902 (Fig. 2.14): The *Catalogue of Active Volcanoes of the World* (volume 20: 35-36) gives the death toll in the May 8 eruption as 28 000. Heilprin (1903: 112; 118-120) notes about 30 000 killed in this eruption. Hill (1902a: 235) gives the population of St Pierre as 25 792. An article in *Nature* (1903, January 15: 258) attributes the deaths in St Pierre to the same causes as those around Soufrière the day before (see above) with the addition of crushing beneath falling walls and other objects and cremation in burning buildings. The article also notes that in both eruptions the lack of respirable air probably caused many deaths. Certainly, witnesses on the perimeter said they could get no air to breathe but were not suffocated by ashes or detectable gases (Hill, 1902a: 257). An interesting paper written by G.N. Cole in 1919 postulated that heat, dynamically generated by pressure, and the pressure changes themselves, killed the population of St Pierre. In refuting Cole's argument and various lines of evidence Jaggar (*Volcano Letter,* March 11, 1926) wrote: "The heat of a burning building disembowelled bodies. There was no mystery about the deaths to the medical officers; at the end of the town near the volcano the rock dust was incandescent and falling by tons [?]. People and houses were burnt. At the far end of the town the dust was scalding hot in a steam emulsion. Thin clothing was spared while skin was burned. The awful cases of bronchitis in the wounded were due to nothing but hot dust. The population of St Pierre died chiefly from instantaneous pneumonia". Although several people and animals were found in frozen positions (for example, a man holding a struggling frightened horse, both dead in that position) it is not true that all the inhabitants died instantly from asphyxiation in positions in which the cataclysm found them. Many bodies were found in positions indicating flight. The captain of the *Roddam* reported that he saw people running along the water's edge for several minutes (Hill, 1902a: 258). Some of those killed were burned by steam; others were killed by force; still others were singed by flame. On shore, a Dr Riley reported, human remains were horribly burnt to the quick (Hill, 1902a: 258). This report was confirmed by Major J. Will of the Royal Army Medical Corps and leader of a medical relief expedition despatched from Barbados and arriving in St Pierre on the morning of May 11:

> At every turn scenes of indescribable horror met the eye. Dead bodies lay scattered about, singly and in groups, most with the face to the ground, and many with the face resting on the fore-arms or covered with the hands. With one exception all the victims were devoid of clothing. This exception appeared to be a white man, probably a sailor, who had escaped from one of the burning ships. He was lying on the face with his jersey pulled over the head.
>
> A brief examination was made of a number of the victims. In many there were no signs of actual burning, they appeared shrunken, dried up, and mummified. On the other hand some were terribly charred, the flesh being in parts entirely consumed, exposing the bones of the limbs and skull. There was no stench, only a faint odour of burning flesh.

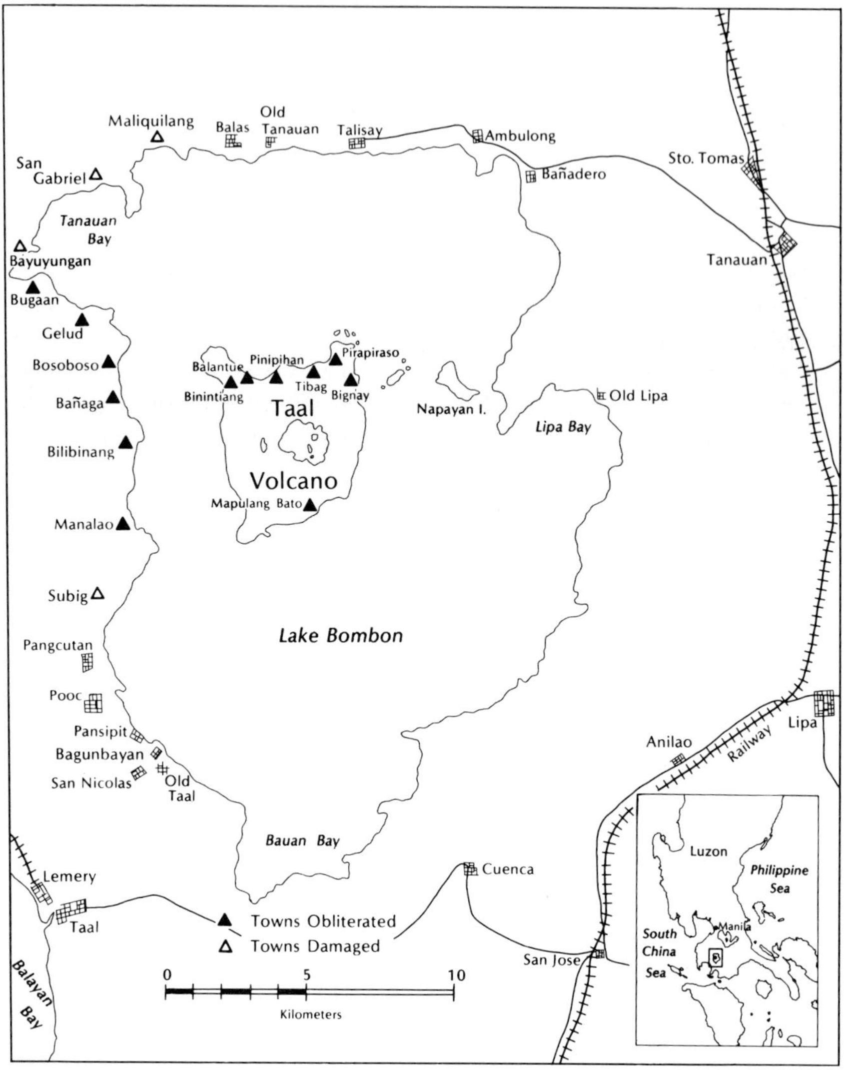

FIG. 3.13 Map of the Taal area showing towns damaged and obliterated during the 1911 eruption (after Masó, 1911).

For a town of 26 000 inhabitants of whom not a single soul escaped, the number of bodies seen in the streets was comparatively small, and even in the houses where, no doubt, the majority had sought refuge from the burning ashes, few bodies were seen. It seemed evident that the majority of the victims had been buried under the fallen and burning roofs, and that large numbers must have been entirely incinerated.

[Will, in *HMSO*, 1903: 9].

Another earlier visitor to the ruins of St Pierre noted that the bodies of victims were greatly reduced in size: "It was as if the moisture had suddenly been extracted from them. But as they were not burned, and as trees and other pieces of wood were barely charred, I concluded that either the duration of the heat had been extremely brief or that the heat had failed to ignite anything because of an absence of oxygen" (Koster, 1956: 414). Certainly, the survivors suffered severe burns. Ciparis, the famed survivor from the dungeon, was severely burnt on his back, beneath his shirt. Another survivor, Compere-Léandre, was burnt on arms, legs and body (Bullard, 1976: 128). On the *Roddam* nearly all survivors had first degree burns affecting only the epidermis but there were also some second degree burns. The eyes of the wounded were unaffected and the eyelashes intact. On the *Roraima,* surviving children moaned for water but were unable to swallow because of the ashes which had clogged and burnt their throats (Hill, 1902a: 258). The various accounts presented here indicate considerable variation in opinions about the causes of deaths in St Pierre. Despite these variations the accounts have a certain consistency in that most observations are confirmed by more than one author. Variations presumably result from difference in the modes and intensities of destruction experienced in even adjacent parts of the city. Such variations would be in accord with the data presented in Figure 2.14 (Fisher *et al.*, 1980).

Taal, 1911 (Fig. 3.13): About 1335 people were killed, chiefly by pyroclastic flows, but only 732 bodies were found. On the island itself 12 or 13 people survived but all were badly wounded (Pratt, 1911: 81-82; Martin, 1911: 89). On the island, only charred and blackened bodies marked the sites of some villages, lighter materials having been washed away by tsunami (Fig. 3.14). At Gelud and Bilibinang villages, in the obliterated zone, no one survived (Masó, 1911: 18). Outside the obliterated zone, at Bayuyungan and Subig villages, very few people were killed. The total number treated for injuries was 175 to 199. The injuries were burns caused by hot sand and ash with the exception of a few fractures. Sulfurous wet ash followed the dry tephra. The survivors say that people suffocated. Some survivors claim to have escaped by getting under mats. Certainly the attitude of many of the corpses, crouching with faces downward, indicate efforts to get away from suffocating gases (Rivers, 1911: 16). Colonel William Rivers of the Philippines Constabulary shared a tent during much of the disaster aftermath with Dr Victor Heiser. It is interesting, therefore, to also present Heiser's observations, published 25 years later:

> The sight was horrible. Not only had all life been wiped out on Volcano Island itself, but also for miles away from the shore of the lake. The dead could wait; our first care was for the wounded and the homeless strays who were without food, drink, and shelter. Temporary hospitals were set up; acres of gauze were used to dress the frightful burns, bruises and fractures. Many women and animals were found to have aborted, doubtless from the terrific concussion.
>
> The number of wounded was in remarkably small proportion to that of the dead, and the evidence showed that the majority of the fatalities had been caused by the explosion

FIG. 3.14 Victims of a pyroclastic flow and tsunami lying amongst house wreckage in the village of Bignay on Volcano Island, Taal, 1911 (© National Geographic Society).

> of gases. When nature runs amok, she often becomes pranksome in a ghastly way. The Filipinos are normally so modest that they are never caught with their clothes off. Even in bathing they are well covered. But among the ruins we found Filipinos stark naked, standing upright and embedded in ashes to their knees, still holding protecting umbrellas over their heads. They had apparently been attempting to get away when the ashes rained over them and the vacuum caused by the exploding gases had torn off their clothes.
> [Heiser, 1936: 69-70]

Lamington, 1953: Pyroclastic flows killed 2942 persons (Fig. 3.15). All 35 of the Europeans living at Higaturu and Sangara Mission died (see Fig. 2.15). At the Administration School (between Higaturu and Sangara) 39 or 40 local people crowded into a small house which was demolished by the pyroclastic flow leaving the dead occupants piled on the floor. In the zone of marginal devastation some villagers were killed; others were burnt with varying degrees of intensity (Fig. 3.16). Many of those who survived had almost their whole bodies burnt, with skin peeling in strips. Of 40 survivors from the inner zone of total devastation who reached Issivita Mission Hospital, 22 died within 24 hours. Of 70 casualties evacuated by air to more distant hospitals, only three died. The survivors had experienced pains in the mouth, throat and eyes followed by a burning sensation in the chest and abdomen and then a feeling of rapid suffocation, and usually sustained extensive burns on uncovered parts of the body and internal burns. These victims had an excessive thirst but were unable to swallow. Internal damage caused a mucus discharge and sometimes bleeding from the nose and

FIG. 3.15 Victims of the 1951 Lamington pyroclastic flows, near Higaturu (see Fig. 2.15) (Photo: G.A.M. Taylor, Bureau of Mineral Resources, Australia, M/1745).

mouth (Taylor, 1958: 48-49). Taylor also made some important comments on the causes of death:

> The exact cause of death in the glowing cloud was, unfortunately, not clarified by autopsy; putrefaction was too advanced when the medical services were free from their urgent obligations to the living...
>
> The appearance of the bodies in the devastated area at Lamington suggested that rapid damage to the respiratory system might have been an important factor in the cause of death. It was extremely difficult to distinguish between Europeans and natives 48 hours after the eruption. This may have been partly due to an intense post-mortem lividity which is characteristic of death from asphyxia. In this form of death the blood does not readily coagulate, but migrates to the extremities of the body. Severe internal burns were suggested by the presence of dried blood around the nose and mouth of some victims. Rigidity was a notable feature of many bodies and was mostly due to the well known effect of heat stiffening brought about by coagulation of the albuminous material in the muscles . . . In some cases, however, sudden death had caused the instantaneous rigidity characteristic of cadaveric spasm and the bodies remained, after death, in sitting or kneeling positions. Lacroix attributed similar occurrences in St Pierre to inhibition, a term which describes the depression of the heart's action by stimulation of the vagus nerves. Few people appeared to have been killed by flying debris or by being crushed by falling trees or buildings. There were no dismemberments such as were common in St Pierre.
>
> [Taylor, 1958: 49]

Mount St Helens, 1980: Table 3.9 indicates the number of bodies recovered versus the number missing presumed dead following the May 18, 1980, eruption. The final death toll reached 57, with only 61% of the bodies recovered. Figure 3.17

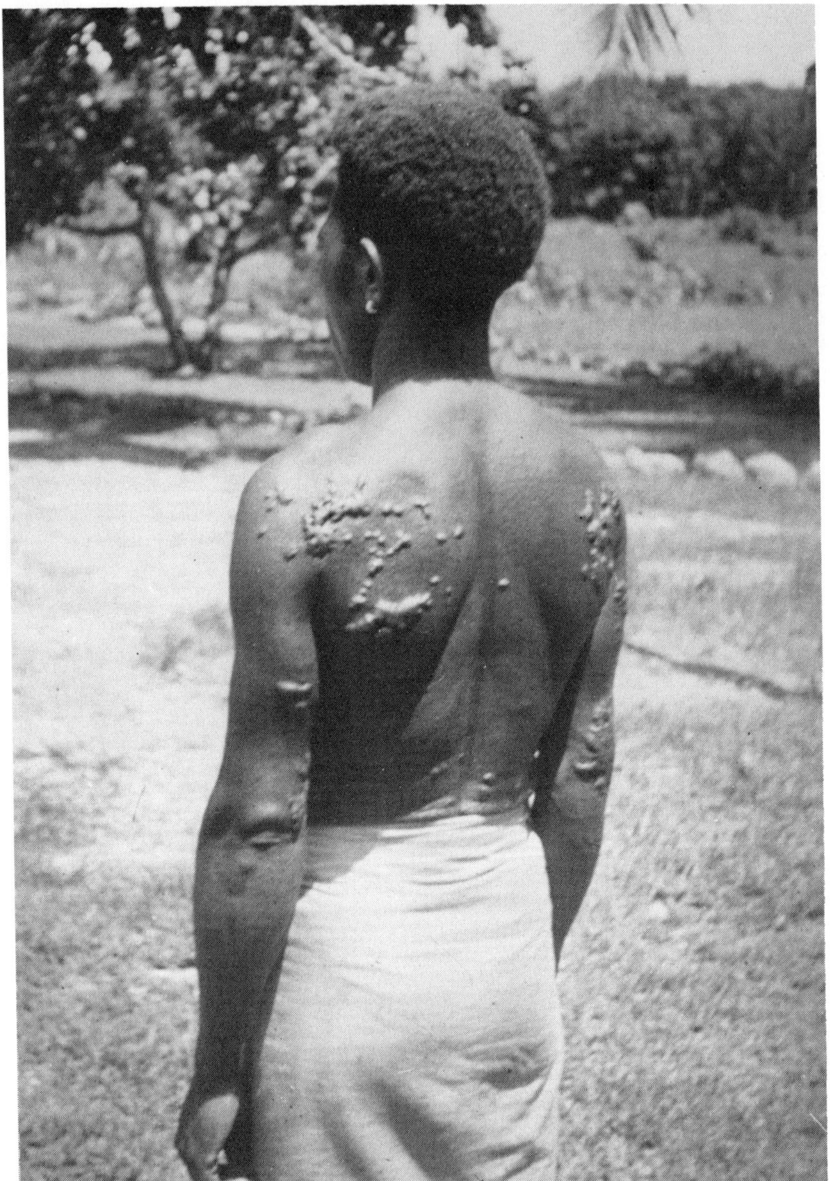

FIG. 3.16 Scars left on burn injuries inflicted by pyroclastic flows, Lamington, 1951 (Photo: G.A.M. Taylor, Bureau of Mineral Resources, Australia, GB/2971).

Table 3.9 Confirmed dead versus missing presumed dead, Mount St Helens, May 1980

Date	Confirmed	Missing	Reference
May 24, 1980	18	72	*The Columbian,* May 26, 1980: 11
May 26, 1980	21	76	*The Columbian,* May 26, 1980: 12
May 29, 1980	21	71	*The Seattle Post-Intelligencer,* May 29: A18
June 27, 1980	25	43	CDC No 10
August 1, 1980	31	34	CDC No. 16
August 22, 1980	31	32	CDC No. 19
December 4, 1981	35	32	Baxter *et al.,* 1981: 2586
April 1, 1983	35	22	*The Oregonian,* April 21, 1983: C5

shows the relationship between the locations of those killed and the missing and the survivors in relation to the areas of pyroclastic flows. Helicopter pilots rescued 113 people around the mountain on Sunday, May 18. Altogether 128 survivors were found in the first few days (EH & WS, 1980). Fifty-three survivors were found within 1.6 km of the tree destruction zone. Many of the other 47 found within 8 km of the vent were located south of the volcano (Fig. 3.17). Of 11 survivors on the edge of the seared zone, six were essentially unharmed, two sustained second-degree burns, two suffered third-degree burns, and one sustained a fracture. Three of these survivors, including two with second-degree burns, were inside a vehicle. Two out of six survivors in the tree destruction area suffered second to third-degree burns. The other four escaped unharmed. Of the 36 survivors within 1.6 km of the tree destruction area, none sustained any serious injuries (CDC No. 19, August 22, 1980). Autopsies were performed on a number of bodies. In 17 cases asphyxia by tephra was the primary cause of death, most victims having an occlusive plug of ash and mucus in the upper airways. Death probably occurred within a few minutes. Some of these victims were inside automobiles. Asphyxia was a contributory cause of death in two further cases. Three victims died from physical injuries, two were killed when their tent was struck by a falling tree 22.5 km north of the mountains and the third died from head injuries received when a large rock penetrated his vehicle 16 km from the vent. In many, if not all, of the above cases the skin showed a tanning or mummification with dark discoloration and an increased leathery consistency. Exposed muscles appeared cooked rather than desiccated and charred, while those portions of the bodies buried in the flow showed skeletonisation and disarticulation. These changes appeared to have occurred after death. Singeing of body hair, which occurs at around 120°C, was found only on those victims within 18 km of the volcano (Eisele *et al.,* 1981: 933-935). Thermal burns accounted for three deaths 15 km from the summit and contributed to two more. "There was extensive thermal destruction of these bodies. The thermal changes differed from those seen after a fire; instead of the usual blackened,

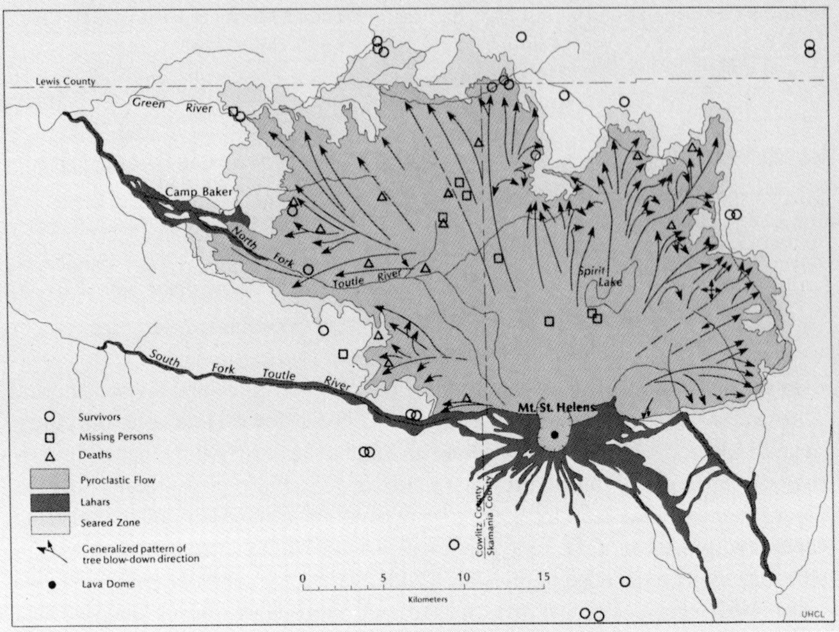

FIG. 3.17 Locations of dead, missing, presumed dead and survivors in Mount St Helens area on May 18, 1980. The area mapped as pyroclastic flow includes areas variously shown elsewhere as debris avalanche, directed blast devastation, downed timber and/or pyroclastic flow. The pyroclastic flow and seared zones are, respectively, approximately equivalent in character to the areas of total and partial destruction shown on Figure 2.15 for the 1951 Lamington eruption (Mount St Helens pyroclastic flow and lahar data modified after Kieffer, 1981; death-survivor data from CDC No. 22).

charred appearance, these bodies appeared dried and baked. The skin was mummified and friable. The exposed muscles were dried and frayed and the ligaments had lost most of their tensile strength. Several fractures involving ribs and extremities were identified, but if they were present before death they were not believed to have contributed substantially to death. Internally, the organs were markedly shrunken and hardened but maintained their usual shape, thus producing miniatures of the originals. The blood formed dry, dark-red casts of the vessels and cardiac chambers. Although the bodies were totally covered with ash [still warm when recovered six days after the eruption], examination of the airways showed no appreciable amounts of ash, leading to the conclusion that the deaths were thermal rather than asphyxial". (Eisele *et al.*, 1981: 934-935). Two victims who died in hospital after walking some distance had second-degree and third-degree burns over exposed portions of their bodies and a purulent trachobronchitis resulting from inhalation of tephra and gas.

These findings are not at variance with most of the less precise information reported following earlier pyroclastic flows. While at least six causes of death were identified in the Mt Pelée and Soufrière, St Vincent, pyroclastic flows, asphyxiation resulting from the inhalation of (hot?) ash and gas was probably the most important. However, this conclusion was based largely on analysis of the bodies recovered. As Table 3.10 suggests it is common for only 25%-60% of the dead to be recovered. The bodies not recovered are usually buried within pyroclastic flow deposits. Although the causes of death in these victims are not known, asphyxia could be important. However, burns or scalds may kill more people in some eruptions. Air has a low heat conductivity so that the human body gains very little heat from hot air (Rivers and Brown, 1979: 20). Thus, dry air at 93°C can be breathed for about half an hour without serious damage occurring. At 500°C dry air can be breathed for only about three minutes (Foster, 1976: 164). Because water is an extremely good conductor of heat, hot wet pyroclastic flows are likely to cause scalds with only a few seconds' exposure.

Few protective measures can be suggested against lightning strokes and ballistic missile impacts other than a sturdy earthed building. Sections of large diameter concrete sewer pipes, walled in at the ends, might offer protection from low volume pyroclastic flows (UNESCO, 1979). It is at least possible that asphyxiation can be slowed by use of one of the masks described in Table 3.7. Slowing the process may be enough as exposure to pyroclastic flows will frequently be only for a few minutes (Chapter 2, Section 3). Similarly, protective clothing is likely to decrease the likelihood of burn and scald injuries. Some examples from Australian bushfire studies illustrate the value of protective clothing. A heat loss of about 1200 watts is a typical maximum cooling rate for a breathing, sweating human. Vigorous exercise will generate about 630 watts internally, and radiation from the sun could add up to 100 watts (if wearing brief clothing). Radiation and convection from a 60 cm high flame at a distance of 1.5 m adds about 1500 watts. This total heat development of 2230 watts is about 1000 watts above the capacity of the human heat regulation system. Under these conditions collapse quickly ensues. However, the wearing of overalls, hats and gloves reduces the heat developed and absorbed to below 840 watts, well within the capability of the body's regulatory system. Studies of radiation heat transmission through clothing materials indicate that woollen and heavy cotton materials transmit less than 5% of radiant heat, cotton shirting 16%, and light nylon shirt material 60% (King, 1962; Foster, 1976: 164). However, these figures refer to heat radiation alone; they have only a limited bearing on the situation in a pyroclastic flow where both conduction and radiation are important.

Table 3.10 Proportion of bodies recovered in eruptions where pyroclastic flows caused most deaths

Eruption	Total killed	Bodies found	Proportion	References
Hibok-hibok, 1951	500	266	53%	*Time,* December 17, 1951: 36
Taal, 1911	1335	732	55%	Pratt, 1911: 81-82
Taal, 1965	189	51	27%	Moore, 1966: 1
Bandai, 1888	461	117	25%	Sekiya and Kikuchi, 1889: 137-138
Torishima, 1902	125	0	0%	Kutschera, 1903: 439
Mount St Helens, 1980	57	35	61%	Eisele *et al* 1981: 932

Finally, it is clear from the data presented in Chapter 2, Section 3 that relatively small pyroclastic flows follow topographic lows. In general, the safest (least dangerous?) areas should be the lee slopes of high ground distant from stream valleys. However, where large topography-mantling flows are produced, the protection afforded by such locations will be minimal.

3.7 LAHARS

As indicated in Table 3.2 lahars and related phenomena have been responsible for a significant number of deaths in volcanic eruptions. Although eruptions such as Galunggung, 1822, and Kelut, 1919, have produced lahars which killed large numbers of people, relatively small death tolls (less than 50) are quite common (Table 3.11). Consequently, popular accounts of death and destruction are much less common than they are in the case of pyroclastic flows. Brief examples illustrate both the range of phenomena and the range of deaths. As with earlier examples many of the estimates of the numbers killed by lahars are merely "guesstimates" based on data set out in the *Smithsonian Institution Volcano Reference File*.

White Island, 1914: Eleven workers at the sulfur mine were killed by a lahar which carried boulders 10 m x 5 m and deposited around 1 m of mud in the workmen's huts (*CAVW,* 22: 42). According to Coleman (1946: 124) only the pet cat survived.

Ruapehu, 1953: At about 8.15 p.m. on December 24 a lahar generated by the discharge of water from Ruapehu's crater lake weakened the central span of the Whangaehu River railway bridge on the North Island main trunk line. Five minutes later an express locomotive, tenders and five carriages plunged into the river killing 151 people (Fig. 2.21). This represented the world's sixth worst railway disaster (*CAVW,* 22: 102; Neall, 1976: 234).

Galunggung, 1822: At a distance of perhaps 50 km from the volcano on October 8: "The rivers, obstructed with these enormous masses of burning lava, were transformed into boiling streams, the overflowing of which arrested at every step the wretched fugitives, and doomed them to a painful and cruel death, in the midst of rivers of liquid fire, or of boiling water. The rivers above named, and the Tji-losse, and the Tji-Konnir floated down in great numbers the bodies of men and animals, and the inhabitants of the lower districts saw whole houses, with their tenants still living, floating down these swollen and heated streams, without the possibility of giving them relief... Further from the seat of the volcano the scene was still more dreadful. There were to be seen the lifeless and half consumed bodies of men, women and children, who had escaped wounded from the burning flood, or had been intercepted in their flight, and the still living who were seeking in vain for an asylum amidst the general destruction. On the 9th, 10th and 11th it rained without interruption, and the wretched inhabitants who had

Table 3.11 Deaths in lahars

Eruption	Number killed by lahars	Total killed	Reference
Vesuvio, 1631	1 750	3 500	*VRF*
White Island, 1914	11	11	*CAVW*, 22: 42
Ruapehu, 1953	151	151	*CAVW*, 22: 102
Galunggung, 1822	3 600	4 011	Anon, 1832: 415
Kelut, 1919	5 110	5 110	*CAVW*, 1: 133
1966	210	210	*CIRV*, 79: 801
Semeru, 1976	40	40	*BVE*, 16: 32
Agung, 1963	200	1 184	*VRF*
Awu, 1966	10	39	*BVE*, 6: 8
Mayon, 1766	46	46	*VRF*
1814	720	1 200	*COMVOL*, 1975: 6-9
1968	3	3	*BVE*, 8: 8
Sakura-jima, 1914	25	62	Kotô, 1916: 95
1974	8	8	*BVE*, 14: 40
Asama, 1783	550	1 151	Sapper, 1917: 25
Tokachi, 1926	144	144	*CAVW*, 11: 311
Usu, 1977-1978	3	3	Kadomura *et al.*, 1979: 42
Mauna Loa, 1868	31	77	Hitchcock, 1909: 107
Fuego, 1963	7	7	*BVE*, 4: 5
Irazú, 1963	30	30	Tazieff, 1965: 19
Ruiz, 1845	700	700	*Bull. Volc.* 38 Supp.
Cotopaxi, 1877	1 000	1 000	Coleman, 1946: 81
Villarrica, 1964	25	25	*BVE*, 6: 5
1971	15	15	*BVE*, 11: 42
Mt Pelée, 1902 (May 5)	25	25	*CAVW*, 20: 35-36
Öraefajökull, 1727	3	3	*Bull. Volc.* 38 Supp: 57

fled from their houses, were left without shelter, and were intercepted in their flight by the overflowing of the rivers, and the destruction of all the bridges" (Anon, 1832: 36-37).

Kelut, 1919: There were 5110 people killed on May 19 and 20, with 104 villages damaged or destroyed. Lahars covered more than 130 km^2 (*CIRV*, 1979: 798). The death toll is given as more than 5110 by *CAVW* (1: 133) and about 5500 by Francis (1976: 322).

Kelut, 1966: On April 26, 210 were killed and 86 were injured. Several villages were totally destroyed and many were partly damaged. The main lahar entered the River Badak, covered a maximum distance of 31 km in a southwest direction and an area of 45 km^2 (*CIRV*, 1979: 801). The *Bulletin of Volcanic Eruptions* (*BVE*, 6, 1966: 7) gives the death toll as 208 killed and 74 missing. Jennings (1969: 2) notes more than 300 killed.

Agung, 1963: Most of the deaths attributed to lahars occurred in Subagan village (Fig. 3.18) 16 km southeast of the cone. Here 140 people were killed and buried in their houses, while others were swept along to the sea. Other deaths occurred in rivers where people were not careful enough in crossing lahar areas during the rainy season. Some people were killed up to 25 km from the volcano. Surjo (1965: 24-26) also reports 26 people injured by lahars.

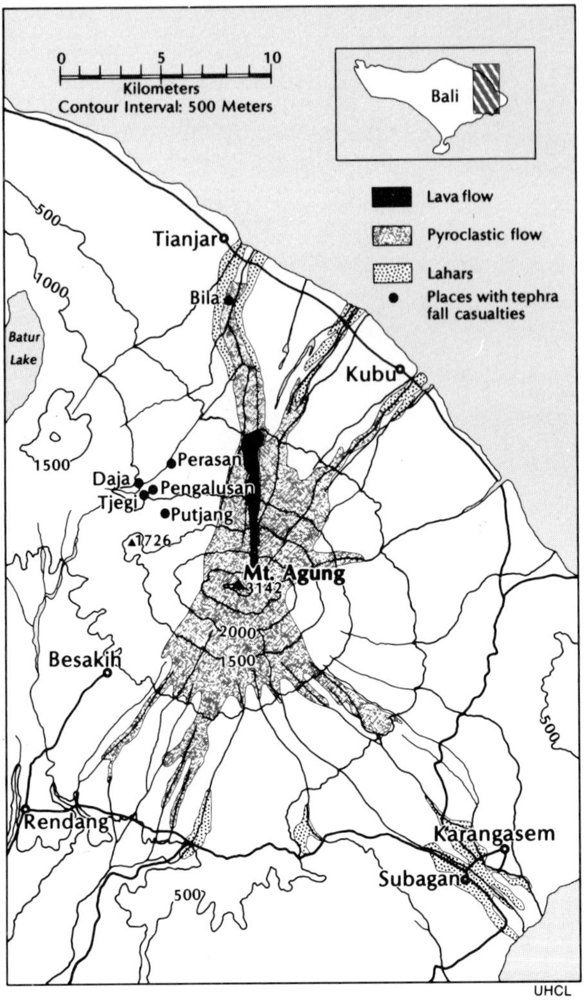

FIG. 3.18 Map of Agung showing areas affected by 1964 pyroclastic flows and lahars and locations where people were killed and injured by tephra fall (after Surjo, 1965: 25).

Mayon, 1814: Most of the deaths in this eruption resulted from lahars through the towns of Cagsawa and Budiao. In the church at Cagsawa (photos of the tower appear in innumerable postcards and textbooks), several hundred people who had sought refuge were killed by the lahar which destroyed the church (Bullard, 1976: 183; *COMVOL*, 1975: 9).

Tokachi, 1926: On May 4, a hot avalanche combined with snow to form a lahar which extended 28 km down the Hurano Valley and 25 km down the Biei Valley (Fig. 2.20) devastating an area of about 25 km^2 and killing 144 people. Of the deaths, 137 occurred in the Hurano Valley where 207 people were injured (Tanakadate, 1926; 1927; *CAVW,* 11: 311).

Mount St Helens, 1980: Possibly six of those killed on May 18 died in the lahars down the Toutle Valley. One newspaper reported that a family of three was swept away as a rescue helicopter watched (*The Columbian,* May 26, 1980: 4). The same newspaper also gave an account of how two young campers survived the Toutle River lahar; Venus Ann Dergan with a possible cracked wrist and looking as though she had been in a washing machine, and Roald Reitan, bruised but unharmed:

> "...I woke up and the river just didn't sound the same", said Reitan, who had pitched his tent only 30 feet from what was to become a raging, timber-choked waterway before he and Ms Dergan could escape. "All of a sudden, I see this wall of logs coming straight toward us."...
>
> "I started throwing stuff in the car and I looked down and saw the water around my ankles. Then I looked up and saw what looked like a piece of railroad bridge. It was holding all the logs back in one huge pile. And then it started to move.
>
> "Man, that thing was snapping trees when it came down the river. It was pure havoc. Havoc and a half."
>
> Terrified, the couple scrambled to the top of their car to escape the rising water. Seconds later the currents swept the vehicle down a steep embankment, and they jumped for their lives...
>
> Seconds later the 5-foot-2 [1.55 m] Ms Dergan was caught under a log.
>
> "It just sucked her under," said Reitan. "Her head was caught between two logs. All I could see was her nose."
>
> Ms Dergan, her hair and face caked with dirt, said she thought she was going to die.
>
> Reitan said he straddled a log and pulled himself along to reach his girlfriend. "I had her hair," he said. "Then I pulled her up. She said, 'You aren't leaving me, are you?' "
>
> Surrounded by the deafening roar of the sea of logs, the two somehow managed to reach the edge of the splintered mess. There they ran in mud up to their knees to reach a road, said Reitan.
>
> Fearing the waters soon would wash out the roadway, they fought their way up a steep hillside.
>
> [*The Columbian,* May 26, 1980: 6]

Mt Pelée, 1902 (May 5): A mudflow in the River Blanche (Fig. 2.14) "entombed 150 persons it is estimated but officials give a smaller number" (Hill, 1902a: 248). Bullard (1976: 123) gives the death toll as 30 or more workers in the sugar mill, but the official figure is 25 dead (*CAVW,* 20: 35-36).

Öraefajökull, 1727: Three people were drowned when a jökulhlaup swept away a chalet (*Bull. Volc.,* 38 Supp.: 57): "Two women and a boy took refuge on roof of the highest; but they had hardly reached it, when, being unable to resist the force of mud that was forced against it, it was carried away by the deluge of water, and as far as the eye could reach, the three unfortunate persons were seen clinging to the roof. One of the women was afterwards found among the mud of the jökulhlaup, but burnt, and as it were parboild [sic]; her body was so damaged and tender but it could scarcely be touched" (in Thorarinsson, 1958: 32).

In general, large death tolls have been produced by primary lahars resulting from the emptying of crater lakes and the melting of snow and ice. Small death tolls often occur in secondary (rain) lahars but have seldom been reported in the international literature (or media). Unfortunately, no details are available concerning the modes of death in lahars. Presumably, drowning, asphyxiation/burial and crush injuries are the main causes of death, with scald injuries being important in some cases. High ground away from stream valleys affords some protection when lahars threaten. Warning systems have been developed for some situations (see Chapter 5, Section 5).

3.8 EARTHQUAKES AND GROUND DEFORMATION

Several points of major interest arise from a consideration of deaths and injuries in volcanogenic earthquakes. As indicated in Table 3.2 very few deaths in eruptions are attributable to earthquakes. A rather different list appears in Table 3.12. Each of the examples in the table is worthy of discussion, each illustrating important points about the nature of eruptions and the nature of the fatality record:

Campi Flegri, 1970: Three people were killed in the collapse of houses by falling masonry and rocks. Although the ground deformation and earthquakes which caused house collapse were believed to be volcanogenic and precursors to an eruption (Imbò, 1979: 511), no eruption has occurred as yet.

Etna, 1169 and 1669: Both Pichler (1970: 266) and Coleman (1946: 188) report that 15 000 were killed by an earthquake in Catania in 1169. Similarly, Coleman (1946: 188) reports that 20 000 perished in an eruption in 1669. Etna was certainly in eruption during both these years (Simkin *et al.*, 1981: 35-36) but there is no substantive evidence that people were killed by either eruption or earthquake even though lava destroyed much of Catania in the 1669 eruption. The Earl of Winchilsea (1669) in a 38-page treatise, *A true and exact relation of the late prodigious earthquake and eruption of Mount Aetna*, details the destruction of vineyards, houses and the very walls of Catania but mentions no deaths. In short, there is no evidence that either eruption or earthquake killed anyone — some people may well have been, but the figures of 15 000 and 20 000 are certainly apocryphal.

Table 3.12 Deaths in volcanogenic earthquakes

Eruption	Number killed by earthquakes	Reference
Campi Flegri, 1970	3	Imbo, 1979: 511
Etna, 1169	15 000	Pichler, 1970: 266
1669	20 000	Coleman, 1946: 188
Borawli Complex, 1631	50	*VRF*
Api Siau, 1974	4	*BVE*, 17: 37
Sakura-jima, 1914	25	Kotô, 1916: 95
Colima, 1806	2 000	*CAVW*, 6: 10

Colima, 1806: An earthquake is reported to have killed 2000 in Zapotlán, 10 km west of Colima (*CAVW,* 6: 10). Sperrey (1903: 487-488) notes that 2000 were killed when the church in which they sought refuge collapsed during the earthquake. Although Colima was in eruption in 1806 (Simkin *et al.*, 1981: 89), the connection between eruption and earthquake is not well founded. Furthermore, Servando de la Cruz wrote in 1974 (*Bull Volc*. 38, Supp.: 69) "No person is known to have been killed by a volcanic eruption in the history of Mexico". At best the 2000 fatalities must be regarded as doubtful.

Api Siau, 1976: Earthquakes in February, March and April associated with ongoing eruptive activity (January, 1972 to October, 1976; Simkin *et al.*, 1981: 61). On April 27, earthquakes caused landslides which killed four persons (*BVE,* 14, 1976: 37).

Borawli Complex, 1631: Fifty fatalities were caused by an earthquake (*VRF*). No other information is available.

Sakura-jima, 1914 (Fig. 3.19): A major earthquake, perhaps the largest volcanogenic earthquake recorded (see Chapter 2, Section 5), occurred at 6.28 p.m. on January 12, about 8.5 hours after the eruption began (Shimozuru, 1972: 25). Five accounts offer varying interpretations of the number killed and the number of deaths attributable to earthquake activity. Jaggar (1924: 447; 455; 466) notes that the actual eruption killed no-one but that 35 people were crushed by falling buildings or buried in landslides during the earthquake. Shimozuru (1972: 24) records two people killed and 27 missing on the island. Nakano (1974: 238) states that 140 were killed by lava streams and ash fall. Kotô (1916) records 35 killed including six on the island, and a further 23 with fate unknown. Of the 29 killed on the mainland, various portions of Kotô's text suggest two were buried in a lapilli and ash fall in Osumi Province, 12 were buried in a landslide and 13 were killed in building collapses during the January 12 earthquake. Of the 127 wounded, 111 were injured in the earthquake and one was injured on the island. Omori (1922: 511) reports only two deaths and 27 missing from the island population. The two dead were village clerks drowned attempting to swim to the Osumi shore after the eruption began (Omori, 1914: 13). Of 16 killed in Kagoshima District about a dozen were killed in a landslide. Thirteen were reported killed in Kagoshima City (Omori, 1914: 53). However, Omori (1914: 475) writes "of the nine cases of the direct mortality in the city none was due to house collapse, all the victims having met their death on the street by the overthrowing of the stone fence walls". Omori (1914: 513) also records four people killed in the Kimotski District of Osumi Province. Because Kotô's various accounts offer the most detail, it is assumed here that 25 were killed and 111 were injured by volcanogenic earthquakes during the 1914 Sakura-jima eruption.

Thus, of the total number of 37 582 deaths listed in Table 3.12 only 79 seem to be

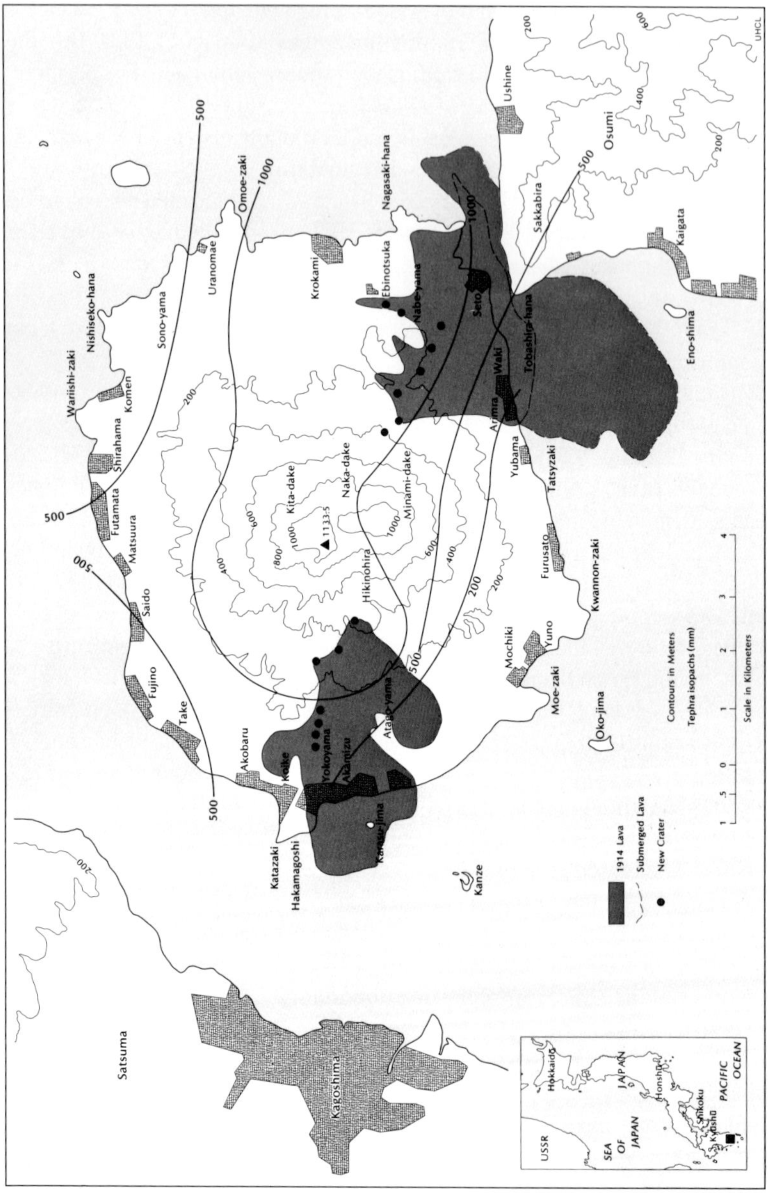

FIG. 3.19 Map of Sakura-jima island showing the extent of the 1914 lava flows which blocked the Seto channel and joined the island to the mainland. Tephra thicknesses are in millimeters (after Omori, 1916a,b; Kotô, 1916).

definitely associated with volcanogenic earthquakes. As no information is available about events leading to 50 deaths in the 1631 Borawli Complex eruption it seems there is worthwhile data on only 29 deaths. This record, certainly incomplete, is improved a little by noting that 10 persons were injured by house collapses resulting from an M = 6.1 quake with focal depth 10 km occurring at 11.19 p.m. on January 23, 1975, at Aso (*BVE*, 15, 1977: 36).

As volcanogenic earthquakes cannot be predicted and as fatalities and injuries generally result from the collapse of buildings, few protective measures can be suggested. Clearly, wearing a hard hat and/or relocating to an area free of falling masonry would improve personal safety. However, increasing the seismic resistance of buildings and other structures is the measure most likely to reduce deaths resulting from volcanogenic earthquakes (see Chapter 5, Section 6).

3.9 TSUNAMI

Table 3.2 suggests that tsunami are the second most important direct cause of deaths in eruptions (after pyroclastic flows) or, if calculations are made for the twentieth century only, one of the more insignificant causes of fatalities. Table 3.13 indicates that this variation in status is almost totally dependent on the 1883 eruption of Krakatau and, to a lesser extent, the eruptions of Unzen (1792) and Tambora (1815).

The areas inundated by tsunami generated during the Krakatau eruption are shown in Figure 2.24. In all, 165 villages were completely, and 132 partly, destroyed (Verbeek, 1886 quoted in Simkin and Fiske, 1983). While Verbeek (1886), the Royal Society of London volume edited by Symons (1888) and Furneaux's (1965) historical novel contain considerable information on the fate of the few hundred Dutch present on the Java and Sumatra coasts (of whom 37 died) little comment is made about the 726 Chinese and about 35 600 Javans and Sumatrans who died.

Dr Sollewyn Gelpe's description of his arrival in Tyringen (Fig. 2.24) where more than 12 000 people died provides some idea of the tsunami's effects;

Table 3.13 Tsunami deaths

Eruption	Number killed by tsunami	Reference
Ritter, 1888	many	*CAVW*, 5: 16
Krakatau, 1883	~ 33 000	*CIRV*, 1979: 796
Tambora, 1815	~ 4 600	Neumann van Padang, 1971: 60
Taal, 1911	few	Pratt, 1911: 81-82
Sakura-jima, 1781	15	Omori, 1916a: 67
Unzen, 1792	4 300	*CAVW*, 11: 43
Oshima-o-shima (Honshu), 1741	1 475	*CAVW*, 11: 281
Komagatake, 1640	700	*CAVW*, 11: 285
Kilauea, 1975	2	*BVE*, 15: 46-48
Mauna Loa, 1868	46	Hitchcock, 1909: 107

> As far as the eye can reach, the only thing that remains standing there is a solitary tree, a gigantic durian, maimed, branchless and leafless. It forms the gravemark of a heap of corpses and carcasses lying under roofs, houses, and trunks of trees. Hundreds of such graves, though of smaller dimensions, may be seen over and over again on the plain. The turned-up earth often merely covers a corpse, alongside which a cocoanut branch or bamboo is stuck upright for the guidance of the authorities. Thousands of corpses of human beings and also carcasses of animals still await burial, and make their presence apparent by an indescribable stench. They lie in knots and entangled masses impossible to unravel, and often jammed along with cocoanut stems among all that served these thousands as dwellings, furniture, farming implements, and adornments for houses and compounds (in Simkin and Fiske, 1983).

Similar accounts exist for many of the small towns along the shores of Sunda Strait.

Little information is available on the deaths of 10 000 people around Tambora in 1815. Neumann van Padang (1971: 60) notes that they died as a result of bomb impacts, tephra falls and seismic sea waves; 4600 deaths as a result of tsunami is a "guesstimate". Sapper (1917: 195) attributes 1200 deaths on Sumbawa to heat. Zollinger (1855) put the death toll at 57 925; 34 000 fewer than Neumann van Padang's total.

The number of deaths resulting from tsunami during the 1792 Unzen eruption is also subject to doubt. The tsunami was caused by an avalanche from the eastern part of Maye-yama falling into the sea; 4,300 were killed along the coast to the east and south (*CAVW,* 11: 43). Sapper (1917: 133) places the death toll from tsunami at only 707.

On November 29, 1975, an M = 7.2 earthquake occurred on the Hilina fault system 39 km east of the Kilauea summit (see Chapter 2, Section 6). The earthquake caused widespread damage, subsidence of as much as 3.5 m and a tsunami with a maximum wave height of 15 m. Two people were killed by the tsunami. Two hours after the earthquake an eruption occurred, interpreted as induced by seismic disturbance to the volcanic plumbing system. The tsunami was also caused by the earthquake (*BVE,* 15, 1977: 46-48). The tsunami which caused 46 deaths during the 1868 Mauna Loa eruption appear to have been generated in the same way.

No warning systems are available for volcanogenic tsunami except at considerable distance from the source. As such tsunami may have a variety of causes (Chapter 2, Section 6) adequate warning systems will be difficult to develop. Evacuation to high ground before an eruption begins seems a sensible precaution. However, as the Krakatau tsunami caused the deaths of people in Merak at elevations 40 m above sea level, very extensive areas on low-lying coastal hinterlands may require evacuation.

3.10 ATMOSPHERIC EFFECTS

Electrical discharges and atmospheric shock waves are most likely to be associated with strong Vulcanian, Plinian, or Peléean eruptions (Table 1.1). As shown in Table

3.2 an insignificant number of deaths are ascribed to either cause. This is not surprising in the case of shock waves as these are more likely to result in injury or mere discomfort. On the other hand, deaths from lightning strikes probably occurred during the Soufrière and Mt Pelée eruptions of 1902, but details are lacking.

Observations of deaths due to lightning strikes are absent from accounts of other eruptions except for three cryptic notes: (i) Sir William Hamilton (1772: 164) wrote, concerning the 1631 eruption of Vesuvio (which he clearly didn't witness!), that men and beasts were struck dead by lightning; (ii) A Philippines *COMVOL* report (1975: 6) commented that some of the 1200 fatalities resulting from the 1814 eruption of Mayon were caused by electrical discharges; and (iii) at the First Point lighthouse in Sunda Strait (Fig. 2.24) on August 27, 1883, a lightning stroke hit the lightning rod, breaking it: "hereafter the lightning hit the entrance of the lighthouse, wounding four of the 10 convicts (only those wearing iron rings; they were burned from the ring down)" (Verbeek, in Simkin and Fiske, 1983). Although evidence is sparse it seems very likely that quite large numbers of people have been killed by electrical discharges during eruptions. Such deaths are most likely to occur and go unreported where strong eruption columns create innumerable lightning discharges and where victims are burnt by hot pyroclastic flows.

Comments on minor injuries or effects on humans arising from electrical discharges are also rare. During the 1965 Taal eruption a municipal official tried to use the telephone but was unable to hold the receiver because it was electrically charged (Carroll and Parco, 1966: 26). Of lesser consequence is the report that a man on the summit of Mt Adams holding an ice axe aloft during the May 18 Mount St Helens eruption felt a jolt through a thick mitten (*Volcano,* 1980: 78).

On the other hand, several interesting reports are available concerning minor injuries resulting from atmospheric shock waves. No deaths can be attributed to shock waves although the following apocryphal account appeared in the *Akron Beacon Journal* (May 22, 1980: B3) during the height of media interest in the Mount St Helens eruption. It refers to atmospheric shock waves from Krakatau, 1883: "More than 1000 miles away in Borneo, a tribe of head hunters were so frightened by the noise that they jumped off cliffs to their deaths".

During the 1958 eruption of Asama, seven persons were injured by flying glass when shock waves broke windows (Murai and Hosoya, 1964: 211).

Although broken windows and other building damage are quite common (see Chapter 5, Section 8) other cases of injury resulting from such damage are not known. However, reports of deafness or earache are quite common. At Santorini in 1925, continual sound and shock waves reportedly kept many people on Thera and Therasia awake during the first months or so of the eruption (Washington, 1926: 365). During the 1537 eruption of Etna (Rodwell, 1878: 88) the noises of explosion were described as "so vibrant people were struck deaf". Near Soufrière, St Vincent, in 1902 a few informants noted that the pressure of the air seemed to be increased and the effect on the tympanic membranes of their ears was such that the sound was acutely painful (Anderson and Flett, 1903: 401). Similarly in Kagoshima, 10 km from Sakura-jima (1914) air concussions gave earache (Jaggar, 1924: 447). These shockwaves were heard and felt far beyond Kagoshima (see Fig. 2.27).

Several precautions can be taken to reduce the likelihood of being struck by lightning during a thunderstorm — presumably similar precautions are effective where volcanogenic lightning is present. In open areas, shelter in a depression away from isolated trees or metallic objects such as wire fences which can conduct current from a stroke some distance away. Stay out of water for similar reasons and avoid projecting above the surrounding landscape. Indoors, seek shelter in a metal-framed building or one with a lightning rod, stay away from windows, the plumbing system, the wiring system and the telephone (Uman, 1971: 21).

3.11 GASES AND ACID RAINS

Gases, like electrical discharges, have probably been responsible for quite a number of deaths in volcanic eruptions, but specific cases are difficult to document. Furthermore, the possibility of delayed or chronic effects has not been adequately studied. Some accounts which ascribe death to tephra fall or asphyxiation may well be describing the effects of gases.

It is even more difficult to attribute deaths to specific gases. However, the only human death during the 1973 Heimaey eruption resulted from the accumulation of gas, 98% CO_2, in a cellar in Westmannaeyjar. The house was partly buried by scoriae and tephra at the time. Several other people were overcome (but not killed) in houses and on streets and large parts of the town had to be closed to traffic (Gunarsson, 1973; Bolt *et al.*, 1975: 113; Williams and Moore, 1973: 15).

In February, 1979, 142 persons were reported killed by CO_2 gas emitted from a fissure in the Dieng Volcanic Complex on Java. One account notes that one group was killed when the gas ponded in a low-lying area near the Batur road; others were killed near an elementary school to which they had retreated after observing the earlier deaths (*SEAN Bull.*, 4(3), 1979, 5). A later account (Le Guern *et al.*, 1982) reports all 142 deaths as occurring along a trail to the south of the school, the victims apparently in single file and appearing to be asleep. These authors conclude that the victims were overwhelmed suddenly by the gas, otherwise the tail-end of the file would have escaped.

Other accounts of death attributable to gases are much less specific. In 1903, 17 people were killed at the 1600 m level on Karthala near the solfataras. There was no eruption in 1903, but the emanations of noxious gases were particularly strong that summer (*CAVW,* 16: 37). Another account concerns the asphyxiation and calcination/desiccation of 16 students of the University of Popáyan on the summit of Purace on May 19, 1949. It is not clear from two accounts whether the students were killed by ballistic impacts or gases (*BVE,* 17, 1979: 74; *Bull. Volc.,* 38 Supp., 1974). However, if the date is correct, the deaths must have been due to gases as the eruption did not begin until May 26, 1949 (Simkin *et al.*, 1981: 95). According to

Le Guern *et al.* (1982: 155) CO_2 streams flowing down the slopes of Tangkuban-parahu (Indonesia) have sometimes killed children.

While the number of confirmed deaths might be small, five interesting reports indicate the effects of gases on both populace and scientists:

Santorini, 1707: A horrible stench was evident at a distance of about 5 km from the volcano. Bourguignon (1708) gives a graphic description (referring to the eruption toward the end of July): "The Sea also seemed much more disturbed and loaded with Sulphur and Vitriol. The boiling of the Water was more fierce and violent: the Smoak thicker, and in greater abundance; and the Fire bigger and more frightful. But above all, a Stench that infected the whole Countery, grew so insupportable, that Persons of the strongest Constitutions could scarce breath in it; others, that were weaker, fell into frequent Faintings; and almost every Body was seized with Vomitings. I could not then but imagine my self on Board some Man of War, where at a general Discharge of all the Guns, the confused stink of the Powder, Tar and Stench of the Ship, especially in foul Weather, often overcomes the strongest Seamen. Just such a nauseous stink we were forced to breath in, without being able any ways to avoid it, or defend our selves from it. But as we were infested with it but at times, and as the wind brought it; so it was sometimes more in one place than another, and more or less tolerable, as the Wind fate" (Bourguignon, 1708: 204).

Soufrière, St Vincent, 1902: An official medical report notes with particular reference to the northeastern part of St Vincent: "It is certain that with the hot dust there was some sulphur dioxide, and it is probable that there was also a little hydrochloric acid gas, but had these gases been present in large quantities, the survivors would have suffered severely from irritation of the respiratory mucus membrane. Except in one or two of the elderly patients, there were no signs of tracheal or bronchial irritation. There is no evidence of the presence of more lethal gases such as carburetted hydrogen, carbon monoxide or carbon dioxide had these gases been the cause of death, it is impossible to conceive how anyone could have survived; moreover, death would have been more sudden" (Will, in *HMSO,* 1903: 14).

Vesuvio, 1906: Perret (1924: 54) reported the presence of CO_2 at the observatory on Vesuvius during the 1906 eruption and noted that it produced a curious sensation of heat about the feet and legs and "an indescribable feeling of oppression" and slight difficulty in breathing. On April 18, 1906, a strong wind forced the gases from the crater down to the observatory. "The air became irrespirable, and several persons were nearly overcome. It was surmised from the symptoms that carbonic acid gas and sulphuretted hydrogen were the principal gases present" (Jaggar, 1906: 108).

Langila, Papua New Guinea 1952: An excellent account by G.A.M. Taylor *et al.* (1957) provides detailed (but incidental) observations of gases in the area of the three Langila craters. Apart from steam the main constituent of the gas appeared to be SO_2. This characteristically choking gas is unmistakeable. Concentrations were always high and vents were rarely approachable without a gas mask. During turbulent conditions in the fissure crater, the filtering medium in the gas mask, made to eliminate concentrations of up to 2%, was quite unable to cope with the existing concentrations. "One fluctuating constituent of the gas caused a stinging sensation in the eyes. It was at first thought to be hydrogen chloride, but a silver nitrate test proved negative. Physiological effects may supply a clue to its composition. It was noted while collecting temperature data that if the observer, without a mask, became enveloped with warm gas he experienced the most acute physiological discomfort. Even with eyes tightly closed and breath held an intense burning sensation in the membranes of the nose and eyes was felt. Condensation on any exposed skin also produced severe irritation. It was concluded from these observations that a higher unstable *oxide of sulphur* was present in the freshly emitted gases; a conclusion which seemed to be confirmed by the fact that copper wire exposed in the vents was often converted to a blue product, which was presumably the sulphate . . . The odour of *sulphuretted hydrogen* was not often detected close to the vents, but was detected quite frequently at a distance from the crater. This could be explained by assuming either the presence of *higher sulphides of hydrogen* or an inability to detect higher concentrations of dihydrogen sulphide prevailing near the vents owing to paralysis of the olfactory centres, a recognized physiological reaction to exposure to higher concentrations of H_2S. When it was observed that stainless steel articles were blackened in the crater atmosphere it seemed more likely that the higher sulphides were present . . . The fact that the observers did not experience any lasting physiological effects from exposure to the gases suggests that concentrations of the very poisonous higher sulphides were low. The extreme irritability and nervous depression experienced after a prolonged period of work in the crater were probably due more to a deficiency in oxygen than to the effects of poisonous gas. The ineffectiveness of the filtering medium in the gas masks may have been a contributory factor" (Taylor *et al.*, 1957: 17).

Masaya, Nicaragua, 1979 and continuing: Degassing has occurred at this volcano at roughly 25 year intervals and lasting for five to 10 years. Measurements indicate that SO_2 levels of 1 ppm occur, significantly above the World Health Organization recommended maximum 0.5 ppm, 24 h mean. Thirty kilometers downwind, sulfate aerosol concentrations have reached 400 $\mu g/m^3$ (Baxter *et al.*, 1982: 150).

Examples given above make it clear that volcanoes do not have to be in eruption for deaths and discomfort to occur. Deaths at Karthala and Purace occurred before eruptions. Similarly, Jaggar's observations at Vesuvio were made three months before the 1906 eruption began. Before the May 18, 1902, eruption of Mt Pelée the fumes of sulfur were so strong in the streets of St Pierre that people wore wet handkerchiefs over their mouths (Heilprin, 1903: 39,62; Hill, 1902a: 256). Taylor (1958:

21,24) reports that both H_2S and SO_2 were noticeable before the 1951 Lamington eruption, not only by people on the ground but also by airplane pilots some kilometers away. In other cases (for example, Hekla, 1947) CO_2 and other gas concentrations may be particularly noticeable after an eruption (Kjartansson, 1957).

The effects of acid rains on humans are rarely commented on. Brief examples from two major eruptions illustrate the range of effects. At Makasar, 370 km from Tambora (see Fig. 2.12c), the 1815 tephra fall "was perfectly tasteless and did not affect the eyes with painful sensation, had a faint burnt smell, but nothing like sulphur" (Anon, 1816: 253). On the other hand, the Katmai-Novarupta tephra fall of 1912 made life unpleasant at Kodiak, 160 km distant. Sulfur fumes made throats raw and eating almost impossible; burning drops of combined gas and water (following a few showers of rain during the night) tasted like vinegar and left large blisters on the lips (Erskine, 1962: 46,65). At Latouche, 480 km northeast of the volcanoes, rain caused stinging burns wherever it touched the flesh. A chemical analysis showed sulfuric acid in the rain (Griggs, 1919: 175).

Another adverse effect on humans resulting directly from acid rains occurred during the 1917 eruption of Boqueron. Jaggar (1925, April 2, *Volcano Letter;* 1945: 120) reported that a mud rain on bare-headed wanderers in the city of San Salvador, 7 km from the volcano, caused their hair to fall out. Eye and skin irritations have been reported following extremely acid rains (pH 2.5-3.5) through the plume of degassing Masaya. Furthermore, this acid rain reacts with zinc galvanising on house roofs, possibly releasing heavy metals into drinking water (Baxter *et al.,* 1982a: 150).

Personal protection measures likely to be effective against gases and acid rains are similar to those discussed for tephra fall (Chapter 3, Section 5.4). However, face masks need to be designed for both toxic gases and respirable dusts. Table 3.14 summarises data on the likely effects of volcanic gases on humans based on the toxicology of industrial pollutants, (Sax, 1963, 1974; Wilcox, 1959). The values in the table refer to threshold limits of repeated, day after day (40 hours per week) exposure, without adverse effects for the majority of workers. However, a small proportion of the population may be adversely affected by concentrations below those given in the table.

Bernstein (1983) has recently summarised recommendations which may help prevent excessive exacerbation of pre-existing noncommunicable respiratory diseases such as destructive airways diseases (for example, asthma, chronic bronchitis and emphysema), interstitial lung diseases (for example, pneumoconioses and hypersensitivity pneumonitis) and pulmonary vascular diseases by acid rains, gases and particulate pollutants, such as tephra:

1. Drink fluids to encourage coughing and to loosen secretions.
2. Don't smoke and stay away from smokers as smoke will only add to breathing problems.
3. Relax . . . take it easy; avoid exertion.

Table 3.14 Toxicology of volcanic gases

Ammonia, NH_3 — extremely pungent colorless gas
Threshold limit (TL) = 79 mg m^{-3} of air; 100 ppm in air
Effects of overdose: causes irritation of eyes, conjunctivitis, swelling of eyelids, irritation of mucous membranes of nose and throat, coughing, dyspnea and vomiting. Irritation of the skin may occur, particularly if it is moist.

Carbon dioxide, CO_2 — colorless, odorless gas
TL = 9000 mg m^{-3} of air; 5000 ppm in air
Effects of overdose: usually regarded as a simple asphyxiant, symptoms appearing only when such high concentrations are reached that there is insufficient oxygen to support life. Symptoms are those which precede asphyxia, namely, headache, dizziness, shortness of breath, muscular weakness, drowsiness and ringing in the ears. Rapid recovery on removal from exposure. For persons with pre-existing cardiopulmonary diseases much lower levels of exposure may be dangerous acutely, as well as over years at low levels.

Carbon monoxide, CO — colorless, odorless gas
TL = 115 mg m^{-3} of air; 100 ppm in air
Effects of overdose: Has affinity for hemoglobin 210 times that of oxygen. With CO-hemoglobin in the blood $<10\%$ rarely any symptoms; 20-30% causes shortness of breath on moderate exertion and slight headache; 30-50% causes severe headache, mental confusion and dizziness, impairment of vision and hearing, and collapse and fainting on exertion; 50-56% unconsciousness results; 80% almost immediate death. Chronic effects of repeated exposure to relatively low concentrations have been described. For persons with pre-existing cardiopulmonary diseases much lower levels of exposure may be dangerous acutely, as well as over years at low levels.

Fluorine, F_2 — pale yellow gas
TL = 0.2 mg m^{-3} of air; 0.1 ppm of air
Effects of overdose; a very powerful caustic irritant; conjunctivitis, skin irritation, bone degeneration, mottling of teeth.

Hydrochloric acid, HCl — colorless gas or colorless fuming liquid
TL = 7 mg m^{-3} of air; 5 ppm in air
Effects of overdose: Irritates mucous membranes of eyes and respiratory tract; concentration >35 ppm causes irritation of throat after short exposure. Concentrations >100 ppm result in pulmonary edema, and often laryngeal spasm.

Hydrofluoric acid, HF — clear, colorless, fuming corrosive, liquid or gas.
TL = 2 mg m^{-3} of air; 3 ppm in air
Effects of overdose: extremely irritating and corrosive to the skin and mucous membranes. Inhalation of the vapor may cause ulcers of the upper respiratory tract; concentrations at 50-250 ppm are dangerous even for brief exposures. Produces severe skin burns which are slow in healing. Subcutaneous tissues may be affected becoming blanched and bloodless. Gangrene of the affected areas may follow.

Hydrogen sulfide, H_2S — colorless, flammable gas with offensive odor
TL = 28 mg m^{-3} of air; 20 ppm in air
Effects of overdose: irritates and asphyxiates; concentrations of 20-150 ppm cause eye irritation; slightly higher concentrations cause irritation of upper respiratory tract and, on long exposure, pulmonary edema. A 30 minute exposure to 500 ppm results in headache, dizziness, excitement, staggering gait, diarrhea and dysuria, followed sometimes by bronchitis or bronchopneumonia. In small amounts acts as depressant; in larger amounts as stimulant; with very large amounts respiratory center is paralysed.

Sulfur dioxide, SO_2 — colorless gas or liquid with pungent odor
TL = 13 mg m^{-3} of air; 5 ppm air
Effects of overdose: causes inflammation and irritation of conjunctiva. Forms sulfurous acid and oxidises to form sulfuric acid (see Section 2.8). Concentrations of 6-12 ppm cause immediate irritation of nose and throat; 20 ppm or more irritating to eyes; 10 000 ppm irritant to moist skin within a few minutes. Chiefly affects upper respiratory tract and bronchi. For persons with pre-existing cardiopulmonary diseases much lower levels of exposure may be dangerous acutely, as well as over years at low levels.

Sulfuric acid, H_2SO_4 — colorless, oily liquid.
TL = 1 mg m^{-3}
Effects of overdose: causes severe burns with rapid destruction of tissue; inflammation of upper respiratory tract; dermatitis.

After Sax, 1963, 1974; Wilcox, 1959

4. In heavy pollution stay indoors, close windows and doors, turn on air conditioner/purifier.
5. Only use commercial masks or (wet) handerkerchiefs for respiratory protection in consultation with your doctor as such devices may cause added stress by increasing the "work" of breathing.
6. During periods of volcanic pollution avoid contact with people who have colds (Bernstein, 1983).

3.12 OTHER CAUSES OF DEATH

Table 3.2 and 3.3 indicate that the largest single category of deaths associated with volcanic eruptions is that of "other causes" and that nearly all the deaths in this category can be attributed to starvation and disease following two eruptions — the Lakagigar fissure eruption of 1783 and the cataclysmic 1815 eruption of Tambora.

Figure 3.20 illustrates population change in Iceland for the period before and after the Laki eruption. The death toll following this eruption has generally been accepted at about 10 000 (*Bull. Volc.*, 38, Supp. 1974: 57), amounting to about 20% of Iceland's population. Jackson's (1982: 47) analysis based on natural birth and death rates concludes with a most reliable estimate of about 9350 deaths, but it is not possible to be more specific with the currently available data. Presumably, most of the human deaths resulted from starvation following massive stock losses as a result of the eruption. Stock losses occurred as a result of lack of feed. Stephensen (1785) provides some grim accounts of starvation but the symptoms he describes could also occur, at least in part, as a result of fluorine poisoning.*

> Diseases of the most inveterate kinds, in the form of scurvy, broke out in sundry places, and those even far distant from the fire: as, for instance, in the districts of Guldbringue, Borgefiord, and Myhre, especially in the first. The district of West Skaptefield was, however, the chief seat of this distemper; and in only six parishes there, no less than one hundred and fifty person were carried off between the commencement of the new year and the month of June following; but some of these perished by famine. The same symptoms shewed themselves, in this disorder, in the human race, as among the cattle. The feet, thighs, hips, arms, throat, and head, were most dreadfully swelled, especially about the ankles, the knees, and the various joints, which last, as well as the ribs, were contracted. The sinews, too, were drawn up, with painful cramps, so that the wretched sufferers became crooked, and had an appearance the most pitiable. In addition to this, they were oppressed with pains across the breast and loins; their teeth became loose, and were covered with the swollen gums, which at length mortified, and fell off in large pieces of a black or sometimes dark blue color. Disgusting sores were formed in the palate and throat, and not uncommonly at the extremity of the disease, the tongue rotted entirely out of the month. This, dreadful, though, apparently, not very infectious, distemper, prevailed in almost every farm in the vicinity of the fire during the winter and spring; but, happily,

*Baxter *et al.* (1982b: 271) note: "Consumption of water with a fluoride concentration of 2–10 ppm would not be expected to cause ill health if the contamination lasted only a few days, though it would be prudent for susceptible people, mainly those with chronic sickness, to use uncontaminated water. Acute exposure to higher concentrations can cause gastrointestinal illness. Consumption of water containing greater than 1 ppm fluoride over long periods could lead to dental mottling in children and, at higher concentrations, osteofluorosis" (see also Prasad, 1978: 60).

> its extreme horrors were confined to the district of West Skaptefield, beyond which it was attended with less melancholy consequences. Many of the unfortunate inhabitants, who resided in the vicinity of the place of eruption, and who could not procure either medicine or assistance, were starved to death; from an utter incapability of swallowing during the prevalence of the disorder any portion of food, even if they could obtain it, which was not often the case.
>
> [Stephensen, 1785: 239-241]

Figure 3.20 shows that it was more than 20 years before the population returned to pre-eruption levels.

Less information is available about deaths following the Tambora eruption. Neumann van Padang (*CAVW*, 1: 166; 1971: 60) reckons 38 000 died on Sumbawa and 44 000 on Lombok as a result of starvation and disease, in addition to the 10 000 actually killed by the eruption *per se*. As mentioned earlier, Zollinger (1855) puts the total number of fatalities at 57 925. Even if Zollinger's figures are correct, starvation and disease remain the greatest single cause of death during and after volcanic eruptions, at least for the period since AD 1600.

The only moderately well documented account of epidemic disease following an eruption concerns the 1902 eruption of Santa Maria: "One of our greatest troubles was that of sickness, owing to the balance of Nature having been upset by the eruption, which, having killed all the birds for some hundreds of miles, enabled the flies, mosquitoes, and rats to multiply to such an extent that life to man became nearly unbearable. The immediate consequence was an epidemic of malaria, which cost more lives than the eruption itself — many times more. It has passed, happily; the birds having come again, the breeding of these pests is checked, and the district again enjoys the excellent reputation for health that it deservedly had before" (A.H. Gehrke, quoted in Anderson, 1908: 488). The death toll from malaria is estimated at about 3000. Malaria was also believed to be responsible for 31 deaths around Galunggung in December, 1982, even though it was previously reported eradicated.

Epidemic diseases frequently increase where survivors are crowded together in refugee camps (see PAHO, 1981). Around Galunggung, for example, the high level of turberculosis endemic in the population was probably increased further by the crowding of about 750 000 evacuees into a variety of camps. Stommel and Stommel (1979) make the plausible suggestion that the first cholera pandemic, which spread from India in 1816 to New York in 1832, resulted from famine, in turn produced by the anomalous weather conditions generated by the 1815 eruption of Tambora.

The relationship between volcanic eruptions and epidemic outbreaks of endemic or previously well-controlled diseases has been a popular topic since the mid-nineteenth century. For example, Parkin (1851) noted that volcanic action and epidemic diseases have three things in common: (i) they progress along particular

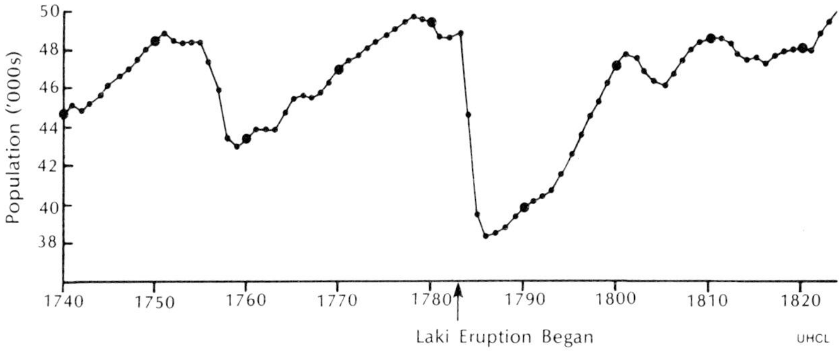

FIG. 3.20 Graph showing population decline in Iceland as a result of starvation and disease following the Laki fissure eruption of 1783–1784 (after Jackson, 1982: 45).

lines; (ii) they are more marked toward the sea and near large expanses of water; and (iii) they are of limited duration and periodical return, with total cessation in a particular locality after certain definite periods (Bardswell, 1901: 321). In his Parkin Prize winning essay "Volcanic action is a cause of outbreaks of epidemic disease" Bardswell (1901: 325) is more cautious, concluding: "It does not, however, seem to me to be unlikely that volcanic action may be one of various determining factors of the high mortality, especially in towns possessing anything like modern water supply and drainage systems". In fact, Bardswell was so cautious that it is difficult to know whether this states 'yes' or 'no'!

Johnston-Lavis (1914: 63), "author of many works on medicine and volcanology", in a 67-page treatise with the superb title *On the Effects of Volcanic Action in the Production of Epidemic Diseases in the Animal and in the Vegetable Creation, and in the Production of Hurricanes and Abnormal Atmospherical Vicissitudes,* concludes that volcanic ejecta are incapable of *directly* producing epidemics because the earth's surface and atmosphere are "antisepticised" by their agency. However, he cites three *indirect* ways in which eruptions cause or accelerate epidemic diseases: (i) the irritant and depressing effects of poisonous fumes lower the resisting power of the ocular, respiratory and upper digestive systems to microbial invasion; (ii) water courses and wells may be contaminated by organic impurities, sewage, etc., and ventilation of houses may be prevented — all of these being malignant influences, favoring diseases, epidemic or otherwise; and (iii) moral depression from fear, a condition well recognised as rendering living beings less immune to infection.*

Contamination of lakes by organic impurities (bacteria in this case) led to an

*Some of the psychological effects of eruptions are considered in Chapter 4.

interesting situation following an eruption in 1980. A number of scientists who worked around lakes in the devastated area at Mount St Helens after May 18 developed symptoms of Legionnaire's disease of the Pontiac fever type. Concentrations of *Legionella* species in some lakes were found to be about an order of magnitude higher than the highest values reported elsewhere. No deaths occurred (Dr D.L. Tison, microbiologist, personal communication).

Other minor causes of miscellaneous deaths include drowning and transport accidents. During the 1779 Sakura-jima eruption the Seto Channel (Fig. 3.19) was filled 1.5 m deep with floating pumice. Two islanders fled on foot to the Osumi coast. A third, following the two pioneers, drowned (Kotô, 1916: 32). In 1781, a tsunami upset three boats, drowning 14 men and one woman (Omori, 1916a: 67). During the 1914 eruption two island clerks were drowned attempting to swim to the Osumi shore after the eruption had started (Omori, 1914: 13). Similarly, some drownings may have occurred during the eruptions of Rabaul in 1937 and Taal in 1911. When Taal erupted in 1965 some islanders were drowned as overloaded boats capsized in high waves on the way to the lake shore (Carroll and Parco, 1966: 7; *Newsweek,* October 11, 1965).

Transport accidents have also accounted for a few deaths during eruptions. A P-39 pilot was killed when his plane crashed while flying over the eruption of Cerro Azul in the Galapagos during 1943–1944. To what extent the death resulted from the eruption is not known (VRF). Perhaps more directly related to the eruption was the death of the pilot of a crop-duster when the plane hit powerlines in Central Washington State during the May 18 Mount St Helens tephra fall (*The Columbian,* May 26, 1980:3; *Holocaust,* 1980: 24). In 1982, two Boeing 747s *en route* to Australia with hundreds of passengers lost power on ingesting tephra into the engines while flying through plumes from Galunggung, but managed to land safely after restarting one or more engines (see Chapter 6, Section 6.2). At least one traffic accident during the Mount St Helens tephra fall also resulted in a death; in Yakima (Fig. 3.8) the driver of a pickup truck was killed at an intersection, evidently as a result of poor visibility (Dr Bob Attwood, Yakima, personal communication, June, 1980). In another traffic accident on the same day 18 people were admitted to hospital as a result of a freeway pileup (CDC No. 14).

Exposure and/or shock may also result in some deaths during and after an eruption. A few persons died of exposure and shock following the 1937 Rabaul eruption (Cilento, 1937: 47). During the second phase of the 1972, Piton de la Fournaise eruption, three visitors died after 48 hours in the cold mists on the rugged lava fields (*SICSLP* 71-72/1472). The extent to which these deaths are connected with the eruption are not known; certainly they did not die as a direct result of the eruption, but perhaps they would not have been on the mountain if it had not been in eruption!

Other indirect deaths result from cardiac arrests. For example, two elderly people with heart conditions died during the tephra clean-up phase in Yakima, evidently over-exerting themselves while shovelling tephra from their properties (Dr Bob

Attwood, Yakima, personal communication, June, 1980). The question still remains as to what extent these deaths are attributable to the Mount St Helens May 18 eruption?

More exotic deaths during eruptions, or at least in relation to volcanoes, include suicide, sacrifice and murder. Suicides, undoubtedly, have taken place at a number of volcanoes, but the most popular volcano for such deeds must be Oshima at Mihara-Yama. The number of suicides in the crater of this volcano is unknown, but Coleman (1946: 151-153) maintained that there were more than 200 suicides in 1933 alone by Japanese who believed that Mihara-Yama was a sacred place in which eternal rest could be found in the flames of the volcano. Many of those who jumied were lovers. Before World War II, 80 or 90 suicided each year, but since the war the death rate has fallen dramatically (*Time,* June 6, 1965). Interested readers are referred to the *Illustrated London News,* May 19, 1934, which includes photos of a gondola descent into the crater and bodies draped over various crater wall projections.

Sacrifices to volcanoes are probably rare. Around Cosiguina in 1835 some of the inhabitants held the belief that the volcano could be bribed into keeping quiet only by the sacrifice of a three-month-old infant into the crater every 25 years (Coleman, 1946: 73). If that example should seem remote, *Volcano* (1980: 20) reports that prior to May 18, 1980, the Cowlitz County Sheriff's office received a phone call from Cleveland, Ohio, offering a young woman as a sacrifice to Mount St Helens (see also, *Westways,* 72(9), September, 1980: 38).

Murders during eruptions are also rare. However, Winchilsea (1669: 25-26) reports murders and executions in Cantania during the 1669 Etna eruption. Following the 1965 Taal eruption, there were stories current, both in the disaster area and in the newspapers, of fights and killings on the island shore over possession of boats (*Newsweek,* October 11, 1965; Carroll and Parco, 1966: 15-16).

3.13 CONCLUSIONS

Several important points arise from this great catalog of death:

1. There is little relationship between the frequency of occurrence of various types of volcanic activity (Table 1.4) and the frequency of deaths caused by the hazard (Table 3.2). Since AD 1600, about 40% of all deaths as a result of volcanic eruptions have been caused indirectly, principally by starvation and disease. If we exclude such indirect deaths from consideration, pyroclastic flows and tsunami have been the principal hazards. However, these figures are much influenced by the few events which resulted in more than 8000 fatalities (Table 3.3). Using arguments such as big events should/should not be included, starvation and disease are not relevant in the twentieth century, and the last 80/380 years is too short a time period to have any statistical validity, it is possible to argue for or against the significance of almost any agent of death. Nonetheless, two conclusions stand out: (i) in the absence of large events in the (short-term) future, pyroclastic flows and

lahars are likely to remain the most frequent causes of death, and (ii) if large-scale events do occur in the (short-term) future, pyroclastic flows and tsunami are likely to remain the most common agents of death.

2. There is remarkably little known about the clinical causes of death. As pyroclastic flows are the most frequently occurring agent of death, the evidence from Soufrière (St Vincent), Mt Pelée and Mount St Helens suggests that asphyxia is the most common, and burn injuries the second most common, clinical cause of death. Nonetheless, remarkably little is known about the processes leading to deaths in volcanic eruptions. Medical teams are initially involved in rescuing the injured; in tropical countries putrefaction may often be too far advanced for worthwhile autopsies by the time the rescue phase and the recovery of bodies are completed. We should also recall that significant proportion of bodies (Table 3.15) might not be recovered.

The general lack of autopsy information has inhibited the formulation of a sounder understanding of the clinical causes of death and, in turn, the development of effective protective mechanisms. For example, to what extent do the burn injuries reported in pyroclastic flows result from heat conducted from the fast moving gas-solid dispersion, from lightning strikes, or from ignition of buildings and other shelters? Would small particle and toxic gas masks prevent asphyxiations in pyroclastic flows? Similarly, does the high incidence of irritation of the eyes and respiratory tract reported in many tephra falls result from the tephra particles themselves or from associated acid coatings and aerosols?

3. A third point also concerns the paucity of data. For many eruptions where more than one source of information is available there is considerable conflict in the data on deaths and injuries. Part of the disparity results from changing death tolls with time after the eruption as more bodies are discovered, missing persons are found (Table 3.9) and the injured die, but others are due to "guesstimates" in reporting, statistics from different official agencies, and differing bases of calculation. For most of the death tolls cited in this chapter it is not known whether the totals include indirect deaths. If indirect deaths are included, is it possible to produce a death toll even for Mount St Helens, 1980? It is difficult to even state which deaths should be included in the indirect category. Should deaths attributable to gaseous emissions in noneruption periods even be included in the statistics? What marks the start of an eruption — could it be increased gas emission?

These questions, of course, serve only to emphasise the problems with the available data. They do not alter the major conclusions about the most common agents and causes of death.

4. In cataloging the problems with death statistics we might also note that such figures tell us only a little about survival. While it is not a trivial matter that authoritative reports represent the 1902 Soufrière, St Vincent, death toll as 1585 and 1680 killed (see Chapter 3, Section 6), the debate has a different significance when we recognise that of those at risk in the eruption about 62% survived (based on 1680 killed). In many ways the import of a death toll in an eruption is only fully compre-

Table 3.15 Survival rates

Eruption	Total Killed	Survival rate	Comment
Vesuvius, AD 79	3 300	83%	Pompeii only; tephra fall and minor pyroclastic flow
Laki, 1783	9 350	81%	Whole of Iceland; starvation
Asama, 1783	477	15%	Kambara pyroclastic flow only
Tambora, 1815	38 000	65%	On Sumbawa } starvation and disease
	44 000	77%	On Lombok } starvation and disease
Soufrière, St Vincent, 1902	1 680	62%	May 7 only; pyroclastic flow
Mt Pelée, 1902	28 000	0.01%	May 8 only; pyroclastic flow
Torishima, 1902	125	0%	Everyone on island killed; pyroclastic flow
Santa Maria, 1902	50	95%	Helvetia plantation only; tephra fall
Taal, 1911		9%	Taal island only; } pyroclastic flows only
		3%	Gelud (Fig. 3.13) } pyroclastic flows only
Sukara-jima, 1914	29	>99%	Island population only; nearly all population evacuated
Rabaul, 1937	441	95%	Mainly tephra fall
Heimaey, 1973	1	>99%	Tephra fall and gas; nearly all population evacuated
Mount St Helens, 1980	57	16%	Assumes 11 survivors on margin of free blowdown area

hensible when coupled with a survival rate (or death rate). Table 3.15 presents some examples. The data in this table need to be interpreted carefully. For example, in Sakura-jima in 1914 and Heimaey, 1973, evacuation of almost all the populace occurred before the eruption reached a climax. Some early evacuation may have occurred in other examples. It could be argued that the survival rate in St Pierre in 1902 is low because of political pressure to remain in the city despite the warning signs that suggested a visit to the southern part of the island might be expedient. Other physical and social factors may have operated in other eruptions.

Table 3.15 suggests that survival rates in pyroclastic flows are generally less than 20%, while the chances of surviving major tephra falls are better than 90%. Even where starvation and disease have been the dominant hazard, survival rates are still in the range 65% to greater than 80%. Where evacuation has occurred early in the eruption the likelihood of survival is better than 99%.

5. Section 3.12 demonstrates that some volcano-related fatalities have such exotic causes as murder and, possibly, human sacrifice. Similarly, consideration of Yakima Emergency Room records following the Mount St Helens tephra fall suggests that some cases of asthma and bronchitis may be psychosomatic in origin. As noted previously, evacuation of either whole communities or individuals sometimes occurs before the onset of the eruption climax. These examples indicate that perception of volcanic hazard varies and that an eruption may have a variety of impacts which influence human mental and social, as well as physical, well-being. These aspects of volcanic hazards are considered in Chapter 4.

4
SOCIAL ASPECTS OF VOLCANIC ERUPTIONS

4.1 INTRODUCTION

Although death and physical trauma are the most dramatic and the most commonly described effects of volcanic eruptions on humans, a wide range of social effects also occurs. Factors which influence the social effects of an eruption on a community include: (i) the nature of the volcanic hazard; (ii) the deaths, injuries and damage incurred; (iii) the reactions of the community leaders to the threat and the aftermath; (iv) the nature and extent of assistance from state, national and international sources; and (v) individual and community perceptions of the threat. Interactions among these factors are complex and the range of possible social effects is very large.

Human perceptions of the threat and the aftermath pervade social aspects of volcanic eruptions but it may be difficult to separate cause from effect. For example, attitudes toward evacuation prior to an eruption are conditioned by perception of the threat posed, the perception itself being a function of previous experience. Similarly, disaster myths about panic, looting and shock influence individual perceptions but stem largely from the media. Clearly, the reactions of public officials are influenced by their perceptions as well as their training. At the same time, the extent and nature of disaster relief is strongly influenced by the media. All, in turn, may be influenced by religious beliefs — or, perhaps, in some cases, religious beliefs stem from volcanic activity!

The social fabric of society may be affected by eruptions. Some alterations may be transitory, lasting only for the duration of the eruption or for the period of evacuation. Others, such as psychological trauma may endure for months or even years in a small fraction of the afflicted population. Still other effects may be permanent, resulting, for example, from the enforced relocation of an entire community. In this chapter the effects described are arranged roughly in order of the time span over which they occur.

Few carefully formulated studies of the social effects of eruptions have been made. Only a sample of the range of likely effects can be provided here — in some cases these reports are little more than sketches of an anecdotal nature. However, where possible, the available information is related to the broader theories and results of natural hazards research.

4.2 HUMAN PERCEPTIONS OF ERUPTION RISK

Human perceptions of natural hazards are a very complex area of study because of the multiplicity of influences that shape attitudes, awareness, insights and beliefs. The media, particularly the electronic media, play an important role in shaping perception by emphasising some of the effects of hazard impacts as well as (sometimes) conveying warnings and providing information about hazard mitigation. Religious beliefs also shape perceptions, some religions, for example, promoting fatalistic attitudes, notions of punishment or acceptance of 'God's will'. Middle-aged people usually have more accurate perceptions of hazards than do young adults; similarly, rural folk, being 'closer to the earth', generally perceive natural hazards more realistically than do townspeople. Prior experience of hazards also improves perception, while the magnitude and recency of the last impact has an important influence. As we might expect, perceptions before and after a hazard impact may be quite different (Burton *et al.*, 1978: 100-102).

Few studies of perception of volcanic hazard have been made, and all of those are American. While these studies are neither strictly comparable nor universally applicable, they do provide some interesting data on social attitudes.

The 1974 study of volcanic hazard perception in the Puna district of Hawaii by Murton and Shimabukuro must be placed in the context of the relatively frequent but generally non-explosive volcanism characteristic of the East Rift Zone (Fig. 4.1), the ethnic and religious mix of those interviewed, and the fact that 85% of the respondents had experienced a number of eruptions. Most of the 101 interviewees did not perceive volcano eruptions as hazardous, more than half the sample emphasising that it is not people, only land and property, that are at risk. Some 79% of the sample expected future eruptions in the East Rift Zone, but 12% denied that future eruptions would occur. Fifteen per cent said that there was no way of knowing when an eruption would occur, while 61% claimed various 'signs' indicate when an eruption is imminent. However, these signs included 'funny' rains, unusual heat, dry spells and calm oceans as well as earthquake activity. Many newcomers believed that the volcanologists could predict eruptions while a few Hawaiians indicated that a priest (*kahuna*), or the volcano goddess *Pele,* would tell them when an eruption was imminent (Murton and Shimabukuro, 1974: 154-6).

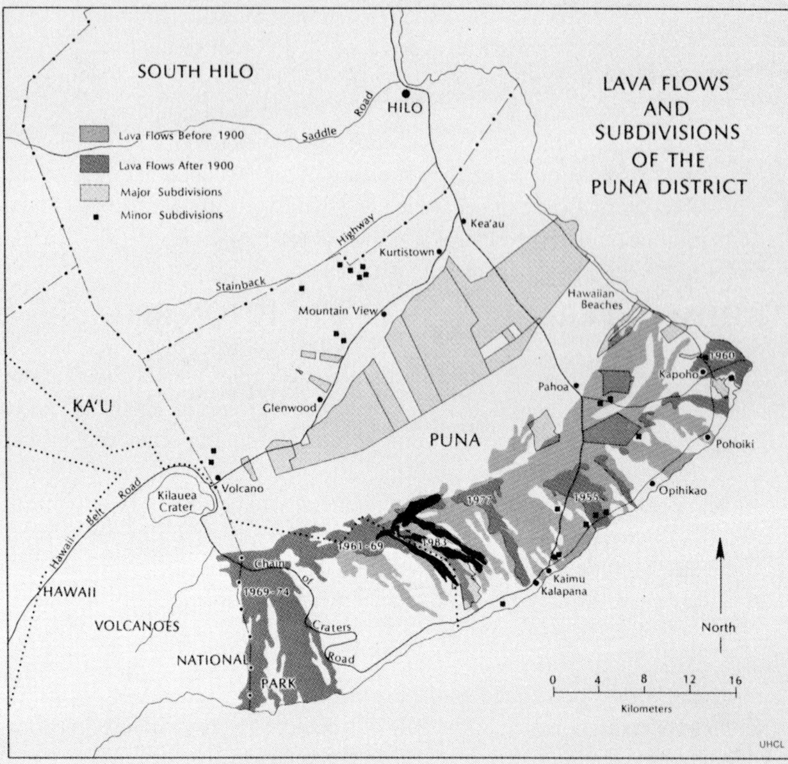

FIG. 4.1 Lava flows and subdivisions in the Puna District of Hawaii. Note that some subdivision on 1955 lava flows has already occurred (modified after Sorenson and Gersmehl, 1980: 134; Macdonald and Hubbard, 1982: 51).

Evacuation and prayer in the event of an eruption were the most popular adjustments cited by interviewees. Awareness of the possibility of diverting the lava flow by wall construction was low (20%) given the serious (and partially successful) attempts made during the 1960 Kapoho eruption. On the other hand, it is important to note that virtually the entire population of Puna knew about the emergency evacuation procedure (Murton and Shimabukuro, 1974: 157). In large part these results are reinforced by those presented by Hodge *et al.*, (1978, 1979).

Data collected immediately before and after the May 18, 1980, eruption of Mount St Helens provide a useful contrast with the Hawaiian studies. The perceptions of residents, visitors and officials can be viewed in the light of the clear perceptions of the volcanologists Crandell and Mullineaux (1978: 25) published two years earlier: "The volcano's behavior pattern suggests that the current quiet interval will not last as long as a thousand years; instead, an eruption is more likely

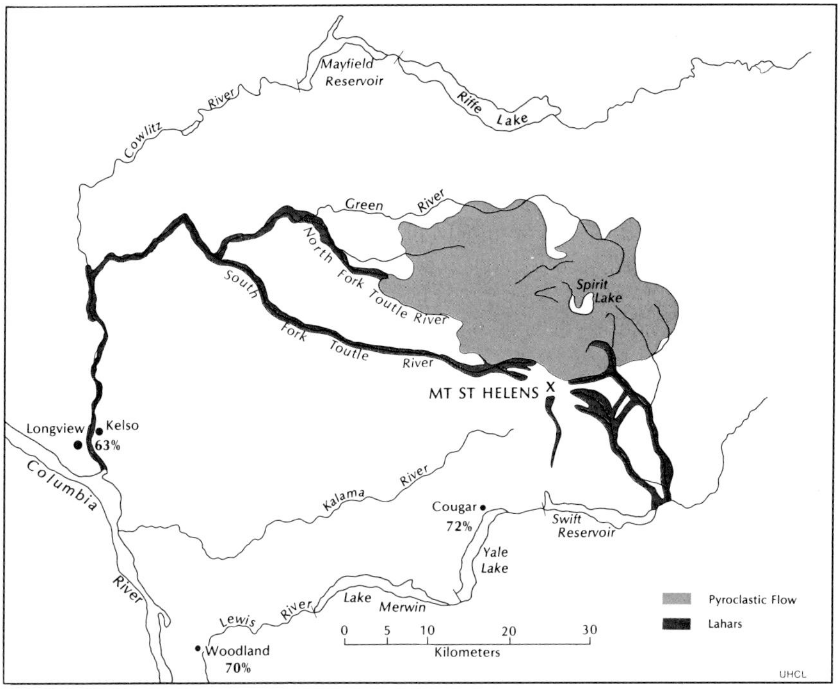

FIG. 4.2 The area west and south of Mount St Helens showing the extent of lahars and the proportion of samples in three communities indicating knowledge of the potential for volcanic eruptions in the Pacific Northwest before the March 1980 activity of Mount St Helens (data from Greene *et al.*, 1981: 52).

to occur within the next hundred years, and perhaps even before the end of this century... Because of the variable behavior of the volcano [in the past], we cannot be sure whether the next eruption will produce lava flows, pyroclastic flows, tephra, or volcanic domes, or some combination of these..."

Batelle Human Affairs Research Center staffers (Greene *et al.*, 1980, 1981; Perry *et al.*, 1980) collected information by telephone from 174 respondents in three communities between April 5 and 7, 1980; that is, after the initial eruptive events beginning March 27 but prior to the 'cataclysmic' Mount St Helens eruption of May 18. The three areas sampled can be referred to as Cougar (about 22 km from the volcano), Woodland (40 km) and Longview–Kelso (64 km). The numbers in Figure 4.2 show the proportion of the sample in each area expressing knowledge, prior to the onset of activity at Mount St Helens, of the potential for volcanic eruptions in the Pacific Northwest. More importantly, Table 4.1 indicates a dramatic change in perception of the potential threat once activity had begun. The 'before' data support the view from research elsewhere, that infrequent hazards are perceived to have low risks (Greene *et al.*, 1981: 53). The 'after' data perhaps suggest a relationship between distance from the volcano and perception of the degree of risk. Certainly, the Longview–Kelso sample had the highest proportion (57%) expressing the view that it was unlikely the volcano represented a threat. As it turned out, these communities suffered physical damage whereas Cougar and Woodland did not.

Interviews conducted in early April, 1980, helped determine the 'before' and 'after' perceptions of 14 officials from a variety of federal, state and local organisations. These interviews indicate that officials found it difficult to perceive an eruption, and there was a high level of uncertainty about the volcano's future behavior. Furthermore, it seems that most officials had a very poor perception of what might happen or of the risks to life and property (Sorenson, 1981: 35-37) despite the efforts of US Geological Survey personnel (Miller *et al.*, 1981: 791, 797). On the other hand, Table 4.2 indicates that lahars or floods and tephra fall were perceived by citizens to be the most hazardous types of activity. The importance attached to tephra fall in the Cougar area undoubtedly stems from the light tephra falls experienced there before the survey date. Only one of the 174 respondents believed that an explosive eruption posed a threat. Despite the uncertainty on the part of officials, the great majority of respondents thought that they had adequate information to guarantee personal protection (Greene *et al.*, 1981: 58). There seemed to be little variation in this confidence with distance from the volcano. It is difficult to interpret these results fully. As it turned out on May 18 and during subsequent eruptions during 1980-1981, Cougar was most affected by tephra fall and the other two areas most affected by lahars or floods. This suggests excellent respondent awareness, but these perceptions were formulated before it was known that the north side of the volcano was bulging; if, in fact, the ominous swelling and subsequent events had occurred on the south side of the mountain, these perceptions would have been sadly awry.

Table 4.1 Perceived likelihood that volcanic activity at Mount St Helens would affect safety of respondents' family or property

Likelihood of an effect	Cougar		Woodland		Longview-Kelso	
	Before	After	Before	After	Before	After
	recent activity		recent activity		recent activity	
Likely	13%	22%	6%	17%	9%	19%
Even odds	6%	39%	6%	37%	6%	25%
Unlikely	81%	39%	87%	46%	85%	57%

After Greene *et al.*, 1981: 52-53.

Table 4.2 Responses to "In what way will your safety be affected?", Mount St Helens area, April 5–7, 1980

Safety affected by	Cougar	Woodland	Longview
Lava flows	6%	3%	3%
Lahars/floods	29%	70%	70%
Tephra fall	65%	26%	27%
Explosive eruption	0%	1%	0%

After Greene *et al.*, 1981: 54.

Sorenson (1981) also interviewed 25 residents of Cougar and 50 visitors to Cougar and Yale Reservoir (Fig. 4.2) to determine their perception of risks involved in the eruption. These interviews were also conducted in early April, 1980. When respondents were asked whether they wanted the volcano to have a major eruption, a series of small eruptions, or a cessation of activity, 56% of the locals interviewed wanted the eruptions to cease, but this category was preferred by only 6% of the transients. On the other hand, 58% of the non-local population desired that a major eruption occur. Sorenson (1981: 55) interprets these responses as evidence of the glorification of the volcano as a media event and an indication that many people have weak or inaccurate images of the consequences of a major eruption. Those interviewed also believed that the threat to personal safety at the interview location was relatively low. This finding probably relates to the common conclusion in hazard perception studies, that hazards happen to someone else (Burton *et al.*, 1978: 106).

Another study of volcano hazard perception focused on the May 18 tephra fall area (see Fig. 3.8) northeast of Mount St Helens. In Ellensburg, 150 km northeast of the mountain, only a few of the more than 30 people interviewed claimed any knowledge of the risks or consequences of an eruption prior to May 18. Despite more than 60 articles in the local newspaper before May 18, from some of which inferences could be drawn that Ellensburg was in the hazardous zone, the prevailing belief in the city was that it was not subject to any of the effects of volcanism. Despite the media publicity one agricultural official had associated volcanic activity only with lava flows! Not one respondent was aware of the *US Geological Survey Bulletin* 1383-C "Potential hazards from future eruptions of Mount St Helens Volcano, Washington", or of the maps therein which clearly show Ellensburg to be in the tephra hazard zone (Fig. 4.3). According to Saarinen (1980: 20) 200 copies of this publication were shipped to the Pacific Northwest after the March 20 activity for distribution to appropriate persons.

Similar low perception of volcanic hazard was apparent in the other cities. In Ritzville, about 40 respondents indicated no prior awareness of the potential threat and had no prior knowledge of US Geological Survey hazard studies (Warrick *et al.*, 1981: 36). In Cheney, local (Spokane) newspapers published 152 articles on Mount St Helens in the March 20-May 18 period. Only eight articles suggested that tephra from Mount St Helens had fallen as far east as Spokane, and these articles did not hint at any real threat. In Missoula, Montana, several dozen articles appeared in local papers before May 18. Nearly half were front page stories but the possibility of a local tephra fall was not mentioned. None of those interviewed by Warrick *et al.* (1981) had perceived any local threat.

Despite the difficulties involved in comparing sample size, sample characteristics and experience, and variations in methods of data collection the Mount St Helens and Hawaiian perception studies form important contrasts. The Puna study can be regarded as indicative of hazard perception amongst a population 'conditioned' by relatively frequent and non-explosive eruptions. On the other hand, the

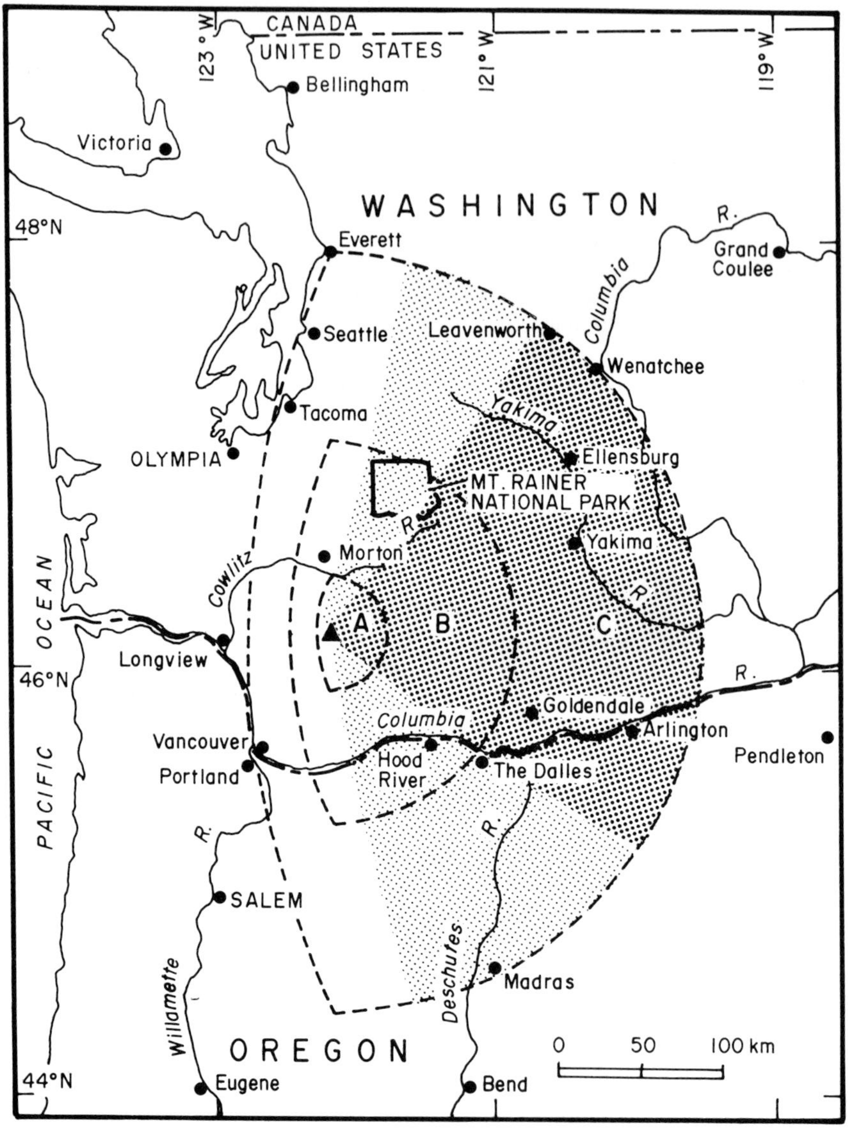

FIG. 4.3 Areas likely to be affected by tephra fall from Mount St Helens (Crandell and Mullineaux, 1978).

Mount St Helens studies represent the low level of hazard perception found in communities where hazard occurrences are infrequent but of high magnitude. These studies also suggest that the US Geological Survey (and other groups of volcanologists) need to spend more time interpreting eruption phenomena and risks for other government officials and the media. Although the May 18 eruption may have dramatically altered Pacific Northwest perceptions of volcano hazard, a few decades without eruptions will certainly see a gradual return towards a pre-1980 level of perception.

Although there is ample evidence that United States perceptions of disasters and disaster behavior are profoundly influenced by the media, especially the electronic media (Wenger *et al.*, 1975), there is little evidence that similar misconceptions (see below) are as pervasive around other volcanoes. We might surmise that the inhabitants of volcanic hazard areas in Asia, central America and the Pacific, with their varied cultural and religious backgrounds and with a variety of eruption experiences, will perceive risks offered by volcanism in different ways to their North American counterparts.

4.3 MENTAL HEALTH PROBLEMS

The types of mental health problems encountered following a natural disaster may cover a wide range. They include psychosomatic problems (for example, vomiting, headaches, rashes), bereavement and separation, social stress (for example, divorce, alcoholism), sequelae of physical trauma, and psychological problems (for example, nightmares, unreasonable fear of a return disaster, depression, disorientation, hysteria and psychosis). Less extreme problems found by counsellors include replacing lost eyeglasses or medication, providing help with income tax forms, and the need for a sympathetic ear (NIMH, 1978).

Nearly all the available information on mental health problems following a volcanic eruption stem from the 1980 eruption of Mount St Helens. At a counselling workshop in Longview "numerous participants expressed that they had never been so depressed before. Many did not want to get up in the morning nor go to work. Psychosomatic illnesses, especially headaches, prevailed. Certainly the quality of life had greatly decreased for the general population. Additional factors of stress included: relocation, families bunching up in one living quarter and entertainment of frightened children. Many left the area, many more questioned whether or not to stay" (*Project Ashlift,* 1981: 10).

Bill Peckham, a spokesperson for the Washington State Department of Social and Health Services reported that they had seen: "evidence of high levels of anxiety related to the ongoing nature of the disaster. Particularly concerned are those who live near rivers that have already been raised because of the mudslides . . . The increased tension has resulted in reports of child abuse, spouse abuse, drug abuse and alcohol problems. In some already marginal families, this has been the last straw. It has given rise to family problems and marital conflicts in a number of cases" (*APA ψ Monitor,* September 1980).

Such problems were not confined to the areas experiencing lahars from Mount St Helens. Of 78 clients counselled for mental health problems in two counties experiencing only tephra fall, 58% of the sample had experienced marital difficulties, 31% alcohol/drug abuse problems, 23% life crisis problems and 22% had problems with parenting effectiveness. In two counties which experienced lahars the respective proportions among a sample of 70 persons were 39%, 30%, 30% and 30% (*Project Ashlift,* 1981: 23).

Several points arise from such data. First, mental health problems may not arise until some time after the eruption and may continue for some time. Pennebaker and Newtson (1983) observed that police calls in Longview for interpersonal conflicts were 50% lower following the May 25 ashfall than in the same period a year previously. Such conflicts remained at a low level for two weeks. On the other hand, one year after the May, 1980, eruption, delayed stress reactions were blamed for "a tremendous increase in all kinds of problems — nervous breakdowns, marital friction, drug abuse, alcohol problems" (*The Wall Street Journal,* May 18, 1981). Even three years later some Toutle River residents were described as living in constant fear of flooding (*Paradise Magazine,* March 6, 1983: 17).

Second, data presented by Leik *et al.* (1982) indicate that problems did not necessarily lessen with time after the eruption (Table 4.3). While economic problems did not increase with time, health and interpersonal problems frequently increased even in areas such as Yakima and Pullman where tephra falls were relatively light (9 mm). It is surprising to find that nearly 20% of the Yakima sample attributed health problems to the volcano a year after the eruption. Certainly psychosomatic problems have occurred in the aftermath of eruptions. Examples presented in Chapter 3, Section 5.2 from the eruptions of Usu (1977), Karkar (1979), Soufrière, St Vincent (1979) and Mount St Helens (1980) suggest that at least some of the reported cases of asthmatic bronchitis, headaches and/or hyperventilation were of psychosomatic origin. However, such problems would normally be shortlived in places such as Yakima and Pullman where the tephra cleanup was effective and rapid. In Longview, where the lahar hazard continued, psychosomatic problems might also be expected to remain at elevated levels. Furthermore, some data pre-

Table 4.3
Proportions of sample attributing problems to Mount St Helens six and 12 months after the May 1980 eruption

Problems	Pullman		Yakima		Longview-Kelso	
	6 mo	12 mo	6 mo	12 mo	6 mo	12 mo
Economic problems	0%	2%	5.8%	6.4%	16%	16.3%
Health problems	4%	12.5%	9.6%	19.1%	16%	11.6%
Interpersonal problems	2%	6.3%	1.9%	2.1%	8%	14.0%

After Leik *et al.*, 1981: 33.

sented by Pennebaker and Newtson (1983) seem to be in conflict with those in Table 4.3. A survey of letters in the Yakima newspaper one month and 12 months after the May, 1980, eruption showed that the proportion of letters with a positive viewpoint had increased from 54% to 89%. In Longview the proportion increased only from 31% to 47%. These data suggest that the people of Yakima recovered psychologically more rapidly from the effects of the eruption than did newspaper letter writers in Longview.

A final point arising from Table 4.3 concerns attribution. Although the data presented by Leik *et al.* (1982) indicate that many respondents attributed their problems to the eruption and the aftermath, misattribution may occur. A survey by Pennebaker and Newtson (1983) in seven towns of the Pacific Northwest indicated that 63% of respondents reported some degree of at least one plausibly ash-related symptom (nasal congestion, eye irritation, coughing, breathlessness, or wheezing sounds in chest) but 36% reported at least one supposedly ash-irrelevant symptom (headaches, queasy or upset stomach, tense muscles, insomnia, or general tiredness). The issue of attribution of symptoms is a complex one. As Pennebaker and Newtson (1983) note: "...common feelings, emotions, or symptoms that undoubtedly existed before the volcano that were not particularly noteworthy (feelings of anxiety due to job or family stresses) are now attended to and organized within the context of the volcano. In other words, these symptoms or emotions that may have been there all along are later attended to much more closely because the effects of the volcano now fit into a general psychological schema that is pertinent to the volcano".

Research results from the aftermath of Mount St Helens suggest that mental health problems following a volcano disaster are little different to those found after other natural disaster impacts. Two provisos must be attached to this statement. First, most natural disasters are short-lived and there is little problem in determining when the hazard impact has finished. Tornadoes last but a few minutes, hurricanes a day or so, and even flood waters have usually receded within a week. Mount St Helens, however, has presented a continuing threat at least to the people of the Toutle and Cowlitz river valleys. Frequent media references to the possibility that the volcano will continue erupting for five to 30 years may also have exacerbated mental health problems.

Second, most of the information on mental health problems following natural disasters stems from North America investigations. Because such problems are integrated with human perceptions, attitudes and cultures it is difficult to generalise these conclusions to other cultural mileu. However, in a general sense it is possible to identify those persons most prone to mental health problems following natural disasters. The following profiles are believed to have at least some cross-cultural validity.

Psychological stresses created by natural hazards span a wide range but, in general, three groups of people can be defined (Wolfenstein, 1957; Parad *et al.*, 1976; Golan, 1978; NIMH, 1978). During the period when the threat seems remote most people (at least in Western societies) deny the existence of the threat either by not thinking about it, by believing it will not happen, or by deciding that it will not affect them. On the other hand, a minority group who do think about remote dangers may be predisposed to emotional disturbance. Nonetheless, anticipation prepares the individual, to some extent, for the effects of the impact.

When the threat is more immediate many people continue to deny the threat, tending not to believe warnings as a defence against anxiety and inconvenience. Those who believe in life after death may take fewer precautions than those who do not. Those who enjoy the social contact offered by an evacuation center may be more willing to evacuate than those who find such centers claustrophobic. The most adaptive group during this period are those who consider the threat seriously, make preparations, take precautions and develop possible scenarios in their minds, thus becoming used to the idea of death and destruction. Another minority group develops monstrous fantasies which may prevent adaptive behavior. Panic is only likely to occur amongst those who perceive an immediate and severe threat and a limited number of closing escape routes.

During impact those who denied the existence of the threat have an illusion of centrality, experiencing a withdrawal of interest in the rest of the world. Those who had monstrous fantasies may be overcome with terror while those who took precautions retain a strong belief in their own survival.

In the immediate disaster aftermath many victims experience feelings of abandonment, even the deaths of loved ones being seen as desertion. Many victims suffer from the so-called "disaster syndrome" in that they appear docile, stunned, shocked, dazed with an absence of emotion and a lack of response to stimuli and an inhibition of outward activity. Some may experience the "counter-disaster syndrome" exhibiting a euphoric identification with the community, physical overexertion and low efficiency with an uncritical acceptance of "leaders" who emerge during the rescue. Some may be unaware of injuries to themselves. A minority of victims remains calm, lucid and practical. Another minority group may be disoriented, cry hysterically and/or be paralysed by anxiety.

The counter-disaster syndrome represents an active denial of the fear, anxiety and grief created by the disaster. Both syndromes particularly affect those who denied the threat — it is a response to the unreality of the impact. Unawareness of personal injuries results from an inability to take in any more stimuli. Both syndrome should be seen as conditions of severe suffering despite the lack of an outward display of stress. The effects may last from a few hours to many days.

In the post-disaster period psychological trauma may not manifest itself for several days or even for several months. Early symptoms of stress such as insomnia, nervousness and digestive upsets tend to disappear fairly quickly but psychosomatic symptoms may persist. As mentioned previously these may be indistinguishable from the physiological effects of tephra fall. A wide range of responses is possible but those who are most adaptive are those who can relive the experience, talk about it, and turn it over in their memories. These victims can assimilate the event and eventually look back on it with equanimity. On the other hand, those who cannot assimilate the event, will not talk about it and cannot bear the revival of the experience in their memories, are the most likely to suffer continuing stress. In many cases these victims would benefit from mental health counselling. The elderly, middle-aged and children are more likely to suffer mental health problems than teenagers and young adults.

While these profiles of natural disaster victims are believed to have some universality and cross-cultural applicability (Quarantelli and Dynes, 1977) mental health counselling is largely confined to developed societies. For most of the world's volcano disaster victims psychological trauma is likely to be a less immediate problem than the issue of day-to-day survival.

4.4 EVACUATION

The term evacuation can mean many things — flight, eviction, formal evacuation by authorities, or abandonment of an area and resettlement. Formal evacuation by authorities in the face of natural hazards is certainly the most common in developed countries. Although it seems to be a widely held belief that the public co-operate and evacuate quickly when so requested by the authorities, various studies in the United States indicate that the majority of people does not move (Quarantelli and Dynes, 1972; Wenger *et al.*, 1975). For example, Quarantelli and Dynes (1979: 67) note: "Those who do flee are primarily transients and tourists — not the people who live there". Nonetheless, a Delaware survey showed that more than 80% of the sample respondents believed that the majority of the public willingly and rapidly complies with requests to evacuate. The extent to which these American findings might apply in other countries is not known, but some of the examples presented below suggest that such generalizations do not apply universally.

Evacuations as a result of volcanic eruptions are common — numbers and conditions range from a few people for a few days to the permanent resettlement of tens of thousands. Some recent examples are presented in Table 4.4. Whether these evacuations constituted forced evictions, formal requests to move to a safer area or spontaneous decisions by individuals is not always known. Conditions may be different because of differing hazards (or perceptions of hazards) in towns on opposite sides of the volcano, or even within the one village. Nolan's (1979:302-303) account of the evacuation of villages around Parícutin (Fig. 4.4) indicates some of the complexities of the evacuation process. Although many people fled the immediate area shortly after the onset of the eruption, many of these subsequently returned home. About three months after the start, however, a more permanent exodus began:

Table 4.4 Examples of evacuations under threat from volcanic activity

Volcano	Reference
Karthala, 1977	4000 evacuated (*BVE*, 1979, 17: 13).
Piton de la Fournaise, 1977	2500 evacuated homes including 800 in Bois Blanc and 600 from Piton Ste Rose (*BVE*, 1979, 17: 15-16).
Ulawun (Papua New Guinea), 1980	2000 evacuated to temporary shelter (*PNG Post Courier*, October 9, 1980: 1).
Tinakula, 1971	160 evacuated by ship (*BVE*, 1973, 11: 16).
Galunggung, 1982	75 000 in temporary shelters (*Galunggung*, 1982: 13).
Api Siau, 1976	1800 evacuated (*BVE*, 1978, 16: 35).
Una Una (Indonesia), 1983	Entire population evacuated to other islands (*SEAN Bull.*, 8(8), 1983: 2).
Pagan (Marianas), 1981	Total island population of 53 evacuated by Japanese merchant ship (*SEAN Bull.*, 6(4): 5).
Usu, 1977	Inhabitants of Toya and Sobetsu Spa evacuated for about one month (*BVE*, 1979, 17: 50-52).
Mt Hudson, 1971	Most of settlers evacuated by sea from mouth of R. de los Huemeles (*BVE*, 1973, 11: 43-44).
Santiaguito Dome (Santa Maria), 1982	Hundreds of residents forced to evacuate (*SEAN Bull.*, 7(9): 12).
Beerenberg, 1970	39 evacuated by plane from Jan Mayen (*SICSLP*, 89-70).

> It was composed mainly of those who had kinsmen in other communities or skills allowing reasonable employment in other places and/or resources that could be liquidated to pay for relocation. In other words, it was generally the richer, better-educated, more exogamous, Spanish-speaking mestizos who first left their home communities. However, some of the richer families stayed because their wealth was in land and homes that could not be disposed of without sacrifice, and they felt they had too much to lose by leaving. A few of the early migrants were among the poor but ambitious Spanish speakers who had already been thinking of leaving their villages to seek better opportunities elsewhere.
>
> Hardly anyone left the Indian-oriented communities of Parícutin and Angahuan before mid-June 1943, but by then the situation was desperate...
>
> Accounts of the evacuation of Parícutin village by the Mexican army range from stories of people forced into government trucks at gunpoint to dramatic accounts of a peasant population gladly following General Lazaro Cardenas to a new land, a new destiny, and a better future. All the accounts apparently contain some truth, perhaps because many of Parícutin's younger citizens did welcome new opportunities, whereas many others, especially among the older people, "did not wish to leave and... preferred to die covered by lava rather than abandon their homes".
>
> [Nolan, 1979: 302-303]

Numerous factors may influence the willingness of those at risk to evacuate. For example, alternative accommodation, food and employment may, or may not, be provided. In some cases political considerations may also be involved. While such details are not always known, the following examples illustrate various aspects of evacuation.

Actual eviction seems to have been rare. Apart from the Parícutin example above, eviction also occurred in Pozzuoli near Campi Flegrei in 1970. Although no eruption has yet occurred, seismic phenomena and ground deformation were interpreted by some as possibly heralding volcanic activity; on March 1, 1970, a decree

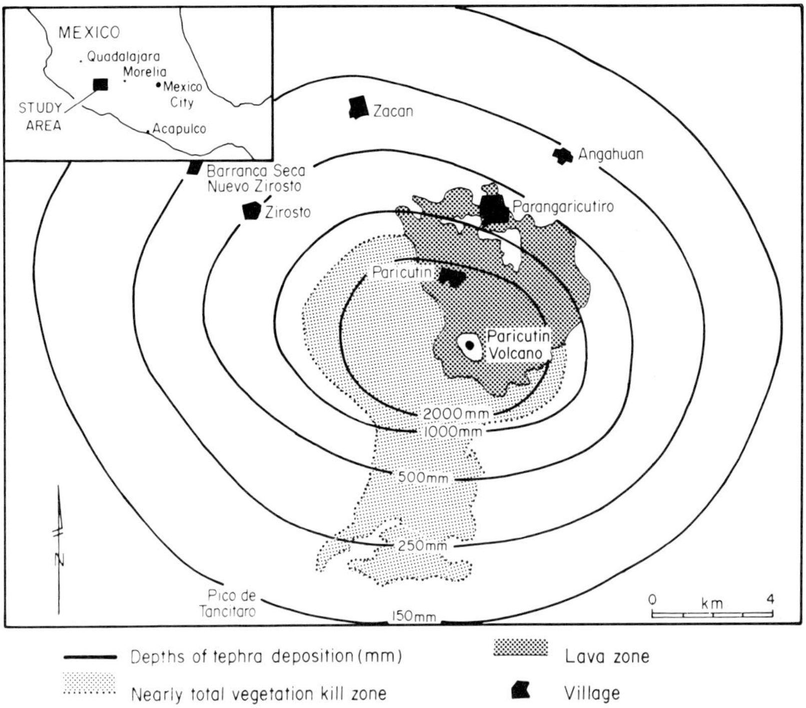

FIG. 4.4 Map of Parícutin showing 1943–1952 deposits and villages (modified after Rees, 1979, and Nolan 1979).

was issued ordering the evacuation of tottering buildings as a precaution. According to Hoffer (1982: 125) 6000 people were evicted by the police and a further 30 000 evacuated on their own accord. This eviction possibly saved lives as buildings collapsed in 1970 and 1972 (Imbò, 1979: 511).

Temporary abandonment, particularly of islands, has occurred on several occasions. An eruption on Bam, Papua New Guinea, probably in the 1870s, resulted in abandonment of the island for several years (*CAVW*, 5: 6). Similarly, a series of severe eruptions between 1771 and 1786 caused Vulcano to be left uninhabited for a number of years. At Santorini in 1925, sufficient people left the island as a result of the eruption, that a large school had to be closed (Washington, 1926: 367; 383).

Following the 1982 Galunggung eruption, 10 475 people from around the volcano were resettled in south Sumatra and the Riau Islands (H. Aboeng Koesman in *Report and Recommendations of the National Workshop on Mt Galunggung Volcanic Risk Management*, September 1982, volume 3: 5). Although more evacuees will migrate to other areas including Kalimantan and Sulawesi the majority will resettle around the volcano. Transmigration resulting from the 1963 eruption of Agung was less successful. *Keesing's Contemporary Archives* (July 13, 1963:

19 521) notes that 70 000 people were homeless and 400 000 were destitute. A total of 332 234 were evacuated from Bali (*BVE,* 3: 7), with resettlement occurring in Sumatra, Sulawesi, Sumbawa, and the Moluccas. Jennings (1969: 37) suggests that (only) 140 000 people required immediate evacuation, but that many evacuees returned as soon as they could to Bali because none of the transmigration areas had the conditions necessary for wet rice cultivation, the agriculture that the Balinese knew best.

Several other interesting examples of evacuation and abandonment with eventual return can also be cited. The South Atlantic island of Tristan da Cunha was abandoned in 1961 as a result of earthquakes, ground deformation, rockfalls and destruction of the water supply prior to an eruption near the only settlement. The 270 islanders moved briefly to an adjacent island and then to industrial southern England. Their new lifestyle was totally different to the isolated 19th century style they had left. By the end of 1963, over 250 islanders had returned to their homeland. Some subsequently returned to England and then to Tristan again; by 1972 few Tristans remained in England (Lewis *et al.,* 1972).

The northernmost Tongan outlier, Niuafo'ou, was abandoned after an eruption in 1946. Political decisions forced the abandonment of the island, unnecessarily it seems, with the passing of the 1946 Evacuation Ordinance. Even so, between two and 30 of the 1300 or so Niuafo'ouans refused to leave despite proposed penalties of a £50 fine or two years imprisonment or both. The evacuees were moved to the vicinity of the capital Nukualofa. Some village groups were relocated again and again, finally being resettled on 'Eua, a sparsely populated island 20 km east of the main island Tongatapu. By 1958, the Niuafo'ouans, scattered in many areas of Tonga and still pressing for permission to return to their homeland, finally won that right (Rogers, 1981).

The evacuation of people resident on the southeast and west coasts of Ambrym, Vanuatu, because of tephra falls in 1950 and 1951 also provides an interesting case study. Some residents left voluntarily because the tephra made life uncomfortable; another 701 were subsequently forcibly evacuated by the colonial authorities to Epi, a nearby island. A few weeks later 48 of these evacuees were killed in a tropical cyclone. The majority of the survivors of Maat village were evacuated to Efate, seat of the capital, Vila. Eventually those remaining on Epi also went to Efate to work with their fellow islanders on a copra plantation and to establish the new village of Maat Efate. Although these Ambrymese regarded this migration as temporary and intended to return to their home island, access to the cash economy and an improved living standard caused them to remain even though other evacuees had returned home by 1954. Strong links were maintained with those on Ambrym, the residents of both Maats visiting each other. By the end of 1977, 40% of the people of south-east Ambrym were absent from the island; many of the absentees were regarded at home as permanent expatriates, but, 25 years after the eruption, still said they would return to Ambrym some day (Tonkinson, 1979).

These examples provide some evidence contrary to the conclusions drawn from United States studies that the majority of the population at risk does not evacuate when requested to do so. Evidence from other evacuations in the face of volcanic eruptions is more difficult to interpret because various authors have provided different viewpoints. The 1906 eruption of Vesvius, for example, caused "a great exodus of tourists with much confusion and a greatly augmented train service early on in the eruption" (Hobbs, 1906: 640). Evidently the exodus was large enough to cause many of the grand hotels to dismiss their staffs. Perret (1924: 47) reckoned that about 100 000 people left the city in the first five days of eruption; many of these were local people.

Varying viewpoints are also available concerning the 1914 evacuation of Sakura-jima island and, subsequently, the city of Kagoshima (see Fig. 3.19). Premonitory earthquake activity on January 10, 1914 warned most people to leave the island. Kotô (1916: 59) noted that 24 577 of the 27 116 inhabitants left for the nearest part of the mainland on January 11. Jaggar (1924: 445) wrote: "Every available sampan sculled with frantic speed back and forth across the channel all day Sunday, January 11, moving the natives of the island, their bedding, mats, rice bags, and canary birds to the mainland. By Monday the army, navy and other government departments, railways and steamship lines were all helping". Jaggar (1956: 105) placed a general of the army in command of evacuation with private and public vessels called into service. On the morning of January 12, 2493 were quickly evacuated by steamboat, these inhabitants waiting for rescue on the shore by the time the rising plume was visible on Kagoshima at 10 a.m. A further 46 people were evacuated during the next three days (Kotô, 1916: 62-63, 95).

In the early afternoon of January 12 tephra began to fall in Kagoshima (thickness not shown on Fig. 3.19). This, rumors of poisonous gases and tsunami, the detonation of volcanic explosions, air concussions powerful enough to cause earache, the rattling of *shoji* (paper house partitions) and the trembling and shaking of the ground caused the populace of Kagoshima to leave, to the accompaniment of clanging fire bells and the alarm bugling of the 45th Regiment (Kotô, 1916: 63-64, 69; Jaggar, 1924: 447). Although Kotô inferred that the evacuation of Kagoshima was disorderly and based on individual initiatives, Jaggar (1956: 105) wrote that all except public officials were ordered to leave for the back country after the 6.28 p.m. January 12 earthquake. Young men's clubs were ordered to receive refugees along the back roads leading into the interior of Satsuma Province, with temples and schoolhouses opened to house them. "The migration of more than 50 000 people with packs on their backs and with handcarts bearing household goods demonstrated how easily the Japanese people took to the nomad existence" (Jaggar, 1956: 105). Jaggar (1924: 455) cites an American schoolteacher who "walked through the deserted streets of Kagoshima without meeting more than five people in a mile; shops were nailed shut; soldiers took up their quarters on street corners and guards walked their beats". Jaggar also reported that there were no ghouls and no looters.

By January 19, Kagoshima city was full of life again and in every way restored to its former state although another large explosion and ash fall two days later caused about 1000 people to again leave the city (Kotô, 1916: 76-87).

While there are contrary opinions in this information as to whether the evacuation of Kagoshima was ordered or spontaneous, orderly or disorderly, reports from the Caribbean indicate that two evacuations during the 1970s — one ordered, one spontaneous — were both orderly! On April 12, 1979, 22 000 people evacuated themselves from the slopes and surrounds of Soufrière, St Vincent. There had been no time for an evacuation order as seismic data provided only a few hours warning of the ensuing eruption. The authorities, however, did enforce the evacuation for several weeks (Fiske, 1980: 152; 1981). According to Shepherd *et al.* (1979) 15 000 people were evacuated in 24 hours; Starbird (1979) notes that 20 000 people fled southwards to Kingston, all returning to their homes within two months.

On the other hand, the evacuation of 73 000 people from the slopes of Soufrière de Guadeloupe was ordered by civil authorities following some months of eruptive activity. "Early Sunday morning, to the sound of church bells, the evacuation began. Small pickup trucks, cars and buses clogged the roads to the adjoining island of Grande-Terre, some residents were taken out by sea. And some others, participants in the Tour de Gaudeloupe bicycle race, left on their bikes. 'Most people didn't wait to pack . . . they left with just the clothes on their backs'." (*Times,* August 30, 1976: 7). Everyone was evacuated from Basse-Terre except one elderly couple who refused to leave the slopes of the volcano. "Emptied of human life, Basse-Terre now resembles the post-atomic spectacle of the movie *On the Beach*. French tricolors flutter unattended in the breeze. The traffic lights are still on but jammed on red or green. Goats wander the streets and chained dogs howl in hunger. Others, having broken the bonds, forage in the grounds of the Saint-Claude Hospital. A large poster in Basse-Terre announces that the Tivoli Cinema is showing *Hell is Empty*." (*Time,* August 30, 1976: 7). A week after the evacuation, farmers were complaining that the livestock they had left behind would starve. Some were let through police barricades to feed their animals or to lead them to safety. Housewives also protested about the sudden increase in the price of vegetables which used to be harvested on the slopes of the volcano (*Time,* August 30, 1976: 7). Complaints also broke out in other quarters as to whether the massive evacuation was justified, whether it lasted too long (3.5 months), and whether there was panic or complacency regarding assessment of the hazard (Fiske, 1979: 212; see also section 4.7.)

Within six hours of the onset of the 1973 Heimaey eruption nearly all 5300 residents of the island had spontaneously evacuated. The speed and efficiency of the evacuation was a result of the farsighted evacuation plan previously prepared by Iceland's state civil defense organisation, the fortuitous presence of the fishing fleet in port and the population's experience with the Surtsey eruption a decade earlier (Williams and Moore, 1973: 15; Gunnarsson, 1973: 41) and, no doubt, the strong community identity of the islanders. Despite the orderliness and efficiency of the evacuation other problems arose as evacuation of material possessions proceeded 20 hours a day. Tempers flared as men watched their houses destroyed by the lava and falling pumice. Other problems occurred on the mainland as permanent homes were hard to come by, children had to be put into schools, and boats had to find space in other ports. Families used to six to seven room houses now had to crowd into two to three rooms. Although the islanders wanted to stay together as a community, this proved difficult in practice (Gunnarsson, 1973: 62f). However, by May, 1974, more than 3000 of the islanders had returned to Heimaey; a late 1973 survey by the Vestmannaeyjar Town Council indicated that only 8% of the island population would not return (*The American-Scandinavian Review*, 1973, 61(4): 338 in Bruner, 1974).

Although little specific information is available in the volcanological literature, most evacuees move as family groups. This is important as post-evacuation problems are likely to be minimised; numerous sociological studies show, for example, that concerns in the post-impact period are initially for the welfare of family members, then neighbors, relatives and friends, and later for members of the wider community (Wolfenstein, 1957). We may infer that in most orderly and ordered evacuations, people leave as family groups. Evidence presented by Carroll and Parco (1966) concerning the evacuation of Taal island in 1965, suggests that this is also correct in at least some cases where flight is spontaneous and rapid.

The Taal eruption began at about 2 a.m. on September 28, 1965, with a series of flashes and explosions. Survivors on the island knew immediately what had happened but many of those on the lakeshore and in the Taal–Lemery area (see Fig. 3.13) had other thoughts, including a storm, a fire, a plane crash, the war in Vietnam getting closer, the communists dropping a bomb on Lipa air base, or robbers invading the place.

A 27-year-old male refugee from the island said. "I woke up my mother, brothers and sisters. I took the family aboard the volcanologist's speedboat but they were made to get off as the boat was overloaded.* My brother saw another boat in which he wanted to load the family but only my unmarried sister was able to board because they started the boat at once. Luckily, my mother and the rest of my brothers and sisters were able to take another boat. I was left at our place and I got some clothes and money. I was able to take the surveyor's boat. The next morning I learnt that my sister who took the first boat died because their boat sank." (Carroll and Parco, 1966: 12-13). Out of 221 refugees interviewed by Carroll and Parco, 144 fled as a household group. A further 23 fled as part of a household group. Fifty-three fled with non-household members but in most cases they were with specific neighboring households. The town of Lemery which was most distant from the volcano, and where neither the precise danger nor what should be done were evident, revealed the greatest tendency for households both to break up and to join

*According to Moore *et al.* (1966: 955) this boat, with a normal capacity of six people, left the island at 2.13 a.m. hours with 21 people aboard.

with non-household members. However, there were frequent mentions in the interviews of awakening others in households and in neighboring households where relatives frequently lived (Carroll and Parco, 1966: 13-15). Despite the harrowing experiences reported by survivors it is clear that social roles continued to operate and to guide a substantial proportion of individual behavior.

In all, about 13 000 families were evacuated and refugee camps were quickly set up. Carroll and Parco's (1966: 33) interview data show the following reasons for evacuation to a particular place: 28%, presence of family members in that place; 24% sufficiently high or distant to be safe from the volcano; 18%, friends, relations or acquaintances there; 9%, followed other evacuees; 7%, followed advice of other family members; and 15%, a variety of reasons. Thus, the existing ties of family and friendship had far more to do with structuring the flight of refugees than did public officials.

Although some of the above examples provided considerable information about evacuation during volcanic eruptions, the data are rather piecemeal. Carroll and Parco's study is the only one which looks specifically at sociological aspects of evacuation in the face of volcanic threat. Clearly, the success of an evacuation attempt will depend on the immediacy of the threat, the cultural background of the potential evacuees, their perceptions of the risk, the inducements offered and a host of other factors. Although some doubt has been cast on the relevance to other areas of United States research findings regarding evacuation, a summary checklist is provided from Perry's (1979) conclusions based, particularly, on experience with floods, hurricanes and human-caused disasters; at the least this list suggests points which need to be considered in the design of an evacuation plan:

1. Enforced evacuation is relatively inefficient; enforcement means convergence of enforcement officers and mass arrests of recalcitrants, when the aim is evacuation.
2. Voluntary evacuation often requires incentives.
3. Evacuation plans must be formulated and communicated to potential evacuees long before the hazard impact occurs. A warning should then stimulate recall of the evacuation plan.
4. Evacuation is dependent on effective communication and transportation.
5. Evacuation routes must be specified in warning messages and must remain clear until evacuation is complete.
6. Most potential evacuees seek confirmation of an evacuation order from neighbors, relatives or officials.
7. Separation of family units during evacuation creates anxiety and attempts to return to the evacuated area.
8. A large proportion of evacuees do not use public shelter facilities, but stay with friends and relatives.
9. Evacuees worry about the security of their property.

4.5 PANIC, RUMORS, CRIME

Fear, panic, rumors, looting and increased crime rates are the subject of frequent media reports in the aftermath of natural hazard impacts. Examples of reports of each following volcanic eruption are found in the literature but it is worthwhile stating the results of American research first.

Numerous American investigators have observed that panic flight is rare in natural disaster situations in the United States.* Quarantelli and Dynes (1972: 67) believe "the most widespread myth about disasters is the belief that people will panic in the face of great danger". "Panic" here refers to extreme agitated behavior that increases the danger to themselves and others. Despite the accumulated evidence to the contrary, 84% of the Delaware sample of Wenger *et al.* (1975: 42) believed that a major problem officials confront when faced with a natural disaster is controlling the panic of people fleeing from the danger area.

Similarly, 65% of the Delaware sample disagreed with the statement "looting rarely occurs after the impact of natural disasters", while 51% believed "the crime rate of a community usually rises after it has experienced a natural disaster". A further 41% were undecided or made no response to the latter statement. Despite these views, there are very few verified cases of looting in the aftermath of natural disasters in the United States (Wenger *et al.*, 1975: 35). Furthermore, it appears that disaster victims are so anxious about possible looting that they tend to believe and spread unfounded rumors that confirm their fears (Quarantelli and Dynes, 1972: 69). Various studies indicate that the crime rate usually falls during the emergency period, at least in the United States.

A questionnaire survey undertaken in Sydney, Australia, produced similar responses to those found in Delaware (Blong, in preparation). Clearly, widely held views about crime rates and panic are at variance with the conclusions reached by detailed investigations. As with widely-held views about evacuation, these items of misinformation almost certainly stem, in large part, from rumors and errors reported in the media (Quarantelli and Dynes, 1972, 1977; Wenger *et al.*, 1975; National Academy of Sciences, 1980).

Before embarking on the detail of examples concerning each of these matters it is worth pointing out that many statements about panic, looting and crime are the product of individual opinion, perhaps biased by observations in only part of the area suffering from hazard impact, or by observation for only a limited part of the disaster aftermath. Opinions will also be colored, of course, by racial, political and religious backgrounds, and by prior experience. Two statements about the behavior of the local populace in the aftermath of the 1906 eruption of Vesuvio illustrate part of the range of possible perspectives. Perret, an American volcanologist working at

*It must be emphasised that this observation applies to *natural* disasters and not necessarily to human-caused and/or civil disturbances, race riots, etc.

the Vesuvio Observatory at the time the eruption began, reported that the behavior of the Neapolitans was admirable "and a greater patience, resignation and 'savoir faire' could hardly have been expected of any race. An interesting feature noted by the writer was that the children had, so to speak, taken charge; it was they, rather than their elders, who were directing the flight, suggesting destinations, urging on the beasts" (Perret, 1924: 47-48). On the other hand, Lieutenant-Colonel Delmé-Radcliffe, military attache to the British Embassy in Rome, made several observations imputing a different character to the affected communities. On April 11, he noted, there was some disturbance in Naples owing to the mob endeavoring to break into the Prefecture. The whole town was almost immediately occupied by troops and there was no further disturbance. The troops' manner with the people was quiet and tactful (Delmé-Radcliffe, 1906: 1-2). The military attache also commented on conditions in the Terzigno–San Guisepe area (Fig. 3.3): "The troops had cleared the road partially, and then civil labour had been requisitioned, but the country folk did not display a good or self-reliant spirit. They remained looking on as long as they could at the soldiers working to save *their* houses, and asking how much they were to be paid for helping them. The officers several times told me that they showed no disposition to help themselves and no appreciation of all that was being done for them. Still the troops everywhere continued to work for them, and to feed them, with great consideration and patience. Some of the natives, however, had been made to work in Terzigno, clearing the roads &c" (Delmé-Radcliffe, 1906: 3).

Although the observations of both gentlemen may well be accurate they certainly create differing impressions of the attitudes of the afflicted population. Without prejudicing the veracity of Delmé-Radcliffe's comments, it is interesting to review these comments alongside the epigraph to this volume.

This example, and several others presented by Blong (1982: 172-176), indicate that comments in the literature about social aspects (and no doubt, other aspects of volcanic eruptions) should not be accepted uncritically as encompassing the total reality of a situation. Unfortunately, this comment must also be borne in mind throughout this chapter!

4.5.1 Panic and fear

Five examples illustrate the range of comments available in the literature.

Taal, 1965. According to *Newsweek* (October 11, 1965): "In dozens of mainland villages people ran panic-stricken from their homes, faces swathed in towels to protect them from the searing hot ashes that floated down like snow. On the island itself, the few terror-crazed survivors fought, shot and hacked each other to death for the few boats still intact. In one case, five armed men mowed down a family of seven to get possession of their canoe . . ." There is little reason to doubt that this report is highly exaggerated and sensationalised. However, Carroll and Parco (1966: 11, 15-16) note that there were cases of genuine panic, defined as highly emotional behavior excited by the presence of an immediate severe threat, and which results in an increased danger to that person and to others. They also suggest that there may have been more real panic at Lemery (see Fig. 3.13) on the mainland than there was on the island, because there, neither the precise danger nor what should be done were evident (Carroll and Parco, 1966: 13-15). However, as reported in Chapter 4, Section 4, most people fled in family groups and normal social roles continued to operate.

It is interesting to compare these comments with Colonel W.C. Rivers' (1911: 17-18) description of the survivors some days after the 1911 Taal eruption. He noted that the people accepted their misfortune without complaint, so far as appearances went, showing no signs of emotion at all. They seemed to take the disaster as a matter of course, and made no request for assistance (see also Heiser, 1936: 71).

Katmai-Novarupta, 1912. Erskine (1962: 165) reports that the Indians in Kodiak, where 40–50 cm of tephra fell, gave up, stunned with fear and beyond hysteria. Everyone turned to the prominent citizens (including Erskine's father!) for guidance. Many people who lost their way in the total darkness during the tephra fall, just lay where they stumbled and fell; they could not see, they could barely breathe and they were prostrate with fear. Continued tolling of the church bell, ordered by the local priest, gave hope and a sense of direction. Many people apparently experienced vertigo/dizziness/disorientation during the darkness, but those who spent several hours guiding others to safety reported an increased sense of touch (Erskine, 1962: 92, 100). The presence of the revenue cutter, *Manning*, which would have allowed rapid evacuation in the event of another ashfall, provided a boost to the town's morale; it seems likely that Kodiak would have been abandoned if the cutter had not been present (Erskine, 1962: 198-199).

Soufrière, St Vincent. During the first two or three days of the 1812 eruption the Caribs settled at Morne Ronde at the foot of the Soufrière, abandoned their houses, livestock and possessions and fled toward the town. The blacks became confused, forsook their work, looked up to the mountain, and, as it shook, trembled with dread (*Evening News,* June 30, 1812, quoted in Huggins, 1902: 28). Similar reactions are accorded the 'native' survivors of the 1902 eruption. On the day after, every noise struck terror into their minds and they generally showed a great aversion to revisiting the scenes of their suffering. They seemed to adapt readily to their new surroundings on the southern part of St Vincent. Although the (British) government wished the survivors to emigrate to Jamaica, a public meeting appealed against the edict stating that there were abundant funds and stores to feed them and that they loved their island and did not want to leave. Gradually, they lost their fear and returned for a few days at a time to their northern homes (Anderson and Flett, 1903: 413, 418; HMSO, 1903: 84-5, 104-105).

Mt Pelée, 1902. On the night of May 5 (three days before the main eruption) terrible detonations were heard at short intervals in Fort-de-France some 25 km south from Mt Pelee. Electric lights in the town failed but the town was lit by flashes of flame from the mountain. According to Hill (1902a: 248), terror-stricken inhabitants rushed for the hills screaming and wailing. People in night clothes, carrying children and lighted by any sort of lamp or candle they had caught up in their haste, ran out into streets wailing and screaming and running aimlessly about the town. The mental strain became unbearable. The *Topaz* made ready at 3 a.m. with refugees aboard and left at 5 a.m. for St Lucia. Jaggar's (1904: 220) report on reactions in Fort-de-France on the morning of May 8, 1902, conveys a quite different impression. Before the eruption a singularly oppressive stillness was noticed and then the eruption cloud with a lightning display was observed: "When the phenomenon was first observed in the streets, some slight disturbance akin to panic had been audible, but the people as a whole had grown so callous to these happenings, that after a few moments very few individuals seemed to take any note . . ." It should be understood that neither Hill nor Jaggar was on the island at the time of the eruption. Perret (1950: 151-153) made an interesting comment on the effect of his presence as resident volcanologist during the 1929 eruption series on Martinique: "A science-controlled psychology eliminates panic fear, and I am glad to testify that, everywhere, I have found the population intelligent and avid for knowledge. The people are to be trusted."

Krakatau. Reports from around the volcano during eruptions in 1928 and 1972 provide an interesting contrast. Jaggar (*Volcano Letter,* March 8, 1928) notes a report in the *Paris Times* of January 24, 1928: "the refugees from the southwestern coast of Java [40–50 km from the volcano] were coming into Batavia [Jakarta], that the government had sent a naval vessel for rescue and investigation and that an enormous column of steam and fire was rising 200 to 300 yards above the top of the crater. Terrifying rumblings were heard under Sunda Strait on all sides of the Krakatoa island-ring. Natives were said to be standing in terrified awe, assembled in groups along the coast of Java and Sumatra, panic had caused the inhabitants of Coneet Bantam in southern Sumatra to flee into the high regions inland, the men venturing back to the coast only during daylight, and local insurance companies have increased their rates for volcanic insurance". In June, 1972, there were weak tephra eruptions but the reactions of the populace were evidently different: "People living on the coastal areas of western Java and southern Sumatra find the volcanic activity more or less routine and no cause for fear" (SICSLP, 1972: 139). It is difficult to draw many conclusions from these examples. The contrasts in the 1928 and 1972 reactions may result from differences in eruption magnitude — the 1972 eruption was a minor affair — or from the fact that the memory of the 1883 cataclysm was still fresh in 1928 whereas by 1972, no survivors of the nineteenth century events remained; the effects of the 1883 eruptions were now just stories passed down by parents and grandparents. In some other examples, contrasts in reported observations suggest different times and different observation points or, perhaps, differing interpretations of the same scenes. Some reports contain overtones of white superiority. Some suggest that only a small proportion of the population actually exhibited panic behavior. The latter, at least, are in accord with disaster studies which show that: "Solo or collective panic flight, in fact, is so rare as to be an insignificant practical problem" (Dynes and Quarantelli, 1976: 235). Public beliefs in widespread panic in natural disasters are but one of several disaster myths.

4.5.2 Rumors

Rumors, misinformation and contradictory statements abound during volcanic eruptions. Of course, rumors are not spread only by the media, but the media has the potential to misinform large numbers of people. As Rosenblum (1979: 25) noted, if a UPI or AP correspondent gets a story wrong, more than a billion people are likely to be misinformed.

Rumors are no doubt common during volcanic eruptions but only a few are presented here. Two examples originated with eruptions of Sakura-jima. In the 1779 eruption the population of Kagoshima was said to have been thrown into confusion by unfounded rumors of a tsunami (Kotô, 1916: 46). In 1914, rumors of scorching poisonous gases and more tsunami helped persuade many of the citizens of Kagoshima to evacuate inland (Kotô, 1916: 63-64). On the other hand, the collapse of a basalt cliff during the major earthquake damaged the railway to the north and the telegraph. This effectively cut Kagoshima off, causing rumors around the world that 73 000 had been buried alive and the city annihilated (Kotô 1916: 63-64, 69; Jaggar, 1924: 447).

Some sensational and entirely erroneous accounts appeared in American newspapers following the 1883 Krakatau eruption, suggesting that part of Government House in Batavia (now Jakarta) had collapsed (Tenison-Woods, 1884). "The first accounts brought to Europe stated that Thwart-way Island, situated at the eastern end of the Strait [between Anycr and the southernmost point of Sumatra — Fig. 2.24], had been split into five portions; that the Cape of Anjer had been sundered by a great fissure; that a number of small islands had subsided; and that no fewer than sixteen volcanoes had burst into eruption within the Strait" (Judd, 1888: 2). In 1899 the *Baltimore Sun* (July 30) noted that it saw as exaggerations reports that there were grounds for fearing the crest of Mauna Loa (4170 meters above sea level) had been blown off, that the sea had flooded the blazing crater, and that the island of Hawaii was shattered!

In the immediate aftermath of the May 8, 1902, eruption of Mt Pelée in the Caribbean, the *U.S. Dixie* sailed from New York with American disaster relief and a group of volcanologists aboard. Both before the *Dixie*'s departure on May 10 and *en route* to Martinique rumors abounded. Hill (1902a: 225-226) reported the following rumors: the top of the mountain had been blown away (partly true); the island had decreased in area by a half; the entire north coast had disappeared; the sea bottom had sunk 1000 feet [300 meters]; streams of lava were flowing; the configuration of the entire island had been changed; Fort-de-France (25 km distant) had been burned; volcanoes had been found in Nebraska and upon sedimentary islands in Alaska; and floating bodies and debris had been picked up far from the scene (possibly true).

It might be uncharitable to note that Hill also records that 150 were killed on May 5 in the River Blanche lahar (Hill, 1902a: 248), whereas most other commentators put the death toll at around 30. It would be fairer to note that the recording of a certain amount of misinformation and rumor is almost inevitable given the confused, chaotic excitement of a disaster aftermath and the frequent need to meet news deadlines.

Serious misreporting of death tolls occurred in the month after the 1965 eruption of Taal. An investigation by a University of Philippines Institute of Mass Communication team (Anon., 1965) produced the casualty estimates from six Philippines' newspapers on which Figure 4.5 is based. Tolentino (1973) gives the death toll as 150, Moore (1967: 337) at 189 and Carroll and Parco (1966: 7) at about 250. The very high estimates in Figure 4.5 apparently all stem from the one source — the Governor of Batangas Province (where Taal volcano is located), Feliciano Leviste. The University of Philippines investigation concluded that: (i) sensationalized reports and conflicting data erode the credibility of newspapers; (ii) the problem of conflicting estimates appears in other areas such as fire and typhoon damage, and the sizes of crowds; (iii) newspapers should study ways of checking casualty data as soon as possible; and (iv) newspapers should err on the side of caution, perhaps withholding the release of casualty figures until they have been checked (Anon., 1965: 2).

It should be noted that very similar problems with casualty estimates occur in the United States (see Dynes and Quarantelli, 1976) and, no doubt, most other countries.

Although the evidence cannot be presented here, it seems likely that Mount St Helens will win the prize for the volume of misinformation created. This is less a reflection on Americans and American journalism than it is on the sheer volume of information produced on the eruption. Much of the misinformation stemmed from scientists who really knew very little about volcanic eruptions and their effects but were given the opportunity by the media to have their say. With so many media outlets in attendance, it was easy for a large number of people, who should have remained silent, to have their opinion heard.

Table 4.5, reproduced from *The Spokesman Review,* Spokane, contains a list of contradictory remarks released in the media in the week following May 18, 1980. Some such contradictions were probably of little direct consequence. For example, citizens of Ritzville were told to wear wet rags as makeshift masks. Later they were

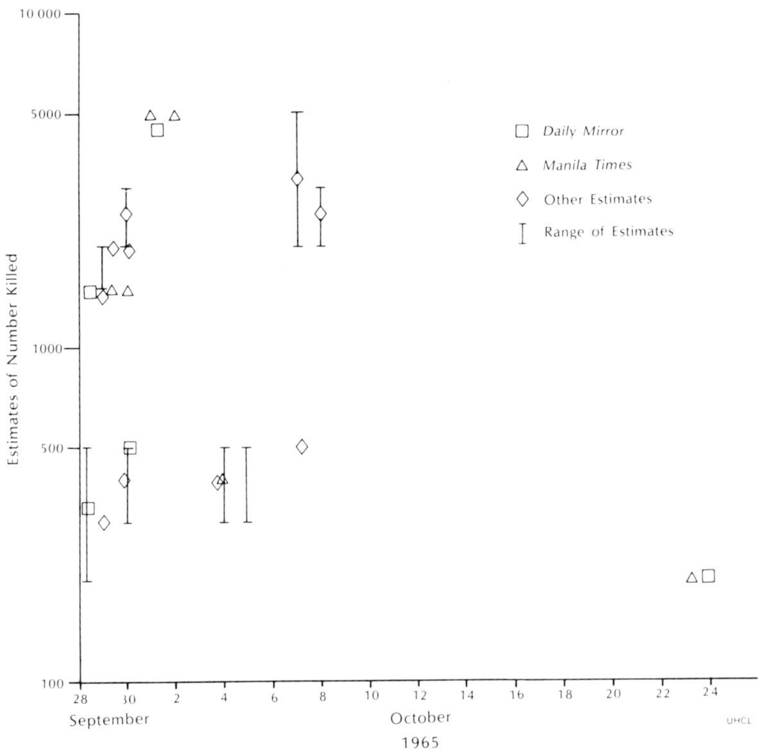

FIG. 4.5 Estimates of death tolls presented in various Philippine newspapers following the 1965 Taal eruption (data from Anon, 1965). The actual death toll was probably 150–250.

told that dry rags were better, but later still it appeared that wet rags were better after all (*Volcano,* 1980: 68). Directly, the consequences may have been slight, but indirectly, such contradictions added considerably to the frustrations of the cleanup campaign.

Radio announcers and citizen's band (CB) operators contributed their share of misinformation and rumor as well. Rumors that the south side of Mount St Helens was bulging (Miller *et al.,* 1981: 797) or that a major eruption had started again were evidently the work of a crank Citizen's Band radio operator on the night of May 22 (*The Columbian,* May 26: 10). Around Ellensberg (see Fig. 4.3) confusion about highway conditions stemmed from erroneous reports by CB and ham radio operators. These reports proved difficult to control and counteract (Warrick *et al.,* 1981: 21). In the same area, a radio announcer unilaterally urged residents to conserve water. It took 24 hours to countermand his announcement and to convince citizens that using water to clean up the tephra would not jeopardize domestic supplies (Warrick *et al.,* 1981: 25).

Table 4.5 Contradictory remarks published in the media within three days of the May 18 Mount St Helens eruption 1980

Questions
Scientists have answered these questions several ways:
Will Spokane receive any more ash after Sunday's initial flurry?
(1) No. It's all over for now.
(2) Yes. Spokane can expect more ash fallouts from time to time for: (a) several days; (b) weeks; (c) months; (d) years.
Is the ash a health risk?
(1) No. There's no danger to people.
(2) Maybe. We're not sure about the health problem. Protect your lungs.
(3) Perhaps. If you have a breathing problem, stay out of it.
(4) Get out of town fast.
(5) None of the above.
How do I protect my lungs?
(1) Use a damp cloth.
(2) Don't use a damp cloth, the moisture and ash will create a dangerous acidity. Use only a dry cloth.
What about plants?
(1) Spray the ash off with water.
(2) Don't try to wash dust from lawns or crops with sprinklers. Moisture will damage plants.
What about livestock?
(1) Leave 'em out. The ash is "not an irritant" to their skins.
(2) Clean an area and move animals to it. Feed them only clean hay protected from ash fallout.
Is the hay crop I just cut ruined?
(1) Yes. Don't bale it.
(2) Perhaps.
How about trees and shrubs?
(1) Wash them.
(2) Don't wash them. Wait until the wind blows it off.
(3) One scientist (this is for real) advised shaking trees to knock the ash off.
Should orchardists turn on wind machines to blow away the dust?
(1) Yes. It's the best way.
(2) No. It'll just ruin the motor.
And the house?
(1) Dust and vacuum inside.
(2) Don't vacuum, dust will clog the machine's filters. Dust only.

From *The Spokesman Review,* May 21, 1980: C1-C2.

More serious instances of misinformation surfaced with rescue efforts west of Mount St Helens. Officials referred people around and around. Some anxious relatives became so frustrated they launched their own searches without permission — and without success (EH & WS, 1980). One family was twice told their parents were safe, once that they might be dead and once that only one was dead. The family became so angry and frustrated that one member joined a helicopter search and another two took part in a ground search (*Volcano,* 1980: 65).

The misinformation amongst officials was also serious. During the first three days after May 18, various agencies were still unsure who was co-ordinating rescue efforts. At first Washington State Department of Emergency Services thought that it was in charge of responding to the disaster. It took three days to establish that the US Forest Service was in charge of the rescue effort (Kilijanek, 1981: 72f).

Late in 1983 the headline "Volcano rumors cause panic" appeared in the Papua New Guinea national daily *Post-Courier* (November 4: 2). The following article, based on the increasing instrumental evidence that an eruption might occur in

Rabaul (Dr P. Lowenstein, Rabaul Volcano Observatory, personal communication), indicated that various organisations had received dozens of phone calls requesting further information and noted that: "Some people said they were told the eruption would be between November 11 and 22". No other evidence was presented to support the leading paragraph: "Speculation on the date of the anticipated volcano eruption in Rabaul has caused widespread confusion and panic among the people". On page 19 of the same issue it was reported: ". . . the people of Rabaul, especially Tolais living around the Gazelle Peninsula do not seem to be alarmed at talk of another major eruption. They still go about their normal day to day activities even though an eruption has been predicted".

Although it might seem that rumor, misinformation and a "credibility crisis", as *The Spokesman Review* called it, are an integral part of disaster aftermath, we might note that *The Daily News* of Longview won the Pulitzer Prize for the local spot news category for its Mount St Helens' coverage. That prize was won, in part, because of the accuracy of its scientific reporting — apparently every scientific statement in the paper was checked with multiple sources before and after publication. Media-spread rumors and misinformation may be a necessary adjunct of disaster reporting, but they can be minimised.

4.5.3 Crime and Looting

As noted in the introduction to this section a majority of the population believes that looting occurs and that the crime rate in the afflicted community increases after a natural disaster impact. Such myths are sufficiently widespread that Perry (1979), for example, suggests leaving a proportion of the populace in the evacuated areas in order to allay such fears and to free enforcement agencies for more important tasks, even though such vigilance seems unnecessary (Wenger *et al.*, 1975: 35; Quarantelli and Dynes, 1972).

Assertions that looting and other crimes have occurred following eruptions appear in several publications. Delmé-Radcliffe (1906: 2) notes that the troops experienced some difficulty in keeping evil-disposed persons away from abandoned houses in Boscotrecase and Boscoreale (see Fig. 3.3) following the 1906 eruption of Vesuvio. Similarly, around Taal in 1911, Rivers (1911: 15) reported that constables rushed to villages to guard property in houses from which the populace had fled. Again in 1965 villagers, returning home temporarily to help others, found some evidence of stealing so they decided to remain in order to look after their possessions. Two water buffalo were also reported stolen (Carroll and Parco, 1966: 31). On the other hand, Jaggar (1924: 455) reports that there were no looters in evacuated Kagoshima when Sakura-jima erupted in 1914 although shops were nailed shut, soldiers stood on street corners and guards patrolled.

The most exciting account of looting and its punishment comes, however, from the Earle of Winchilsea's account of the eruption of Etna in 1669. The portion of the account reproduced here refers to the city of Catania on Thursday, March 14, 1669:

> Whilst the People were thus busied in their Devotions, and astonisht by their Feares, News was brought to the Magistrates of the City, that a considerable number of Thieves and Robbers had taken the opportunity of this general Distraction, to make a Prey of the already distressed People, and that they had Murdered several of them for their Goods; and that is was to be fear'd, that the City of *Catania* itself might run some danger from the great numbers of them which were about the Country, and from thence took their opportunities to get into the Town. Whereupon, consultation being had for the prevention of further mischief from them, the Commander of the Castle was Ordered with a considerable number of Horse, and a Party of *Spaniards,* to secure the Country and City against these Robbers; who immediately sent out several Parties with his Provost-Marshall, with Order to seize on all suspected Persons, and such as were not able to give a good account of themselves: and, for such as were taken in the Fact Robbing, to Execute them by Martial-Law, without any further Tryal; and accordingly caused three pair of Gallowes to be set up for their speedy Execution; one before the Gate *Di Aci,* a second in the *Market-Place,* and a third before the Gate *Della Decima,* setting strong Guards upon the Gates of the City, and causing all suspected Houses to be searched, an Account to be given in of all Lodgers, and such persons to be secured, and could any wayes fall under a Suspition.
> [Winchilsea, 1669: 25-26]

Another account detailing looting concerns the Tyringin area of west Java after the 1883 Krakatau eruption and associated tsunami:

> The fields, strewn with corpses, offer an unusually animated scene in spite of their horrors. Controleur Tromp has taken vigorous action for burying the corpses at the rate of 5 guilders apiece. For this purpose, men are called up from each village separately. But along with the gravediggers, thousands of people come provided with handspikes or sharpened stakes to be serviceable to them as implements either in searching for their own property or in robbing what belongs to others. For robbery is the only object for which hundreds of all ages and both sexes are disposed to venture going to Tjuringin, where but few police are available. On my approach a group of them dispersed, to return forthwith, like hyenas scared from a battlefield, should the check on them be removed. Many gangs remained, robbing on as if they were owners of the articles which they dug up. Some of the robbers are aware of the whereabouts of the missing Government cash boxes. At least this must be inferred from the considerable amounts of money found on several persons who were arrested near Pandeglang.
> [Dr Sollewyn Gelpte, 1883, quoted in Simkin and Fiske, 1983]

Evidence of looting and crime during and after the May 18, 1980, eruption of Mount St Helens have not, to my knowledge, been reported. However, looting was evidently at least suspected by some owners of damaged property in the Toutle Valley. One farming family in the Toutle Valley with more than 80% of their land buried under lahars reportedly moved into a motor home on the property to stop looting and vandalism (Gamroth, 1980: 10). The house shown in Figure 4.6, damaged by a lahar on May 18, had the sign shown in Fig 4.6b attached.

Some data on crime are available from the city of Yakima (see Fig. 3.8). One burglary suspect was caught on Monday, May 19, after detectives were able to follow his clear footprints in the freshly-fallen tephra all the way to his mobile home and to match the prints with a pair of shoes (*The Yakima Herald-Republic,* May 21, 1980). According to this newspaper, crime in Yakima on Sunday, May 18, was much lower than usual for a Sunday but that crime was back to normal by Monday,

A

B

FIG. 4.6 (top) This house in the South Fork of the Toutle River Valley was half-buried by lahars on May 18, 1980. (bottom) A sign attached to the house. The photo was taken in May, 1981.

except for car thefts which remained above normal for three days. The above-average incidence of car theft has been attributed to the notion that the frequent warnings against driving one's car in the dust suggested that it was better to drive a vehicle belonging to someone else! However, Yakima City Police Chief La Rue indicated that there was no evidence of an increased rate of car theft after the tephra fall, and that there was a general decrease in street crime and misdemeanour crime (perhaps a 60%-75% decrease), perhaps lasting nearly a month (La Rue, personal communication, May, 1981). In fact, a general decrease in crime makes good sense in the tephra fall area as everyone in the community was involved in the ash clean-up.

Officers of the Central Washington Comprehensive Mental Health Service had a sense (but no hard evidence) of increased drinking and domestic violence, occasioned by the long periods of time families spent together under trying circumstances during the cleanup period (Fran Wilson and Su Warren, personal communication, June, 1980). While police documentation presents no support for this contention (La Rue, personal communication, May, 1981), neither does it deny the possibility. In general, Yakima and other communities in the tephra fall zone (for example, Cheney — Anderson, 1982: 187) experienced an excellent community cameraderie during this period. Such cameraderie and altruism is typical of the immediate post-disaster impact period (Dynes and Quarantelli, 1976; 1977).

4.6 DISASTER RELIEF

Disaster relief has two broad aims: (i) to save life and salvage what can be salvaged from the immediate crisis, and (ii) to increase disaster resistance by improving infrastructure (that is, ameliorating the impact of the next disaster). The second of these aims is considered, in part, in Chapters 5 and 6 — our current concern is only with the immediate salvage operation.

Disaster relief is a complex topic. The nature of the relief effort changes with time after the disaster. There is nearly always a mass convergence of people and goods into the disaster area, even though neither are required, or at least not in the numbers/amounts that arrive. Survivors are frequently cast in the role of "victims", whether or not they desire assistance. Relief agencies and politicians may compete with one another in the provision or offer of assistance, both groups vying for media attention. Folk from outside the impacted area may arrive and compete with genuine refugees for relief aid. Corruption and/or race relations may affect the distribution of aid. Where international disaster relief is involved, various situations may be exacerbated. International relief may: (i) arrive too late; (ii) not be usable or relevant; (iii) supplant rather than complement local efforts; and/or (iv) be incompatible with societal aims. Past experience indicates that donor countries need to make much more serious efforts to assess international disaster needs (Harrison, 1980: National Academy of Sciences, 1979). There are indeed, many faces to disaster relief.

Table 4.6 Expenditure of Soufrière Eruption Fund, 1902-03 (in English pounds)

Items	Expenditure (£)			
	Oct. 31 1902	Feb. 28 1903	May 31 1903	July 31 1903
1. Burying dead bodies	142	155	158	163
2. Burying dead stock	2	2	37	37
3. Medical aid	516	533	587	587
4. Clothing	212	344	496	497
5. Food and doles, and distribution expenses	5170	7035	7239	7239
6. Temporary shelter	236	572	591	598
7. Landing and storing supplies	110	110	110	110
8. Transport of supplies	226	226	226	226
9. Purchase of land for persons who lived near the volcano	1937	2404	2429	2644
10. Purchase of materials for houses on acquired lands and cost of erection	5974	6910	8592	11927
11. Relief work, roads, etc.	1986	4039	4224	4224
12 Ditto, Georgetown Water Works	10	10	10	10
13. Pay of special constables	15	15	15	15
14. Compensation for losses	2398	2485	3470	3534
15. Settlement of claims for losses		6011	6094	6098
16. Compensation to Crown Land holders			1605	1606
17. Erection of Camden Park bridge			18	203
18. Gratuities for special services	84	113	194	258
19. Sum withdrawn to be expected on experimental cotton cultivation				500
20. Miscellaneous	439	591	1082	1098
	19464	31562	37187	41583

From data in *HMSO,* 1903.

Rather than attempting to illustrate all aspects of disaster relief following volcanic eruptions eight sketches are presented here to indicate as wide a range of disaster relief experience as possible.

Soufrière, St Vincent, 1902. Six months after the major May 7, 1902, eruption, 2000 pounds per month was still being spent on relief, "a large sum compared with what would have been circulated in wages if no disaster had happened, but it must be remembered that all the people had provision grounds which helped out their existence, and unfortunately those provision grounds have been destroyed" (Governor of St Vincent to Prime Minister of Great Britain, in *HMSO,* 1903: 75). Table 4.6 itemises the various categories of expenditure at four different dates in 1902–1903. Food aid, the rebuilding of houses, relief work, and purchase of land and compensation for losses are easily the largest items. The provision of clothing, medical aid and shelter, and the cost of burying the dead total to a relatively trivial amount. The table also emphasises that various forms of disaster relief were still required more than one year after the eruption began.

Mt Pelée, 1902. "Soon after the calamity occurred, M. Maurice Bloch, chief accountant of the Ministries of the Colonies, was despatched to distribute £20 000 to the sufferers, and his accounts were all duly passed by the Colonial

Minister of the day, M. Decrals. But they have been reopened, and strange things now appear. M. Bloch allowed to himself £8 a day expenses, £2 to one assistant and £2.10s to another and treble and quadruple the allowances of Generals or Admirals. Not content with this excessive scale, the allowances dated from leaving Paris instead of from arrival in the colony, the voyage having been made on a warship, where everything was provided gratis. The upshot of it all is that this place knows M. Maurice Bloch no more" (from an undated clipping, *London Globe,* 1905?).

Vesuvio, 1906. Delmé-Radcliffe (1906: 5) assessed those requiring disaster relief as: (i) those rendered homeless; (ii) those dependent on the dead; (iii) those dependent upon the 2700 ha rendered permanently unfit for cultivation; (iv) those dependent upon the 5000 ha around Ottavianio (see Fig. 3.3) and other areas around Vesuvio which had been rendered temporarily unproductive. In all, longer term assistance was required for about 20 000 people and temporary assistance for about 30 000 people. Delmé-Radcliffe's observations of the troops at work assisting the populace led him to comment for future situations: "Any relief measures that may be adopted should encourage the people to work for themselves. At present they seem to think that they can leave all efforts for making good the damage done to the Government, to the troops or charity" (Delmé-Radcliffe, 1906: 5).

Agung, 1963. Although international disaster relief poured into Indonesia for up to 360 000 people requiring assistance, the greater part was siphoned off in Jakarta and Surabaya for sale. According to Jennings (1969: 53,57) only a small proportion of the aid reached the victims.

Irazú, 1963–1965. An airport sweeper, donated to assist in cleaning tephra from the streets of San Jose, was useless as it was too wide to manouver in the narrow streets. Another donation, a 1923 tractor with a truck loading bucket, was delivered in a state of disrepair (Clark and Lee, 1965: 3).

Lamington, 1951. A number of fascinating comments concerning disaster relief after this eruption can be found in the Australian magazine, *Pacific Islands Monthly,* a journal noted for its forthright opinions. There were problems with the local inhabitants who were not allowed to go back to their own villages because of the continuing danger and they "do not want to go anywhere else and if they are forced into new villages they will probably not stay there. Later on they may be persuaded to set up new villages in suitable country outside the volcanic area, but until then they are costing the Australian taxpayers a tidy sum in foodstuffs. It takes a lot of rice to feed 3100 natives [sic], and a lot of close supervising to keep them healthy under camp conditions" (*Pacific Islands Monthly,* March, 1951: 11). Three months later the journal was able to report that the Ilimo refugee camp had been closed at the end of May — an excellent achievement. "Anyone who knows natives [sic], does not need to be told their thousand-and-one objections to settling in even 'kinship' territory, nor the long story of enthusing the young and active men to the job of clearing new areas for homes and gardens. But they have been won into accepting temporary substitute lands" (*Pacific Islands Monthly*, June, 1951: 27). Strong feelings developed over the relief appeal partly because the Territory administration insisted that there should be no appeal for the help of Europeans only but that the local inhabitants should share equally with the Europeans in any public subscription. As a result the town of Lae (second largest in New Guinea) did nothing as a community about relief (*Pacific Islands Monthly*, March, 1951: 85). However, in June, *Pacific Islands Monthly* was able to report that 10 000 pounds had been given for the needs of widows and children of Europeans who died and 5000 pounds for the relief and rehabilitation of the indigenous population. This latter money was given to government officers for the purchase of tools and other equipment. If more money became available, consideration was to be given to causes of hardship amongst Europeans absent from Higaturu (see Fig. 2.15) but who lost possessions (*Pacific Islands Monthly*, June, 1951: 50). These financial matters seem all the more remarkable when it is remembered that only 45 Europeans died, but nearly 2000 of the local Orokaiva people perished and more than 3000 had been in refugee camps.

Taal, 1911 and 1965. In 1911, a house-to-house assessment of losses was made: widows, orphans, destitutes and others who sought out supplies were issued daily with stocks of rice and dried fish. About 500 men took relief work on the roads using funds allocated by the government. All the relief workers had a commendable spirit; they did not spare themselves in their efforts to help the suffering. No salaries or administrative expenses were paid to the relief workers who were all government employees or local volunteers, many from church groups. However, while personal property losses were assessed at about $65 000 only about $16 000 were subscribed to relief funds (Rivers, 1911: 17-18). The most detailed study of the numerous facets of disaster relief following a volcanic eruption has been provided by Carroll and Parco (1966). Nearly all of the following section stems from their work. The 221 refugees they interviewed were probably broadly representative of the 55 000 evacuated from the disaster impact area. Nine per cent of those interviewed had lost a member of their family. About 2.3% had had a family member injured, 67% had suffered losses to crops, 56% to livestock, 40% to houses and furnishings and 15% to fishing boats and nets In all about 13 000 families were displaced. Refugee camps were quickly set up to treat the injured, and to distribute food, clothing and propaganda material for the election campaign then in progress (Carroll and Parco, 1966: 7). *Time* (October 8, 1965: 29) reported that President Diosdado Macapagal left his birthday party to supervise the rescue operation. The refugees needed transportation, food and clothing, money for minor purchases, shelter and medical treatment. The dead 'needed' coffins, disinfectant and gravediggers. Although many studies divide those involved in disasters into 'victims' and 'helpers', this division is far too simplistic. As Carroll and Parco (1966: 22) point out, nearly all of the refugees interviewed had provided assistance to other refugees. As might be expected informants who were community leaders provided more assistance than 'ordinary' refugees. Of those refugees who had returned home at the time interviews were conducted, 11% had stayed one night or more in public evacuation centres; the remainder had stayed with relatives or friends. While it seems axiomatic that the larger the area devastated the smaller will be the proportion of victims having friends and relatives in the immediate hinterland, this figure of 11% is probably also representative of experience in the United States and elsewhere. Carroll and Parco suggest, in fact, that this figure would have fallen to below 10% if disaster relief goods had been more widely available outside the public evacuation centres. Unlike 1911 when it was evidently some days before officials in Manila became convinced that a major disaster had taken place, in 1965 outside aid arrived very quickly. However, Carroll and Parco and their informants expressed several criticisms of the disaster relief effort. Outside aid, including collections by relief organisations and newspapers, exceeded all expectations but lack of coordination between different bodies developed. There were limited communication facilities within the disaster area and between it and the Manila headquarters of various organisations; there was a lack of reliable information as to number of refugees, their whereabouts and their needs. In the absence of any overall authority, developments included a considerable amount of inter-organisational rivalry, vehement differences of opinion, and a tendency a month before the national elections to interpret almost any action of a candidate or government agency in political terms. Organised relief agencies may have tended, because of the lack of information as to the number and whereabouts of refugees, to concentrate too much on the evacuation centres and to overlook those in private homes. Many private citizens and civic groups, on the other hand, insisted on giving relief directly to any and all comers, and jeep loads of real and spurious refugees (*falsificados*) were observed following the convoys of relief goods, their occupants ready to jump out and fall into line whenever the convoy stopped and distribution began. Sometimes, it was asserted, the genuine refugees were too weak, sick or proud to go out and compete for relief with the *falsificados*; it was also asserted that beggars disappeared from the streets of Manila, having found their Eldorado around Taal. It was also thought that many of the genuine refugees were stockpiling relief goods in the belief that the bonanza could not last beyond election time (Carroll and Parco, 1966: 40-41). Refugees interviewed the second week after the eruption were asked "what are your most immediate needs?": 81% replied food; 62%, money; 52%, clothing; 25%, shelter. Carroll and Parco (1966) comment: "It is difficult for one who watched the relief convoys passing through Taal town to interpret this insistence on food as indicative of a genuine shortage in the disaster area at that time. It seems more likely that it should rather be attributed to one or both of the following: the desire of the refugees to stockpile for lean days ahead; and the haphazard system of distribution which resulted in the overlooking of many genuine refugees (and in particular of those who were living with relatives rather than in evacuation centres) or their being beaten out by 'falsificados'". There was an emphasis on money in that refugees needed small amounts of cash for cigarettes which were seldom included in the relief packages and male refugees needed jeepney (bus) fares to commute back and forth from Taal or Lemery (see Fig. 3.13) where they spent nights with families, to lakeshore homesites where they spent their days attempting to salvage and repair what they could. It may also reflect that by the second week the needs of the refugees were becoming quite specific and of a kind which people in Manila would not be likely to think of: feed for the animals which had survived; shovels; brooms; and other tools to clear the ash and make repairs. Carroll and Parco also asked "how should the giving of aid be facilitated?" Table 4.7 indicates that the respondents clearly preferred house-to-house distribution of relief aid. There was also an almost violent insistence on the part of some refugees that politics and politicians should be kept out of aid distribution (Carroll and Parco, 1966: 43).

Heimaey 1973. Those remaining in the town after the evacuation lived very well as everyone had emptied their refrigerators and given the food to the Red Cross mess hall where the volunteers were fed free of charge (Gunnarsson, 1973: 75; G.P.L. Walker, personal communication, 1983). Under Icelandic social welfare laws the evacuees had no doubts that the cost of their food, clothing and temporary housing would, if necessary, be met from public funds. On February 7, 1973, the Althing passed an additional bill which aimed to "share the burden equally among the nation" in providing assistance to the Vestmannaeyar islanders. An emergency fund of US$20 million was created by raising sales tax by 2%, property tax by 30%, and municipal taxes by 10% *inter alia*. Finland, Norway, Denmark, Sweden, West Germany, France, Israel, Holland, the Peoples Republic of China and the American-Scandinavian Foundation also provided financial assistance (*Atlantic and Iceland Review*, 1973, No. 1 Supp.: 1, quoted in Bruner, 1974). The evacuees had little difficulty in finding jobs but housing proved more of a problem. Unused apartments in Reykjavik were quickly renovated, 500 prefabricated houses were purchased from Scandinavian countries and a new home for the aged was built. Children were absorbed into schools in Reykjavik and

Table 4.7 Opinions on how relief distribution could be better organised

Opinions	Male (104)	Female (115)	Key Informants (57)
House to house distribution	70%	68%	35%
Identify real victims	6%	7%	38%
Up to the one giving	8%	18%	0%
Through village officials	8%	4%	23%
Through municipal officials	1%	2%	16%
Through distribution centers	2%	3%	2%
Through parish priests	0%	0%	5%
Through civic organisations	1%	0%	4%
Others	6%	1%	7%
Total	102%	103%	130%

From Caroll and Parco, 1966: 43.

more than 900 children accepted summer holidays in Norway offered by several organisations. As families returned to Heimaey late in 1973 relief fund monies covered many of the transport costs, wages for relief workers and compensation for damaged or destroyed homes (*American-Scandinavian Review*, 1973, 61(3), quoted in Bruner 1974).

Galunggung, 1982. The initial eruptions of Galunggung in 1982 caused the evacuation of about 75 000 people in the regencies of Tasikmalaya and Garut (Fig. 4.7). Because many more evacuees were expected with the widespread development of lahars during the 1982–1983 wet season intensive local and international relief efforts were made to assist perhaps as many as half a million people (*Galunggung*, 1982: 13). The information presented below summarises the situation up to September 1982 — it is intended to illustrate the complexity of a modern international volcanic disaster relief effort. Assistance came from a variety of United Nations organisations (United Nations Disaster Relief Organization [UNDRO], United Nations Development Program [UNDP], United Nations Children's Fund [UNICEF], World Health Organization [WHO], World Food Program [WFP], and the International Labour Organization [ILO]), numerous countries (including the United States, Japan, Australia, Canada, the ASEAN countries, New Zealand, Holland, Sweden and Switzerland) and a number of international agencies (Catholic Relief Services [CRS], CARE, Church World Service [CWS], OXFAM *et al.*). Monetary donations and aid including medicines, sanitation facilities, powdered milk, clothing, emergency shelters and technical assistance totalled more than US $4.4 million. Eighteen emergency shelters were built in Garet and 334 in Tasikmalaya for a total cost of US$0.8 million. Refugees were equipped with rice and other foods, medical and health supplies, clothing and blankets, tents and straw mattresses, kitchen utensils, water pumps, gas lamps, farming tools, sports equipment and religious materials such as the Qur'ān. A daily rice ration, in varying amounts, was issued at first but then general field kitchens were established. When a coupon system was instigated the number of patrons dropped from 33 000 to 22 000 (presumably other refugees were staying with friends and relatives). This coupon system also discouraged the sale of donated goods in the markets. A September, 1982, census established the number of evacuees in the camps at 17 500. It was believed that 45% were capable of looking after themselves or could be relocated to other villages where their merchant and laboring services were needed. The remaining 55%, mainly poor peasants, were encouraged to transmigrate to southern Sumatra, Kalimantan and Sulawesi. It was hoped that this relocation process would be complete by the end of October leaving the emergency shelters ready to receive those evacuated from areas experiencing secondary lahars during the wet season (Sapardjo, Minister of Social Affairs, in *Report and Recommendations of the National Workshop on Mt Galunggung Volcanic Risk Management*, Bandung, September, 1982, volume 2: 7-12). While the Galunggung

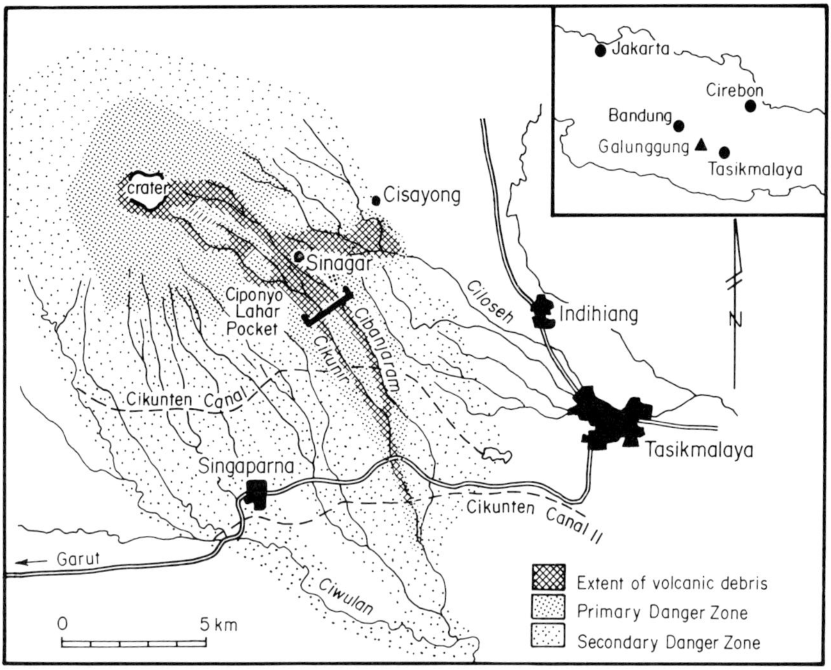

FIG. 4.7 The area around Galunggung showing the extent of lahars in November, 1982, and indicating primary and secondary danger zones (after *Galunggung,* 1982: 24).

disaster relief effort seems to have been quite successful it is also worthwhile looking at it from the perspective of one of the victims:

> We spent one night in Ciamis and then returned to Tasik where we spent two days and nights in my office. Realising that it was not safe to go home, we came to Bandung where we are fortunate enough to have close family.
>
> We stayed in Bandung for two months and waited and waited. Eventually we could no longer bear the separation from our village and rice-fields and so we went home.
>
> They say no one should inhabit this area — we are in the highest risk area [see Fig. 4.7]. Nonetheless, almost everyone has returned, I suppose, now that the soldiers have built a dam to protect this area from mudflows. Although we know this is a high risk area, we feel that there is still some small hope that our house will stay standing.
>
> We now sleep here overnight and only leave if the rains get severe. We have built some temporary shelters on a nearby hill in case of emergency.
>
> There is not much to do in the daytime. I go to collect the meal from the rice kitchen everyday and then I grow what little I can in the garden. Of course our rice-fields and fish ponds are ruined through lack of water, but I manage to grow a little cassava. The men tend to sit around the house or help building the dam.
>
> All our neighbours seem quite healthy. If we go to the barracks near our house, every Monday and Thursday, our children can be given prophylactic injections and check-ups. Those under five years of age are given milk and *bubur kacang jijau* (green bean porridge).

> The Government has said evacuation is compulsory, but we are obstinate people – but where can we go to make a new life? Besides, we feel we will have time to escape any mudflows that come our way.
>
> We have been informed of the safer areas and the Government has told us to move there. We can rent land for 5000 rupiahs a year and build our own houses. I'm not sure whether the Government's advice is valid or not.
>
> There are places in the Government barracks but they are too far away for us to be able to visit our fields regularly and transmigration has no attraction for me. They say transmigration to Sumatra has been successful. I have met people who have come back to take their families to live there with them. They are doing well.
>
> But my house is still here and so I do not want to leave it. Last week Government officers came around to cut down my coconut trees. They need the roots and earth to build the dam. They gave me Rp. 1500 for each tree. I think it's quite a good idea because I can still use the actual trees for whatever I want. I think I will keep them for building materials. I would like to extend my kitchen if we survive this disaster.
>
> [Galunggung, 1982: 17]

4.7 THE EFFECTS ON RESCUERS, OFFICIALS, POLITICIANS

It is not possible to document in any complete way the effects of volcanic disasters on rescuers, emergency service personnel, scientists and politicians. For all of those involved, disaster impact is a stressful time; important decisions need to be made quickly and rationally in the face of a myriad other activities and frequently with neither the necessary information nor adequate sleep. Disaster impact may be a time when careers are enhanced or ruined, and when reputations are made and lost. While all these effects cannot be documented here, some examples indicating the range can be presented.

Rescue workers get tired and depressed. Even well-organised experienced teams bound together by a strong camaraderie and faith, such as the British Salvation Army group which took part in rescue efforts at the Italian earthquake in 1980, reported the surfacing of petty jealousies between those laboring at sorting clothing, those working in the soup kitchens, and those toiling amongst the debris of fallen buildings (Burlison, 1981). Not all those involved can be seen as "heroes".

Volunteers may experience the so-called "Magna Mater complex" in which the volunteer attempts to take on all the problems of the world as his or her own, quickly becoming tired and discouraged with the victim's uncooperativeness and ungratefulness. Volunteers may also develop feelings of omnipotence (the Jehovah complex), with the ability to solve all problems and, thus, reinforcing the victim's feelings of helplessness and passivity (NIMH, 1978: 35).

Similar minor problems occur during volcano disaster rescue efforts. During a visit to an evacuation centre near Mount St Helens, United States President Carter stopped to talk to a woman wearing a Red Cross jacket. She turned out to be a reporter for *National Geographic*. Naturally, the real Red Cross volunteers, who had been working at the centre for days, were rather annoyed (*Volcano,* 1980: 68). In Heimaey in 1973, a few days after the eruption began, a fashion model and film crew arrived from a *haute couture* magazine.

> Dressed in the latest fashion she had a photographer spinning around her like a satellite, snapping pictures for which the spouting volcano provided the backdrop. No doubt the contrast was striking, but few could appreciate the gimmickry. Some young men wanted to give her a dead cat as a farewell present, just to impress upon her that more was at stake here than a fashion magazine feature.
>
> [Gunnarsson, 1973: 76]

Heiser (1936: 71) reported a not dissimilar incident during the relief operation after the eruption of Taal in 1911:

> Newspaper reporters came from all over the East to cover this catastrophe. Among them were two Frenchmen who arrived after most of the rescue work had been accomplished, and who remarked superciliously that it had hardly been worth their while to come. In the endeavour to make it worth their while, fifty stalwart constabulary were selected, who eagerly entered into the plan. The next night we waited until the Frenchmen were asleep, and then cast a rope over the roof of the nipa shack in which they were housed. The soldiers took hold of either end, and stood poised. When the customary *tremblement de terre* came, they began to sway, pulling back and forth on the rope in accelerating tempo. The bamboo creaked and groaned under the stress as the hut lurched dangerously on its stilts from side to side. The soldiers gathered momentum, and it swung faster and farther, until the terrified Frenchmen catapulted out the window, hit the ground, like Antaeus gained new strength, and sped away into the darkness.
>
> We learned later that they had not stopped until they had reached the nearest town, many miles distant, and we read with great interest and amusement their account of the frightful earthquake at the scene of the Taal volcano.

It is likely that a variety of officials get hassled during disaster situations. At Mount St Helens, both before and after the May 18 eruption, officials had to deal with problems associated with road closures and restricted access. It proved impossible to keep people out of the danger zone because of the network of logging roads. Skamania County Sheriff Bill Closner was reported as saying: "It didn't matter what we did. People were going around, through and over the barricades". Closner also reported that law enforcement officers were the target of verbal and written abuse, some of it from Spirit Lake cabin owners. In the days after the May 18 eruption people with property or relatives in the devastated area became increasingly vocal in demanding to get into the area, a Forest Service spokesman noted (*The Columbian*, May 26, 1980: 9,10).

The Galunggung disaster relief example included in Chapter 4, Section 6, illustrates that there are bound to be a lot of organisations involved in various activities during and after a volcanic eruption. Almost unbelievably, Kilijanek (1981: 21) reckoned there were 246 volunteer organisations in Washington State — explorers, civil air patrol, search and rescue dog units, ski patrol, four wheel drive units, diving teams, cave rescue units, and a host of others — which might become involved with country, state and federal officials in search and rescue efforts. Given the number of organisations involved, it is not surprising that leadership conflicts and territorial disputes can erupt. In part such conflicts (and lack of leadership in certain organisations) explain why it took four days after the May 18 eruption for a unified search and rescue effort to evolve. Kilijanek (1981) provides an interesting portrait of inter-organisational jealousies, tensions and conflicts during the aftermath.

Scientists wanting to get into the devastated area around Mount St Helens also became increasingly vocal in their demands. Between April 20 and September 17, 1980, the Mount St Helens Research and Educational Coordinating Committee issued 121 permits covering nearly 500 scientists from 55 academic and 23 other organisations. Some permits were used for five to 15 visits (Allen, 1980; 1982). The number of permit applications rejected was very small. However, in early June, 1980, various scientists at the University of Washington, for example, believed that the US Geological Survey was preventing non-government scientists gaining access to the mountain and generally hindering research efforts. This denial of access supposedly occurred despite an even-handed policy which guaranteed both groups access. On July 26, 1980, an article appeared in *Science News* with the title "The right to research – a government agency may have robbed the US scientific community of a chance of a lifetime" (West, 1980). This paper presents both the Survey and academic viewpoints. Although the Mount St Helens Research and Educational Coordinating Committee had been appointed to review permit applications by researchers, scientists still complained that, even with permits and the required safety and emergency equipment, they were still turned away at the edge of the restricted zone (West, 1980: 61). Survey personnel pointed out that it was the US Forest Service who were in charge of access, not the Survey. However, one researcher said: "They [the Survey] say they don't want duplication in the field. So you try to find out what's being done and they won't tell you. Then you suggest something and they say it's being done. Then you ask for the results and they won't give them to you" (West, 1980: 62). This and other comments from researchers indicate the depth of feelings generated, feelings that had not altogether subsided three years after the initial eruption.

Earlier eruptions have also had their share of scientific controversy. A great deal of controversy stemmed from the evacuation of 73 000 people from the area around Soufrière de Guadeloupe in 1976 for 3.5 months (Fiske, 1979; 1980). As Fiske (1981: 93) noted "On some days, more media attention was focused on disagreements among the scientists than on what was actually happening at the volcano". Scientists disagreed with scientists, leading to acrimony and, eventually, proposals for a deontological code (Tazieff, 1979a; Bostok, 1978). An editorial in the *Journal of Volcanology and Geothermal Research* maintained that scientists must be prepared to take risks and accept criticism (Bostok, 1978). This editorial (written under a pseudonym) also commented on the overdesign of nuclear and other industrial installations because of overestimation of possible hazards. Three volcanologists involved in the original debate made trenchant criticisms of the Bostok editorial in the same journal. Each was particularly incensed by Bostok's (1978: 1) comment referring to Haroun Tazieff: "He was correct of course. Nothing happened. Nevertheless, the intervention of self-appointed experts and the confusion generated by conflicting interpretations resulted in wholesale disruption of the life and economy of the island". There is no space here to assemble all the counter-arguments presented by Sigvaldason (1978), Fiske (1979) and Tomblin (1979), but the latter's final comment is worth repeating:

> A final responsibility is that the volcanologist should remember, and if necessary remind the civil authorities, that the decision to evacuate involves not simply the numerical assessment of the hazard probability but also the offsetting of this against the economic and social consequences of evacuation. Most volcanologists have little or no expertise in sociology or economics and are therefore not in a position to provide the best judgement as to what constitutes the limit of acceptable risk. This is emphatically not a case of avoiding a responsibility: it is a case of limiting one's opinion to one's field of professional competence. An analogy is that in a court of law, an expert witness is allowed to give opinions only on subjects within his field of expertise.
>
> This indicates to me the need for a multidisciplinary team to decide upon the limit of acceptable risk, preferably in advance of any volcanic crisis. On the part of volcanologists, a readiness to present and discuss hazard estimates in quantitative, probabilistic terms will serve the public interest far more directly and will be a subject on which reasonable consensus can be achieved far more easily than the creation of a deontological code.
>
> [Tomblin, 1979: 215]

Whether the debate between volcanologists is public or private, attempts to determine the future activity of erupting volcanoes that are little understood and inadequately instrumented, are likely to prove worrisome for the responsible volcanologists, especially where a large population is at risk. At Mount St Helens the resident population at risk was relatively small but at Guadeloupe in 1976 and on Karkar in 1979 the populations were large. On Karkar, after the deaths of two volcanologists monitoring the activity, plans were drawn up for the evacuation of 24 000 residents. Although activity eventually died down, the volcanologists in charge were under enormous strain for more than six months. Similarly, Sigvaldason (1978) commented on the 1976 Gaudeloupe affair: "It must further be remembered that both scientists and the authorities were working under extremely high tension. When men are strained, inexperienced and faced with an unexpected situation they are likely to make decisions which later prove to be overdimensioned. That is a simple fact of life". Privately, some volcanologists involved in the prediction of future activity at Mount St Helens prior to May 18, 1980, admit that they would have assessed some aspects more accurately, if they had not been so tired, so involved and if they had had time to stand back and assess the whole situation.

Politics and politicians also frequently become involved with volcanic eruptions. Around Taal in 1965 many of the refugees evidently stockpiled relief goods because they believed that the distribution would not continue beyond election time. As mentioned earlier, there was an almost violent insistence on the part of some refugees that politics and politicians should be kept out of aid distribution (Carroll and Parco, 1966: 41, 43). It also seems likely that the very high estimates of casualties during this eruption (Fig. 4.5), all attributed to Governor Feliciano Leviste, were designed to increase the flow of outside aid into Batangas Province.

Naturally, volcanic eruptions give politicians the opportunity to enhance or damage their reputations. During the 1983 Etna eruption volcanologists evidently advised the Italian civil defense that lava flows posed no immediate threat to villages. However, village mayors thought defensive measures were necessary and persuaded the minister, who faced an election, to spend seven times the annual Italian volcano research budget on lava diversion (see Chapter 5, Section 2; MacKenzie, 1983). In Japan local mayors have the responsibility of issuing and rescinding evacuation orders although such orders can have little effect unless coordinated with police traffic control. When Usu erupted in 1977 the evacuation of Abuta (see Fig. 3.4) began about one hour after the eruption on August 7, continuing on the next two days (Iida and Ishimoto, 1978). Although it was 73 days before all evacuees returned home, only 10 days after the eruption began the Toyako-Onsen Tourism Association began pressing the town manager to rescind the evacuation order (Hirose, 1979: 58). The town manager organised a compromise between this viewpoint and those of the Governor, the police and the scientists who thought rescission was premature given the potential for further eruptions. He initiated an evacuees' association and instituted emergency drills, then one hour and three hour visits to the town and, finally, daytime visits on August 23. The rescission order which allowed most (but not all) evacuees home was given on September 7. The police were pressured by the town manager and those running the tourist industry — the situation was thus viewed firstly as an economic crisis and secondly as a threat to public safety (Hirose, 1979: 59). As no further eruptions threatened the Abuta area, the decision was seen as a wise one. The Abuta town manager also commented: "Toyako-Onsen suffered 50 million dollars in damages, but got 150 million dollars worth of free advertising". Certainly he achieved maximum political mileage from the eruptions. Because the town manager listened to the people and sympathised with their desires he gained the near-unanimous support of the town council. In the subsequent election he was re-elected town manager in the first uncontested election in 20 years (Hirose, 1979: 60-61).

The Abuta town manager also made some interesting comments on responsibility, following his decision to rescind the evacuation order: "Heads of municipalities should always be prepared to commit *hara-kiri*. We obey the dictates of conscience and do what we think is right. If we are wrong and the people suffer as a result, then we should be prepared to atone for the error. If you can't accept the responsibility, you shouldn't accept the position" (Hirose, 1979: 59).

Two examples can also be given of international politics and volcanic eruptions. The first example concerns the routing of the Panama Canal. At one stage in the United States Senate debate the Panama route was considered superior to the Nicaraguan route, the chief argument being that the latter was prey to volcanic eruptions. On the morning of the crucial Senate vote one lobbyist sent every Senator a Nicaraguan five-peso stamp picturing an active volcano. The volcano could have been Momotombo, near the proposed canal line and active in January, 1905 (*Time,* August 22, 1977: 28).

The second example concerns an agreement signed on July 1, 1964, between the Indonesian Government and the World Food Program of the United Nations concerning international aid to Bali following the Agung eruption in the previous year.

Under this agreement, the Indonesian Government would undertake the complete rehabilitation of Bali's social and economic infrastructure including: (i) restoration of irrigation works; (ii) clearing of buried land; (iii) rehabilitation of the transport system (including bridge construction); (iv) rehabilitation of the destroyed forest to improve water catchments and to stabilise slopes; and (v) resettlement of displaced persons. Despite the importance of the items in the agreement, the political and economic turmoil in Indonesia in 1964–1965 ensured that little was done (Jennings, 1969: 57).

In many ways politicians exert their greatest influence on volcano affairs through budget strategies. No better example can be found than the annual budgets of the US Geological Survey Volcano Hazards Program before and after 1980. The 1980 and 1981 fiscal year allocations were enormous compared with those for 1978 and 1979. However, politicians have short memories, at least as far as natural hazard impacts are concerned, so many programs in the fiscal years 1982 and 1983 had to be drastically reduced or terminated. This occurred at a time when Long Valley, California, was showing signs of renewed activity, so that many programs on the Cascades volcanoes were cut. It is probably fair to say that politically-enforced budget levels mean that of the seven Cascades volcanoes (St Helens, Long Valley-Mono-Inyo, Lassen, Shasta, Rainier, Baker, and Hood) which can be expected to erupt in the short term only Mount St Helens and Long Valley are adequately monitored. Similarly, in Hawaii, Hualalai and Haleakala, both with recorded eruptions since European 'discovery' of the island chain, are inadequately monitored. Far worse situations exist in most countries where volcanoes have repose period of more than a few years. Nonetheless, politicians with influence on budgets affecting volcanic hazard studies should reread the statement above attributed to the Abuta town manager!

4.8 VOLCANIC ERUPTIONS AND RELIGIOUS BELIEFS

It seems likely that twelfth century Cistercian monks in Iceland deliberately spread rumors that Hekla was the gateway to Hell, if not Hell itself. This tangible proof of the fiery depths was adduced in order to deter sinners from their heretical ways (Thorarinsson, 1970: 5). In the sixteenth century, such beliefs persisted:

> Out of the bottomless abyss of Hekla Fell, or rather out of Hell itself, rise miserable cries and loud wailings, so that these lamentations may be heard for many miles around. Coal black ravens and vultures hover around this mountain and have their nests there. There is to be found the Gate of Hell, for people know from long experience that whenever great battles are fought or there is bloody carnage somewhere on the globe, then there can be heard in the mountain fearful howlings, weeping and gnashing of teeth.
>
> [Thorarinsson, 1970: 6]

A pagan view of an eruption can be cited from the report by Cassius Dio (AD 150-235) of the AD 79 eruption of Vesuvio:

> The day was turned into night and light into darkness. Some thought that the Giants were rising again into revolt (for at this time many of their forms could be discerned in the smoke and, moreover, a sound of trumpets was heard), while others believed that the whole universe was being resolved into chaos or fire... [In Rome] no little fear was occasioned, that lasted for several days, since the people did not know and could not imagine what had happened, but, like those close at hand, believed that the whole world was being turned upside down, that the sun was disappearing into the earth, and that the earth was being lifted to the sky...

Although these early Icelandic and Roman examples are perhaps typical of the role of volcanic eruptions in religious beliefs (see also Vitaliano, 1973), beliefs and superstitions continue to the present day. Moreover, such supersititions are not confined to semi-literate peasants and 'primitives'. Behavioral scientists hypothesise that in times of stress and uncertainty many individuals seek security in supernatural beliefs, rituals and related behavior. Several examples are included to indicate the range of cultures involved. To emphasise that such views are not confined to less developed societies, the first examples come from the United States.

Around Mount St Helens after the May 18 eruption in 1980 some local radio preachers portrayed the death of Harry Truman as God's warning of his anger toward those who cuss and drink liquor (*The Christian Century,* July 2-9, 1980: 695). Even in 1983, two ministers of religion were reported as saying that God was using the volcano to tell Kelso and Longview residents to be more religious, charitable and caring of their families (*The Columbian,* March 31, 1983).

In Hawaii, a lady in a long flowing red robe had a permanent booking on the first special flights to see new eruptions. She claimed to hold conversations with *Pele,* the volcano goddess, from the aircraft (*Aviation Week and Space Technology,* September 9, 1968). During the Kapoho (1960) eruption Lachman and Bonk (1960: 1095-1096) reported that many of the evacuees openly acknowledged a belief in *Pele,* probably reinforced by the apparent inability of science and technology to cope with the destructiveness of the lava flows. Rituals and offerings were made to the volcano goddess. A photograph in their paper showed a Chinese-Hawaiian man reading a Hawaiian language chant to *Pele.* The traditional offerings of breadfruit, bananas, pork and tobacco were at his feet. The Xmas gift wrappings indicate Western influence. The throwing of the offerings onto the lava flow was accompanied by the singing of a Chinese song. Similar small offerings to *Pele* were left on the margins of the 1983 lava flows in Royal Gardens subdivision.

Lachman and Bonk (1960) indicated that such behavior and beliefs were not related to any one religious creed, ethnic group, age level or degree of educational achievement. Believers in *Pele* included some highly educated individuals and some prominent citizens.

The Christian God has often been called to halt eruptions. In 1572, 1591 and 1605–1611 Augustinian priests exorcized Taal island to "tranquilize the natives" and to accustom them to "place their trust in the only God" (Masó, 1911: 6). During the 1897 eruption of Mayon the people of Alby took *Divino Rostro* on a procession while the citizens of Legaspi prayed to St Rafael, their patron saint, to save them (Anon., 1897). In Italy, religion has always played an important part in halting volcanic eruptions. Hamilton (1772) commented that the list of eruptions which he gave could not have been compiled but for the curious fact that sacred images and vestments of the church had been used by priests to halt volcanic fury. A record of saintly interference — the veil of St Agatha around Etna and relics of St Januarius (Saint Gennaro) around Vesuvio — had been carefully recorded by the priests (Bullard, 1976: 211).

Around Etna the efficacy of the veil of St Agatha was tested as early as AD 253 when lava flowed in the direction of Catania. The saint had been martyred only the year before. When the citizens saw the stream of lava approaching they rushed to the tomb and removed the veil which covered her body. This was carried to the edge of the lava flow and is asserted to have at once arrested its progress (Rodwell, 1878: 83-84).

In 1669 the relics appeared again:

> As the Fire approached the Religious everywhere appeared with much Devotion, carrying in Procession their Reliques, especially those of *St Agatha,* the famous Martyr of *Catania,* in which they reposed no small confidence, followed by great Multitudes of People, some of them mortifying themselves with Whips and other signes of Penance, with great Complaints and Cryes, expressing their dreadful expectation of the Events of those Prodigious Fiery Inundations.
>
> [Winchilsea, 1669: 16, 25]

A similar account comes from the 1811 eruption of Etna:

> But when the lava had arrived with two hundred yards of this deserted habitation, the eruption ceased, to the great joy of the natives, who attributed this mercy to the merits and inteference of the Patron Saints, whose images were brought daily from Castiglione (a distance of three miles) in procession, during the progress of this calamity, and placed, while mass was performed, amidst the tears of a wretched multitude, a few yards in front of the slowly advancing fire. This procession was composed of the miserable and ragged natives of both sexes and all ages, crying and sobbing, beating their breasts, tearing their hair, and flogging their backs in penance, whilst their priests were calling on all their saints to assist them. On their way to the lava, they stopped at the baron's house, from the balcony of which the chief priest, with the most violent gestures of grief, delivered a short sermon, in which he told them the eruption was a judgement upon their sins, and recommended them to mend their lives, and to pray to all the saints to intercede for them. Every pause of their discourse was filled with a general burst of tears, beating of breasts, tearing of hair, and flogging of backs.
>
> [Anon., 1811: 219]

The religious influence has continued to the present day. When Etna erupted in 1971, nearly every resident of Sant'Alfio, a small winegrowing village, joined a procession up the mountain. "Praying, singing hymns and carrying relics of their patron saints, the villagers advanced to within a few yards of the glowing smoking wall of lava. As his flock knelt before the threatening stream, Sant'Alfio's parish priest, Don Francesco Parisi, tilted his head skyward and implored God to 'send away this menace from us and from our homes" (*Time,* May 31, 1971: 44). Religious relics have also played their part around Vesuvio (see Hoffer, 1982: 116-120). During the tephra fall from the 1767 eruption the Neapolitan mob set fire to the Cardinal Archbishop's gate because he refused to bring out the relics of St Januarius (Hamilton, 1772: 32). During the 1906 eruption, Delmé-Radcliffe (1906: 5-6) was quite scathing about the population's religious fervor: "... beyond running away, the utmost the people did to help themselves, was to carry about images of their saints in front of the lava in processions which largely consisted of women and girls, with their hair loosened, wailing and singing. This strikes the feminine spectator as very 'touching' and picturesque, but it is really a pitiful exhibition of supersitition, a mixture of vanity, and hysteria on the part of the women and indolence and ignorance in the men. It is also deplorably unpractical".

The 1781 eruption of Sakura-jima greatly frightened the local people causing them to vow to the God of the mountain (Gongen) that they would refrain from fishing on the eighteenth day of every lunar month. More than 130 years later, Kotô (1916: 55) heard people talk of this sacred day which had lately not been strictly observed thus causing the death of Gongen and the 1914 eruption.

This theme is quite a common one. For example, Hibok-hibok, on the island of Camiguin, has caused death and destruction on several occasions "always, said the elders because the God had been displeased by younger Camiguenos, who grew lax in their churchgoing, forgetful of the feast days and neglectful of the sign of the cross. When his children did wrong an elder would glance fearfully toward the horizon and mutter 'The volcano will get angry'" (*Time,* December 17, 1951: 36). Similarly, when Agung erupted in 1963 the Balinese knew that the gods were angry. Once every 100 years festivals are held in the temple of Besakih to appease the gods; 1963 was the time for the festival but the Balinese priests feared that the Gods would be angry because the festival had been started without the usual rites (Mathews, 1965). Although the priests were the first to die the temple of Besakih was still standing after the first eruption (*Newsweek,* April 1, 1963: 36). Balinese mystics evidently interpreted the eruption as foretelling the political demise of the late President Sukarno. Some Javanese mystics believe the 1982 eruption of Galunggung signifies that President Suharto is losing his divine source of power (*Washington Post,* 1982, November 25: G2).

God's punishment also figures amongst the reasons for the eruption of Lamington in 1951 (Keesing, 1952). Some Orokaiva (the local people) thought the eruption "was God's visitation because they had disobeyed the bishop's instructions to build new churches. Others said that God had punished them because they had not helped the allies sufficiently during the war and because some of them betrayed missionaries to the Japanese" [others mentioned the lack of cooperation in mission and government plans for development] (Belshaw, 1951: 242). However, Schwimmer's (1969; 1977) much more detailed analysis provides further insights into Orokaiva beliefs and interpretations of the eruption. It appears that Orokaivans are capable of holding three or four explanations at once, depending on the context.

One explanation was that the eruption of Sambaripa, as Lamington is called by the Orokaiva, was the result of the mountain's anger at noisy disrespectful acts such as the throwing of handbombs near the crater during World War II and the later acquisition of guns which led to hunting. Thus, to some, the eruption indicated the potency of the old order — it fortified ideology. This view was reinforced by the fact that while the Sangara, Isivita and Sasembata populations lived equally close to the volcano, it was mainly the Sangara (who had many hunting rifles compared with the others) who were killed.

Another explanation saw the destruction of the government station at Higaturu (see Fig. 2.15) as payback for hanging Orokaivans judged responsible for killing missionaries. This view probably comes from the Yega people on the coast and not from the Orokaiva as the deaths of 45 Europeans and nearly 3000 Orokaivas is not likely to be seen as payback by the Orokaiva. However, the Yega, traditional enemies of the Orokaiva, suffered only small losses (Schwimmer, 1977: 313).

A third explanation for the eruption was that it was God's punishment. This had been reinterpreted by the mid 1960's as many children had died and God would not have made them suffer. The eruption was then seen as an indication of what the end of the world would be like. As Schwimmer (1977: 317-318) notes, this explanation makes no tribal divisions and omits to mention the deaths of European Christians.

Many years after the eruption a fourth explanation appeared — that the eruption had a geophysical cause. This view was evidently put forward by the missions to counteract criticism from white planters and public servants that the church, by preaching that the eruption was due to God's wrath, was pursuing power politics. It was rumored that the bishop had threatened to blow up the mountain again unless he was obeyed. The bishop denied these allegations and said that it was the Orokaiva themselves who had this idea of retribution. However, the matter is not that simple as an acknowledgement of guilt in the Orokaivan cognitive system places some obligation on God or Sambaripa to forgive them. Thus, there was a transactional element in this sincere state of contrition (Schwimmer, 1977: 322).

Evidently, more than one of these beliefs can be held at the one time. For example, the eruption punished the Europeans for punishing the Orokaivans and the Orokaivans for not being good Christians (Schwimmer, 1977: 327).

Orokaivan beliefs about the causes of the eruption also produced some side effects. Because the government distributed quite a lot of aid immediately after the eruption, this was seen by the Orokaivans as part of an exchange transaction thus placing them under some obligation. This probably promoted native–government and native–mission cooperation (Schwimmer, 1977: 327-330). As we will see in the next section the eruption also had other profound effects on Orokaivan lifestyle.

4.9 LONGER TERM SOCIAL CHANGES RESULTING FROM ERUPTIONS

Apart from the short term social changes resulting from psychological trauma, evacuation and disaster relief some longer term changes may also occur. These changes, although resulting from eruptions, need to be seen in the context of existing social changes. Eruptions, rather than causing such changes, act as catalysts, altering the rate at which adjustments in social and political institutions occur. Because the effects of changes are generally to 'westernise' areas, all four examples presented here are from less-developed countries.

Lamington, 1951. The 3000 refugees were allowed back into the devastated area in 1953. Although villagers were generally able to use the land they had previously occupied, some features were permanently changed including, for example, the location of the administration patrol post, hospital, mission and the network of roads. Relatively large villages were also created, encouraged by the government, so that schooling, clinics and roads could be more easily provided. These larger, relocated villages displayed more innovations (advocated and sponsored by the government) such as coffee-growing, and had higher levels of school education than did surrounding villages. The differences could have been the result of government contact and guidance but they also may have been ecological in that the new villages were closer to wage employment and markets and had a climate suitable for cash cropping (Schwimmer, 1977: 306). One other factor also influenced the receptiveness to innovations. The Sangara people mainly fled in small groups. As it happened, almost the entire older generation was killed by the eruption. This meant that the survivors were not fully instructed in traditional religious knowledge; it also meant that innovation proceeded without the obstruction of the older generation. On the other hand, the Sasembata people mainly fled in village groups led by the village constables. This weakened the leadership of the Sasembata traditional leaders and enhanced the authority of the village constables. This authority was also strengthened by the constables' dealings with the administration in food supplies and house building matters (Schwimmer, 1977: 306-12). For the Orokaivans separation from their land meant separation from their ancestors who were left unattended while the survivors were in refugee camps. At the same time birth rites for infants and the dead became impossible as such rituals must be performed on the taro field associated with certain deceased relatives. Furthermore, absence from the land was dangerous as it could reactivate residual rights to the land by other groups. Fear that the land would be taken by others was sufficiently strong among the Orokaiva in 1966–1967 that they had begun to accept the idea of converting land titles to sole tenure. Acceptance of this idea, with its numerous consequences for kinship organisation, had been hastened by the separation from the land during the eruption aftermath and by the advent of cash cropping (Schwimmer, 1977: 334). By the late 1970s cattle projects had become viable economic propositions in the area; land that was 'borrowed' after the eruption was thus being demanded back by the original owners (Richard Jackson, University of Papua New Guinea, 1979, personal communication). While it appears, then, that the eruption had a great influence on some aspects of Orokaivan life, Ingleby (1966) reported that 15 years after the event few children knew anything about the eruption and few people held any fear of the volcano. However, the eruption added to legends, songs and history. Schwimmer (1977: 317) noted that the songs admit the guilt of the dead but that the strongest emphasis was on the suffering of the living through the loss of 'brothers'.

Parícutin, 1943–1952. The social and economic effects of the 1943–1952 eruption of Parícutin have been documented by Rees (1970, 1979) and Nolan (1979). Around Parícutin, population declined, partly absorbed by larger villages, partly as a result of migration of young men to the United States as laborers (*brazeros*). The local people came into far more contact with urban Mexicans and North Americans in their roles as refugees, tourist guides and research assistants. They also became part of the planned social change co-ordinated by officials of the modernising Mexican nation. Outside aid poured in to the resettlement villages; water storage and piping and electricity were installed, and medical clinics and poultry-raising projects were begun so that the new villages were 10 to 20 years ahead of the other Tarascan villages (Rees, 1970; Nolan, 1979). In this case, the eruption provided a catalyst for forces that were already at work. Nolan (1979) provides considerable detail on the longer term changes in each of the five affected communities (Fig. 4.4). The new settlement of Caltzontzin was formed by Tarascans of Parícutin. They were given new houses that they did not like, and new shoes that most were not accustomed to wearing. Although well-meant efforts were made little attention was paid to their non-economic needs; and culture shock stemming from rapid change plagued the settlement for more than a generation (Nolan, 1979: 303). Between October, 1943, and January, 1944, Miguel Silva was founded by the people of Zirosto and San Juan. By the time a survey of the land was completed more refugees had arrived than the land could support. In early 1944 the local rancheros killed the surviving livestock and the secular leaders of the refugee settlement and threatened the priest who was recalled to another post. With local leadership gone the refugees from Zirosto and San Juan began to fight among themselves. Many of the seeds they brought failed to grow. The water was bad and most of the refugees were sick with intestinal orders and malaria. As many as one in 10 died, especially the elderly who were said to have lacked the will to live under such conditions. Within a year many of the survivors had left Miguel Silva and by 1946 the population had decreased to about 300. Most of the San Juan people went to New San Juan after its founding in May 1944. Zirosto families either went to the United States or returned to the volcanic zone. The few from Zirosto who stayed cut themselves off emotionally from the mother community and came to view themselves as the pioneer founders of a new settlement on a difficult frontier (Nolan, 1979: 305). By February, 1945, much of the drama at the volcano was over and adjustments were under way in the refugee settlements. Calzontzin's citizens began rebuilding their government-designed homes to suit their own tastes; Miguel Silva got a pure water system and favorable government intervention in the feud with the local rancheros. The population stabilised as the people began to think of themselves as pioneers rather than as refugees. As the eruptive activity declined hard times fell on Angahuan which had become economically dependent on the tourist trade. Some men were forced to leave the town as temporary laborers because the lands were still too unproductive. Eventually the tourist trade picked up as people came to see the volcano they had read about in the school textbooks. Zirosto was most affected by social disruption. The town fragmented into two groups as the men fought among themselves for the few available resources. Some moved to New Zirosto, others stayed put. Brothers made different choices placing severe strains on family relationships. It is said in New Zirosto that the move would have been truly successful if all had gone together, taking with them the old bells that symbolised the community. In Old Zirosto, those who left are blamed. It is argued that if all had stayed, the government would eventually have rerouted the road, provided electricity, water, six years of school, and all the other benefits (Nolan, 1979: 311-313). As fertility returned to the areas of moderate ashfall aggression became more intense as formerly useless lands regained agricultural value. The conflicts were particularly difficult to settle because the eruptions had destroyed landmarks traditionally used as boundaries. The most serious conflicts involved New San Juan, Caltzontzin, and the people of ranches founded by both refugee settlements on community lands in the volcanic zone. Court battles were in process by the early 1950s and incidents of aggression were reported in 1959 and 1960. In 1965 the feud intensified and reports of hired gunmen, killings and substantial property destruction were filed with government agencies. In 1967 a government investigator documented destruction of harvests, killing of livestock, and the burning of 128 houses (Nolan, 1979: 313-314). It is also of considerable interest to record the general views of the community years after the event. In general, people in San Juan and Miguel Silva expressed positive views based on success in meeting the challenges of the eruption years. In Zacan there was a strong expression of educational achievement thought to have resulted from an widening of horizons during the volcanic years, but there was also concern about the future because few of the educated young could work in this small, still relatively remote community. In both Old and New Zirosto, emphasis was placed on the 'death' of the old community, sometimes followed by a comment, blaming the other settlement. There was ambivalence in Caltzontzin, but in spite of individual achievements and what seemed to be a new sense of hope and community spirit, there was also a negative side to many comments. In contrast to the new San Juan and Miguel Silva, there was a tendency to state that the old town had been better than the new.

Tristan da Cunha, 1961–1962. The social system of the people of Tristan da Cunha was also severely affected by volcanic activity in 1961 when the entire population of 270 was forced to abandon its isolated island home and nineteenth century lifestyle. They resettled in industrial southern England, changing from being property owners to paying rent and becoming dependent on buses, punctuality and jobs. On Tristan, time had never been much of a factor in their lives; life had been straightforward and simple. In England, life was complicated. "In the welfare state, which they took for granted, there were many authorities to deal with: the labour exchange, the police, welfare office, post office, and so forth, each with its own set of bewildering forms" (Lewis *et al.*, 1972). Not all the Tristan's problems resulted from social upheaval following the eruption but it was then that the problems became manifest.

> Life within the Tristan community, it turned out, results in some lack of personal initiative, very limited competitiveness, and self acceptance. The children were free of anxiety, had a strong sense of belonging together distinct from the 'outsiders', and a certain fatalism or passivity. They are not a talkative people so it is not surprising that the children had diminished linguistic skills, limited imagination, lack of curiosity, and a certain immaturity of thought. Although some were of normal intelligence, there were many among the older children epecially who appeared to be of limited intellectual capacity. They were all slow in the basic school subjects.
>
> [Lewis *et al.*, 1972]

As mentioned in Chapter 4, Section 4 most Tristans returned to their island home in 1963.

Niuafo'ou, 1946. The evacuation of 1300 people from Niuafo'ou in 1946 seems to have been entirely for political purposes (see Chapter 4, Section 4). There was little danger from the eruption that could have threatened the whole island. Just what the political motives were is difficult to fathom but the discussion presented by Rogers (1981) makes it clear that the Tongan Royal Family, the Tongan Government and the British Agent and Consul were not all in agreement. When the refugee Niuafo'ouans arrived in Nukualofa, the capital, in 1946 the Government ordered local citizens to supply them with foodstuffs. Resentment at these demands spilled over on to the Niuafo'ouans. Their stay in old United States Army huts was not a happy one — they remember especially the shortage of food, the cold, and the humiliation of living on charity. After another move most of the refugees were resettled on the island of 'Eua. Land had to be cleared before villages could be built, and the Niaufo'ouans had to work long days for the 'Eua residents to purchase house poles, thatch and seed crops. Antagonism developed between the newcomers and the residents but in 1950 the first families settled into their permanent homes. By 1955 most of the Niaufo'ouans had settled on 'Eua although some people were still waiting for allotments in 1964. Gradually the Niaufo'ouans established their independence on 'Eua with their own churches, schools and shops. Social contact with the 'Euans had also improved.

> They eventually built nine new villages with the same names and in the same geographical sequence as in their homeland, and they set about achieving their independence from local residents and from government aid by working in and through village committees. In Kava circles on 'Eua in 1967 the people recalled the cold, wet winters of 'Eua and the inadequate shelter and poor food; the influenza, bronchitis and other respiratory diseases which carried off a distressing number of old people and babies; the long wait for their bush allotments; the hostility of local residents who measured every stick of wood and coconut in terms of money, labour and economic gain. But with the increase of neat mounds of gravel and sand in village graveyards, there grew up a new Niaufo'ou history in 'Eua and a feeling that this place was home.
>
> [Rogers, 1981: 162]

Not all Niaufo'ouans have settled on 'Eua. Others scattered throughout the kingdom continually pressing for permission to return to their homeland. Permission was granted in 1958, on the condition that everyone lived in the one village. The people readily assented to all government stipulations before boarding the homeward bound vessel but upon arrival at the island they rapidly spread out to reestablish their own villages. Government services on the island were only gradually reestablished over more than a decade. "Thus despite the argument that the people were removed in 1946 because of the threatened suspension of government services the first decade of permanent resettlement vigorously survived without most of them" (Rogers, 1981: 163). By 1976 there were 678 Niaufo'ouans back on the island, and 2108 on 'Eua.

4.10 CONCLUSIONS

In many ways it is difficult to draw adequate conclusions from such a disparate body of information. Insufficient detailed case studies of each aspect have been made by trained observers. Much of the available data reflects strongly the bias of the reporter. Perhaps that leads to the first conclusion — statements can be found that will support almost any interpretation of the social consequences of an eruption.

Sadly, nearly all the data on human perceptions of eruption risk and information on mental health problems stem from the United States. The results seem to conform well with American data on other natural disasters, but leave unanswered the question of applicability to other areas of the world. In general, we might believe that western society is furtherest removed from its agricultural or rural base and therefore understands nature least. While this may be fair comment with reference to relatively frequent meteorological hazards, it may be quite erroneous for societies and areas where eruptions occur but once a century or even less frequently. Some support for this notion comes from the recognition that many of the differences between Pacific Northwest and Hawaiian perceptions of volcanic hazards stem from differences between the two areas in the frequency and character of volcanism.

The studies of evacuation suffer from much the same problems — the general theory is largely American in conception and we are unsure of its general applicability. However, Carroll and Parco's (1966) study of the social consequences of the Taal eruption suggest that there are numerous comparisons; on the other hand, Perry's (1979) comments presented at the end of Chapter 4, Section 4 probably have only limited applicability. Evacuation in the face of volcanic hazards has one major difference when compared with evacuation from other natural hazard impacts — the duration is much less certain. Evacuation from a hurricane or a flood impact is unlikely to be for more than a week but numerous examples presented here indicate evacuation periods of a month or more are not uncommon when volcanoes erupt. Whether knowledge of this influences the willingness to evacuate or evacuation behavior is unknown.

The information presented on panic, rumors and looting indicates that few people panic, that little looting occurs, and that rumors and misinformation are inevitable. Although the available data are limited, they do suggest that volcanic hazard impacts produce reactions similar to those found in other natural hazard impacts. These examples also indicate the prime influence of the media, the role the media plays in perpetuating disaster myths, and that control of rumors and misinformation lies in 'control' of the media. In turn, 'control' of the media implies a strong, well-organised emergency operations centre with informed media spokespersons conversant with media desires for information and media deadlines. Disaster officials and the media alike have been well aware for some time that the quality of disaster reporting leaves much to be desired. The 1980 United States National Academy of Sciences report *Disasters and the Mass Media* makes several suggestions for improving the situation — despite media participation in and agreement to these recommendations there is, so far, little evidence of improved standards.

Disaster relief includes a diverse field of activities. Although relief may be focused on the provision of food, clothing, medical care and shelter the examples given indicate *inter alia* relief work may be required, gravediggers and special constables paid, gratuities presented, transportation provided, corruption and freeloaders dealt with. Aid may still be required over a period of months or years even though media attention has been diverted elsewhere. Furthermore, it should be noted that some prominent authorities regard the current provision of international disaster relief as quite unsatisfactory in many aspects (Harrison, 1980; Anon., 1983).

Relief workers, officials, volcanologists and politicians all stand to gain or suffer as a result of an eruption and its aftermath. The fact that volcanic disasters are a stressful time involving long hours and little reward, or only criticism, for such officials should be recognised by all concerned.

Religious beliefs obviously play an important part in social attitudes for at least a section of the populace. Such beliefs may inspire a fatalistic attitude in the afflicted populace, promote seemingly irrational behaviour, or give some the strength to assist fellow sufferers in the face of danger or adversity. Such beliefs are likely to affect perceptions of the hazard, willingness to evacuate, the acceptability of disaster relief, the apportioning of blame and a host of other social attitudes.

Some eruptions promote long term changes in afflicted communities. These changes, probably greatest and most traumatic in less developed communities, are frequently part of a longer term social and economic evolution; eruptions and their aftermath hasten rather than alter the course of development. The examples presented here also indicate a very strong attachment to 'place'. Evacuation and relocation may have spanned more than a decade, but many evacuees still have a strong desire to return home.

With the available information and space it is not possible to do more than sample social aspects of volcanic eruptions, or to present more than a series of portraits. Although disaster theory is rapidly evolving, few case studies have been made of volcanic hazard impacts, and, as indicated, there may be some dangers in applying impressions gained in other natural disaster studies. Although Mount St Helens has produced a great deal of valuable studies related to disaster theory, investigations outside North America are urgently needed.

5

VOLCANIC HAZARDS AND BUILDINGS

5.1 INTRODUCTION

Volcanic eruptions effect buildings through a variety of damage mechanisms. Buildings may suffer collapse, burial, transport, foundation failure, undermining, racking, excessive wall or roof loads, impacts from ballistic projectiles, corrosion, or catch fire. Several of these damage mechanisms may operate at the one time as a result of the action of one or more volcanic hazards. Some damage mechanisms are obviously more effective in the destruction of buildings than others. Furthermore, the occurrence of individual damage mechanisms varies not only in the frequency with which they are associated with the one volcanic hazard but also from one hazard to another. For example, buildings are more commonly buried by lava flows than transported by them, but transport occurs more often by lahars than by lava flows.

Table 5.1 summarises the occurrence of damage mechanisms in relation to each volcanic hazard. The list of hazards is a slightly truncated version of that presented in Table 1.4 and used in Chapter 2. In general, the damage mechanisms are arranged in order of decreasing damage potential. The numbers in the table refer to the relative likelihood of occurrence of the damage mechanism for each hazard. In the almost total absence of quantitative information the table is but a subjective assessment.

Table 5.1 raises several points of interest. The table suggests that ignition and building collapse are associated with the greatest number of volcanic hazard types but that damage resulting from wall loads and burial or flooding are the most likely to occur. On the other hand, pyroclastic flows, lahars and tsunami are each associated with 10 of the 11 damage mechanisms. These hazards are also the most

Table 5.1 Damage mechanisms to buildings by volcanic hazards and their relative likelihood of occurrence

Damage	Damage mechanisms	Volcanic Hazards							
		Lava flows	Tephra falls	Pyroclastic flows	Lahars	Seismic activity	Tsunami	Atmospheric effects	Acid rains and gases
Total destruction	Collapse/overturning	3	2	3	2	2	3	1	
	Burial/inundation/flooding	4	2	3	4	1	4		
	Ignition	3	1	3	1	2	1	2	1
Severe damage	Uplift/transport	1		3	3		3	2	
	Foundation failure/deformation	1		1	1	3	2	1	
	Racking			3	2	3	3	2	
	Wall loads	4	1	4	4		4	3	
	Undermining/erosion			1	2		2		
Minor damage	Roof loads/		4			1		2	
	Projectile impacts	2	4	2	2	2	2		
Negligible damage	Corrosion/discoloration		2	1	3		2		2

1 → 4: increasing relative likelihood of occurrence
Blank: not applicable or extremely unlikely

likely to cause damage to buildings. Acid rains and gases represent the least hazard to buildings.

Clearly, Table 5.1 presents generalised conditions. The vulnerability of individual structures to volcanic hazards depends on construction style and materials, building age and the quality of workmanship as well as building shape, orientation and aspect ratio. Although buildings in many less developed countries may be of flimsier construction than those common in the western world, they may also exhibit fewer/smaller chimneys, parapets and other vulnerable "decorations", contain fewer flammable furnishings, and suffer less damage as a consequence of ruptured lifelines (for example, gas, electricity and water supplies). Damage resulting from hazard impact may also vary with the time of the day. Even minor volcanogenic earthquakes may lead to severe building damage if ground shaking causes objects to fall when open fires, or gas heaters are alight. Such considerations suggest the general hypothesis that relatively strong structures will survive initial impacts more readily than weak structures, but may suffer more damage as a result of secondary damage to fittings, furnishings and lifelines.

Examples of each type of impact on buildings and their contents are presented in the following sections.*

5.2 LAVA FLOWS

As indicated in Table 5.1 the most significant damage mechanisms resulting from lava flows are excessive forces on walls, burial, ignition and building collapse. Several lava flow characteristics contribute to the destruction of buildings. As flows with high yield strengths exert the largest stresses on walls, thick, slow-moving aa and block flows are more damaging than thin pahoehoe flows. Although the latter flows have higher temperatures (see Chapter 2, Section 1) the temperatures of all mobile lavas are well above the ignition points of flammable construction materials. On many occasions buildings are damaged by projectile impacts as rubble rolls down the front of advancing flows. Impact forces resulting from flow momentum and the hydrostatic pressures exerted by ponded lava may also lead, in a few instances, to building damage through excess wall loads, foundation failure and/or uplift forces. In almost all cases, however, buildings will ultimately be destroyed by fire, collapse or burial.

Other general considerations indicate that thick-walled stone buildings with little wood or other flammable materials in doors, window frames, ceilings and/or furnishings will best survive the passage of a lava flow. Exterior walls diagonal to the encroaching lava may divert the flow whereas walls set normal to the forward movement provide a more serious, and a more vulnerable, obstruction. However,

*In effect, building contents are limited to the furnishings and appurtenances of dwellings. The effects on machinery, electrical equipment, automobiles etc., are described in Chapter 6.

the spartan dwelling set diagonally to the lava flow may prove subsequently to be just as uninhabitable as the luxuriously furnished, lightly-built home that was ignited and consumed!

Table 5.2 provides some examples of building destruction by lava flows. Notably, nearly all examples are of basaltic flows. Further details are provided for some of these examples.

Table 5.2 Examples of damage to buildings by lava flows

Eruption	Damage	Reference
Vesuvio, 1872	One third of Massa, one quarter of San Sebastiano	Palmieri, 1873: 90
1906	Many buildings in Boscotrecase and elsewhere	Jaggar, 1906: 111
1944	Partly destroyed San Sebastiano and Massa	Rittman, 1962: 35
Etna, 1910	Lava destroyed small town of Mascali	Sapper, 1917: 40
1928	Town of Mascali wiped out; village of Nunziata almost entirely destroyed	Coleman, 1946: 190
1983	More than 100 buildings destroyed	*Discover,* July 1983: 13
Nyiragongo, 1977	800 people lost houses	*BVE,* 17, 1979: 13
Piton de la Fournaise, 1977	April 8, 12 houses burned; April 10, two houses and filling station burned; April 13, lava reached church in 4 m high flow, burnt door, penetrated 3 m inside nave and stopped	*BVE,* 17, 1979: 15-16
Niuafo'ou, 1925	Nearly all buildings in village of Futu destroyed	Coleman, 1946: 119
Matavanu, 1905	Four villages	Anderson, 1910: 633
Api Siau, 1974	24 houses destroyed, 40 others saved (removed)	*BVE,* 16, 1978: 35
Miyake-jima, 1983	80% of Ako town buried	*SEAN Bull.,* 8(10): 2
Kilauea, 1955	17 houses	Macdonald, 1962: 250
1960	87 houses	
1983	8 buildings	
Mauna Loa, 1868	37 houses destroyed	Hitchcock, 1909: 197
1926	Buried village of Hoopuloa; not a single building left	Coleman, 1946: 107
Hualalai, 1800-1801	At shore, lava buried Hawaiian villages and Paiea fishpond	*CAVW,* 3: 8
Heimaey, 1973	About 300 houses	Williams & Moore, 1973
Lakagigar, 1783	14 farmsteads, 20 churches destroyed	*CAVW,* mss.
San Jorge, 1808	Town of Ursalina and adjacent country houses, cottages and farm houses inundated	Dabney, 1809: 343-344
Tenerife, 1706	Port town of Garachico totally destroyed, not a single edifice left standing	Hulbert, 1827: 64

The distribution of lava flows produced during the 1906 eruption of Vesuvius is shown on Figure 3.3. The major flow, on April 7 and 8, 1906, advanced through part of Boscotrecase and advanced to the walls of the cemetery of torre Annunziata where the flows bifurcated (de Lorenzo, 1906: 477). In Boscotrecase some of the streets were blocked with high walls of lava, particularly where street directions were parallel, or nearly so, to the flow direction. Roads with walls at right angles to the slope were cut by the flow, but the lava generally also moved along the road in both directions for some meters (Lacroix, 1906: 229). "Houses met various fates according to their situation, some being destroyed. Others were invaded by the lava, which found an entrance through doors and windows facing the mountain. The lava was sufficiently fluid to fill the chambers and courts and to mould itself there. It is hardly necessary to add that although the lava was sufficiently cool after a few days to allow access to the houses, the chambers not filled with it had a suffocating temperature" (Lacroix, 1906: 229-230).

Both Lacroix and Jaggar (1906: 111) comment on the comparative absence of fires in Boscotrecase. Evidently, the outer surface of the lava flow was slaggy and, thus, only weakly conductive so that combustible woodwork did not ignite immediately, but carbonised slowly adding to the gases coming from the lava. Chimneys and gutters gave forth smoke and heated air. The masonry construction of the buildings with limited use of timber also protected adjacent buildings from fire (Fig. 5.1). Nonetheless, the lava flow, 5 meters high at places along its margin, cut walls and even some houses in two, and banked against the stately ruins of the church (Lacroix, 1906: 229-230; Jaggar, 1906: 111).

In 1944 lava from Vesuvio flowed to the northwest toward San Sebastiano (see Fig. 3.3). Milton Bracker reported by wireless to *The New York Times:* "Gradually the stream spread out in the little valley. The last few trees went up in flame — peculiarly outlined in blue — and then the crackling mass crunched down on an eight-foot wall [2.4 m] and began to devour it... Guiseppe Battaglio's house was on the far side of the fence, and for a while it seemed that the stone fence might channel the flow and save the modest stone structure... But as the incandescent mass roared over the fence, it was plain the house was fated. A spear of fire shot up to a corner of the building. Then it subsided, and the house seemed to be winning the battle. The odds were too great, however. The lava ground into the base on the other side, and with a roar the wall fell in. A few minutes later the surging flow literally cracked the house in half. What looked like an iron bedstead twisted into the air... Thus the destruction of the town began. A few hundred yards [500 m] back, but directly in line of the flow, stood the town's best houses and the three-storey, yellow school that the inhabitants cherished. It was estimated that they all were crushed and buried within two hours" (*The New York Times,* March 21, 1944: 5).

During the 1905–1906 eruption at Matavanu, Western Samoa, lava spread out along the coast for 15 km, destroying the villages of Salago, Toiapaipai, Maleola

FIG. 5.1 At Boscotrecase in 1906 the Church of St. Anne was blockaded by lava which formed a wall before the entrance (Photo: F. A. Perrett, Lyman House Memorial Museum, Neg. #28 [copy]).

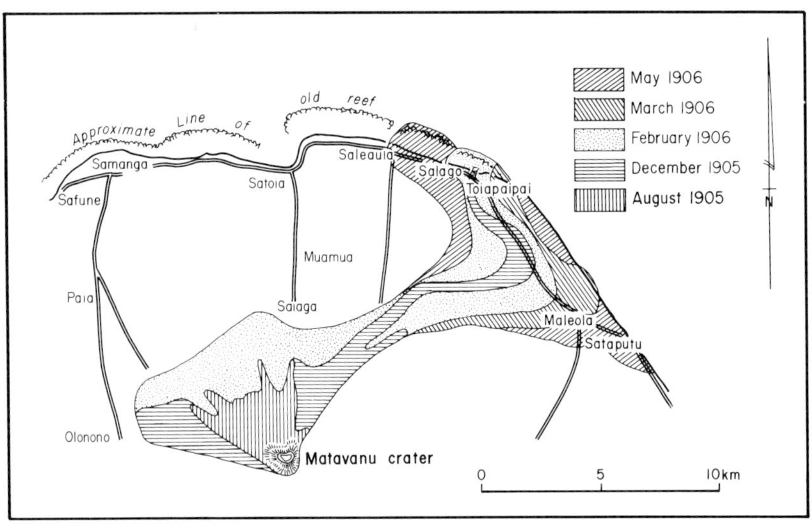

FIG. 5.2 Lava flows from Matavanu buried villages and roads along the north coast of Savaii in 1905–1906 (modified after Anderson, 1910: 622).

and Sataputu and the whole town of Saleaula except for two houses (Fig. 5.2). The Samoan houses in the villages were built of thatched roofs supported on wooden poles, but the churches at Sataputu and Saleaula and Bartley's house near the latter were substantial stone buildings. The lava flowed around these three buildings, partially burying them but without causing the walls to fall. In each case, however, the interior was burned and gutted. At Sataputu the church was buried by 8–11 meters of lava but the tower still stands. At Saleaula, the church was near the edge of the lava which entered through the windows covering the floor to a depth of more than a meter (Anderson, 1910: 633-634).

The area covered by lava flows during the 1914 eruption of Sakura-jima is shown on Figure 3.19. Yokoyama and Akamizu villages on the western side and Seto, Waki and Arimra villages in the south western part of the island were overrun by lava. In these five villages all houses, a total of 1272, were destroyed (*Kobe Herald,* 1914: 12). Although Omori (1922: 513) gives the total number of houses burnt in the eruption as 2067 plus 1790 other buildings, it is clear from his earlier text (Omori, 1916b: 215) that not all these losses can be attributed to lava flows (see also Chapter 5, Sections 3, 4).

The extent of the lava flows and the approximate locations of the buildings destroyed in the 1955, 1960 and 1983 eruptions of Kilauea (Table 5.2) are shown on Figure 4.1. The progressive destruction by fire of the Kapoho school in 1960 is illustrated in Figure 5.3. Other buildings were ignited directly from adjacent vents (Fig. 5.4). The 1944–1945 lava flows from Parícutin also destroyed many buildings in the villages of Parícutin and San Juan Parangaricutiro (see Fig. 4.4). Part of Parícutin village was evacuated in June, 1944, in the face of an advancing aa flow (Fig. 5.5). In San Juan most of the village was buried but the masonry church tower remained intact (Fig. 5.6). The 1973 eruption in Heimaey also buried and/or burnt many houses (Fig. 5.7). By late March, 1973, two months after the eruption had started, 60 to 70 houses had been buried by lava. By early May, 300 buildings had been covered or gutted by fire, though many of the fires resulted from pumice fall (Clapperton, 1973; Williams and Moore, 1973). In all these examples the advance of the lava flow was sufficiently slow that evacuation of nearly all possessions was possible. However, at Heimaey, and no doubt elsewhere, personal property suffered severe damage during evacuation and transportation by ship to Reykjavik. Furniture and white-enamelled refrigerators seemed particularly damage-prone. Some families had up to half their personal possessions ruined by the time they reached the "mainland" (Gunnarsson, 1973: 56-58).

Numerous observations have indicated that lava flows with low yield strengths may be diverted by masonry walls and other resistant structures. Most protective measures involve diversion of the flow to another route and can only be successful if vacant land is available for the diversion, a point well illustrated by Rittman: "The first recorded attempt to divert a flowing lava stream was made by Diego Pappalardo during the greatest eruption of Etna, in 1669; unfortunately he was too late, since the lava had already penetrated into the town of Catania and not much

FIG. 5.3 (top and bottom) Progressive destruction of the Kapoho School following ignition by a lava flow in the 1960 Kilauea eruption (US Geological Survey photos, 2d 443, 2d 450).

FIG. 5.4 Spatter from a vent ignited the adjacent building, Kapoho, 1960 (US Geological Survey Photo File No. P10-79-83 [2d]).

FIG. 5.5 Final evacuation of dwellings as aa lava flow (left of view) invades Parícutin village on June 21, 1944 (Smithsonian Institution Archives, Record Unit 7281, William F. Foshag Papers, 1923-1965 and undated, F214B). See also Fig. 2.1.

FIG. 5.6 Partial destruction of the church in San Juan by lava flow (Smithsonian Institution Archives, Record Unit 7281, William F. Foshag Papers, 1923-1965 and undated, F282, January 24, 1945).

remained to be saved. Under Pappalardo's leadership half-a-hundred men turned out, protecting themselves against the heat with wet hides, and proceeded to hack a breach with picks and axes through the western lateral moraine of the lava flow above the town, so that the molten river should flow away to the side. Had this attempt to divert the lava been made earlier, when it was still some distance from the town, it might perhaps have been more successful, since the lateral outflow would have reduced the pressure at the snout of the flow and thus slowed its advance. At all events, the scoriaceous covering was indeed broken through, and the

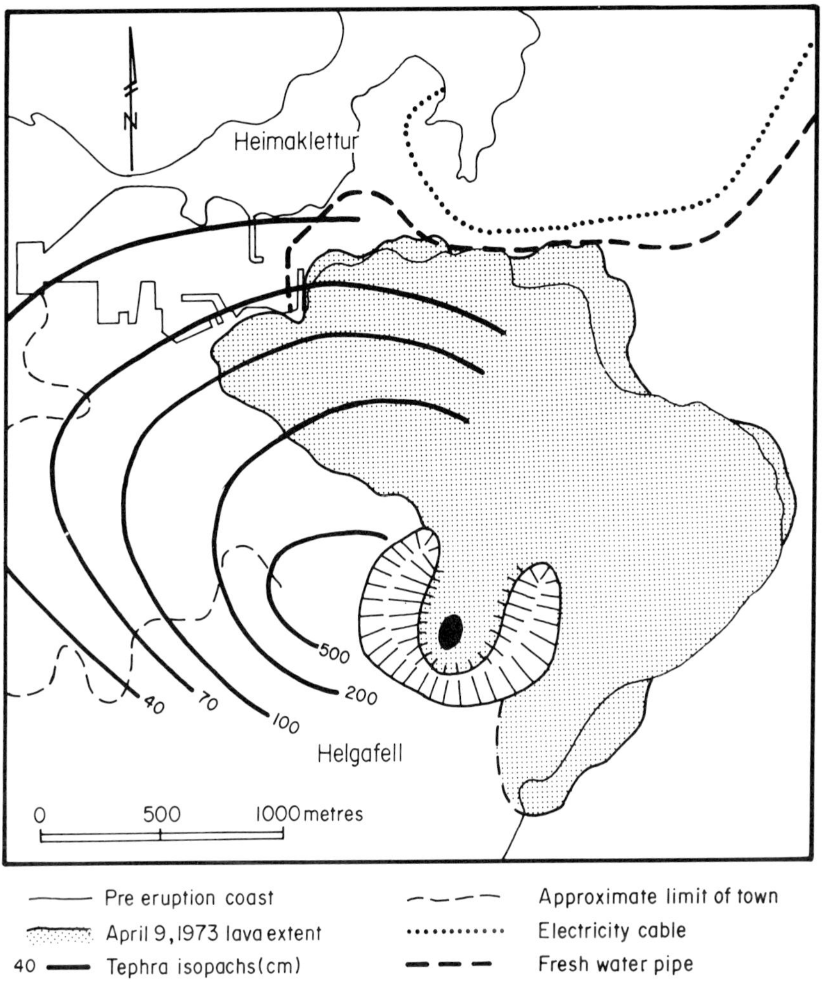

FIG. 5.7 Distribution of lava flows and tephra from the 1973 Heimaey eruption (modified after Clapperton, 1973a,b; Preusser, 1973).

lava induced to flow out laterally as planned. But when the people of Paterno heard of this, they feared that the diverted stream would threaten *their* town. And so, 500 heavily armed men turned out with trumpets and drums, and put Pappalardo and his helpers to flight" (1962: 54-55).

During the same eruption the walls of Catania held up the lava flow for 15 days before they were breached (Clapperton, 1972: 163). Diversons and/or damming of lava flows have been tried in a number of other situations with varying degrees of success. At Kapoho in 1960 bulldozers were used to build six walls with a total length of 4500 m and a volume of more than 300 000 m^3 of old lava (Fig. 5.8) ripped from the surrounding area in 9.5 days. Although there has been some debate as to the effectiveness of these barriers (Macdonald and Abbott, 1970: 103) a great deal was learnt about wall construction techniques. Macdonald (1962: 292-293) concluded:

> 1. The wall must have a broad base. If the flow against the wall is in the order of 10 to 12 m thick, the base should be in the order of 30 m wide.
> 2. The wall must be constructed of heavy materials. Cinder is too light, and the lava tends to push it aside or burrow under it and heave it up.
> 3. The wall should have adequate height to prevent overflow, so far as possible. Even small overflows can be serious, because they may erode the face of the wall and eventually cut a gap through the wall.
> 4. The slope of the wall away from the lava should be gentle so that any overflows that do occur will move slowly down the side of the wall and erode it less than if the movement were rapid.
> 5. Sharp salients projecting into the flow should be avoided. Salients are equivalent to a low place in the wall, and invite overflows. If a salient is unavoidable, the wall should be built somewhat higher there than along the rest of its course.
> 6. If a wall is to act as a diversion barrier, an unobstructed path should be provided along the side of the wall toward the lava. This clear path is nearly as important in directing the course of the flow as the wall itself. All vegetation larger than grass and small bushes should be removed. Even the trails of rock fragments spilled from the edges of bulldozer blades in building the wall constitute a noticeable obstacle to the advance of the flow and should be cleaned away so far as possible.
> 7. Diversion barriers cannot operate successfully unless the gradient in the channel behind them is sufficient to cause the liquid lava to flow readily. The amount of slope necesary depends, of course, on the viscosity of the lava, more viscous flows requiring a steeper slope than less viscous ones. For flows of the sort ordinarily encountered in Hawaii, the minimum slope appears to be about 1.5 to 2 per cent.

In 1973 barriers built by bulldozers in Heimaey slowed the advance of lava toward the town. In 1977, a wall 3–4 m high was built around the diatomite factory at Namafjall near Krafla (Iceland) to divert lava flows (*BVE,* 1979, 17: 80). More recently, various plans for the construction of diversion barriers have been considered in order to protect Hilo from future lava flows from Mauna Loa (United States Army Corps of Engineers, 1980). It is worth noting that suburbs of Hilo requiring protection include those built on 1881 lavas!

Explosives were used in attempts to halt lava flows advancing toward Hilo from Mauna Loa in 1935 and 1942. Although Rittman (1962: 55) and others (for example, Macdonald and Abbott 1970: 64-65) have claimed that these aerial bombings were at least partially successful in diverting and/or stopping flows a detailed analysis by Lockwood and Torgerson (1980) indicates that the bombs had no significant effects. However, experiments on Mauna Loa in 1975–1976 using bombs up to 900 kg with fuse delays of 0.05 seconds and impact velocities of 250–275 m s^{-1} caused maximum flow disruption particularly where spatter cones were targeted.

An attempt on Etna to divert a lava flow using explosives was only partly successful. The main technical problem with this effort, in May, 1983, revolved around

FIG. 5.8 Bulldozers work at constructing a wall of old lava fragments to dam a 1960 Kapoho flow while spectacular fire fountaining occurs in the background (Photo: American Red Cross).

the use of explosives which became unstable at temperatures of about 100°C to remove a rock barrier behind which lava flowed at temperatures of greater than 1000°C. Water jackets were used to cool the explosives but variations in lava flux and accretion on the barrier resulted in a diversion of only about 20% of the flow for a few days. However, other overflows slowed the lava front in the main channel (*Discover*, July, 1983; *SEAN Bulletin*, 865: 6).

Attempts have also been made to halt or slow the advance of flows by spraying cooling water onto the lava. It was noticed both at Matavanu in 1911 and at Kapoho in 1960 that lava advancing into the sea moved along the shore rather than continuing into deeper water, presumably because chilling increased the viscosity of the flow front. "Later in the 1960 eruption, on several occasions the Hawaii Fire Department tried artificial spraying of the flow margin to determine its effects. One example may be given. For several days the edge of the flow had been creeping slowly toward a wooden house. When the lava was about 6 m from the house volunteer firemen, who were not then otherwise occupied, turned two streams of water from 75 mm hoses onto the side of the flow. Within a few minutes the lava stopped moving, and it remained stationary for several hours. In short, it was found that even with rather small amounts of water the spread of the flow could locally be checked for long enough to remove the contents of buildings, or even to move entire buildings to safety" (*UNDRO,* 1976: 7).

A much more substantial experiment was undertaken at Heimaey in 1973 where the harbor entrance was at risk (Fig. 5.7). At first the advancing lava was sprayed with 12 000 tonnes of water per hour from a ship. By late March, 30 high power pumps together with fireboats and ships' pumps were delivering 32 400 tonnes of seawater per hour to the lava surface (Williams and Moore, 1973; Clapperton, 1973: 628). It appears that these efforts were instrumental in saving the harbor entrance and much of the town. Large-scale pumping schemes have also been considered at Krafla and Hilo. In 1983 sea water was sprayed on the Miyake-jima flow front at 40 places. By the following day forward movement of the flow had halted at 38 of the 40 places (*SEAN Bulletin,* 8(10): 3).

While bombing, dams, walls and cooling of the lava provide valuable methods which may halt, slow or divert lava flow, non-structural methods of protection may be simpler and cheaper if applied early enough. The most important of these is land-use zoning. The 1980 estimate of the cost of a barrier to protect much of Hilo was $US3.5 million. Zoning, using the 1881 lava flow as an example, would have been much cheaper. Relocation of buildings may still prove cost-effective in some circumstances.

It is also worth noting that structural protective measures need to be considered within the cultural background of the affected residents. Human intervention may change lava flows from "Acts of God" to "human-caused events" with dramatic effect on insurance policy provisions (Barberi and Gasperini, 1976: 220). On the densely settled slopes of Etna diversion of lava flows is often impractical because other settlements will be placed at risk as indicated above. Attempts to divert a 1971 flow brought to light a seventeenth century law which forbade such efforts (*Newsweek,* 1971, May 17: 43; Booth, 1974: 317). During public hearings regarding the Hilo diversion barriers it became apparent that the goddess Pele (see Chapter 4, Section 8) was important in shaping some local attitudes to control measures: "... many believe it is possible to pray to dead relatives who have been dedicated to Pele to intercede with the goddess to direct a lava flow away from a particular piece of land. Pele beliefs also play a part in behavior associated with evacuation on the part of old Hawaiians. Such people may designate a flow as being Pele who 'can take anything she wants when she will', and when a lava flow approaches their property, they will do nothing, believing that to take any action would anger the goddess. During the 1955 eruption in Puna district [see Fig. 4.1], an old Hawaiian man refused to move his belongings, including a new television set and refrigerator, because of his beliefs. During the recent volcanic eruption and lava flow in the Kalapana area of south Puna in 1977, there was widespread local opposition to any form of experimental waterbombing by the Army to determine the effects on the flow of lava. The degree of opposition finally lead the Mayor of Hawaii County and the Governor of the State to informally request federal authorities to stop their experimentation" (US Army Corps of Engineers, 1980: 89).

5.3 TEPHRA FALL

As indicated in Table 5.1 the most damaging effects of tephra falls result from projectile impacts and roof loads leading to collapse. Secondary effects such as ignition of furnishings or, indeed, the destruction of the entire building by fire may result from penetration of the structure by hot projectiles or shortcircuiting of electrical systems by fine-grained tephra. Corrosion of some surfaces may result from contact with the acidic coatings on grains.

5.3.1 Projectile Impacts

Some of the effects of projectile impacts on buildings have been previously summarised by Blong (1981). Figure 5.9 illustrates the relationship between the penetration resistance of a variety of building materials and the impact energy of volcanic bombs of varying size, shape and density falling at terminal velocities (and hence with varying kinetic energies of impact $= 0.5\ mv^2$; see Chapter 2, Section 2.1). The data presented in Figure 5.10 indicates:

A. Even 20 mm thick sheets of tempered glass will be penetrated by bombs of 10 g mass. Some bombs of only 2.5 g mass but with specific gravities greater than 1.0 g cm^{-3} can break some window glasses. All window glasses except laminated glasses will be broken by projectiles of greater than 25 g mass whatever the projectile shape.
B. Plywood, hardboard, fibre-cement, asphalt and cedar shingles, slate and clay tiles are all likely to be penetrated or at least damaged by projectiles of 50–60 mm diameter.
C. Steel cladding material could be damaged, particularly around fastenings, by projectiles of greater than 40–45 mm diameter (mass 250–500 g). Steel cladding can be penetrated by bombs of greater than 2.5 g cm^{-3} with masses of greater than 200 g (Blong, 1981: 410).

These results need to be treated with some caution. Impact energies have been calculated for normal (90°) impacts on buildings materials fastened in accordance with the codes extant in Australia and the United States. The penetration resistances of most building materials decline with age but sheet steel and asphalt shingle show increased resistance with increasing ambient temperature. Pointed projectiles will achieve higher penetration than spherical or cubical projectiles. The results shown on Figure 5.9 assume that volcanic bombs have cooled and solidified. While the penetration achieved by hot malleable bombs will be less than

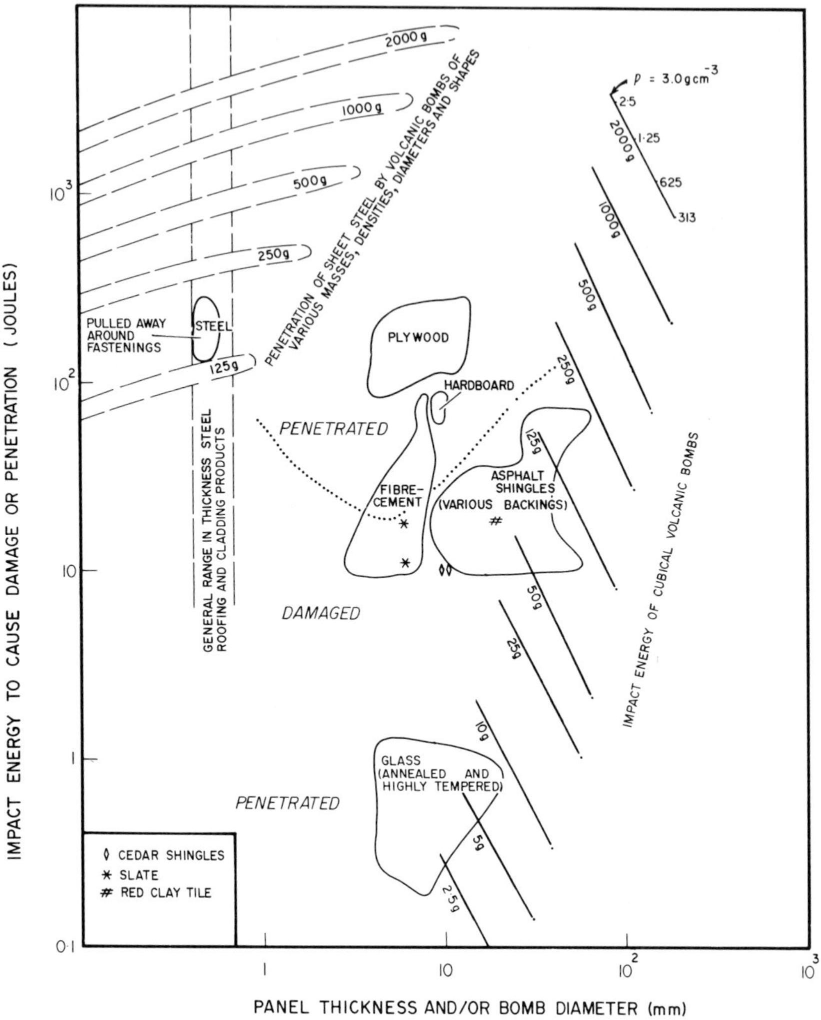

FIG. 5.9 The relationship between volcanic projectile size, density and impact energy and the damage caused to a range of building materials. The sloping lines indicate projectile dimensions and impactenergies while the fields define commercially available cladding and other building materials (after Blong, 1981: 407).

that produced by cold hard projectiles, ignition of ceilings, floors or furnishings by the former may cause more serious damage. Finally, repeated impacts may seriously weaken elements of the structure, even though penetration does not initially occur.

Blong's (1981) presentation, derived from a variety of engineering studies, remains untested from a volcanological point of view as detailed observations of the

FIG. 5.10 Five stages in the progressive destruction of the Lassen Peak lookout by projectile impacts in 1914. In the second diagram the roof has been penetrated by falling bombs. The fourth diagram shows the condition of the building in October, 1914, the last diagram, in August, 1915 (redrawn after Hill, 1970: 213).

effects of bomb impacts on structures have not been recorded. Nonetheless, some interesting observations have been made.

Two excellent examples describe the effects of volcanic bomb impacts on buildings. The first, involving just the impact effects of projectiles without the added destruction of fire, results from the destruction of the lookout on Mt Lassen in 1914. Figure 5.10 (Hill, 1970: 213) illustrates the progressive destruction of the lookout which was located 400 m from the crater and 150 m *above* it. The building was designed to withstand the strong winds common on the 3200 m mountain.

The first diagram shows the lookout before the eruption began, with shutters over most of the windows. In the second diagram the roof and walls have been penetrated by bombs in several places; a rafter was broken. Progressive demolition of the building resulted from continued bomb impacts until little remained (fourth diagram) on October 20, 1914. The last diagram indicates that only a pole remained by August 8, 1915 (Hill, 1970: 213). It is perhaps interesting to note that while the lookout was demolished entirely by successive bomb impacts, the diagrams show no accumulation of boulders!

The second excellent account of the destruction of a building by projectile impact clearly indicates that bombs ejected from Vulcano on August 4, 1888, landed still hot on Mr Narlian's villa, only a few hundred meters from the crater. This account comes from a letter dated August 30, 1888, from Mr Narlian to Dr H.J. Johnston-Lavis:

> Consequently, I spent all the night dressed on a sofa in the drawing-room. Towards morning I was overpowered by sleep, and went to the little bedroom which looks towards the mountain, rested on the bed, and soon was evidently sound asleep. Soon afterwards I was awakened by a *tremendous din* which can hardly be described. As I jumped up from my bed I felt stones falling on the roof as hail — such cannonading going on — I understood what was the matter — ran to the opposite room, where I had made my children sleep that night. They were also up in consequence of indescribable noise of thunder, rush of gases, flames, falling of huge boulders, rock, &c. I took them to the drawing-room, but as soon as the door was opened a big stone *red hot* (all these stones are quite red with heat) fell through the roof, ceiling and floor a few yards from us, smashing all, setting fire to everything. Now I took my children back to the bedroom, which looks on the verandah, tried to gain the terrace by that side. The house doors were shivering and shaking so that it was a difficult matter to open the doors. At last I succeeded, but before we were out in the verandah another stone fell at *our feet,* was smashed to fragments, and burned the feet and legs of my boys. Now we passed through the verandah, regained the house at the top of the stairs; here another stone fell very near us (none of these stones were less than two feet [0.61 m] in diameter). The last stone (which is the fourth that struck the house including the one that fell on the ceiling of my bedroom while in bed) had nearly blocked our way out with the rubbish that it brought from the roof. We passed through over the heap of rubbish and were now out in the open to the north of the house. By this time (not many minutes after all) the whole place was lighted up, woods, grass, buildings, hedges — all was on fire; the huge boulders and stones were literally raining everywhere about us — what confusion! Natali, the faithful boy, had by this time come to the help of my little boys — we all began to run to Vulcanello and away from the dreadful thundering mountain. All the means of communication we possessed at Vulcano consisted in an old, half broken 20-foot boat, and a lighter, both of which the men in their panic and mad despair had taken away with them, leaving us on the sands of Vulcanello. Towards noon, however, boats of rescue reached us from Lipari, and we thus ended one of the most eventful days of our life. How we escaped death I do not know.
>
> On revisiting the spot I saw the whole plains below the mountain, to the distance of 1.5 miles, specially the neighbourhood of the house and the men's habitations, *literally* covered with boulders and rocks of all sizes, which had imbedded themselves in the ground to various depths. The most huge of them is near the well of the house, which is not less than 10 *yards* [9.4 m] in diameter, or better length, and is some 10 to 11 [3.1 to 3.4 m] feet deep in the ground. This is about ¾ of a mile [1.2 km] from the crater. Another, of nearly equal dimensions, is on the shore near the Quarantana at the end of the bay of the Levante! Rocks of one yard [0.91 m] in diameter are as plentiful as can be as far as the middle of Vulcanello, near the Punta Samossa! We have to thank God for going unhurt through this *'hail storm'!*
>
> [Anderson and Johnston-Lavis, 1888: 16]

In some cases it is difficult to separate the effects of ballistic impacts from those resulting from wind-blown tephra fall (see Chapter 5, Section 2.2.2) but building destruction by projectiles seems most often to be accomplished by subsequent fire. At Heimaey in 1973 incandescent blocks ranging from 0.1–2.0 m in diameter caused fires when they crashed through roofs and unprotected windows (Booth,

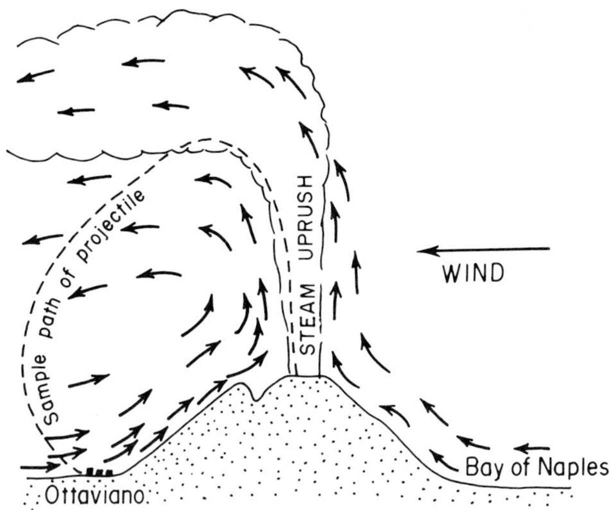

FIG. 5.11 The trajectory of pumice lapilli at Vesuvio in 1906 as portrayed by Hobbs (1906).

1979: 331). The Glacier Hut, 1200 m from Ruapehu's Crater Lake, was severely damaged when hit 35 times by ballistic ejecta during the 1975 eruption. Much of the damage was apparently caused by one large block which formed a 2 m wide impact crater near the hut, broke into pieces, and bounced upward through the hut walls leaving holes up to 0.3 m across. The bombs landed hot and charred the wood where they came to rest. Several bombs burnt through the floor of the hut (Nairn *et al.*, 1975). During the 1783 eruption of Asama, house roofs at Karaizawa, 11 km from the vent, were perforated by red hot pumice blocks up to 0.5 m in diameter. Fifty two of the 162 houses were completely burnt (Aramaki, 1956: 197). During the 1914 eruption of Sakura-jima, buildings in the villages of Yokoyama, Koike, Take and Fujino (see Fig. 3.19) were ignited by falling bombs (Omori, 1916b: 215). The number of buildings destroyed or damaged is not known because some of these villages were also affected by lava and pyroclastic flows.

In other cases, buildings have been damaged but not destroyed by bomb impacts. During the 1906 eruption of Vesuvio, windows in Ottaviano and San Guiseppe (see Fig 3.3) were often broken by falling lapilli. Figure 5.11 presents Hobb's (1906) explanation for the fact that windows facing away from the volcano were often broken. Jaggar (1906: 113) noted that: "Some of the flying pebbles were endowed with such high velocity from their long flight miles [kilometers] up, out, down, and in, that they cut clean holes through the glass".

Some other observations provide evidence that can eventually be used to test some of the results presented in Figure 5.9. At Georgetown, St Vincent, in 1902, bombs up to 900 g from the Soufrière perforated thatched roofs, dented galvanised

FIG. 5.12 A sheet of corrugated aluminum cladding damaged by blocks ejected at Soufrière de Guadeloupe, 1976 (Photo: Grant Heiken).

iron roofs and left shingle roofs undamaged (Anderson and Flett, 1903). In 1903, 60–80 mm diameter bombs penetrated iron roofs (*HMSO,* 1903: 193). Similar damage was done during the 1976 eruption of Soufrière de Guadeloupe (Fig. 5.12). At Manam, Papua New Guinea in 1957 bombs 50 x 75 mm perforated thatched roofs (Palfreyman and Cooke, 1976) and at Stromboli in 1912 boulders up to 30 tonnes in weight were hurled up to 3 km destroying several houses in Ginostra (Rittman, 1962: 37). In 1963, in a zone up to 8 km from Agung, bombs set palm leaf roofs on fire — but did not ignite bamboo roofs (*UNESCO,* 1979).

5.3.2 Tephra falls

As fine-grained tephra deposition is largely controlled by the wind profile around an object, building shape, height, orientation and spacing relative to other objects will influence deposition and drift. From such considerations and a comparison with snow deposition and drift, Blong (1981) outlined some general principles:

A. With no wind the roof load of tephra is uniformly distributed and equal to the ground load (assuming none falls off the roof).

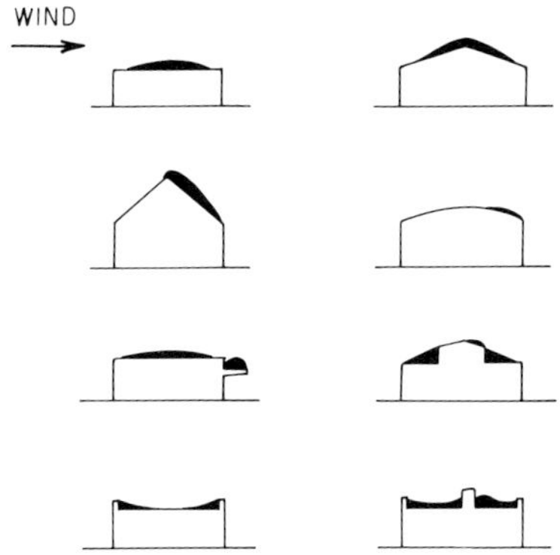

FIG. 5.13 Suggested distribution of tephra loads following drifting on a variety of roof styles (after Blong, 1981: 415).

B. The extent to which tephra accumulates on sloping roof surfaces depends on tephra–roof and tephra–tephra coefficients of static friction. 'Smooth' surfaces such as sheet metal and glass are more likely to shed tephra loads than are 'rough' surfaces such as thatching and asphalt or wood shingles. Unbalanced loads on roofs are more likely to be the result of wind scour and deposition than of sliding.

C. Roof configuration has a very important influence on tephra accumulation and roof live loads. Where gables and other projections lie transverse to the wind, areas of 'aerodynamic shade' are created in which tephra accumulation occurs (Fig. 5.13). If the analogy with snow is realistic, tephra drifts will occur not only on the windward side of the parapet but also along other parapets to form a saucer-shaped depression. Other roof projections such as chimney and ventilation shafts also enhance accumulation leading to unbalanced live loads. Roof loads may reach two to three times ground loads. Sliding of tephra from higher surfaces may also create impact loads.

D. Unbalanced or asymmetrical loads on roofs are important in causing roof failure. Roof support members normally in tension may be placed in compression under a large unbalanced load. Failure of only one member may be required to cause partial roof collapse. Wall failure may occur before roof failure if bracing is inadequate.

Building failure under tephra loads may also occur when tephra drifts against walls and/or foundations. Similarly, tephra loads against chimneys and parapets can lead to failure of these 'obstructions' and, indirectly, to roof failure.

Building damage, rather than failure, can also result from tephra fall. As Blong (1981: 417-418) notes: "(i) continued penetration of fine tephra into the ceiling space via eaves ventilation gaps and cladding joints could lead to eventual ceiling collapse or sagging under the accumulated load; (ii) similarly, penetration of fine tephra could cause a short-circuit in the electrical system, leading to fire; (iii) deposition of acid tephra and associated aerosols could, through corrosion, lead to premature ageing and weakening of cladding and other building materials; and (iv) deposition and redeposition of tephra following rainfall could lead to overloading and collapse of the roof gutter system, clogging of drains, water damage to ceilings, wall and sub-floor spaces, short-circuiting of the electrical system and so on."

These considerations are theoretically based — observations presented in the literature are not of sufficient quality to establish their validity. Nonetheless, there are some interesting examples of the effects of tephra falls on buildings.

Figure 3.4 illustrates the tephra accumulation ellipse around Usu in 1977. The area experiencing a tephra fall greater than 200 mm in depth extended about 8.5 km by 2 km. As the specific gravity of the tephra was about 1.0, a live load of 200 kg m^{-2} corresponds to the 20 cm isopachs. In the western part of Toyako Spa tephra live loads reached 200–300 kg m^{-2} and in the Konomi area on the western edge of Mt. Konpira, 400–800 kg m^{-2}. Buildings in this area of Hokkaido are designed for a snow live load of 100 kg m^{-2} with the critical live load generally being 200–300 kg m^{-2} (Shibata *et al.*, 1978).

In most areas where tephra falls were thick enough to exceed critical live loads there were few buildings. At Konomi, however, the roof of a nursery play room was damaged and roof beams broke in the public housing area. The latter buildings were of concrete block construction. At the two-storey filtration plant, a ferro-concrete building, cracks appeared in the lower edges of roof beams with a span of 9.4 m. Most of the other 34 buildings in the Konomi area were evidently undamaged (Shibata *et al.*, 1978).

At Heimaey at least 60-70 houses were buried by tephra by early May, 1973 (Williams and Moore, 1973: 16). Many strong buildings with flat roofs collapsed under the load of less than one meter of tephra (Booth, 1979: 331). Houses that survived the total tephra fall of more than 4 m had their roofs cleared periodically. However, many of these houses were damaged when water left in heating systems froze and cracked the pipes ruining walls and filling cellars with water (Gunnarsson, 1973: 72, 84).

Buildings have also been damaged around Vesuvio by tephra fall. At Ottaviano (see Fig. 3.3) in the 1737 eruption many houses were crushed by the weight of tephra and bombs that accumulated to four palms deep (d'Aragona, 1739: 244). We cannot be sure whether 'four palms' refers to a tephra fall of 280–400 mm or 720–1000 mm but the lower value seems the more likely as flat house roofs in the same area collapsed under a tephra load of only 100 mm in 1906 (Perret, 1950: 155). In Naples in the same eruption the roofs of the market of Monte Oliveto and a few poorly constructed houses collapsed under a load of only 25–30 mm of tephra (Hobbs, 1906: 640; Lacroix, 1906: 234-235). However, it was the buildings of Ottaviano and San Guiseppe that were most seriously affected during the 1906 tephra fall. Whilst there is considerable doubt about the thickness of the ash and lapilli (see discussion in Blong, 1982: 173) it was certainly more than 700 mm and perhaps as much as several meters. Jaggar (1906: 197, 112-113) wrote: "The overpowering weight of this snowstorm of gravel [the sand and lapilli fall on April 7 and 8, 1906] was too much for the flat and tiled roofs of the Italian stucco houses, and Ottajano [sic.] simply *fell in,* the crumbling walls adding to the volcanic debris... The ruin at Ottajano and beyond it at San Guiseppe was due to a simple, single cause — falling sand and gravel. There was no serious earthquake, fire, lightning, or wind. The sand and *lapilli* at Ottajano reached a depth of from three to 3¼ feet [0.9 to 1.0 m] on the level. The stones average one-third inch [8 mm] but there were many as large as an apple. The roofs have caved in, carrying with them the *debris* of the walls, frequently shifting sidewise and producing puncture, rupture, or total crumbling of the loose masonry. Vaults of sufficient thickness, arches and dome-shaped huts were the only structures which resisted the attack. The narrow streets are piled high with the sand and fallen rubble, the footpaths through them now following a course far above the pavement."

Figure 3.19 outlines the distribution of tephra fall in the 1914 Sakura-jima eruption. In all, 1446 houses and 1331 other buildings were buried under tephra (including pumice lapilli — see Fig 2.13a) and lava (Omori, 1922: 515). However, Omori seems to have included the buildings burnt by lava flows in a separate category so that most of the 2777 buildings mentioned above were probably destroyed by tephra fall. Thus, we may assume, in Krokami village on the east-coast of the island 200 of the 270 houses were destroyed by 1 m of tephra. In the north coast villages of Komen, Futamata, Shirahama and Matsuura, where less than about 500 mm of tephra fell, no houses were destroyed (*Kobe Herald,* 1914: 12) though houses may have been damaged. Although most buildings at Kapoho in 1960 were buried by lava, some collapsed under the load of tephra (Fig. 5.14). Similarly, well constructed timber frame buildings were destroyed by tephra fall during the 1937 Rabaul eruption (Fig. 5.15a) and local material houses were badly damaged by tephra falls around Parícutin in 1943 (Fig. 5.15b).

Table 5.3 presents other examples of the effects of tephra falls on buildings. Several points are worth making: (i) in most cases only a small proportion of the houses at risk suffered damage or collapsed; (ii) it is sometimes difficult to separate the effects of the load of tephra fall from those of projectile impacts; (iii) in most cases we do not know whether the tephra was dry, damp or saturated (certainly rainfall on tephra can easily double the roof live load) (iv) although many of the examples presented do not relate to 'western' communities or 'modern' building practices, occasionally modern buildings/roofs do collapse under small live loads (for example, the one hectare roof of the Hartford, Connecticut, arena collapsed in January, 1978, under a live load of only 120 mm of wet snow! [*Time,* January 30, 1978]); (v) although the thicknesses described in Table 5.3 refer to "uncompacted tephra", tephra compacts within a few days of deposition to about half the freshly-fallen volume (thus, there may be quite large discrepancies among thickness recorded); and (vi) there is considerable variation in the building styles and ages referred to in the table.

FIG. 5.14 (top) Buildings covered with a mantle of tephra during the 1960 Kapoho eruption. Fire fountaining in the background (Photo: Lyman House Memorial Museum 74.30.166); (bottom) The same scene at Kapoho. Two buildings have collapsed under the tephra load. Fountaining activity has moved uprift and the lava flow has advanced to the rear of the buildings (Photo: Lyman House Memorial Museum 74.30.167).

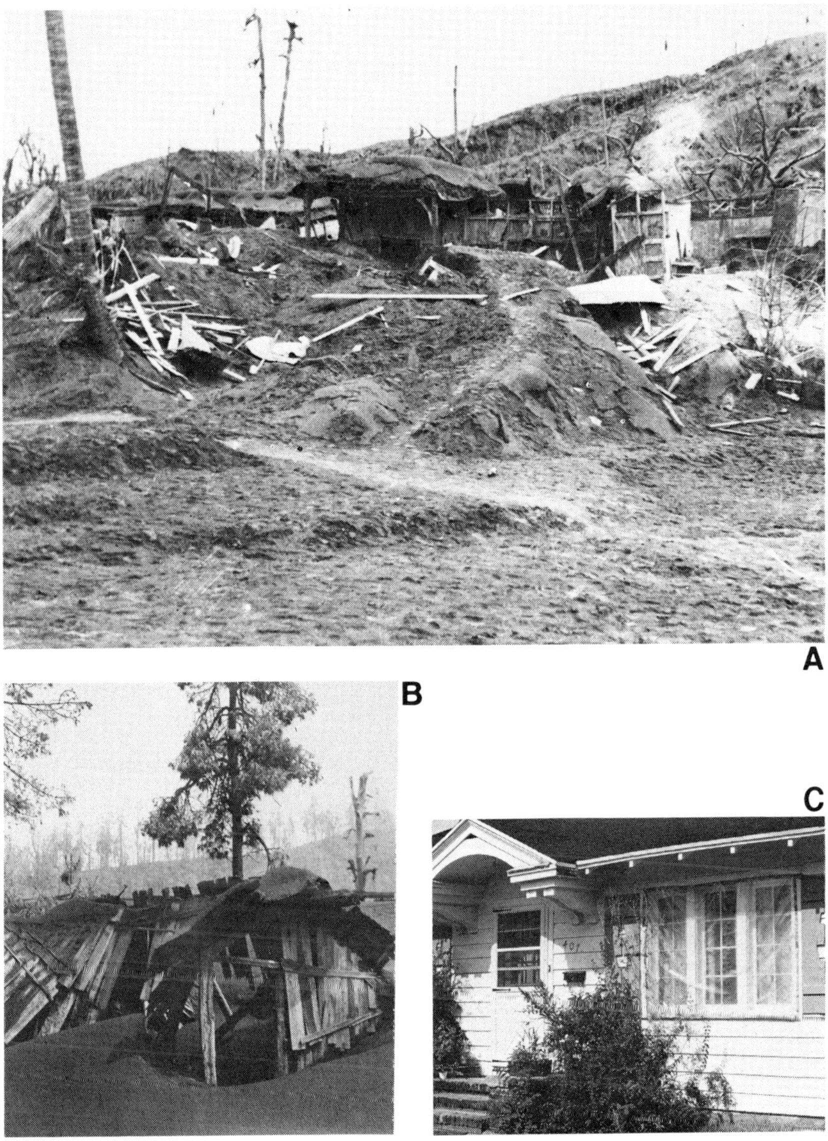

FIG. 5.15 (top) Destruction of the Baden-Jones house on Malaguna Road by the 1937 Rabaul tephra fall (Photo: Bureau of Mineral Resources, Australia, NHF#3); (left) a local-material house partially destroyed by Parícutin tephra fall in June, 1943 (Smithsonian Institution Archives, Record Unit 7281, William F. Foshag Papers, F32); (right) Plastic sheeting taped over windows provides some protection for the house interior from wind-blown tephra. Numerous houses adopted such methods in Yakima in May-June, 1980 after 10 mm of Mount St Helens tephra fell on May 18.

Table 5.3 The effects of tephra falls on buildings.

Eruption	Uncompacted tephra thickness (mm)	Damage	Reference
Ulawun, 1967	10-12	Minor damage	Johnson *et al.*, 1970: 15
Tambora, 1815	95	Roofs collapsed	Anon, 1816: 254
Taal, 1754	1100	Tile roofs of apses of church and kitchen collapsed	Masó, 1911: 10
Asama, 1783	<1100 but including red hot lapilli	82 out of 162 houses crushed (+ 52 burnt – see Chapter 5, Section 2.3.1)	Aramaki, 1956: 197, 198, 200
Santa Maria, 1902	200	Houses and farm buildings crushed, some totally destroyed	Coleman, 1946: 65
Fuego, 1971	300 including 50 mm bombs	20% of roofs in Yepocapa collapsed including those of many public buildings	Bonis and Salazar, 1973: 395; *SICSLP*, 1971: 114
Soufrière, 1812	150-250	Roofs collapsed	Anderson and Flett, 1903: 468
Soufrière, 1902	75-125 460	Roofs collapsed Houses collapsed	Anderson and Flett, 1903: 468

For a variety of reasons, including those just presented, it is not possible to define a critical thickness of tephra above which collapse or damage will ensue. Data in the table suggest that about 100 mm may be critical in many cases but damage to a few buildings can be expected with falls of even less than half that amount, particularly if accompanied or followed by heavy rainfalls.

A tentative conclusion that about 100 mm of tephra fall may be a critical threshold is also supported by three examples where no damage occurred. When Stromboli deposited 60–80 mm of tephra on the flat roofs of houses in nearby towns Perret (1912: 287) reported that no adverse effects were observed. Similarly, the 1937 eruptions in Rabaul harbor deposited about 100 mm of tephra on buildings in the town (Lowenstein, 1982), but most of the buildings were unharmed (*Official Handbook of the Territory of New Guinea,* 1943 supplement). Third, during the 1967 eruption on Deception Island the British Antarctic Station emerged unscathed from a tephra fall of 80 mm, while the Chilean Station was severely damaged, with some roofs caved in, by a fall of 500 mm (Baker, 1969: 119).

Finally, it is worthwhile considering some of the effects resulting from minor tephra falls. Although San Jose experienced a total tephra fall of about 12 mm from the 1963–1965 eruption of Irazú, individual falls rarely amounted to more than tens of grams per square meter. Under these conditions corrugated iron roofs proved to be essentially self-cleaning, but the tephra collected in roof gutters and drains, breaking gutters when it was not periodically removed (Clark and Lee, 1965: 10). In another example, about 10 mm of tephra fell on the city of Yakima during the

May 18, 1980, eruption of Mount St Helens (see Fig. 3.8). According to some accounts (*Agricultural Engineering*, 1980, August: 17) galvanised roofing was badly damaged and even new metal buildings were reportedly rusting badly in areas of heavier fallout. However, four points emerge from discussions with several insurance loss adjustors (in June 1981):

A. There appear to have been no insurance claims for damage resulting from electrical short-circuiting in houses as a result of the tephra fall.
B. the tephra mixed with water leached the color from dark-stained paint on cedar sidings, or placed a grey stain on dark wood stains so that they had to be re-stained. There were also a few claims for the recoating of fresh paintwork that had not hardened when the tephra fell.
C. It proved difficult to clean the tephra from whiterock chips and shingles used on some roofs; consequently, moss started to grow on the tephra on roofs so that a more thorough cleaning was required. Tephra also accumulated in the vesicles of red pumice rock used for garden borders, ornamentation, and so on; as this proved impossible to clean some insurance companies paid out for replacement red pumice rock.
D. Damage to shingle and shake roofs was caused by the use of snow shovels and walking on roofs during the cleanup process. Similarly, some damage was done to roof seams on mobile homes. After the cleanup, some flat roofs were observed to leak, a problem caused by live loads perhaps 100 times those imposed by the tephra fall.

Table 5.4 indicates the damage reported by 900 households in Eastern Washington State. Damage to roofs, gutters, downspouts and swimming pools was regarded as the most severe. However, 48% of respondents indicated no housing-related damage and a further 33% gave the estimated total damage, including injury to landscaping, as less than $250 (Dillman and Roberts, 1982: 192). No relationship between damage and tephra thickness was presented.

Much of the damage reported in Yakima resulted from the cleanup process rather than from the tephra fall itself. This was probably also the case throughout

Table 5.4 Type and degree of damage to buildings and other structures caused by Mount St Helens tephra fall in Eastern Washington, 1980

Damage	Proportion of 900 respondents reporting damage	Proportion of 900 reporting damage		
		heavy	moderate	light
Roofs, gutters, down spouts	33%	14%	7%	12%
Awnings, patio/porch covers	34%	6%	9%	19%
Interior floors, carpets	48%	4%	13%	32%
Swimming pools	20%	8%	4%	8%

After Dillman and Roberts, 1982: 192.

the fallout area. Although average insurance company payments on policies in Yakima were \$300–\$400, 90% or more of such amounts were for the labor involved (see also Chapter 8, Section 5). Thus, in some societies a tephra fall producing almost negligible damage to buildings may still result in significant costs.

The most obvious protective measure that can easily be taken against tephra fall is the removal of accumulated ash from the roof before the safety of the structure is endangered. Providing care is taken, the removal of the tephra should result in only minor damage to gutters. Where buildings are of a flimsier construction, perhaps made of local materials, it may be possible to "bump" the ash from the roof from the inside or to provide additional uprights to strengthen the roof. Such methods are reported in legends about the mid-seventeenth century fall of tephra from Long Island, Papua New Guinea (Blong, 1982: 107-108). Figure 5.13 and the earlier discussion suggest several design precautions which would make roofs less suceptible to collapse under unevenly distributed loads. Protection of house interiors from ingress of minor amounts of tephra can take many forms. One such method is illustrated in Fig. 5.15c.

Whilst protection from projectile impacts is more difficult some precautions can be taken. At Heimaey in 1973, 95 tonnes of corrugated iron sheets were used to cover all the windows facing the crater (*The American-Scandinavian Review,* 61(4): 331 quoted in Bruner, 1974) thus protecting house interiors from hot lapilli. Although it would be impractical to strengthen large areas of roofs, portions of houses constructed of concrete or of 10 mm steel sheet could provide a "projectile shelter". The design of such a shelter would need to recognize that the size of potential bombs within a few kilometers of a crater prevents the construction of a "bomb-proof" shelter and to take into account other likely volcanic hazards.

5.4 PYROCLASTIC FLOWS

As indicated in Table 1.4, pyroclastic flows are the chief agents of death in volcanic eruptions. Not surprisingly, this results in large part from the vulnerability of buildings to the impact of pyroclastic flows as indicated in Table 5.1. Severe forces on walls, projectile impact, collapse, racking, burial and ignition are all highly likely occurrences although no analyses of the causes of failure have been made. Some brief examples of the extent of destruction of buildings are recorded in Table 5.5. More detailed examples indicate that building failures in pyroclastic flows result primarily from flow velocity rather than from the elevated temperatures; however, as the following descriptions indicate, the extent of destruction is quite variable across even small areas (see also, Chapter 2, Section 3).

Excellent descriptions of the destruction of buildings by pyroclastic flows moving at about 33 m s^{-1} during the 1951 eruption of Lamington have been presented by Taylor (1958: 40): "At Higaturu [see Fig. 2.15], six miles [9.6 km] from the crater, only one house remained reasonably intact. The U-shaped residence of the District

Table 5.5 Examples of damage to buildings by pyroclastic flows

Eruption	Damage	Reference
Tarawera, 1917	Steamblast at Frying Pan Flat lifted roof off Waimangu accommodation house 800 m away	*CAVW*, 22: 66
Agung, 1963	More than 2000 houses destroyed, mostly by pyroclastic flows (see Fig. 3.18)	*BVE*, 1963, 3: 7
Ruang, 1874	200 houses destroyed	*CIRV*: 796
Sakura-jima, 1914	50% of houses in Take and 30% houses in Fujino (see Fig. 3.19) destroyed by pyroclastic flows and projectile impacts	Omori, 1916b: 215; *Kobe Herald*, 1914: 12
Usu, 1769	Pyroclastic flow burned down all houses in Osarubetsu (now Nagawa) 6 km S of crater	Katsui *et al.*, 1981: 7
Arenal, 1968	2 small villages completely destroyed	*BVE*, 1968, 8: 12

Commissioner... was pushed fifteen feet [4.6 m] northwards and damaged on the southern side by flying debris. Most of the other buildings at the government station were carried away and in most cases only the floors remained. A group of three steel Sydney-Williams huts on the eastern side of the parade ground was badly damaged and had partly collapsed...; on the opposite side of the parade ground the superstructure of the District office was swept away completely."

Damage to buildings at Higaturu was also quite variable: "The southern and western parts of the settlement sustained the most severe damage [Fig. 5.16a]. At the southern end a few leaning foundation posts were all that remained of some buildings, whole structures having been carried away. On the western side of the settlement, the steel-framed hospital building was completely wrecked..., whereas similar structures on the eastern side sustained lighter damage, and one house remained more or less intact... Some of the damage was undoubtedly caused by materials picked up and carried along by the nuée; corrugated iron and planks were scattered over a wide area..." (Taylor, 1958: 43).

Extreme variations in effects on buildings also occurred at Sangara mission in the marginal zone of devastation. "The mission residence... was undamaged except for a few small holes in the fibro-cement wall on the southern side of the building. Even the glass louvre windows were left intact...; but on the southern side of the house adjacent store buildings were badly damaged [Fig. 5.16b]... These buildings certainly helped to break the force of the nuée, but the apparent anomaly in the relative damage to the buildings was probably due more to relative structural strength than to a shielding effect" (Taylor, 1958: 40-41).

Pyroclastic flows of the base surge type during the 1911 eruption of Taal produced devastation every bit as complete as that around Lamington but at Taal it is difficult to separate the effects of flows from subsequent damage wrought by tsunami. On Volcano Island seven villages were obliterated (see Fig. 3.13). In most places no traces of the villages remained. To the west of the lake six villages which

A

B

FIG. 5.16 (top) The remains of a house in the northwestern part of Higaturu after the pyroclastic flow swept through; (bottom) Although the Sangara Mission House was hardly damaged by pyroclastic flows at Lamington, adjacent store buildings were destroyed (Photos: G.A.M Taylor, Bureau of Mineral Resources, Australia, M1745 and GB/3013).

FIG. 5.17 Houses in the village of Subig damaged by a pyroclastic flow during the 1911 eruption (Photo: *National Geographic,* April, 1911: 341).

contained a total of 543 bamboo houses were destroyed (Fig. 5.17). Debris was washed into the lake by the waves, leaving here and there a tree stump and an occasional house corner post standing above the smooth mud. In two villages elevated above the influence of the tsunami, wrecked houses were strewn everywhere. In the villages of Bayungan and Subig, located just outside the devastated zone, about half of the houses were destroyed (Rivers, 1911: 13-15).

The April 24, 1975, eruption of Ruapehu also produced a pyroclastic surge. At the "bomb proof" geophysics instrument pit on the northwest of the dome about 500 mm of local ground surface was eroded by the blast, and the heavy iron pit covers and the upper course of link logs lining the pit walls were blown away. The pit was completely filled with rocks up to 500 mm diameter and a matrix of coarse ash and mud (Nairn *et al.,* 1975).

The pyroclastic flows of Lamington, Ruapehu and probably Taal evidently destroyed buildings primarily as the result of lateral forces exceeding the strengths of the various structures in their paths. Despite evidence of ignition in St Pierre in 1902, most of the considerable destruction there also resulted from laterally directed forces far exceeding the resistance of structures. Keenan (1902: 189-190) commented: "Away from the main streets there were only the foundations of walls and heaps of rounded stones to show where houses had been; and one could pick one's way across the chaos of wreckage, in almost any direction, over piles of rubble, sheets of metallic roofing, steel braces or girders, masses of tree branches, iron bedsteads, smashed roofing-tiles, tangled telephone wire, burst-open safes and great mounds of ash-plastered building-stones, blocking up and almost obliterating the narrow streets. Even if I had been perfectly familiar with the city, I should have had great difficulty in finding my way about; and without such familiarity I could not orient myself at all. It was often impossible for me to determine whether I was in a street or in the midst of a ruined block of buildings. Of course, in a city that has been so completely wrecked there is little to describe. One can only say that it is a chaotic mass of rubble, plaster, roofing-tiles and shattered walls, with here and there the fire-scorched branchless trunk of a big tree" (1902: 189-190). Anderson and Flett (1903: 486) observed that a large statue of the Blessed Virgin was moved 13 m from its pedestal. A low stone wall nearly half a meter thick was toppled and also moved. Many buildings were levelled; afterwards many remaining walls crumbled at the slightest touch (Heilprin, 1903: 101-102). Similar comments have been made by Jaggar (1902) and Hill (1902a).

Most of the commentary on the destruction of St Pierre indicates that few of the buildings were burnt. Keenan (1902: 192) wrote: "In all parts of the city, and particularly at its southern end, there were quantities of wooden wreckage, in the shape of beams, planks, barrels and fragments of roofing, that had not been burned, nor even singed." On the other hand, a medical team that arrived two days after the May 8 eruption reported that everything combustible had been ignited, including clothing capable of withstanding a temperature of 135°C (Will in *HMSO,* 1903: 10). These contrasts in observations probably result from investigations being made in different parts of the city. Localised fires did consume buildings and their contents perhaps generated by lightning strikes, ignition of particularly flammable objects, or the gases from the large quantities of rum and oil stored in the city (Trechmann, 1938: 435).

Nonetheless, the pyroclastic flow was hot. Perret (1937: 94) provides clear evidence that glass objects were softened and green juicy fruits carbonised where no conflagration occurred. Trechmann considered the effects of the pyroclastic flow, separate from any subsequent ignition, on a number of household and other objects: "Most glass objects are struck and collapsed on one side; lamp shades or bottle necks have been crumpled up while the ash was falling through, and it has been imprisoned in the folds... Smaller iron objects, nails, wire, locks etc., are converted to crystalline magnetic oxide, sometimes with a core of metal left... It seems to be galvanized objects that were most easily oxidized, and around the nails the layer of oxide is sometimes arranged in a spiral form as if the gases had whirled round and completed the oxidation instantaneously. Coins are fused together or become saucer-shaped or are partly melted into blobs or have swollen up, obliterating the design on them; some have a malachite-like coating on them resembling the patina on ancient coins. A gold Portuguese coin in the St Pierre museum is merely distorted; gold melts at about 1075°C... Organic material — biscuits, macaroni, coffee, wafers for religious purposes, etc. — are, when struck by the *nuée,* changed to glistening porous charcoal with little or no loss of outward shape" (1938: 436-437).

These observations indicate that although temperatures reached at least several hundred degrees Celsius in these pyroclastic flows, the temperature effects on buildings and their material contents are relatively minor compared with the effects of the high-density, high-velocity flow itself. The same is true of the Mt Lamington pyroclastic flow in 1951; it is only where the effects of the flow itself were minimal that the elevated temperature and its effects could be determined. Thus, Taylor (1958: 46) estimated the temperature during the pyroclastic flow which devastated Higaturu to have been 200°C for 90 seconds. This temperature was not high enough to ignite or char objects in the interior of the one house that survived. Although a plastic lamp shade was softened and collapsed, a reel of cotton and the leather drive-belt on a sewing machine beneath the lamp were unaffected by heat. In the hospital, analysis of penicillin suggested a temperature of 145°C or greater for one to two minutes.

One further example illustrates both the effects on buildings and contents of elevated temperatures in pyroclastic flows and variations in the extent of damage. The main research station of the Geophysical Institute of the University of Alaska near Augustine in 1976 sustained major damage from air blast and thermal effects accompanying pyroclastic flows even though the two corrugated aluminium buildings had been located in the lee of a small hill to provide some protection from such flows. The station was apparently hit by the turbulent hot gas and dust clouds of two pyroclastic flows on Janury 23, 1976, but was not affected by the dense basal part of the cloud. In many ways this situation seems to be analogous to that at St Pierre. All of the windows in the main building on the side facing away from the volcano were blown out. The aluminum sheathing was also dented on this side (the north side). On the east side the corrugated sheathing was torn loose and two mattresses near the window opening were incinerated. Other mattresses in the building did not ignite. "A plastic towel holder located under the kitchen window sill showed the effects of partial fusion, and a plastic measuring cup on a shelf at window level was also partially melted... Some of the paper wrappings and cartons on these same shelves showed incipient charring. A battery which was sitting on the floor opposite a window also showed the effects of incipient melting" (Kienle and Forbes, 1976: 44-45). The softening of plastic objects suggests a relatively low and transient temperature in this portion of the pyroclastic flow evidently affecting only those objects close to window openings. The ignition temperature of the mattresses is not known but as the roof of the building was penetrated by small blocks and lapilli, their incineration may have resulted from a bomb rather than from the pyroclastic flow.

In all the examples presented so far, the deposits left by pyroclastic flows have been very thin — about 50 mm at the Augustine research station, less than 2–3 m in St Pierre (Fig. 2.14), and 150–300 mm at Higaturu (Taylor, 1958: 52). At the other extreme, the deposits left by pyroclastic flows (especially those of the debris avalanche type) may be very thick; Harry Truman's house at Spirit Lake at the foot of Mount St Helens and Kawakami spa near Bandai in 1888 are both buried by more than 40 m of deposit. Herculaneum was buried by hot mud (pyroclastic flows and lahars — Sheridan *et al.*, 1981: 283) to a depth of 20 meters. There the force of the flows varied greatly: "At some points its pressure felled walls and columns, smashed furniture, and carried away statuary... At other points it rose so gently that it oozed over eggs without breaking them and filled rooms without disturbing the position of pots on a kitchen stove or a cradle in a nursery... Luncheon is still waiting on tables; snatches of poetry, erotic phrases, and scores of names in Latin and Greek — and even in Oscan — are legible on the walls; the case of fine glassware was found half-packed; the candelabrum and bronze Dionysus await repairs; the sick boy in the house of the gem-cutter lies on the bed, his lunch of chicken uneaten; the baby remains in the cradle... In some places [the pyroclastic flow] was hot enough to carbonize foodstuffs and wood; in others, hot enough merely to scorch cloth and papyrus; and in still others, wood, rope, eggs, and even fishnets were left almost totally unburned. In some houses wax tablets remained unmelted" (Deiss, 1966: 18-19).

In many ways the Herculaneum example is more instructive than the others. Complete burial has meant preservation and has prevented the occurrence of secondary effects resulting from subsequent eruptive activity and even weathering. The area excavated is no more than a few hectares and yet, in that area, enormous variations are found in the velocity and temperature of the pyroclastic flow.

The Augustine and Ruapehu examples emphasise the difficulties of engineering structures to withstand pyroclastic flows. In both cases some protective measures were taken as the possibility of pyroclastic flows was recognised — near Augustine the buildings were located in the lee of a small hill and on Ruapehu the instrument shelter was designed to be bomb-proof. In neither case did the structure survive intact.

Where burial depth is more than 10 m the question of above ground or below ground protection is rather academic. In some cases, it may be possible to build structures, similar to those used in avalanche terrain, to deflect pyroclastic flows. Techniques used to flood proof buildings would prevent the ingress of much of the asphyxiating ash. Such techniques mainly involve the sealing of wall openings against water entry and/or damage by floating debris (Flood Loss Reduction Associates, 1981). Where deposits are thinner, it is clearly possible to design and locate portions of buildings that will retain their structural integrity and provide some protection for the occupants. Below ground cellars are obvious examples. In other cases, bathrooms are frequently the strongest portion of a structure as the wall spans are short, window openings are small and the plumbing provides additional stiffness. Human survival would probably require additional water and air supplies as the room would need to be sealed. Relatively minor modifications to such rooms could enhance but not guarantee personal safety.

5.5 LAHARS

Although lahar velocities may (or may not) be lower than those of pyroclastic flows, impact energies may be equal as lahars are denser and may exhibit quite high yield strengths. Lahars may be hot as in the case of those down the Toutle River from Mount St Helens in 1980, or cold with chunks of ice adrift in them as in the jökulhlaups of Iceland and the 1877 mud flows of Cotopaxi. Lahars may violently destroy buildings, forcing boulders, trees or other debris through the upslope ends of the structures, they may push or float buildings off their foundations, or they may quietly ooze into buildings through doors and windows, causing little damage to the exterior shells (Table 5.1). Because temperatures are lower than in the case of pyroclastic flows, say, less than 40°C–50°C, direct ignition of flammable objects is unlikely, though secondary ignition may result from ruptured gas lines, open fires, or short-circuiting of the electrical system.

A good description of the effects of a lahar on a building is given in Chapter 2, Section 4. This lahar, from Lassen Peak in 1914, destroyed houses on four or five other ranches (Hill, 1970: 216). In some cases the buildings were totally removed: "The Jessen place was once a fertile meadow, but now it is a waste of land, with not a vestige of buildings or fences to show where the place used to be. The Lost Camp ranch was also destroyed. A pile of hay was all that could be found on the place" (Loomis, 1926: 46).

Further north, however, Mount St Helens produced more impressive lahars in 1980. Figure 4.2 illustrates the course of the lahars down the North and South Fork Toutle and the Cowlitz Rivers. Camp Baker was reported destroyed about 2.30 p.m., half an hour after the lahar first moved into the equipment and shop areas (Fig. 5.18). Downstream at the Toutle Boys Ranch (Fig. 5.19) more than one meter of mud was deposited in some buildings (Cummans, 1981a: 9,10). The lahar arrived at the mouth of the Toutle River at about 8.30 p.m. on May 18, carrying with it logs, buildings and a pickup truck.

It is difficult to establish the exact number of houses affected by the Toutle and Cowlitz River lahars as various estimates include damage by pyroclastic flows. In July, 1980, 61 homes were reported as totally destroyed, a further 55 having major damage, and 64 rendered inaccessible with 50% of the latter suffering major damage or total destruction (*Washington Geologic Newsletter*, 8(3): 18). An early estimate (*The Columbian*, May 26, 1980: 11) gave 123 houses and cabins destroyed in Skamania County (see County line on Figure 3.17), 23 more with major damage and 11 with minor damage. In Cowlitz County 75 houses and cabins and two businesses were reported destroyed. A later estimate (*Volcano*, 1980: 36) gave 300 houses destroyed and numerous others flooded in the Toutle Valley alone.

In fact, few accounts of the destructive effects of lahars offer precise estimates of the number of buildings destroyed or damaged. Table 5.6 presents a large number of items on the destruction of buildings by lahars. In most cases the number of buildings involved is not known but in a few cases more precise information is

FIG. 5.18 Building damaged by lahar at Camp Baker, North Fork Toutle River (Photo: R.L. Schuster, US Geological Survey, July, 1980).

FIG. 5.19 Cabins buried by lahar at Boys Ranch, North Fork Toutle River (Photo: R.L. Schuster, US Geological Survey, July, 1980).

Table 5.6 Effects of lahars on buildings

Eruption	Effects	Reference
Mt Agua, 1541	Destroyed former Guatemalan capital Ciadad Vieja	Neall, 1976: 235
Yake-dake, 1585	Lahar buried village of Nakao	*CAVW*, 11: 101
Öraefajökull, 1727	Jökulhlaup swept several buildings away	Thorarinsson, 1948: 32
Asama, 1783	Nuée dammed Agatsuma R., formed lahar, destroyed a number of villages	Aramaki, 1956: 198
Cotopaxi, 1877	Pyroclastic flows onto ice and snow caused mudflows which seriously damaged buildings up to 240 km from source	Miller *et al.*, 1978
Aoba, 1879?	Lahars destroyed villages on SE flank	Warden, 1970
Soufrière, 1902	Secondary lahars overturned several small houses behind Georgetown	Anderson & Flett, 1903: 423.
Katmai-Novarupta, 1912	Secondary lahars carried houses off foundations and crushed walls in Kodiak	Martin, 1913: 174
White Island, 1914	All buildings (only three or four) and plant destroyed; boulders 10 × 5 m carried as far as factory buildings; about 1 m mud deposited at site of workmen's huts	*CAVW*: 22, 42
Dieng, 1939	63 houses destroyed	van Bemmelen, 1949: 18
Villarrica, 1964	Partial destruction of Conaripe	*Bull. Volc.*, 38 Supp.: 44
Deception Island, 1969	Lahars from melting snow and ice forced way through British huts, pushing down walls, clogging generators and congealing to an immovable icy mass	Baker, 1969: 126
Hudson, 1971	Houses destroyed in Rio de los Huemules Valley	*BVE*, 11, 1973: 43-44
Ruapehu, 1975	Lahar from crater lake and melting of snow damaged kiosk buildings and undercut chairlift foundations	Nairn *et al.*, 1975
Semeru, 1976	Secondary lahars destroyed 258 houses in Sumber Wulluh village	*BVE*, 1978, 16: 32
Usu, 1977-1978	Secondary lahars on October 24, 1978, destroyed 4 houses, partially destroyed 21 houses, > 100 houses inundated	Kadomura *et al.*, 1980:169

available. Kelut, for example, has produced lahars during 28 eruptions. In 1826, 65 villages were destroyed, in 1848, 800 houses, in 1875, 30 villages; in 1919, 9000 houses in 104 villages, in 1951, 11 villages were partially destroyed and in 1966 several villages were totally destroyed with many other villages partly damaged (*CIRV* 1979; *Bull. Volc.*, 38, Supp.: 15; Sigit, 1965: 3; Neall, 1976: 235).

Elsewhere in Indonesia, damage to buildings by lahars is common on Merapi although pyroclastic flows are better known on this volcano. Similarly, around Agung in 1963 a number of villages (including Subagan village where 140 were buried in their houses — Surjo, 1965: 24) and towns were severely damaged by lahars (Figs. 3.18, 5.20). Lahars developed on Galunggung in 1982 (see Fig. 4.7)

FIG. 5.20 Lahar damage to buildings and streets of Subagan (see Fig. 3.18) southeast of Agung, Bali, 1964 (Photo: *National Geographic,* September, 1964: 439, Robert F. Sisson).

and caused damage to several villages when buildings were filled 1–3 m deep with mud. An earlier major eruption, in 1822, produced pyroclastic flows and lahars which destroyed 114 villages.

In 1926 large very fluid lahars swept down the Hurano and Biei valleys on the northwest slopes of Tokachi (see Fig. 2.20). Significant discrepancies exist in the reported destruction of buildings; Tanakadate (1927) reported several houses in Biei valley and 73 houses in Hurano valley destroyed with a further 80 buildings damaged in the latter area while Murai (1960: 57) and *CAVW* (11: 311) recorded the villages of Kami-Furano, Shimo-Furano and Biei and a total of 5080 houses devasted.

A newspaper cutting of the time* recorded the scene in Kami-Furano: "Of the submerged half of the town not one house or fragment of a house projected above the muddy waters to show where the street had been. All had been battered to bits and washed away. In their place was a huge causeway of trees razed off the mountain and of tree stumps uprooted from the rice fields by the fury of the flood, and carried along in the swirling tide to act as battering rams against houses made of wood and paper. No wonder that the waters won."

*Source unknown — from a collection of such cuttings in the Lyman House Museum, Hilo, Hawaii.

FIG. 5.21 The effects of lahars on the outskirts of Cartago in the 1964 wet season. Most homes were completely destroyed but a few remained intact but filled to the eaves with mud (Photo: H.H. Waldron, US Geological Survey).

At Irazú, numerous secondary lahars developed following lengthy tephra falls during 1963–1965. Figure 2.18a illustrates the rapid rise of stream hydrographs in the Rio Reventado during the 1964 wet season. The most destructive lahar, resulting from a violent rainstorm on December 9, 1963, "reached a height of more than 12 m. Near Cartago the debris flow spread out over the surface of the alluvial fan in a tongue of destruction about 1 km wide, [Fig. 5.21] devastated an area of nearly 3 km^2, and inundated another several square kilometers with muddy waters. More than 20 persons were killed by this disastrous debris flow and more than 300 homes were destroyed" (Waldron, 1967: I16-I17). Lahars caused further damage to buildings during the 1965 rainy season.

Because lahars almost always follow topographic lows the areas of highest risk are more easily defined than is the case with other volcanic hazards. Furthermore, a wide range of protective measures can be taken including diversion, ponding, re-vegetation, drainage of crater lakes and installation of warning systems.

Extensive lahar control efforts have been undertaken around Kelut (Fig. 5.22) since 1905 when an earth dam was built on the Blitar river for sediment storage. This was destroyed by lahars in the 1919 eruption and rebuilt, but by 1925 only 2 m of the 7.5 m storage remained. The 1919 eruption also led to a scheme, completed in 1926, to lower the volume of the Kelut crater lake from 38 million m^3 to 1.8 million m^3 stored water (Neumann van Padang, 1960). The 1951 eruption deepened and widened the crater and filled the drainage tunnels with debris so that the water volume had risen to more than 23 million m^3 before the 1966 eruption (Zen and

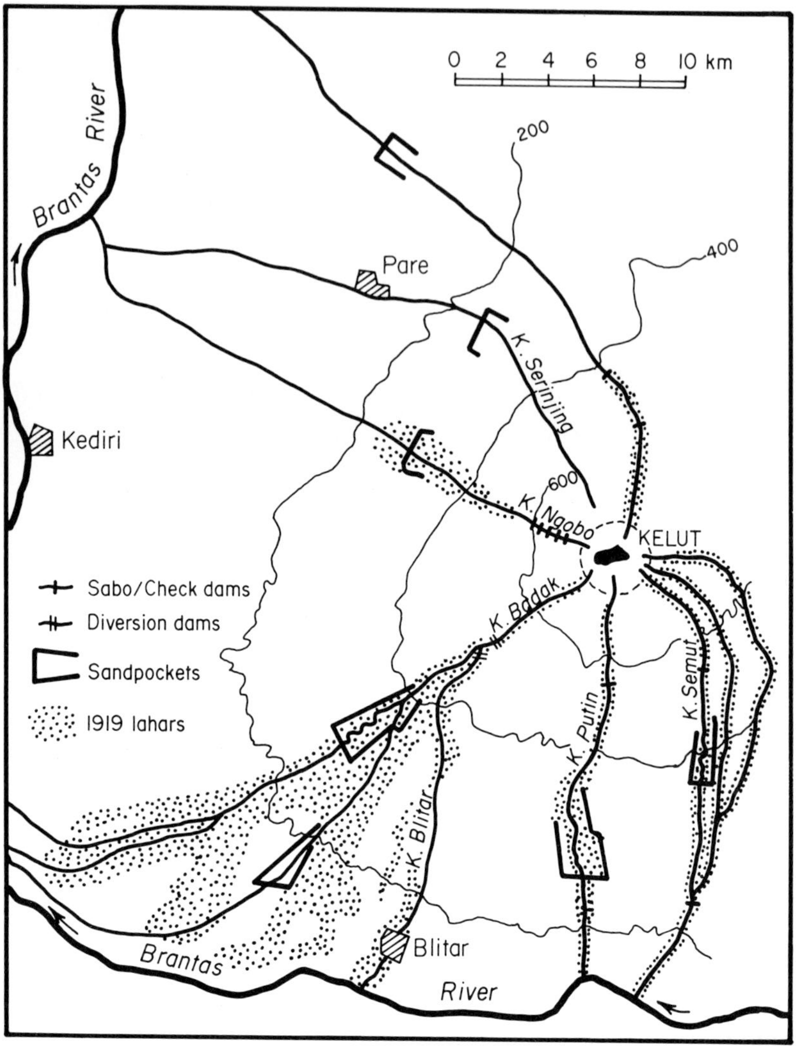

FIG. 5.22 The southwestern slopes of Kelut indicating the location of check dams, sandpockets and other structures designed to protect downstream areas from lahars (after Smart, 1981: 616).

Hadikusumo, 1965; Sigit, 1965). Subsequently, a drainage system was reinstalled and other control measures on the slopes reintroduced (Fig. 5.22) using Sabo check dams (see Ikeya, 1979), diversion dams and sediment retaining structures (sandpockets). The diversion dam on the Badak river is intended to protect the town of Blitar. Other structures are designed to control the points where lahars can leave river channels and the spread of sediments (Smart, 1981).

Similiar efforts at lahar control have been made around other volcanoes. Lahar pockets have been built on the slopes of Merapi (Legowo, 1980). Attempts have been made to lower the level of water in the crater lake of Ijen in Java (Macdonald, 1972: 173) and Neall (1976: 238) has advocated a similar step at Ruapehu. Several kilometers of levees were built along the Rio Reventado to protect Cartago from Irazú lahars. Massive levees were built along parts of the Cowlitz following the Mount St Helens, 1980, eruption. A number of sedimentation basins were also built in the Toutle River and plans were advanced to lower the level of Spirit Lake (Stewart, 1982). Figure 4.7 locates several lahar control structures built on the slopes of Galunggung in 1982. One 1.5 km long lahar pocket alone involved moving 27 500 m^3 of material, much of it without heavy construction equipment. However, efforts to build lahar pockets before the 1982-1983 wet season drew together the largest assembly of heavy plant seen in Indonesia — 90 dumptrucks, 25 bulldozers, 13 loaders, seven backhoes and eight cranes. "The maximum height of the dikes is set at 5 m to avoid engulfment if and when breached. The checkdams, more appropriately called spillways, are there to allow water to pass out of the mud-filled lahar pockets. Once the lahar pocket is filled, the dikes can be raised and the process of containment can begin again. The consolidation dams, which serve to anchor the lahar [pocket] to the earth, are constructed of gabions held fast by piling" (*Asian Building and Construction,* 1983, January: 25). In some cases, existing hydroelectric or water storage dams on rivers draining volcanoes could be used provided the water level was lowered before sediment entered the reservoir. In the United States Pacific Northwest, for example, suitable storages exist on some rivers draining Mounts St Helens (Figure 4.2), Baker and Rainier. Revegetation and other soil conservation works in tephra fall areas may also be effective in controlling secondary lahar formation. Such efforts were made following the eruptions of Irazú in 1963 and Mount St Helens in 1980. More sophisticated techniques have been used to protect important individual structures. Following the 1953 lahar which destroyed the Tangiwai railway bridge on the lower slopes of Ruapehu in 1953, a warning system was designed, which relied on the detection by electrodes of acidic crater lake waters in the river and a visual display panel at a nearby railway station (Neall, 1976: 238). Normal methods of flood detection in streams using telemetered river heights could also be effectively used, particularly where secondary lahars pose a threat.

5.6 EARTHQUAKES AND GROUND DEFORMATION

Building failure under earthquake loads can occur in many ways: (i) foundations or supports may fail; (ii) racking may occur to the point where major structural members fail; (iii) racking severs the (frequently) weak connections between roof, walls, and floor so that the walls fall inwards or outwards leaving the roof (momentarily) unsupported; (iv) relatively weak items such as towers, chimneys and parapets fail, falling through the body of the building, perhaps severing major structural members in the process; and (v) rupturing of gas lines or electrical circuits may lead to fire or broken water pipes may cause flooding of part of the structure.

The seismic resistance of buildings is quite well understood particularly in comparison with building resistance to mass flow phenomena or tephra loads. In general, Table 5.7 indicates decreasing resistance to earthquake failure, based on the relative cost of earthquake insurance for various building styles in California. However, building age, the quality of workmanship, and the quality of construction materials all affect seismic resistance. More importantly, buildings interact with substrate materials during earthquakes, so that the seismic resistance of soils and rocks dramatically influences building behavior. In general, seismic resistance decreases through the series: (i) hard rocks; (ii) soft rocks; (iii) alluvial soils; (iv) lake muds; and (v) reclaimed land. In addition, the earthquake vulnerability of a structure decreases with increasing distance from the earthquake epicenter and increases with increasing ground slope.

Brief examples of damage to buildings resulting from earthquakes are presented in Table 5.8. Despite the title of this table there are strong doubts that the Santa Maria 1902 and Vesuvio, AD 79, earthquakes were of volcanic origin (see Chapter 2, Section 5). For a few other examples more details are available.

During the 1911 eruption of Taal masonry walls in Talisay at the northern end of Lake Bombon (see Fig. 3.13) were cracked and displaced by earthquakes. Even 30 kilometers away from the volcano, on the shores of Balayan Bay, cracks appeared in the ground and the church tower was partly demolished (Fig. 5.23) (Pratt, 1911: 65,69). Other churches and houses in the area were also damaged, but all these structures seem to have been greatly weakened by age, humidity and earlier earthquakes (Masó, 1911: 23).

Earthquakes caused considerable damage to buildings during the 1914 Sakurajima eruption. It will be recalled (Chapter 2, Section 5) that the major earthquake recorded a magnitude, M = 6.7. Table 5.9 indicates damage to buildings in various localities. No information is available as to how this survey was conducted; it should be noted, however, that *The Kobe Herald* (1914: 12) in a special supplement on the eruption reports that in Kagoshima 35 houses were destroyed, 113 seriously

Table 5.7 The relative resistance of various building styles to earthquake damage.

Building style	Relative resistance
Small wooden frame	1.0
Steel-frame with reinforced poured concrete	0.6
Wood-frame with masonry veneer or other liabilities	0.43
Steel-frame and reinforced concrete-frame with segmented masonry walls	0.43
Single-family dwellings with reinforced masonry	0.38
Unreinforced brick and larger reinforced masonry buildings without steel or concrete framing	0.13
Unreinforced masonry, masonry-veneer, and adobe buildings	0.06

After Yanev, 1974: 234.

FIG. 5.23 Damage to Cayaysay church at Taal 1911 by volcanogenic earthquakes (Photo: *National Geographic,* April, 1912: 359).

Table 5.8 Examples of buildings and towns damaged by presumed volcanogenic earthquakes

Eruption	Damage	Reference
Vesuvio, AD 79	Earthquakes of 62-64 AD, and 79 AD caused much damage to buildings in Herculaneum, Pompeii and Campagna generally	Dilke and Dilke, 1979: 743
Etna, 1669	Town of Nicolosi, 15 km from Etna summit destroyed	Rodwell, 1878: 90
Unzen, 1792	April 21, one of many earthquakes destroyed houses of Simbara on east cost of peninsula	*CAVW,* 11: 43
Mauna Loa, 1868	46 houses destroyed by earthquake	Hitchcock, 1909: 107
Soufrière, 1902	At Chateau-belair house walls were cracked and roofs were subsequently noticed to leak when heavy rain came	Anderson and Flett, 1903: 406
Santa Maria, 1902	Town of Quezeltenango destroyed 15 km from volcano	Rose, 1972: 31
Katmai-Novarupta, 1912	At Kanatak, driftwood and sod huts collapsed	Erskine, 1962; 61
Api Siau, 1974	Feb 28, partial collapse of many houses; March 25, complete collapse of primary school	*SICSLP,* 54-74

Table 5.9 Damage to buildings resulting from earthquakes during the 1914 Sakura-jima eruption.

Buildings	Damage	Kagoshima City	Sakura-jima Island*	Other districts	Total
Dwellings	Totally collapsed	29	28	56	113
	Half collapsed	117	38	79	234
	Damaged	9405	1222	921	11548
Other Buildings	Totally collapsed	94	62	93	249
	Half collapsed	182	28	117	327
	Damaged	1058	58	25	1141

After Omori, 1922: 515.
*Includes other portions of Kagoshima District.

Table 5.10 Examples of damage to buildings and contents in Puna District, Hawaii, resulting from November 29, 1975, earthquake.

Locations*	Damage
Volcano	Three wooden water tanks were destroyed, several others were partially damaged.
Kalapana	A wood frame house shifted 1 m from foundation. One water tank damaged. Loose objects fell off shelves.
Kurtistown	Damage in seven residential homes – cracked concrete steps, house and garage moved from foundation, 100 mm wall separation, roof separation, chinaware broken in falls from cupboards. Rock wall damaged.
Opihikao	Two water tanks at church destroyed.
Pahoa	Three homes moved from foundations. Other damage included a broken waterline, collapsed water tank and tool shed.
Glenwood	Loose objects fell off shelves.
Mountain View	Floor of water tank cracked; plexiglass cracked; television set shifted off stand and fell to the floor; rock wall damaged.

After Tilling *et al.*, 1976: 5.
*The locations of these places are shown in figure 4.1.

damaged and 1090 partly damaged. Although earthquakes damaged far more houses than all other hazards during the eruption, earthquakes were responsible for the destruction of less than 5% of the total number of dwellings destroyed.

Some months after the 1957 eruption of Capelhinos (Azores) began, a seismic crisis occurred on the island with more than 450 felt earthquakes in the two to three days beginning May 12, 1958. Many of the earthquakes had Modified Mercalli intensities of X with the epicenters clustered around three centers (Fig. 5.24). A series of fractures developed, with a narrow belt of subsidence, elongated between the eruptive centre at Capelhinos and Fayal caldera. In all, more than 500 homes on the western side of the island were destroyed by the earthquakes in a two day period. There was no major damage in the main town of Horta at the opposite end of the island (Machado *et al.*, 1962: 3523; Parsons and Mulford, 1958: 15).

A great deal more detail concerning damage to buildings is available for the M = 7.2 earthquake and eruption at Kilauea on November 19, 1975. The hypocenter was fixed at 5 km west southwest of Kalapana (see Fig. 4.1) at a depth of 5 km. Several foreshocks and hundreds of aftershocks occurred. Whether the earthquake was volcanogenic in origin or a tectonic earthquake leading to an eruption is not clear; in terms of damage to buildings, the distinction is of little importance. The

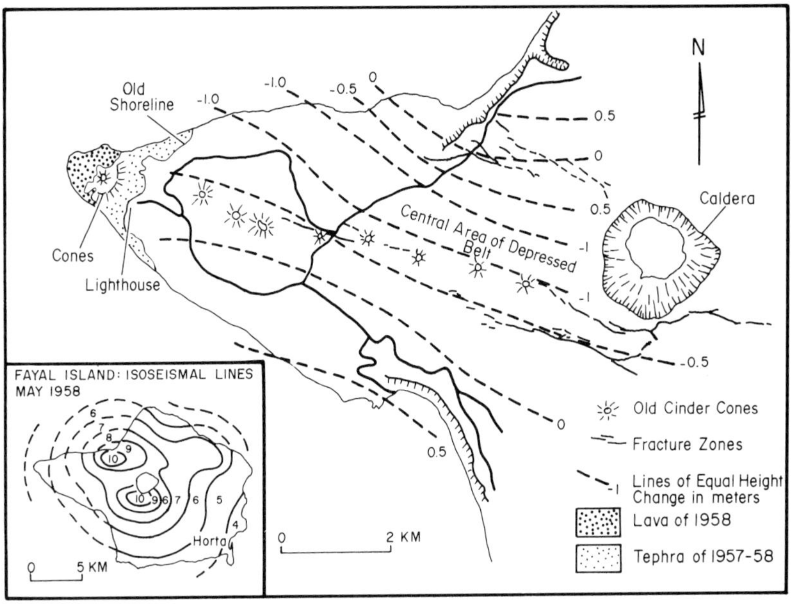

FIG. 5.24 Western Fayal Island showing May 1958 ground level changes and isoseismal lines for May 13-14 seismic crisis (after Machado *et al.,* 1962: 3521).

main earthquake caused minor to moderate damage to five churches (four in Hilo, one in Opihikao), 11 commercial buildings (10 in Hilo, one in Mountain View) and 89 homes (including 51 in Hilo and 23 in Puna). Five poorly constructed or old houses (four in Hilo, one in Ka'u) were completely demolished (Tilling *et al.*, 1976: 1, 3). Table 5.10 indicates the range and the types of damage to buildings and contents in Puna District where Modified Mercalli intensities were described as VII–VIII. Severe earthquakes associated with the 1868 eruption of Mauna Loa caused similar widespread damage. In Kau District (west of the volcano) alone, for example, 46 houses were destroyed by the earthquake (Brigham, 1909: 101-118).

Ground deformation as well as ground shaking may damage buildings. Following the 1902 Soufrière, St Vincent, eruption subsidence occurred in the Wallibou area southwest of the volcano. Huggins (1902: 16) reported that boats now travel over the site of Wallibou village 15 m out from the present beach. Where the land was formerly 6 m above sea level, there is now about 12 m of water. A seismic crisis at Campi Flegrei (near Pozzuoli) produced ground deformation of nearly 800 mm y^{-1} in 1970. Ground deformation and earthquakes opened cracks in the walls of several apartment buildings in 1970 and led to their eventual collapse (Imbo, 1979).

Considerable ground deformation also occurred during the 1944 eruption of Usu. Earthquakes which began on December 28, 1943, caused some damage to

FIG. 5.25 Collapsed portion of reinforced concrete Sankei Hospital building in Sobetsu Spa following ground deformation associated with the 1977 Usu eruption (Photo: June, 1978, by courtesy of H. Kadomura).

water pipes and plastered walls of dwellings at Toya Hot Springs on the northwestern side of the mountain (see Fig. 3.4). In some areas north and east of the mountain, uplift totalled as much as 50 m during the next six months (Minakami *et al.*, 1952: 264-265). In 1977, Usu erupted again. The largest recorded earthquake reached only M = 4.3 and no damage to buildings around Usu was observed. However, ground deformation continued for more than a year along fault lines identified after the 1910 eruption. By August, 1978, 41% of the ferro-concrete and steel frame structures in Toyako Spa and 71% in Sobetsu Spa (a total of 46 out of 91 buildings) had been damaged by earth movements. The two to three storey reinforced concrete Sankei Hospital, built 1968–1974, collapsed in January, 1978 (Fig. 5.25), after showing dislocations of 5 mm per day in November, 1977. A three storey reinforced concrete building, a dormitory for the employees of Manseikaku (Hotel), is shown in Figure 5.26. A fault line ran through the middle of this building. Ground movement was noted in August, 1977, with cracks in the wall becoming conspicuous in mid-October. On November 5 the largest crack was 15 mm. In February, 1978, the vertical difference at the fault line was 150 mm, increasing by mid-March to 350 mm. At midnight on March 17, 1978, the walls and the reinforcing steel gave way, the building shearing completely at the fault. By July, 1978, the vertical difference at the fault line was 1.3 m and continuing. Figure 5.26 shows the

FIG. 5.26 Deformation of the apartment house for workers at Manseikaku Hotel by ground movement along fault line following Usu, 1977, eruption (Photo: April, 1981, by courtesy of H. Kadomura).

condition of the building in April, 1981. Other buildings in the area failed in a similar way or developed cracks in floors and walls and distortion of doors and connections (Shibata *et al.*, 1978).

These examples indicate that ground deformation may cause damage and building destruction every bit as serious as the damage resulting from ground shaking. However, destruction of buildings takes place over a much longer period of time allowing safe evacuation of personnel and the buildings' contents. While the damage to structures may be just as costly as that resulting from earthquakes, the cost in terms of contents should be minimal by comparison. The Usu examples indicate that even seismically-engineered buildings designed to withstand a horizontal acceleration of 0.18 g (Shibata *et al.*, 1978) can be readily destroyed by relatively slow surface movements.

Whilst most of the literature (see for example Seed, 1975; National Academy of Science, 1982; Beavers, 1981). deals with the seismic resistance of multi-storied engineered structures, the safety of individual dwellings can be significantly increased by relatively simple means as P. Yanev (1974) has demonstrated in his excellent *Peace of mind in earthquake country.* Such measures include improving roof–wall–floor–foundation connections, the use of lateral bracing, and anchoring vulnerable objects such as chimneys, parapets, gas stoves, wall cupboards and grandfather clocks!

5.7 TSUNAMI

Table 5.1 indicates that damage to buildings by tsunami may result from flooding, excessive lateral forces, transport, racking and foundation failure. The situation with respect to any particular building will vary with height above mean sea level, distance from high tide level, topography, other obstacles, and wave refraction. Some data suggest that the backwash may be more damaging than the uprush (Vitaliano and Vitaliano, 1971: 94). The first wave may be by no means the largest; the destruction of buildings may be progressive. Brief examples of the effects of tsunami on buildings are presented in Table 5.11.

In general, the extent of damage to buildings is a function of inundation height, h, and the water velocity on land, v (see Chapter 2, Section 6). Studies of the effects of tsunami on the Japan coast indicate that one and two storeyed wooden houses are damaged by inundation heights of 1.0-1.5 m and destroyed as a result of ground floor collapse where water depths exceed 2 m. Houses not tied to their foundations floated with inundation heights of 2 m or more (Hatori, 1964: 189). The proportion of damaged houses, D, may be approximated by $D = 9\sqrt{hv}$ when D is defined as:

$$D = \frac{a + b + c/2}{a + b + c + d} \times 100$$

where a is the number of houses washed away, b the number of destroyed houses, c the number of half-destroyed houses, and d the number of inundated houses, including inundation of underfloor areas (Hatori, 1964).

The best known example of tsunami damage is that resulting from the Krakatau eruption of 1883. Figure 2.24 illustrates the areas inundated by the waves of

Table 5.11 Examples of damage to buildings by tsunami

Eruption	Damage	Reference
Oshima-o-shima, 1741	Destroyed 850 houses	Siebert and Kienle, 1983
Sakura-jima, 1780	168 houses totally destroyed; 630 houses partially destroyed	Omori, 1916a: 66
Unzen, 1792	3284 houses destroyed by tsunami generated by 0.48 km^3 avalanche into sea	Sapper, 1917: 133; *CAVW*, 11: 43
Tambora, 1815	Houses ruined at Bima 70 km east of volcano	Neumann van Padang, 1971: 59
Mauna Loa, 1868	186 houses destroyed	Hitchcock, 1909: 107
Pelée, 1902	May 5, lahar down R. Blanche created wave which flooded lower St Pierre	Keenan, 1902: 207
Taal, 1911	Lake level rose 2.5 m, carried away houses (see Fig. 5.17)	Pratt, 1911: 65
Taal, 1965	Waves on lake reached >4.7m and swept 80 m inshore – houses destroyed	Moore *et al.*, 1966: 960
Kilauea, 1975	Caused property damage of about $1.4 million	Tilling *et al.*, 1976: 9.

August 26 and 27, 1883. Table 2.10 indicates the maximum wave heights for locations shown on the figure. The following chronology of the destruction of towns along the Sunda Strait shores has been pieced together from data provided by Metzger (1884) and Wharton (1888):

August 26, *1800–1900 h:* The first large wave reached Tyringin where it destroyed many houses near the sea. Three fires started as a result of lamps being tipped over. A wave caused considerable damage at Telukbetung, sweeping away houses.

1900–1930 h: The low-lying Chinese camp at Merak was swept away.

August 27, *0100 h:* Sirik village, 11 km south of Anyer submerged.

0630 h: Much damage done in Semanka Bay; a wave swept away nearly the whole of Anyer. At Anyer one of the survivors said "I went out about 5.15. After having talked with several persons, I saw the wave, still far off, rapidly making towards us, I ran away, was followed by the wave, fell down quite exhausted, but happily on a hill, where the water could not reach me. Before my eyes, all the houses along the beach were destroyed". Another person reported: "I was early at the beach... On looking round I saw a high wave which I could not escape; I was lifted from the ground, but caught hold of a tree. Then I perceived several waves, which followed the first; the place where Anjer had been before was covered by a turbulent sea, from which some trees and roofs of houses were still peeping out. After the wave had flowed off, I left the tree and found myself in the midst of the devastation. The Chinese camp was not yet destroyed" (Metzger, 1884: 241).

0730 h: The destruction of Anyer was completed by a second wave. Neale (1885 in Simkin and Fiske, 1983) reported that his companion, who had lived in Anyer all his life, on visiting the town six weeks after the tsunami, could not positively point out even the street in which his home had been, such as the extent of the destruction. Telukbetung was also overwhelmed at about this time. The *Gouverneur Loudon,* a Dutch Indian steamer of 745 tonnnes was under steam. The *Loudon* "turned her head to it, was lifted up, but kept well; now the wave rushed on to the beach, and before the eyes of the passengers and crew of the *Loudon* houses disappeared;... The place where Telok Betong [Telukbetung] had been before was changed into a violent sea (except the buildings on the hills)... [Fig. 5.27] From the barracks at Telok Betong, on the Talang Hill, about twenty-five metres above the level of the sea, an eye-witness wrote: 'At 6.20 I went to Kampong (village) Kankong, about 1400 metres distant from the barracks, to see the destruction which the wave had caused the night before. After I was there I saw a wave rushing on to us; we hastened to the hills, the villagers following us. When I had reached the barracks, I saw Kampong Kankong had disappeared, and so had the other villages near the beach'" (Metzger, 1884: 241-242).

1000 h: An immense wave inundated the whole Sunda Strait foreshores of Java and Sumatra and carried away the remaining portions of Tyringin, Merak and

FIG. 5.27 Present-day Telukbetung viewed from the hill about 25 m above sea level looking toward Lampong Bay. All of the area in this view was inundated and the houses destroyed by the 1883 tsunami.

Telukbetung and many other hamlets and villages. In all, 165 villages were completely destroyed, 132 villages were partly destroyed and two of the five lighthouses on the coast were demolished (Judd, quoted in Symons, 1888: 14; Neumann van Padang, 1971; *CIRV,* 1979: 796). Even in Batavia (Jakarta), tsunami caused considerable damage to buildings on the afternoon of August 17: "Suddenly, without any warning, a tidal wave . . . made itself felt in the city. The Dutch capital has no harbour, and the only approach to it is by a lone canal nearly two miles [3.2 km] in length, lined on either side by massive brick walls. In this channel, leading from the roadstead to the city wharves, the water rose at an alarming rate and burst over the adjoining land. This was the first intimation at Batavia of the terrible wave which (as we discovered later on) was the messenger of death to so many thousand inhabitants on the western shores of Java. Its effects in the city were quite bad enough. Although this great torrent of water had travelled nearly ninety miles [145 km] it dashed up this Batavian canal with great power. In spite of distance, its fury was not then fully spent. In the streets of the capital, adjoining the canals and wharves, the water rose to a depth of several feet [one or two meters], and the people had to run for their lives. Not long afterwards I steamed down the canal in a launch, and saw the destruction which had been caused. In several places the massive brickwork lining the sides had been swept

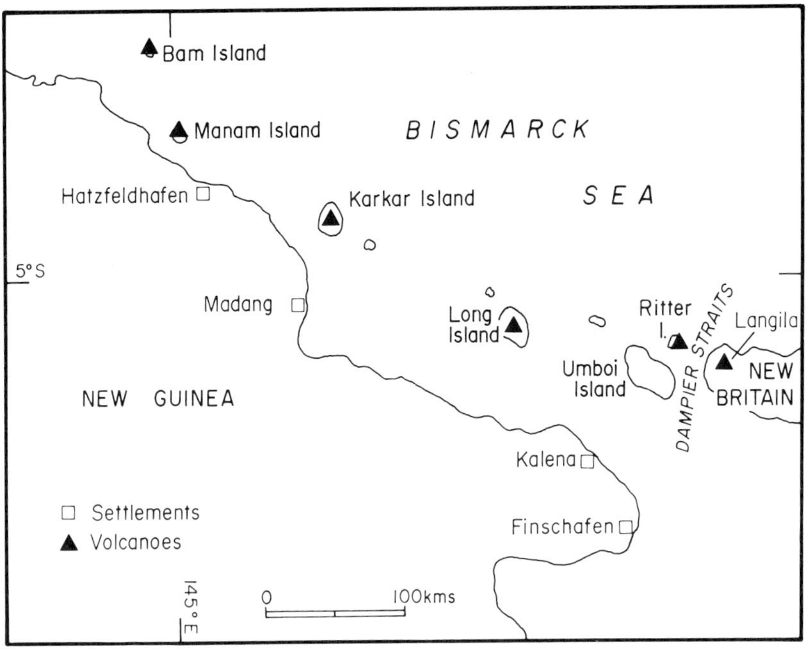

FIG. 5.28 Localities near Ritter Island, Papua New Guinea. Squares represent settlements, triangles represent recently active volcanoes (after Cooke, 1981: 115).

away, leaving huge gaps in the masonry of many feet [meters]" (Neale, 1885, quoted in Simkin and Fiske, 1983).

Although evidence is rather flimsy, the great eruption of Santorini/Thera about 1500 BC, very similar in style to the 1883 eruption of Krakatau, probably also caused massive destruction of coastal settlements by tsunami, particularly on Minoan Crete (Marinatos, 1939; Yokoyama, 1978). Mészáros (1978) has surmised that buildings in northwestern Cyprus, 700 km from Santorini were also destroyed by the Minoan tsunami.

Extensive tsunami damage also occurred during the 1868 Mauna Loa eruption. In Kau District 108 houses were destroyed, the waves in some cases carrying debris 250 m inland. On the Puna coast (Fig. 4.1) entire villages were swept away; even heavy stone houses disappeared (Brigham, 1909: 103, 118).

Significant damage was also caused along the coasts of the Papua New Guinea mainland and New Britain by tsunami generated by the collapse of Ritter Island in 1888 (Fig. 5.28). No comprehensive account of this damage is available but those searching for the missing von Below expedition on the New Britain coast reported: "Amongst the former [beach] campsite a pair of trousers, belonging to von Below; and a pair of boots belonging to Hunstein; a tent; sundry cooking utensils; and a bottle of sauce were found — all covered with mud and fallen trees. Trees which remained standing showed traces of the flood up to about 10 metres" (Steinhauser, 1893). A nearby village was destroyed (Anon., 1888). At Hatzfeldhafen, the tsunami reached 7–8 m destroying a yam store, damaging a boat shelter and carrying away lumber intended for building a bridge. In a village on an island near Kelana less strongly built houses were swept away (Cooke, 1981: 119).

While the type of protective mechanisms advocated for buildings in areas likely to be affected by pyroclastic flows and lahars are also generally applicable to buildings in tsunami-prone areas, there can be little doubt that land-use zoning is the most effective precaution.

5.8 ATMOSPHERIC EFFECTS

In most cases damage to buildings by atmospheric phenomena during volcanic eruptions is likely to be of a relatively minor nature. However, a few examples indicate that structural damage (Table 5.1) or fire may lead to total destruction of buildings. The two main atmospheric agents causing damage are shock waves and lightning strikes.

Some of the most destructive effects of atmospheric pressure waves have been described from around Krakatau in 1883. Shock waves "caused the bursting in of windows, and even the cracking of walls 100 miles [160 kmaway at Batavia and Buitenzorg [Jakarta and Bandung]. Lamps were thrown down, gas-jets extinguished, and a gasometer, under the influence of one of these great waves, leaped out of its wall, causing the gas to escape. Even at much greater distances cracks were produced in walls, and all accounts agree in ascribing the result to air-vibrations and not to earthquakes" (Judd, 1888: 27). Verbeek (1885, quoted in Simkin and Fiske, 1983) wrote of his own experiences in Buitenzorg [Bandung], more than 250 km from the volcano: "Moreover, everything that was loosely attached in houses, especially hanging things, started shaking, with a rattling, squealing or cracking sound, which contributed to the feeling of discomfort among the citizens. At a quarter to seven in the morning [August 27], such a powerful explosion was heard, that even the last who had tried to resume their sleep jumped out of bed. In four houses in Buitenzorg, including mine, which were built with their front to the west, several lamps burst, plaster fell from the walls, doors and windows which were not properly closed jumped open;..."

Similarly, the perceptible shaking of houses in eastern Java, 780 km away, during the 1815 eruption of Tambora resulted from atmospheric shock waves. Shock waves were also felt at Macassar, 370 km from Tambora, the sounds being so heavy during the morning of April 12 that a ship at anchor and houses in the fort were shaken (Anon, 1816).

Eruptions of Asama have also produced shock waves powerful enough to damage buildings. In 1783 shock waves shook nearby houses so violently that stone weights on roofs fell off and windows rattled in Tokyo over 300 km away (Aramaki, 1956: 197-198). In 1958, within an area of 430 km^2 adjacent to the volcano, 2305 houses were damaged by airshocks. Damage occurred up to a maximum distance of 15 km from Asama (Murai and Hosoya, 1964: 211). In 1973 an airshock was felt at Titibu City 70 km from the volcano and many windows were broken in Tigitaki 6 km south of the crater and in other nearby villages (*SICSLP*, 1973: 131; *SICSLP*, 19-73).

During the 1977 eruption of Sakura-jima more than 100 window panes were broken by air shock waves that had an estimated initial velocity of 146 m s^{-1} and an estimated initial pressure of 27.4 MPa. On December 8, the air shock waves broke windows in hotels at Hurisato Spa 3 km south of the crater (*BVE*, 1979, 17: 41).

Atmospheric pressure waves from Italian volcanoes have also caused minor damage to buildings. Hamilton (1772: 29) reported that shock waves from the Vesuvio 1767 eruption burst open doors and windows in Naples, including "a locked door in the Sicilian King's palace" (see also Granville, 1817: 31). Lacroix (1906: 239) believed that air vibrations caused by explosions during the 1906 eruption probably played some part in the destruction of buildings. Explosions during the 1888–1890 eruption of Vulcano broke windows in Lipari 10 km away (Bullard, 1976: 234-235). The explosions accompanying the 1907 eruption of Stromboli were so sharp that air concussions broke nearly every window on the island (Perret, 1912: 286). In the 1915 Stromboli eruption air concussions did not break any windows but the flat house roofs leaked when it next rained (Perret, 1916: 461).

Damage to window glass by air shock waves seems fairly common in volcanic eruptions. More serious damage to buildings is rare. Damage to buildings resulting from lightning strikes during eruptions seems rarer still if the literature is any guide. It seems likely, however, that such damage is frequently attributed to other causes. Lightning strikes and ignition almost certainly played a part in the destruction of buildings in St Pierre in 1902 and around Soufrière, St Vincent, in the same year. The same is probably true of other large Plinian and Peléan eruptions (see, for example, Perret, 1937: 99). During the first evening of Katmai-Novarupta tephra fall, the Woody Island naval station on Kodiak Island was struck by lightning and burnt to the ground (Erskine, 1962: 59, 151). Anderson and Flett (1903: 406) attribute chimney collapses in Chateaubelair near the Soufrière to lightning strikes rather than to earthquakes. Although poorly documented in the literature, lightning strikes probably cause damage to buildings during most large Plinian and/or Peléan eruptions.

Damage to buildings by atmospheric effects accompanying eruptions can presumably be minimised by ensuring that all buildings are equipped with adequate lightning conductors and by the installation of shock-resistant glass.

5.9 ACID RAINS AND GASES

Acid rains and gases are likely to cause much less serious damage to buildings and their contents than the other damaging actions listed in Table 5.1. Although corrosion of electrical fixtures could lead to short-circuiting and fire, this effect does not seem to have been described in the literature. With few exceptions, in fact, damage ascribed to acid rains and gases has been limited to the contents of buildings.

The acid rains and gases created by the eruption in 1912 of Katmai-Novarupta seem to have been particularly damaging. At Seaward, 400 km northeast of the volcanoes, a sulphuric rain ate into the metal work of buildings (Griggs, 1922: 25). At Cape Spencer, 1100 km away, fumes tarnished freshly polished brass in 15–20 minutes (Wilcox, 1959: 427). In Kodiak, 160 km southwest of the volcanoes, every piece of bronze, brass and copper tarnished black even where the object was in a drawer in a sealed house into which little tephra penetrated. In the same town, laundry left on lines overnight was reduced to yellow shreds (Erskine, 1962: 72). In Vancouver, Canada, a month after the eruption acid rains damaged clothes hung out to dry (Wilcox, 1959: 417): "Sulphuric-acid fumes were wafted 1500 miles [2400 km] to Vancouver, where they impregnated the weekly wash hung out to dry. When the laundry was taken off the lines and Canadian housewives applied their hot irons, the clothes turned to shreds" (Herbet and Bardossi, 1968: 68).

Similar effects on laundry and metallic articles have been reported from other eruptions. Forty-five years after the 1931 Aniakchak Crater eruption, Trowbridge (1976: 73) wrote: "My washing which had hung on the line during the eruption literally fell to pieces after about two more washings, eaten by the acid content of the ash". This account refers to a point more than 320 km from Aniakchak Crater! At the other extreme, volcanologists working in the caldera of Nyiragongo reported that tents tore like blotting paper after three days, their clothes were eaten away and that even a wire rope was reduced to half its original diameter by corrosive gases (Tazieff, 1952 : 161-162; 1979b: 130).

Acid rains and gases from the Caribbean eruptions of 1902 attacked metallic articles. In the ruins of St Pierre all silverware was blackened. A silver plate picked up by an officer of the *U.S. Dixie* was black and corrugated. Other metal relics susceptible to sulfur discoloration showed its blackening effect (Hill, 1902a: 255). Tephra from Soufrière, St Vincent, that fell at Kingston (St Vincent) and on Barbados was highly sulfurous; silver articles were attacked by H_2S (Anderson and Fleet, 1903: 401). Similarly, fence and telephone wires and metal equipment on plantations and at a cement plant west of Masaya in Nicaragua have been damaged on several occasions by rases (Bolt *et al.*, 1975: 113).

Such minor effects as the discoloration of metallic objects are quite common in areas surrounding active volcanoes.

6

THE EFFECTS ON UTILITIES AND MACHINERY

6.1 INTRODUCTION

In general, this chapter is concerned with the effects of volcanic eruptions on three broad groups of phenomena: (i) utilities such as water and electricity supplies and communication networks; (ii) transport systems such as aeroplanes and airports, railways, ships and ports, roads and cars; and (iii) machinery such as mechanical and electrical equipment including office machinery and air-conditioning equipment. Table 6.1 presents an example of the diversity of effects that may result from a lahar. Figure 6.1a provides the location of towns along the Puyallup Valley north west of Mount Rainier, Washington State. The impacts listed in Table 6.1 are those estimated by Cullen (in Hodge *et al.*, 1978) to result from a lahar equal in magnitude to the Electron mudflow which swept 56 km down the valley about 600 years ago leaving a deposit near Orting, for example, 5 m thick. Crandell (1973) estimated that such a lahar might occur in the next few centuries. Table 6.1 indicates that the effects of such a relatively small lahar are widespread, affecting a variety of public service delivery systems in different ways. In most cases, these systems are affected during an eruption, but effects may continue after activity has ceased. In a few cases, particularly where volcanogenic earthquakes occur, systems may be affected before an eruption *per se* occurs, or even when there is no eruption at all. For each example considered here, the damaging action, insofar as it is known, is given but the emphasis is on the effects rather than the causes. Details of the ways in which problems have been mitigated are incorporated into each section.

Table 6.1 The impact of a lahar equivalent to the Electron mudflow on public service delivery systems in Puyallup Valley.

Areas of Service	Specific Services	Electron	Kapowsin	Ohop Valley	Orting	McMillin	Alderton	Sumner	Puyallup	Fife	Tideflats
Health Care	Hospitals								§		
	Clinics				*				††§	?	
	Medical labs							*	§		
	Pharmacies				*			*	†§	††	
	Nursing homes				*			*	†	††	?
	Ambulances, etc.							*	§		
Public Safety	Municipal Govt.				*			*	§	††	
	Police				*			*	§	††	
	Fire		‡		*			*	§	††	††
Communications	Radio							?	?	†	?
	Telephone		§	††	*	*	*	*	†	†††	†††
Transport	Freeways							††	††§	†	†
	Other highways		*	†††	*	*	*	*	†††	††	††
	Arterials		*	†††	*	*	*	*	†††	††	††
	Bridges			†	*	*	*	*†	†††	†	†
	Railroad tracks	*	*	††	*	*	*	*	†††	†††	†††
	Railroad bridges			?	*	*	*	*	†††	?	†
	Railroad yards							*	†††		†††
Utilities	Water										†††
	Electricity	*	§	?	*	*	*	*	†††	†††	†††
	Water	‡	‡	††	*	*	*	*	†††	†††	†††
	Sewage	‡	‡	†	*	*	*	*	†††	†††	†††
	Natural gas				*	*	*	*	†††	††	††
	Petroleum pipeline							‡	‡		‡

After Cullen, 1978: 118 * Destroyed. † Damaged (more symbols, greater damage). ‡ Unaffected. § No damage, but service interrupted. ? Insufficient data.

6.2 WATER SUPPLY SYSTEMS

The effects on water supply can be divided into: (i) effects on the reservoir; (ii) effects on the reticulation system; and (iii) effects on water usage. Examples are given of water supplies affected by earthquakes, ground deformation, lava flows, tephra falls, and fumarolic activity. Although no water supply systems were located within the area affected by pyroclastic flows around Mount St Helens this eruption demonstrated some potential effects. Lakes such as Spirit Lake (Fig. 3.17) now occur at a higher altitude, occupy a larger area and are shallower than before the eruption. These lakes, previously oligotrophic, quickly became eutrophic, anaerobic and developed very large populations of bacteria. Recovery will take many years (Dion and Embrey, 1981; Baross *et al.*, 1982).

Reservoirs may be contaminated even when eruptions do not occur. For example, the 1975 (fumarolic) thermal activity of Mount Baker (U.S.A.) contaminated Baker Lake (200 m above sea level) on the south-eastern flank of the volcano, a lake used primarily for hydro-electric power generation but also providing an important fisheries and recreational resource. Increased thermal activity in March, 1975, in the Sherman Crater area of the summit forced closure of several campgrounds and other public access areas.

In spring 1975 a water-filled depression in the Sherman Crater glacier about 2900 m above sea level and 13 km from Baker Lake, had a pH of 2.5 and an acidity of 15 mg L^{-1}. In early summer, 1975, this and other acid water was supplied to Baker Lake via Boulder Creek at a rate equivalent to 0.56–1.09 t day^{-1} of concentrated sulfuric acid, increasing to 3.4–12.4 t day^{-1} in June to August. Although Boulder Creek drains only about 5% of the Baker Lake catchment, down-reservoir dilution and buffering of the density current depends on the temperature and volume of other stream water entering the reservoir. In June, measurements indicated that the pH of Boulder Creek water was in the range 3.8–4.4 while 50 m out into the lake from the mouth of the creek pH ranged 3.8–6.3. In 1975 the median iron concentration of Boulder Creek was about 3700 $\mu g\ L^{-1}$, more than 12 times above the recommended limit (300 $\mu g\ L^{-1}$) for public water supplies. Iron concentrations in other streams ranged from 0–30 $\mu g\ L^{-1}$ while levels below the reservoir ranged from 430–1200 $\mu g\ L^{-1}$ (Bortleson *et al.*, 1977).

The dilution and buffering characteristics of Baker Lake and the short mean water residence time of 0.2 y suggested that Baker Lake could flush the Boulder Creek pollution load relatively quickly. Bortleson *et al.* (1977) concluded that 1975 Baker Lake water was safe for drinking and other human contact. The trace elements selenium, mercury, arsenic and boron were unlikely to be dangerous in the lake water at their then concentrations and pools of acidic water were unlikely to build up because of continued outflow, replenishment, and buffering.

In 1975 acidic waters did not reach as far down-reservoir as the dam wall. However, if Boulder Creek water were to reach the dam wall undiluted — a very unlikely situation — sulfate ions could damage the concrete dam wall at concentrations as low as about 180 mg L^{-1}. Acidic waters are generally regarded as damaging to concrete at pH values less than 6.0 with the corrosion of concrete becoming increasingly severe at lower pH (Bortleson *et al.*, 1977: 29). In such situations high alumina or other sulfate-resisting cement can be used for construction (see, for example, O'Flaherty, 1974: 122).

Reservoir turbidity and acidity may also be affected by even quite minor tephra falls. At Santo, 130 km from Ambrym, a fine tephra fall produced by the 1953 eruption contaminated tank water supplies (*Pacific Islands Monthly,* May 1951: 83). During the 1953 eruption of Mt Spurr a 3–6 mm tephra fall in Anchorage (Fig. 6.1b) caused the pH of the public water supply to fall to 4.5 before returning to pH 7.9 after a few hours. The turbidity of the water rose from 5 ppm to 290 ppm and took six days to return to normal. Soluble sulfate values also varied (Wilcox, 1959: 446).

Similar rises in turbidity for a few days occurred with tephra falls around Mount St Helens in 1980. Where turbidities reached 10 Jackson Turbidity Units (JTU — a common measure of the optical property of water that scatters light), customers were advised to "boil water" (Nielson, 1980: 25). In Montana, where only a few mm of tephra fell (see Fig. 3.8) turbidity increases of 5–10 JTU were observed but the suspended tephra rapidly settled out. At first it was believed that the tephra would cause acid conditions in surface water storages. In anticipation of problems with water quality the Helena surface water storage was shut down for 48 hours but the tephra was only slightly acidic and not at all harmful (Acord, 1980: 71).

Water quality problems produced by tephra fall have generally been minor. Acidity can be reduced by buffering and dilution while turbidity falls as fine sediment settles out even though very fine particles tend to remain in suspension for long periods. Experiments reported by Nielson (1980: 26) indicate that the addition of soda ash to raise the alkalinity to 30–40 ppm with the subsequent addition of 15–25 mg L^{-1} of alum usually produces a good floc. Standard filtration procedures can then be used by systems operating full treatment facilities. However, there appears to be no standard treatment that will work in all cases. Presumably, the flocculation methods mentioned above will also prove effective on the small water tanks used for domestic supplies in many countries. Furthermore, downpipes from collection areas can easily be disconnected before tephra fall begins, and the tanks sealed. One 25 m diameter 1.64×10^6 L conical water tank cover made of 0.91 mm reinforced Dupont 'Hypalon' synthetic rubber survived undamaged a fall of more than 50 mm of wet Mount St Helens* tephra even though only supported by bar joints every 3.3 m (*Public Works,* 112(2), February, 1981: 88).

The long-continued eruption of Irazú which began in 1963 also produced problems with the water supply system in Costa Rica's capital city, San José. Fine tephra

*Figure 3.8, based on the detailed examination by Sarna-Wojcicki *et al.*, 1981, is regarded here as the definitive map of Mount St Helens tephra thicknesses. The "more than 50 mm" reported here is almost certainly an exaggeration.

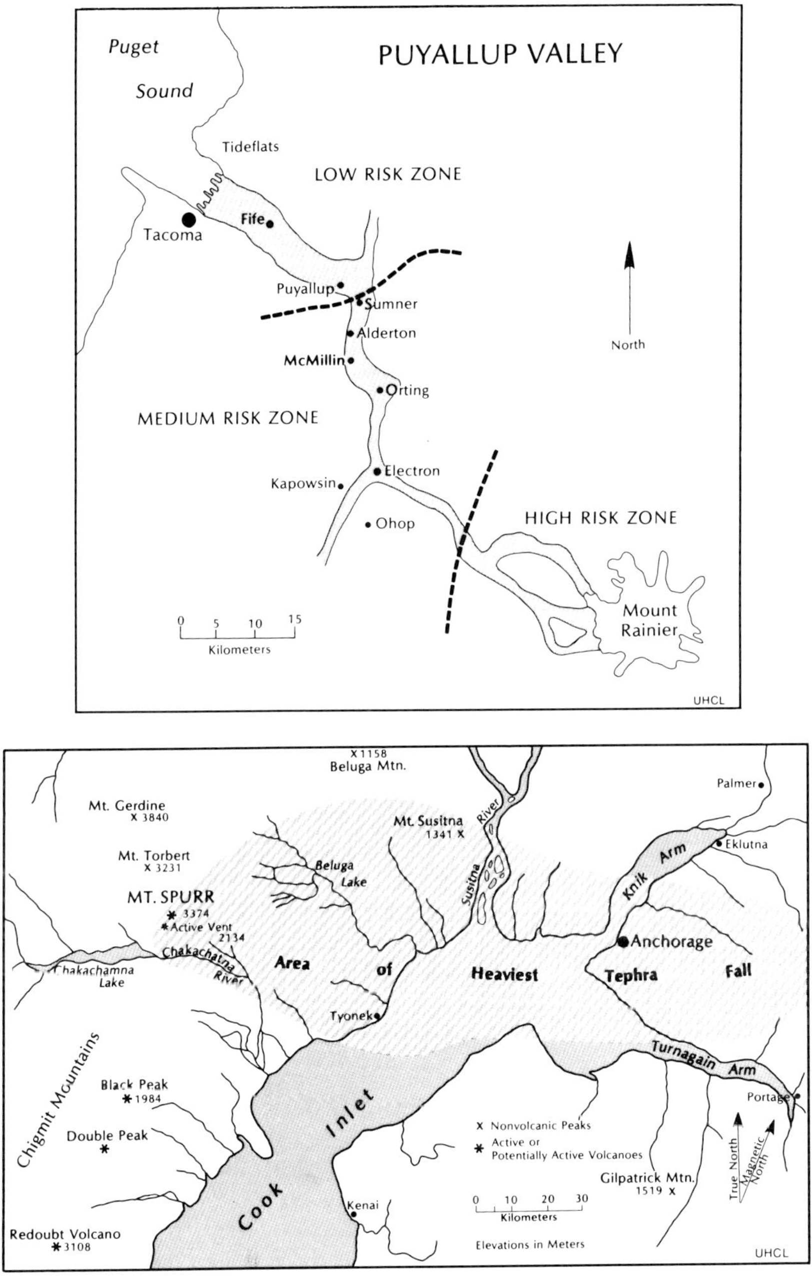

FIG. 6.1 (top) The Puyallup Valley indicating the area affected by a lahar equivalent in magnitude to the 600 year old Electron mudflow from Mount Rainier (after Cullen, 1978); (bottom) Approximate distribution of tephra from the Mount Spurr eruption, Alaska, 1953. The tephra fall in Anchorage amounted to 3–6 mm (after Wilcox, 1959).

suspended in river water clogged the filters in the city's water works. The main water pumping station went out of operation in mid-June 1964, so that water had to be trucked into the capital from surrounding cities (*US News and World Report,* 56, June 29, 1964: 71-73).

Community water supply systems along the Toutle and Cowlitz rivers proved particularly vulnerable to lahars during the 1980 Mount St Helens eruption (see Fig. 4.2). While high turbidities in stream supplies were overcome by closing intake structures and operating from storage or unaffected ground water systems, intake systems in streams were damaged or swept away in some cases. The Toutle community water system was destroyed, buried in mud. Castle Rock lost its intake structure but was able to operate with limited capacity from existing wells while new wells were rapidly drilled (Nielson, 1980: 25). At Longview, the intake system was buried in 5 m of sediment. Utilisation of a temporary intake allowed four pumps to supply 7.3 ML per day, about quarter of the normal requirements. Use restrictions were imposed until May 25 when tephra fell on Longview. Kelso, across the river, but using a different supply system, was also able to help with Longview's supply (*Seattle Post-Intelligencer,* May 29: A11). It is noteworthy that the damaged intake structures were more than 70 km from the volcano.

Water reticulation systems have proved vulnerable in a number of situations. At Heimaey in 1973 the lava flow into the harbor broke one of the two freshwater pipelines from the mainland (Williams and Moore, 1973: 15). However, this created few immediate problems as most of the population had already been evacuated. During the 1957–1958 Capelinhos eruption most of the pipelines carrying Horta's water supply were damaged by earthquakes or ground deformation (see Fig. 5.24) so that a severe water shortage still existed in late 1958 (Parsons and Mulford, 1958). Around Usu in 1944 earthquakes caused damage to water pipes before the eruption began. Ground deformation also caused a number of water supply problems; irrigation canals formed temporary lakes and disturbed subsurface waters. In the upheaved area villagers no longer had water for drinking or irrigation while in adjoining villages wells and fountains overflowed (Minakami *et al.,* 1952: 264-265; Minakami, 1974: 327). At Parícutin in 1944 reworking of tephra caused floods of eroded debris and the silting up and destruction of irrigation systems (Rees, 1970: 13). If ground deformation and earthquakes in the Mammoth Lakes (Long Valley) area of California in the early 1980s lead to an eruption, Los Angeles, nearly 500 km away, will probably lose its Mono Lakes water supply.

Problems with water quality and with intake and reticulation systems inevitably lead to water shortages and rationing. Destruction of Longview's water intake structures by the Cowlitz River lahars during the May 18, 1980, Mount St Helens eruptions led to an odds–evens rationing system based on house numbers and dates (*Seattle Post-Intelligencer,* May 29: A11). While the need for water rationing may be reduced in many situations because of partial evacuation, a tephra fall is guaranteed to increase water demand. Several examples drawn from the area affected by the May 18 Mount St Helens tephra fall illustrate the magnitude of the effect and a

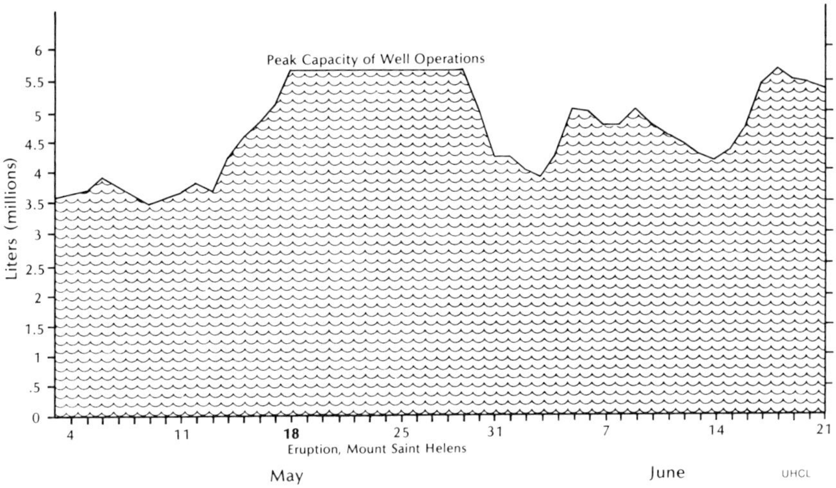

FIG. 6.2 The effects of the May 18, 1980 tephra fall from Mount St Helens on water use in Cheney, eastern Washington. Water use is calculated as a five-day running average (after Warrick *et al.*, 1981: 73).

variety of coping strategies. The relevant thicknesses of tephra and locations are shown on Figure 3.8.

In Spokane water rationing was imposed on an odds–evens basis as residents tried to wash off cars, buildings and streets (*The Columbian*, May 26, 1980: 9). In nearby Cheney the heavy demand for cleanup purposes reached 1.5 times the pre-eruption demand (Fig. 6.2) and remained at this level for 12 days. This demand was more than equal to the city's available capacity to supply as the city's newest and largest well was inoperative. Supply was selectively cut off to Eastern Washington University, the city's largest individual customer, and an emergency well was used to assist in meeting demand (Warrick *et al.*, 1981: 72).

In Ellensburg, the tephra cleanup caused the demand for water to leap 2.5 times, with water consumption rates remaining above average for four days. Some water quality problems were also experienced as the pH of Yakima River, the primary source of supply, fell to 5.8 and turbidity increased. This problem was overcome by shutting down the infiltration wells and shifting supply to tube-pumped wells (Warrick *et al.*, 1981: 25).

In Ritzville demand also increased to 1.5 times pre-tephra fall requirements. Water quality was not a problem as water is drawn from deep wells. As Warrick *et al.* (1981: 48) note: "Since the city water system is designed for peak daily summer averages . . . there were no real supply problems. Yet, the water department did take the precautionary step of restricting residential use by an odd/even address arrangement during peak demand hours. This precaution was deemed necessary in order to keep line pressure up in the event of a fire or some other high requirement situation".

These examples illustrate several important points. Even relatively minor tephra falls, 10–20 mm or less, will generate greater than 50% increases in user demand during the cleanup phase, at least in Western societies. Although simple rationing methods can help supply authorities to meet demands, much greater problems would have been experienced in eastern and central Washington State if: (i) uncontaminated ground water sources had not been available to most communities; and (ii) the cleanup period had coincided with the peak summer water demand.

6.3 WASTE WATER SYSTEMS

Waste water disposal systems are affected in several ways by volcanic eruptions. Both reticulation systems and treatment plants may suffer. The most common hazard is tephra fall, although presumably sewerage and storm water pipes may also be broken by earthquakes and ground deformation.

Because large amounts of tephra fall on roads, roofs and other impervious surfaces, tephra is easily washed into storm water drains. San José experienced this problem during the 1964 Irazú eruption, the clogged sewers causing local flooding during the rainy season. When citizens reported blocked basins city crews removed the tephra with a 250 mm "vac-all" suction pipe and storage tank, taking about one hour per basin (Clark and Lee, 1965: 22). In Ellensburg (Fig. 3.8) the May 18 tephra from Mount St Helens overloaded storm drain grates, pipes (lines) and catch basins but the prompt use of high pressure water jets and "Line Slick" (Gem Chemical Company) largely prevented tephra from settling in the lines (Warrick *et al.*, 1981: 26; State of Washington Dept of Ecology Memo, June 18, 1980).

Tephra falls are particularly detrimental where sewage and storm water are collected into the one pipe network. In this arrangement there are many possible entry points and large quantities of tephra are likely to arrive at sewage treatment plants.

Although Ellensburg had separate collection lines for storm water and sewage through most of the town, in the central business area the lines were combined (Warrick *et al.*, 1981: 25). This meant that significant quantities of tephra reached the sewage treatment plant which was a conventional activated sludge plant except that primary clarifiers were absent. The facility consisted of a raw sewage pump station, headworks (including grit removal and comminution), two aeration basins, secondary clarifiers, a sludge pump station, a waste sludge thickener, primary and secondary anaerobic digesters and two chlorine contact chambers (Department of Ecology Memo, June 18, 1980).

The first significant impacts of the tephra fall on the Ellensburg Sewage Treatment Plant were observed on May 20 by which time the tephra was being washed from the streets in the cleanup campaign. Flow increased from about 18.2 ML per day prior to the eruption to 34.6 ML per day on May 20. Volatile solids in the secondary clarifier dropped from about 70% to 28% on May 20 and to only 8% on May 24. The sludge volume index also declined during this period. Values had still

Table 6.2 Damage to equipment components at Ellensburg Sewage Treatment Plant resulting from Mount St Helens tephra fall, May, 1980.

Equipment	Number	Components	Estimated cost(US$)
Raw sewage pumps	2	Impellers, rings, shaft sleeves	3000
Comminutors	2	Oscillating, stationary	1300
Aerators	8	Impellers, shafts	24000
Sludge recirculating pumps	2	Impellers, shaft sleeves, wear plates	2100
Sludge waste pumps			
duplex	2	Pistons resurfaced, seals	2400
simplex	2		
Flow control valve	1	For sludge recirculating pump	2500
Check valves	2		850
Jet hose	1		1300

Department of Ecology Memo, June 18, 1980.

not returned to normal by the end of May (Department of Ecology Memo, June 18, 1980).

Most of the damage to the Ellensburg treatment plant resulted from tephra-induced wear of pumps and pumping components. Table 6.2 lists the damaged components and their estimated replacement costs. This estimate of $37 450 did not include hire of flushing equipment or overtime.

Much more severe damage occurred at the City of Yakima waste treatment facility which had a collection system of 446 km length (of which 5% is combined sewage and storm water lines) and the capacity to provide secondary treatment of 69 million liters of wastewater per day (Day and Fisher, 1980: 2082). On May 19, more than 2830 L or about 15 times the normal amount (White *et al.*, 1980: 5) of grit was removed in the pretreatment processes but sand was observed in the raw sludge indicating that some tephra was passing into the primary clarifiers. On May 20 it was evident that the equipment was suffering; "vibration occurred in the grit classifiers and the gear box of the mechanically cleaned bar screen. Raw sludge lines were becoming plugged, and pumping difficulties were being experienced" (Day and Fisher, 1980: 2084). On the following day equipment failures and shut-downs occurred. On Wednesday, May 21, the comminutor failed completely, the biofilters were virtually stripped of their growth and secondary treatment was curtailed. At 9 a.m. city officials issued the following press release and construction of a bypass to isolate the treatment plant, temporary chlorination facilities, and a sludge storage lagoon began (Day and Fisher, 1980: 2085):

> The City Manager's Office announced this morning a decision to bypass the Yakima Wastewater Treatment Plant with the sewage flow to the Yakima River, due to the disaster [sic] conditions from the volcanic ash. This decision was arrived at through consultation with the Washington State Department of Ecology.
>
> We anticipate the bypass to be effective May 22, 1980, and to last indefinitely.
>
> The specific reasons for the bypass are as follows:
>
> Volcanic ash and grit accumulations in the system form unpumpable masses forcing equipment and operational failures.
>
> Trickling Filter rock media is stripped by grit which destroys its ability to remove pollutants.

> Abrasive wear on equipment may put the plant out of operation indefinitely and also cause damage to a major investment in new equipment installed as part of the current plant upgrading.
>
> The conditions of the bypass include the following:
>
> Regular testing of influent to determine when the grit content is low enough to handle in the plant.
>
> Lines and equipment will be cleaned and readied for service.
>
> We will assess equipment damage and take any available measures to prevent additional damage.
>
> We will monitor river conditions which at this time are most favorable with high flows.
>
> We will provide full documentation of problems, discussions, tests, remedial efforts and other actions.
>
> [Office of City Manager, City of Yakima, May 21, 1980]

The decision to put essentially untreated sewage into the Yakima River was not taken lightly, nor was it accepted lightly by many people as the following anecdote related by a Department of Ecology official indicates: "Another incident that came to me, I think Monday [probably Wednesday, May 21] afternoon, when I got a call from a lady over in Yakima Indian Reservation. She happens to be the wife of a minister to the tribe over there and in a mission situation. Her problem was that she had been listening to the radio and there was a story on it that the city of Yakima was about to start dropping its sewage, raw, in the Yakima River. Well, she was very upset by this from the standpoint that the salmon resource of the Yakima River was something that the Indians had depended on in all history and that the state should not be a party to that. It shouldn't be done. It should be stopped . . . Of course, I tried to tell her what the facts of the situation were. That, if we did not allow the city to go ahead with the procedure of putting the sewage in the river, the situation would come about shortly that the machinery wouldn't work. Then we would really be fouled up. The sewage would be going in untreated anyway. Even if the machinery was still there and we were running the sewage through the plant, we wouldn't be getting any treatment on the sewage as it went through . . . Well, she didn't understand that really, and then I tried to tell her the river was running at 6000 cfs [170 m^3 s^{-1}] and the effluent from the Yakima plant was of a range of 10 to 20 cfs [0.28–0.56 m^3 s^{-1}]. Very evidently, she didn't understand what a cfs was, or how much volume that amounted to. I asked her, finally, 'Just how much sewage do you think that is?' She said, 'Oh you ought to be able to move it in about two truckloads in a day' . . . Well, of course, we were looking at something more like one 5000-gallon truck [18 927 L] per minute. To try to find that kind of trucking capacity was totally unrealistic. We were looking at this history of the Yakima drainage where, for many years, Yakima and many other communities in that valley had previously, before environmental clean-up times, put sewage directly in the river for decades. It wasn't all that novel a situation . . . But she was serious enough about it. In fact, she said that this was a plot and what it really amounted to was genocide committed against the Indians. It's a little hard to be rational in responding to that kind of an attack" (Hallauer, 1980: 62).

Again, the timing of the eruption of Mount St Helens was important. Spring is a time of high snow-fed streamflows so that downstream dilution of the effluent was rapid. Sampling was carried out along the Yakima River and in some irrigation canals. Community health staff went from door to door in the area where effluent entered the river explaining the position. As a result, no cases of water-borne illness showed up although the monitoring showed up previously unknown water quality problems unrelated to the entry of untreated sewage (Dr Bob Attwood, Yakima Valley Community Health Service, personal communication, June, 1980).

Although the treatment plant bypass was initially implemented for an indefinite period, cleanup and repair of the plant progressed sufficiently rapidly that it was possible to recommence primary treatment on May 25, 1980. During the next few weeks teams cleaned sewer lines at an average 20 km per day (Day and Fisher, 1980: 2086). Damage to the waste disposal system included: comminutor; mechanical bar screen; degritters; grit pumps; gritty clones and classifers; clarifier drive mechanisms; air compressors; trickling filters; trickling filter recirculation pumps; primary and secondary clarifier sludge pumps; trickling filter pumps; all electrical controls; sludge transfer pumps; 125 hp waste gas boiler for heating sludge; sludge recirculation pump; plus controls and pumps in all sewage lift stations (Department of Ecology Memo, June 10, 1980). In late May, 1980, damage to the treatment plant was assessed at US$700 000, to the sanitary collection system US$290 000, and to the storm sewer system US$90 000. Damage to Phases I and II of the new, partially completed plant, was estimated at US$1 479 000 with contingencies totalling US$1 500 000. While this total damage estimate seems very high at US$4.06 million, damage was minimised by halting treatment for four days (White *et al.*, 1980: 7).

Damage to other Washington State treatment plants was reviewed by Kish (1980). The amount of damage seems clearly related to the extent of combined sewage and storm water lines. Although Yakima had only 5% combined lines, it seems likely that illegal connections to the sewage system existed. Furthermore, reports suggested that citizens raised manhole covers to dump tephra (Day and Fisher, 1980: 2085).

The Yakima experience led public works officials in that city to compile a list of things to do in the event of another tephra fall with a view to protecting waste water disposal equipment and to helping other cities that might experience similar situations in the future:

1. Cover all external equipment with plastic.
2. Shut down all equipment not absolutely required.
3. Shut down ventilation equipment where possible.
4. Place all pre-treatment equipment into operation and adjust for maximum removal rates.
5. Place all primary clarifiers into operation and increase pumping rates.
6. Pump all grit and ash into one digester until conditions improve, then dump the contents.

> 7. Shut down the biofilters and cover with plastic.
> 8. Monitor all processes for introduction of grit and ash.
> 9. Monitor torque or current on all motor driven devices.
> 10. Grease everything subject to a dusty atmosphere, and those subject to pumping gritty material.
> 11. Filter or change out gear lube weekly, flush with solvent recommended by the manufacturer.
> 12. Alert the public to the consequences should they introduce this material into the sewer system.
> 13. Notify regulatory agencies and public health officials.
> 14. Formulate a plan to bring the plant processes and equipment back on line.
> 15. Contact equipment manufacturers for their assistance.
> 16. Step up preventive maintenance scheduling until the system is completely purged of contaminants.
>
> It must be realized that it is difficult to ascertain the total effect of a particular ash problem until it occurs because its effect when mixed with water can be different, depending upon its origin and physical/chemical characteristics, i.e., toxicity, pH, settlebility [sic], etc. One must also consider how long it will be airborne, whether it's metallic, and to what degree the plant is designed to resist its entry into interior spaces. This includes electrical cabinets, gear boxes, ventilation equipment, etc. Many preventive measures may not be identified until after the fact, but someone should be assigned solely the task of analyzing the fallout immediately to provide input to operations and maintenance personnel.
>
> The regulatory agencies should be consulted immediately, and their authorization to shut down processes affecting your discharge should be mutually agreed upon by both the city and the agency.
>
> [White *et al.*, 1980: 8]

Given the occurrence of a tephra fall the Yakima team suggested that it was prudent to increase the frequency of performing preventative maintenance to equipment. In turn this would affect maintenance costs, equipment longevity, replacement costs and staffing levels (Day and Fisher, 1980: 2089). With the possibility of repeated tephra falls, White *et al.* (1980: 8) also suggested some alternative equipment designs:

1. House all equipment.
2. Pressurise all structures.
3. Specify air filters capable of removing fine dust.
4. Specify dust-free motors.
5. Specify seals that can withstand dust penetration.
6. Cover the biofilters (a side benefit of this is the savings in energy as a result of being able to adequately treat greater amounts of wastewater during cold months).

6.4 ELECTRICAL SUPPLY SYSTEMS

Electric power generation, distribution and consumption may all be affected by volcanic eruptions. It is probably rare, however, for all three aspects to be seriously affected by the one eruption.

Generation systems have suffered in a number of eruptions. In 1945, tephra-laden floods around Parícutin washed out several power dams (Wilcox, 1959: 454). On Heimaey in 1973 the building housing the island power plant in the north-east part of the town was destroyed by tephra fall and lava flow. The 30 000 volt submarine cables were also severed (Williams and Moore, 1973). During the 1975 phreatomagmatic eruption of Ruapehu floods filled the tunnel and aqueduct of a hydro-electric power scheme still under construction (*BVE*, 1977, 15: 16). Before the May 18 eruption of Mount St Helens water levels in the three hydro-electric dams on the Lewis River (see Fig. 4.2) operated by Pacific Power and Light Company were lowered 15 m as a precautionary measure against possible over-topping by lahars and/or landslides. Lahars eventually raised the level in one reservoir by about 0.6 m (Munro, 1980: 36; Bob Zike, Operations Manager, Pacific Power and Light Co., Yakima, personal communication, June, 1980). The need to keep the reservoirs at a lower level raised the possibility of some loss of generating capability the following winter. However, no generators were affected by Mount St Helens tephra. Tephra fall from the 1976 Augustine eruption, on the other hand, had quite unexpected effects: "Electric utility company officials, particularly on the Kenai peninsula, noticed increased efficiency in their natural gas-fired turbines. The ash contained in the large quantities of air ingested by the turbines scoured a layer of corrosion off the turbine blades. The utility officials became concerned, however, about what would happen when the corroded layer was gone and the abrasive ash began to wear down the blades themselves" (Miller, 1976: 20).

Electricity transmission systems have frequently suffered during eruptions, particularly as a result of tephra fall. Gunnarsson (1973: 76) reported that electricity lines on Heimaey were heated by the rain of cinders and sagged under their own weight so that they lay close to the ground. The effects of wet tephra can be appreciated from the May, 1924, eruption at Kilauea. A severe electrical storm began about noon one day accompanying the rise of the tephra cloud from the first explosions. Flashes of lightning usually accompanied each explosive spasm and thunderstorms and mud rains frequently occurred. Eight kilometers from the vent 21 consecutive power poles were destroyed by lightning (Stearns, 1925: 4,8).

Quite a number of relatively minor reticulation problems occurred in the tephra fallout area from Mount St Helens both on May 18, 1980, and during the subsequent clean-up period. In Yakima the tephra probably caused four or five transformers to catch fire, particularly the older style transformers with pocket-type bushes — these seemed quite susceptible to tephra entry. In the few days after the tephra fall no rain fell and a windy period helped remove the tephra from the poles, insulators and transformers. Maintenance crews were also busy. When it did rain, five or six poles caught fire either as a result of lightning strikes or shorting-out by tephra across contacts (Bob Zike, Pacific Power and Light Co., Yakima, personal communication, June, 1980). In Ellensburg similar problems were experienced. Maintenance crews began work on May 19 blowing tephra off substation transformer insulators and out of switch panels. Strong northwest winds on May 20 removed the dust from pole-mounted stepdown transformers. Momentary power outages occurred as loads were shifted off substations so that they could be cleaned (Warrick *et al.*, 1981: 22). In Cheney, five transformers were lost and two pole fires reported. The only power outages were brief and electricity problems were under control by May 23. Essentially standard procedures for eliminating dust accumulation were used. Insulators and wires were cleaned by pounding poles to shake off as much of the tephra as possible, and then the rest was blown off with compressed air (Warrick, *et al.*, 1981: 72).

Ritzville was not so lucky. Here the tephra fall was both thicker (see Fig. 3.8) and finer. About 200 customers in southern Adams County lost power for six to eight hours on the day of the eruption. Maintenance crews were delayed by poor visibility and tephra-caused vehicle breakdowns. Warrick *et al.* described the other electricity problems in Ritzville: "It was discovered that the ash tended to cake like cement, making it difficult to remove from electrical equipment. While a heavy rain would pose little problems — washing the ash from lines, substations, and other electrical apparatuses — a light rain would set up a conductive layer of ash on the equipment, causing numerous shorts and widespread outages. Weather forecasts for the upcoming week indicated that there was a good chance of light rain for eastern Washington. It then became evident to the power companies [Washington Water Power Co. and Big Bend Electric] that a large effort would have to be made to clean the ash off their equipment . . . The cleanup operation proceeded uneventfully for the next few days. Air compressors and water trucks were pressed into service to blow off substations, wash poles and remove the ash from insulators and other apparatuses. The clear weather pattern continued to hold and, despite long hours and the omnipresent clouds of dust, the line crews completed about three-quarters of the cleanup before the sky turned overcast early Thursday and a light rain began to fall, compacting the ash on electrical equipment. Almost immediately, numerous outages were reported. Pole fires, flashovers, and shorted transformer fuses plagued the power system. These shorts, in turn, tripped automatic shutdown devices — leaving still more customers without power. Problems also developed when lightning arrestors and cutoff skirts were blown off and when current arced across cracked insulators. Emergency crews rushed from place to place, but initially seemed to be losing ground . . . Washington Water Power reported many small outages, due mostly to pole burnings and blown transformer fuses. Fortunately, the city of Ritzville experienced only minor interruptions in service; final repairs were completed by 6–8 p.m. on Sunday, May 25. Customers in rural areas of Adams County were not so fortunate. Big Bend Electric reported six days of straight outages, with almost all of its 3000 accounts affected at one time or another. Three substations (out of a total of 15) which had not yet been cleaned presented the most serious difficulties. Five pole fires were reported. The last domestic re-hookup was completed on Tuesday, May 27, but most irrigation accounts still remained out at week's end (1981: 49-50).

On the wider scene the May 18 tephra fall affected about one-third of Bonneville

Power Administration's (BPA) four state transmission system (Tollefson, 1980: 16). The extensive backup system of the BPA network meant that power could be switched to secondary circuits while cleaning and maintenance of their 54 substations progressed. Thus, BPA customers experienced fewer outages than those common with local utilities. Nonetheless, the cleanup was a massive effort; "in many cases a lot of people who work for the Bonneville Power Administration and who had never come in close proximity to the high voltage electric power system gained some experience in housekeeping work around the [54] substations" (Munro, 1980: 35).

In general, the area affected by Mount St Helens escaped fairly lightly. The problems described for Yakima, Ellensburg, Cheney and Ritzville are typical of those experienced in the area of tephra fall. Additional problems experienced by electric utilities included vehicle-related problems (see Chapter 6, Section 11) and the disruption of meter-reading schedules. The tephra fall of May 25 caused problems in the Portland area less severe than those associated with an ice or electrical storm (Mosebar, 1980: 10). Despite the seriousness of various effects, the Pacific Northwest was lucky — lucky that little or no generation capacity was lost; lucky that no major transmission lines passed through the blast area; lucky that the weather remained fine throughout most of the fallout area for a week or more; and lucky that the eruption occurred at a time of the year when electricity demand was relatively low.

That luck was with the Pacific Northwest as far as electricity demand was concerned can be illustrated from the 1953 Spurr eruption. When tephra began to fall in Anchorage (Fig. 6.1b) at about noon, the automatic switching system turned on the street lights. With office and domestic lights also being switched on, an unanticipated peak demand occurred, and overloading of the network was only averted by radio appeals to economise on power (Wilcox, 1959). By comparison, in Yakima, the automatic switching (which is done by photo cells) occurred at about 11.15 a.m. on Sunday May 18. The street lights switched off at the normal time the following morning. The electricity system was easily able to cope with this demand because it was a warm day and because it was Sunday, always a low-demand day (Bob Zike, Pacific Power and Light Co., Yakima, personal communication, June, 1980).

6.5 TELECOMMUNICATIONS

In this section the term "telecommunications" is used to refer to communication by radio or telephone. From a consumer point of view telecommunications travel via wires or by radio wave propagation, but the distinction is not simple as telephone or telex messages may be transmitted via satellite or microwave repeater stations. Examples of effects presented here consider both the impact period and the immediate aftermath. Some theoretical aspects are also presented.

Telephone communication has been affected in a number of ways in a variety of eruptions. During the 1956 eruption of Bezymianny telephones rang spontaneously and loudspeakers burnt out (Gorshkov, 1959: 86). The tephra cloud over the village of Ust-Kamchatsk during the 1964 Shiveluch eruption was so charged with electricity that telephone and radio communication with other villages was not possible (Gorshkov and Dubik, 1970: 268). A municipal official on the shore of Lake Bombon tried to telephone Taal town during the 1965 eruption but he was unable to hold the receiver because it was electrically charged (Carroll and Parco, 1966: 26). In the 1902 eruption of Pelée the telegraph office in St Pierre and its contents were burnt, presumably by pyroclastic flows. Some fragments of apparatus were moved as much as 100 m. Marconigraphs were made useless in subsequent eruptions (Hill, 1902a: 255). During the 1888–1890 activity of Vulcano, the Lipari to Milazzo (Sicily) submarine cable which runs 5 km east of the island was broken five times (Bullard, 1976: 234-235). In 1907 a submarine eruption at Monaco Bank near the Azores was only detected by the break in the submarine cable (*CAVW*, 21: 52).

The 1980 Mount St Helens eruptions also affected telephone systems. In Yakima, where only 10 mm of tephra fell (see Fig. 3.8), the dust invaded electromechanical systems. In exchanges, during the tephra fall, air cooling systems were shut down, windows were taped up and entries and exits were limited. Nonetheless, the abrasive dust, partly magnetic and highly conductive when wet, proved costly to remove from electromechanical systems. Outside equipment suffered wear. Public telephones in exposed locations subsequently required about twice the normal level of maintenance. On the other hand, identical semi-public phones such as those in restaurants and gas stations required no additional maintenance (Tom Maher, Pacific Northwest Bell, Yakima, personal communication, June, 1980).

While these diverse examples illustrate that volcanic eruptions may adversely affect telephone systems, human usage during and after eruptions may have even greater impacts. For example, during the fall of about 3–6 mm of Spurr tephra in Anchorage in 1953, telephone facilities were strained by neighbors calling each other to compare notes. Radio appeals to cut down telephone use eased the situation (Wilcox, 1959: 448). The eruption of Mount St Helens produced similar effects. In Yakima, for example, the telephone service was able to cater for 10%–13% of telephone owners making local calls at the one time. Local calls reached 278% of normal during and immediately after the tephra fall. According to interviews conducted by Dueker *et al.*, (1980: 37), many of these calls were emergency calls, others were substitutes for travel, and other calls were made to family, friends, and employers, and to stores to check whether they were open. The local call network was so overloaded many potential users could not even get dial tone.

The Yakima telephone system was also able to cater for rather less than 5% of users calling long distance at once. In fact, the toll network was designed to handle with little delay traffic on the busiest days — Mothers Day (in 1980, the Sunday before May 18) and Christmas Day. In Yakima the tephra fall resulted in 70% more attempts to make calls than on Mothers Day. On the Sunday night about 40 telephone operators were unable to handle the trunk traffic usually dealt with by less than five operators. Five days after the tephra fall the toll network was still 150% above the usual demand (Tom Maher, Pacific Northwest Bell, Yakima, personal communication, June, 1980). In similar circumstances and with no telephone network reserved for emergency only use, emergency personnel could be hampered from performing their duties adequately.

It is also interesting to note that the increased telephone maintenance costs generated by the tephra fall in Yakima may well have been balanced or exceeded by the revenue generated from near-capacity use of the trunk network for a week or more. Certainly Mount St Helens created some additional telephone revenue before maintenance costs went up on May 18. As early as March 28, at the United States Forest Service command center in Vancouver, Washington, a mobile trailer with nine pay phones was wheeled into a nearby parking lot to satisfy reporters (*Holocaust,* 1980: 12).

Theoretical considerations of the effects of tephra falls on radio communications have been outlined by Priest and Blong (in preparation). Table 6.3 provides a summary. At less than ultra-high (UHF) frequencies, and hence in the range of public radio, the radiowave propagation distance oveø which tephra particles are present in the atmosphere will be only a relatively small number of wavelengths. Thus, signal attenuation due to absorption and scattering will be small. This conclusion is supported by experimental data on path loss gathered by J. N. Tylor (General Manager Telecommunications, Post and Telecommunications Corporation, Papua New Guinea, personal communication, 1981) during the 1980 eruption of Langila. However, ionisation of particles in eruption columns leads to frequent lightning strikes (see Chapter 2, Section 7.2); even during the intervals between lightning discharges, ionised gaseous columns will cause backscattering of radio wave energy. Ionisation effects and electrical discharges will be important at up to very high frequencies.

Table 6.3 Summary of effects of tephra falls on radio wave propagation

Type of system	Frequency	Likely cause of interference
MF (AM public broadcasting)	535–1605 kHz	Ionised gases
HF (short wave)		Continuous lightning discharge
VHF (FM public broadcasting)	88–108 MHz	Super refractive ducting
UHF		Direct damage to aerial structures
Troposcatter		
Microwave	$> 1GH_2$	Super-refractive ducting Tephra fall into transmission path Tephra fall onto aerial surfaces Direct damage to antenna or repeater structures Heavy/acidic local rainfalls

From Priest and Blong, in preparation.

Atmospheric ducting (super-refractive ducting) occurs when temperature in the atmosphere increases with height and humidity decreases. Turbulence results in inhomogeneities in the index of refraction and inscattering of waves. The radio signal received by scattering from the local disturbance will be combined with that received from the non-turbulent atmosphere; the resultant signal will thus undergo considerable fading depending upon the amplitudes and phases of the multi-path components. At typical troposcatter frequencies the intense local turbulences created by thermal disturbance may cause power to be scattered and to fall outside the receiving beam.

At frequencies in the microwave region, but where the wavelength is still much greater than the size of the tephra particles, scattering is small compared with attenuation of the signal due to absorption, particularly close to the erupting vent. Here, also, super-refractive ducting will be most serious. At small wavelengths (in the low centimeter and millimeter bands), tephra particles can be comparable in size with the wavelengths and both scattering and absorption are likely to strongly attenuate the signal. At the very short wavelengths (the millimeter band and optical systems) scattering becomes more important than absorption (Priest and Blong, in preparation).

Not all of these effects, or those additional ones listed in Table 6.2, have been reported. Certainly, radio communications have been hindered on several occasions. At Deception Island in 1967 continuous thunder and lightning in the eruption cloud produced enough static to make radio communications difficult. However, the British Antarctic base was able to maintain contact with several ships (Anon., 1968: 23,24). At Lamington, in 1951, a violent night eruption caused an electrical disturbance which made radio transmission impossible (Taylor, 1958: 26).

These data provide only partial accounts of the likely effects of a moderate eruption on telecommunication links. It is instructive therefore to consider the likely damage to a network. In 1937 the Rabaul eruption produced the tephra fallout pattern shown in Figure 6.3. At that time, communication links with the New Guinea islands were rudimentary. However, by the late 1970s, Rabaul Exchange was the focal point of all telecommunications traffic into and out of the islands north of the Papua New Guinea mainland. It provided all switching and analysis for dependent terminal exchanges on New Britain and New Ireland. Furthermore, all circuits to the Papua New Guinea mainland passed over the microwave system on Rabaul Ridge and through the coaxial cable between Rabaul Exchange and Rabaul Ridge terminal. On the basis of an eruption similar to the 1937 eruption the performance of the telecommunication system was assessed (Tylor, 1981):

A. The Rabaul Ridge microwave link would fail by power failure and/or cable breakage.
B. The microwave links Bainings–Rabaul and Mt Tomavatur–Rabaul would fade due to ducting and refraction, during the eruption.

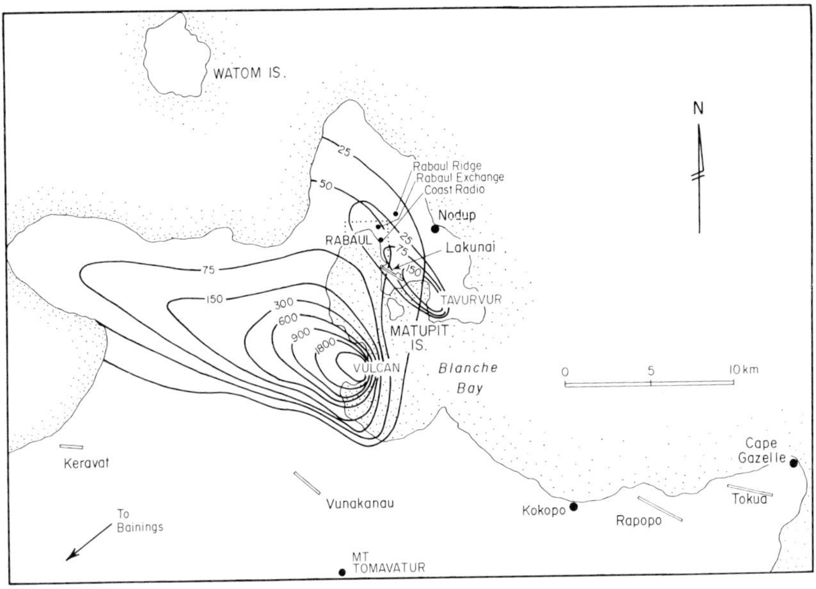

FIG. 6.3 Tephra isopachs from the 1937 Rabaul eruption and present-day telecommunication links in the Rabaul area (tephra map after McKee *et al.*, in prep.; telecommunication links from J. Tylor, *personal communication*).

C. Rabaul exchange would experience power loss and possible flooding and mud damage (less than 0.5 m).
D. Coast Radio Station would probably receive shallow water damage (less than 0.5 m).
E. Kokopo (35 km down coast) could be isolated by cable damage. Power would fail.
F. Transmitters would experience power loss and probable faults due to conductive ash deposits on aerials, feeders and transmitters. The transmitter roof could receive up to 100 mm tephra and could fail.
G. The exchange roof would receive up to 75 mm of ash. It could support this weight without damage if the roof load was evenly distributed. Box gutters might block with ash and pumice causing water to inundate the equipment room.

Tylor's (1981) analysis and the prospect of eruptive activity in the Rabaul caldera (McKee *et al.*, in press) resulted in changes in the telecommunication network, making it more immune to failure. This was achieved by terminating radio systems at Mt Tomavatur and making Rabaul a spur, so that failure at Rabaul would only affect Rabaul (John Tylor, General Manager Telecommunications, Papua New Guinea, personal communication). Analysis of other telecommunication nodes close to potentially active volcanoes would almost certainly indicate similar problems of network survival. Tylor (1981) also suggested the following measures to maintain sensitive telecommunication areas after the fall of conductive, corrosive and magnetic tephra:

A. Traffic into critical equipment areas should be restricted.
B. Critical areas should be sealed off, with adhesive tape if necessary.
C. Removal of dust should be done with damp rags. Sweeping is destructive.
D. Airconditioning plant to be tested. Filters to be cleaned daily. Plant to run on recirculate to extend filter life.
E. Tephra should be removed from roofs and gutters of exchange buildings.
F. Dust will enter emergency power generators causing loss of output power, scouring of sliprings and brushes and cutting out bearings. Portable plant should be cleaned and placed in as clean a location as possible.

It is difficult to overemphasise the importance of telecommunications in natural disasters. Wenger *et al.* (1975: 41) report that a survey conducted in Delaware indicated 74% of respondents viewed the electronic media as their main sources of disaster information. Similar results have been recorded in Australia by Blong (in preparation). A survey conducted in the Pullman, Washington, area after the Mount St Helens tephra fall determined that 56% of the sample would use the radio as the main source of information in the event of another big eruption, 24% television and a further 12% radio and television (Dillman and Roberts, 1982). Where towns and cities have their own radio and TV channels these sources of information generally prove satisfactory, but where broadcast information applies more correctly to other localities, problems can arise. In Ritzville, for example, after the tephra fall, emergency messages were transmitted via police bullhorns. Similar problems arose in Cheney: "During the week following the eruption . . . there was no effective means of disseminating information to Cheney citizens. The city believed that it did not have access to the Emergency Broadcast System; access to Spokane radio stations was very limited under the emergency situation. With the exception of a campus FM station (which is seldom, if ever, listened to by townspeople), Cheney does not have a local station. During the ashfall, Cheney residents tuned to outside stations located in Spokane, but these stations focused on their own local problems. This proved unsatisfactory. With no alternative means of communication at hand, the Mayor was forced to have his May 19 emergency message to the public hand-delivered — door to door" (Warrick *et al.*, 1981: 58,80).

Other forms of communication may also be damaged during eruptions. Telemetry devices employed by volcanologists are no exception! Pyroclastic flows and lahars produced during the 1975 Augustine eruption destroyed the seismic network on the island (*BVE*, 1978, 16: 53-5). At Ruapehu in the same year ballistic blocks chopped a heavy seven-pair communication cable into pieces resembling spaghetti (Nairn, 1975b).

Finally, even that least destructible form of communication — the mail — may suffer adversely. Many Pacific Northwest residents reacted to their first experience of tephra fall on May 18, 1980, by mailing a sample of the stuff to their friends all over the United States. The mail service in some areas ground (literally) to a halt. Postal officials in Pasco said that if residents wanted to mail samples to friends and relatives, they should do so in sealed plastic bags. Envelopes stuffed with ash had burst open and damaged postal machinery (*The Columbian,* May 26, 1980: 9).

6.6 AIRPORTS AND AEROPLANES

Although the great bulk of information on the effects of volcanic eruptions on airports and aeroplanes stems from the May, 1980, eruptions of Mount St Helens, it is probably correct to say that few new effects were discovered. Given the effects of earlier eruptions, little happened to airports and aeroplanes in May, 1980, that was not predictable.

6.6.1 Airports

All of the examples cited in this section concern the effects of tephra fall. The airfield at the Kilauea Military Camp was seriously affected by bombs during the May, 1924, eruption (Fig. 6.4a). In 1951, a tephra fall of only 1–2 mm from Lamington was sufficient to close the Port Moresby airfield (Taylor, 1958). In Anchorage in 1953, 3–6 mm of tephra was sufficient to close Merrill Field, International Airport, and Elmendorf Air Force Base. Numerous planes were diverted to unaffected airfields (*Anchorage Daily Times,* July 11 and 13, 1980). In 1963 the eruption of Agung closed Surabaja airport with a tephra fall of less than 10 mm (Jennings, 1969: 26). Catania airport, 19 km from Etna, was closed for a while by a tephra fall in 1979 (Guest *et al.,* 1980: 156). The May 18, 1980, eruption of Mount St Helens closed airports as far east as Missoula in western Montana where about 1 mm of tephra fell (Warrick *et al.,* 1981: 91). The May 25 eruption which deposited only 2–3 mm of tephra in the Portland area (see Fig. 3.8) closed that airport for 21 hours (*Aviation Week and Space Technology,* June 2, 1980: 19). A tephra fall of 100 mm from Miyake-jima closed the island airport for four days in 1983 (*SEAN Bulletin,* 8(10): 4).

These examples illustrate that only a few millimeters of tephra fall are sufficient to close airports. A more detailed consideration of the closure and clean-up of Yakima airport on May 18 indicates the extent of the effects of a 10 mm tephra fall.

It took five days to satisfactorily clean the tephra from the 420 ha of Yakima Airport. Twenty thousand tonnes of tephra were removed from about 40 ha of hard surface area. The ash was not removed from the adjacent grass but was watered in 24 hours a day for seven days, particularly along the critical areas 20 m each side of the runway. Up to 150 people worked on the clean-up using 40 pieces of equipment

FIG. 6.4 (top) Bombs from the May, 1924, eruption on the airfield at Kilauea Military Camp (Photo: Lyman House Memorial Museum, Album 6 Strip 164 #1); (bottom) Light aircraft sprinkled with Mount St Helens tephra, May, 1980 (Photo: P.W. Francis).

— front end loaders, sweepers, graders, dump trucks and flushers (water trucks). The airport clean-up cost about US$75 000 for labor, including overtime, equipment and fuel. The absence of planes for seven days meant a loss to the city of US$20 000–$25 000 on the fuel flowage fee of two cents per (US) gallon (1 US gal = 3.8 L). A further loss occurred to the airport restaurant where the customer level had still not returned to normal three weeks after the eruption (Yakima Airport Manager, personal communication, June, 1980).

The airport, and the airspace above, was closed for seven days. Actually, a twin-engined Cessna landed illegally at the airport on May 19 and let off two passengers. Airport staff tried to detain the plane which swerved to avoid their truck and took off, the plane and pilot being in violation of numerous regulations by landing with inadequate visibility at a closed airport (*The Yakima Herald-Republic,* May 21, 1980). Otherwise, closure meant the cancellation or diversion of eight Hughes Airwest, 18 Cascade Airways and numerous private flights each day (Yakima Airport Manager, personal communication, June, 1980).

Grant County Airport at Moses Lake took 15 days to reopen after 30–40 mm of tephra fell on May 18. Officials estimated clean-up costs at US$200 000 but other revenue losses occurred when Boeing moved their training program for airline crews to Glasgow, Montana. Japan Air Lines also temporarily closed their jet training base, staffed by 28 (*Aviation Week and Space Technology,* May 26: 21; June 9: 25; June 23: 31).

Similar closure and clean-up problems occurred at other airports throughout the Pacific Northwest. At Missoula runways were washed down and operations were back to normal on May 22 (Warrick *et al.,* 1981: 91). In most cases the tephra was swept up or damped down and then swept up. Evidently truck-mounted jet engines used at Elmendorf Air Force Base and International Airport to "sweep" runways clear of Spurr tephra in 1953 were out of favor (*Anchorage Daily Times,* July 11, 1953).

Airport closures also meant flight cancellations. Cascade Airways, a commuter airline operating out of Spokane, was hardest hit with virtually its entire area of operations covered with tephra. On May 18 the airline was able to complete only 5% of its scheduled flights; by May 20 this had risen to only 30%. Ground transport problems also kept airline staff strength to about 30% of normal. Hughes Airwest had to cancel its operations at five airports: Spokane; Yakima; Pasco; Lewiston (Idaho); and Kalispell (Montana). United Airlines cancelled 18 operations a day from Spokane. On May 25 when Portland Airport was closed, United diverted flights to Seattle and San Francisco, but the commuter airline Air Oregon had to close. Most route cancellations were for only a few days but Cascade Airways, flying into airports such as Moses Lake, had schedules seriously affected for nearly a month from May 18 (*Aviation Week and Space Technology,* May 26, 1980: 21; June 2: 19). On the other hand, the eruption generated an enormous increase in sightseeing flights from nearby airports (see below and Chapter 8, Section 4).

Even threatened eruptions may have important effects on airports. In 1979, the possibility that an eruption of Karkar (Papua New Guinea) would necessitate evacuation of more than 20 000 people led to plans to extend the island's airstrip so that it could be used by Hercules transports. Similarly, a threatened eruption of Rabaul in late 1983 resulted in upgrading of the Vunakanau strip 16 km from Rabaul so that it would meet operational standards for Dash-7s, Hercules and Caribous (PNG *Post-Courier,* 1983, November 10: 11).

6.6.2 Aeroplanes

Tephra falls or plumes have frequently caused problems for aeroplanes. Seventy Allied planes were damaged by a tephra fall from Vesuvio in March, 1944. Thirty five were damaged beyond repair but 20 aircraft were back in service within a month. The eruption evidently occurred so quickly that it was not possible to remove the planes from airfields near the crater (*The New York Times,* April 25, 1944: 4). In November, 1983, following evidence that an eruption might occur in Rabaul, Air Nuigini altered flight schedules so that no planes remained on the ground in Rabaul overnight, thus reducing the possibility that aircraft would be stranded or destroyed (PNG *Post-Courier,* 1983, November 10: 11).

An unusual incident occurred in 1935 when United States Army Air Corps B-3s and B-4s, fabric-covered biplane bombers, were used to bomb lava flows from Mauna Loa. One pilot reported direct hits with 275 kg bombs on a lava channel; he observed that molten rock was thrown to about 60 m elevation, some flying fragments making holes in the lower wing of his plane! (Lockwood and Torgerson, 1980: 729).

In 1953, a United States Air Force F-94 all-weather jet that flew through a dense cloud of Spurr tephra for only a moment or two had paint sandblasted away from the leading edges of wings and its plexiglass canopy frosted (*Anchorage Daily Times,* July 11, 1953; Juhle and Coulter, 1955: 199). However, military aircraft were not caught on the ground as the tephra cloud approached Anchorage but set off for other strategic bases in Alaska where they remained for several days until Elmendorf Air Force Base was again operational (*Anchorage Daily Times,* July 10, 1953). In 1976, when Augustine erupted, civilian and military aircraft flying through a tephra cloud sustained damage. Two F-4E Phantoms, flying in close formation about 10 m apart, entered complete darkness: "Had the lead plane not immediately turned on its lights, the following pilot would have lost contact; he could barely see the lead plane 10 meters away with its lights on. Upon landing in King Salmon the canopy of the aircraft was [sic] scoured, and the paint at the wing tips was sandblasted off. Very fine jewelers rouge-colored material was ingested in the cockpit through the engine air intake. The material was sticky and was found in every nook and cranny of the planes" (Kienle and Shaw, 1979: 150-152). A Japan Airlines cargo DC8 suffered a scoured center windshield and slight abrasion damage to external radio parts, landing gears and the air-conditioning system. Only the windshield needed replacement. A Japan Airlines 747 and another DC8, both carrying passengers, suffered minor damage, less extensive than that to the cargo DC8 (Kienle and Shaw, 1979: 152; Miller, 1976: 22). Between 1977 and 1979, ejecta from Usu and Sakura-jima pitted or shattered outer windshields on six Lockheed L1011 and three McDonnell-Douglas DC8 aircraft (ICAO, 1983).

A few hours after the Mount St Helens eruption began on May 18 a Hughes Airwest McDonnell-Douglas DC-9-30 *en route* from San Francisco to Calgary flew into the plume at about 10 700 m between Pasco and Spokane: "The Seattle air route traffic control center (ARTCC) had advised the pilot of the cloud, and he thought he could fly above it, FAA [Federal Aviation Administration] said. Shortly afterward, the pilot reported entering cirrus clouds, which suddenly began to grow darker while a sulfuric odor was noticed in the cockpit. The pilot turned to the east, toward Lewiston, before resuming the flight to Calgary, the FAA said. There, it was noticed that the windshield had a 'milky appearance,' the FAA said... The return flight was completed, and after the DC-9 landed at Phoenix, all three windscreens were changed, as well as the sliding windows, the FAA said. The wing leading edges showed the sandblasting effect of the particles, and the engine compressor blades showed some abrasion, FAA said. Engine oil was changed as a precaution, and when analyzed, the old oil showed abnormally high concentrations of sulfur, the agency said. The FAA pointed out that these effects resulted despite exposure to the ash for no more than 4 min... The FAA cited this incident in its general warning about the hazards of ash contamination, which it said could lead to 'multiple engine failures,' as well as contaminating pressurisation/air condition systems and damaging pitot static systems. In its warning to general aviation, the FAA said 'the gritty dust will peel finish off leading edges... clog air filters and convert engine oil into a destructive abrasive compound and ruin engines' " (*Aviation Week and Space Technology,* May 26, 1980: 20).

On May 25, a turboprop transport, a Lockheed L-100-30, incurred a four to five minute exposure to heavy tephra shortly after a pre-dawn takeoff from McChord Air Force Base near Tacoma. Contact with the tephra occurred while climbing through 3400 m. The number 2 and 4 engines began surging badly and had to be shut down. The other two engines were low on power. Damage to the Transamerica Airlines plane amounted to about US$500 000. Two turbines were destroyed and two badly damaged; there was some compressor damage. Two center windshields and one side windshield had to be replaced. Later, Boeing reported damages to two 727 trijet transports. Both planes and engines were subsequently returned to service (*Aviation Week and Space Technology,* June 9, 1980: 25,26; June 23, 1980: 31). Several light aircraft made forced landings because of engine failures due to tephra in the induction systems (*FAA General Aviation News,* Sept.-Oct., 1980: 10). On the other hand, a Lockheed U-2 which flew a zig-zag pattern through the tephra plume at 18 000 m suffered no damage. The lack of damage was attributed to the fact that the U-2 flew at higher altitudes where fallout was lighter (*Aviation Week and Space Technology,* June 9, 1980: 26).

In late June, 1982, a British Airways Boeing 747 with 240 persons on board, flying 150 km west-south-west of Galunggung at about 11 300 m encountered a tephra plume which stalled all four engines and abraded the cockpit windshield and wing surfaces. The aircraft lost 7500 m altitude in 16 minutes before the engines could be restarted. A safe emergency landing was made on three engines at Jakarta. In mid-July, 1982, a Singapore Airlines Boeing 747 with 230 passengers at 9000 m stalled three engines in a tephra plume from the same volcano. This jumbo jet lost 2400 m altitude before restarting one engine and landing safely in Jakarta on two engines (*SEAN Bulletin,* 7(6), June 29, 1982: 6; W.S. Smith, *FAA information release,* March 18, 1983). As noted in Smith's (1983) information release: "Post flight analysis for each of these incidents has revealed that a glassy ceramic layer or material was deposited on the surfaces of the turbine section. Additionally, it was also found that some ablation of compressor and turbine blades and stators had occurred. The combination of these effects can change the fundamental airflow characteristics and efficiency of modern turbojet and turbofan engines".

In November, 1982, an Air Nauru jet carrying 39 passengers and flying at about 3000 m entered a tephra cloud from Sakura-jima, six minutes after leaving Kagoshima Airport. Lapilli impacts produced hairline cracks in three cockpit windows. The plane returned to Kagoshima and landed safely (*SEAN Bulletin,* 7(11), 1982: 11).

These brief examples place Mount St Helens in a lengthy history of contacts between tephra falls or plumes and aircraft damage. Following the 1953 Mt Spurr eruption the United States Civil Aviation Administration Office of Aviation Safety issued warnings indicating that pilots should take special precautions to prevent abrasive damage to aircraft engines and instruments and noting that filters, screens and pumps needed to be checked daily (*Anchorage Daily Times,* July 11, 1953). On May 21, 1980, only three days after the main Mount St Helens event, the Federal Aviation Administration issued an alert to air carriers and general aviation warning of the volcanic ash hazard. This alert, reproduced *in toto* below, makes it clear that aircraft on the ground exposed to tephra fall (Fig. 6.4b) also required special maintenance:

> 1. *Introduction.* The purpose of this special alert is to provide safety information and recommended actions to preclude possible airworthiness problems associated with aircraft exposure to volcanic ash either in the atmosphere or on the ground. The following recommendations are based on the best available information at this time and will be amended as further information and facts are gathered.
>
> 2. *Background.* Volcanic ash from Mount St Helens has been analyzed to contain abrasive and corrosive materials such as sulfuric acid and floride and chloride salts and acids. Depending on the location of the volcanic ash fallout, the particle sizes range from as small as .5 microns to 100 microns [0.5 μm to 100 μm]. Most aircraft screens will filter out material down to 15 microns but will pass particles that are smaller.
>
> The ash will probably be encountered as a fine powder, like talcum powder, light grey in color. When dampened it has been reported to set similar to concrete. Due to adhesive action of the sulfuric acid, the acid tends to adhere to the interface between the particles

and the aircraft structure causing corrosion. Although volcanic ash may not be visible on the structure, sulfuric acid may still exist causing corrosion and it is recommended that all aircraft that have been exposed to volcanic fallout be given a test to determine the acidity levels. This can be accomplished by the application of nitrazine paper available from many pharmacies. If the Ph [sic] factor is 4 or below, it is recommended that the following corrective actions be taken as soon as practicable.

3. *Airframe.*
(1) Safety precautions including safety glasses, gloves and protective clothing should be adhered to.
(2) Aircraft should be cleaned in the following sequence:
A. The aircraft manufacturer's maintenance manuals should be followed regarding the protection of aircraft systems during the cleaning process. The procedures recommended by the manufacturer for the inspection and cleaning or purging of pitot static systems, instruments systems, etc., should be followed.
B. If there is any volcanic ash coating on the aircraft structure, it should be removed by hand brushing, air or vacuum cleaning prior to performing any washing actions. If the aircraft is washed before removing ash, it will form a corrosive paste.
C. The aircraft should be thoroughly cleaned, inside and out, before washing.
D. The aircraft should be rinsed thoroughly with fresh water without scrubbing to ensure that all parts of the aircraft have been amply rinsed.
E. The aircraft should be given a test at the completion of each wash cycle to ensure a Ph factor of above 4. The Ph factor may be performed by taping nitrazine paper strips on various parts of the structure and wetting with distilled water.
F. It may be necessary to repeat the wash procedures on a continuous basis in areas where fallout continues. To ensure that the sulfuric acid is neutralized, complete the wash cycle using a petroleum base solvent.
G. It is recommended that regardless of the Ph factor, aircraft exposed to volcanic ash be given a water wash as assurance against corrosion.
H. Close inspection for external signs of damaged seals, especially landing gear and landing gear actuators.

4. *Systems.* For all aircraft exposed to volcanic ash on the ground or air, system checks should include inspection of pitot-static probes, static ports, air conditioning outflow valves and filters, generator cooling tubes and filters, vacuum lines and filters, and externally mounted sensors, such as long wire antenna, and angle-of-attack sensors, to remove any ash contamination. Precautionary inspection should be done on a random basis of electronic equipment subject to cooling air to assure need for equipment removal for removal of ash contamination.

5. *Powerplant Considerations.*
(1) *General.* As noted above the nature of the volcanic ash is that it is both abrasive (in the form of fine powder) and corrosive (in the form of acid content). Both turbine engines and reciprocating engines may be affected. Compressor and turbine blades suffer erosion by abrasive particle impact. Lubrication and other fluid systems are subject to contamination by solids and chemicals, while moving parts are subject to abrasive wear.

The abrasive nature of the material causes rapid mechanical damage to moving parts. Experience has shown that engines operated with oil contaminated with the ash has caused them to fail in as few as 20 hours after exposure.

The acids associated with the ash are soluble in oils and as such attack engine parts resulting in rapid deterioration. Engines can also be attacked externally by the corrosive action.
(2) *Maintenance.* Engines which have been operated or subjected to a volcanic ash fallout need:
A. Thorough external cleaning.
B. Cleaning or changing of all oil, fuel, and other systems screens and/or filters and draining all sumps. Flushing and cleaning of contaminated fluid systems should be accomplished following the engine manufacturer's recommended procedures.

C. Cleaning and scavenging of any open tanks where the ash or chemical action could collect and concentrate.

D. Oil and fluid draining and change.

E. Close inspection for external signs of damaged seals.

F. Cleaning and inspection of accessories and components for contamination such as vacuum filters and regulators.

G. Following cleaning, inspection, and fluid changes the proper operation of the engine should be verified by run-up. Frequent oil and fluid changes should be scheduled. During subsequent operation temperatures and pressures should be monitored closely for changes which may signal problems.

H. Spectrographic oil analysis is an indicator of engine wear and contamination. Comparison of analysis of early samples after cleaning with previous analysis reports can serve as an indicator of an engine's stability.

6. *Recurrent Inspections.* Aircraft owners and operators are urged to closely monitor aircraft and engines, including systems, on a continuing basis and take action as deemed necessary for aircraft exposed to volcanic ash as assurance against corrosion (FAA, 1980).

Other advisory information was issued by airframe and engine manufacturers (see *Aviation Week and Space Technology,* June 9, 1980: 41, 43-45, and *FAA General Aviation News* September-October 1980: 10-11).

Galunggung tephra posed a rather different problem as it was not always known when the volcano had created rising tephra plumes. Furthermore, the intercontinental jets on the South-east Asia–Australia routes were flying at altitudes where the plumes were dispersed containing particles only a few micrometers across and, hence, "invisible" to weather satellites. Australian scientists evidently solved the problem of tephra plume recognition using the United States NOAA 7 satellite and special signal processing techniques on two infra-red images (*The Australian Physicist,* 19, September 1982: 163).

6.6.3 Other aspects of air safety

The Mount St Helens search and rescue operation involved 242 aircraft from United States Air Force regular and reserve units, the United States Army, Washington Air National Guard, Coast Guard and Civil Air Patrol. Aircraft employed by the United States Geological Survey, United States Forest Service and numerous news media were also in the area for at least part of the search period. In all 549 sorties were flown totaling 882 hours (*Aviation Week and Space Technology,* June 9, 1980: 26). Because search helicopters and fixed-wing planes belonging to many agencies took part communication problems quickly arose; many aircraft could not communicate directly with one another. Eventually a Lockhead C-130, with a jeep inside equipped to communicate with the ATS-3 satellite, was brought in to circle the mountain at 6000 m. By using the satellite this unit was able to maintain communications with helicopters working on any side of the mountain – impossible with conventional equipment. This capability, coupled with the C-130's radar equipment, permitted it to maintain lateral separation between as many as 65 helicopters. "We have absolute control" a rescue official said. "Without it, I don't know how many mid-air collisions we would have had" (*Aviation Week and Space Technology*, June 2, 1980: 19).

Restrictions on aircraft approaches to Mount St Helens were emplaced before the May 18 eruption. Special clearance was required to enter the area in a 5 km radius to 18 000 m and a 32 km radius to 6000 m. While rerouting of commercial flights to avoid this zone added no more than 15–20 minutes to flight times (*Aviation Week and Space Technology*, July 28, 1980: 19) problems arose with sightseeing aircraft. As early as March 30, 1980, small planes filled the sky around the mountain and Vancouver's (Washington) Pearson airpark looked like a Trailways station (*Holocaust*, 1980: 12). *The Columbian* reported on May 26 (page 1) "Good weather brought out hundreds of planes, buzzing the peak like angry bees. Despite a closure to the general public, the airspace was jammed because most claimed to represent some press outlet. The five-mile [8 km] limit around the peak shrank and shrank, until it appeared that there was a competition to see who could get closest". In late July, *Aviation Week and Space Technology* reported (July 28: 19) that the Federal Aviation Administration was concerned about the estimated 150–200 sightseeing aircraft that swarmed around the volcano. The FAA issued special warnings to the pilots to exercise extreme caution. But near-miss incidents occurred in the restricted area despite frequent patrolling by FAA and United States Forestry Service aircraft. "A recent incident involved a National [sic] Geological Survey helicopter which was hovering 100 feet [30 m] above the mouth of the crater to allow scientists to scrutinize the interior. Suddenly, the scientists were startled to see a fixed-wing aircraft pass beneath them and disappear in a cloud of ash and steam . . . Some observers are certain that the first aviation casualties [but see Chapter 3, Section 12] will occur as a result of a large number of pilots recklessly flying around in the restricted area around Mt St Helens without prior approval or radar separation by air traffic control. There are approximately 40 cases of illegal penetration being investigated at the present, many of which are expected to result in pilot certificate suspension" (*FAA General Aviation News*, September-October 1980: 10).

6.7 RAILROADS

Railway lines and stock and overhead cableways have been affected in a number of volcanic eruptions. Ground deformations disrupted railroad lines around Usu in 1944 where spectacular ground upheaval moved the lines laterally by 20 m. Ground movement also accompanied the 1924 Kilauea eruption. At Kapoho, 45 km east of the caldera (see Fig. 4.1), about 200 earthquakes felt on April 22 and 23 were accompanied by ground cracking and the sinking of a graben between two parallel faults 800 m apart. Figure 6.5 shows the irregular lowering of the railway line from the Kula quarry used to supply rock for the Hilo breakwater. The maximum deformation was 2.5 m (Macdonald and Abbott, 1970: 78).

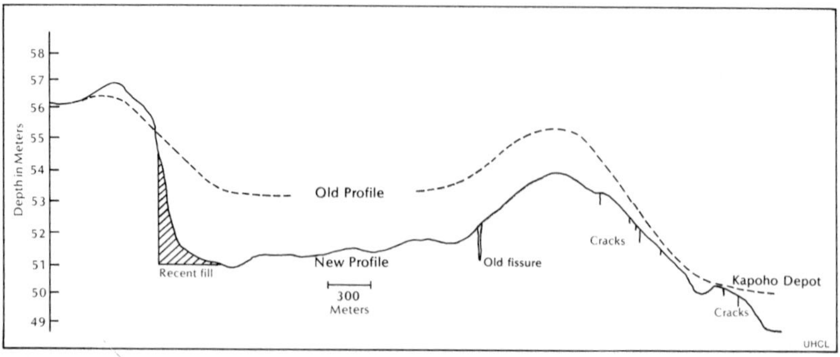

FIG. 6.5 Deformation of the railway line in the Kapoho area, 1924 (after Finch, 1924).

Lava flows have often blocked railways around Etna and Vesuvio. In the 1928 eruption of Etna a flow 30 m wide blocked the Messina–Catania railway and the road parallel with it, cutting off all land communication between the two cities. In 1971 lava high on Etna destroyed the terminal station of the cableway to the observatory (*BVE,* 1973, 11: 9) and surrounded four of the pillars supporting the cable, breaking the cable on April 15 (*SISCLP,* 33-71: 1173-74). The rest of the cableway was destroyed in 1982. The 1906 Vesuvio eruption cut the Circum-vesuvian Railway at a number of points (Fig. 6.6). A very fluid flow during the 1944 eruption of Vesuvio overran a 30 m section of the "tourist" funicular railway to "Valle dell Inferno" (*The New York Times,* March 20, 1944: 3). In 1917 the railway from Quezaltepeque to Ciudad Arce in San Salvador was covered for a length of more than 2 km by aa lava from a flank eruption of Boqueron (Roy, 1957).

Tsunami have damaged railway lines and rolling stock in at least one instance. The area around Merak (see Fig. 2.24) was amongst the worst hit by Krakatau, 1883, tsunami: "The strong railway line, used for conveying the stone to the neighbouring jetty, was torn up for many hundreds of yards, twisted and bent just like wire. The fish-plates connecting the lengths of rail had held securely, and when the metals had been torn from the sleepers by the rushing water, the latter had been curved and bent in serpentine fashion, and carried a great distance from their original position. One of these lengths of torn-up rail must have measured a quarter of a mile. The railway trucks had fared very badly, having been dashed in all directions, and greatly damaged. Two of the locomotives employed on the quarry line, in spite of their great weight, did not escape so well as one would have imagined. One of them, a six-wheeled tank engine, was washed off the rails and thrown completely over on its side. The other, of similar size and construction, was more damaged, and had actually been carried right out to sea. There it lay, a battered wreck, some fifty yards from the beach, with the waves surrounding and breaking over it" (Neale, quoted in Simkin and Fiske, 1983).

FIG. 6.6 The circumvesuvian railway blocked by a 1906 lava flow (Photo: F.A. Perret, Lyman House Memorial Museum, Neg. #27 [copy]).

FIG. 6.7 The destroyed engine, a carriage and a bridge section at Tangiwai, New Zealand, in 1953 (Photo: *Auckland Star*).

FIG. 6.8 Log rack cars tipped over by a lahar in the North Fork of the Toutle River, May 18, 1980 (Photo: R.L. Schuster, US Geological Survey).

Lahars and landslides have also caused major damage to railways. Landslides (and probably lava flows) destroyed much of the Vesuvio funicular railway in 1906. Above the foot of the cone the railway was almost obliterated; stations had gone, but here and there a bent rail was sticking up. One section was still tied together by the sleepers but bent down the mountain at right angles as though the whole roadbed had moved (Jaggar, 1906: 109). Similarly, although no details are available, lahars destroyed a railroad during the 1968 eruption of Mayon (Moore and Melson, 1969: 616).

Railway bridges have proved very susceptible to lahars. On Ruapehu, "The Tangiwai disaster" occurred on Christmas Eve, 1953, when a lahar from the crater lake rushed down the Whangehu River and carried away one of the piers of the railway bridge only a few minutes before the Auckland–Wellington express arrived at 10:20 p.m. The locomotive tender and 5 carriages plunged into the river killing 151 people (Fig. 6.7). The steel chassis of one carriage was transported 2 400 m downstream by the lahar (*CAVW*, 22: 102).

Lahars down the Toutle River west of Mount St Helens in May 1980 had serious effects on railways. The Weyerhaeuser Company's private logging railroad from Camp Baker to Longview was severely damaged. Twenty four kilometers of the 676 km track were buried in the lahar and six railway bridges in the Toutle Valley were destroyed including one with a 120 m span. Although the diesel locomotives working the line were unharmed, Weyerhaeuser lost 86 log rack cars (Fig. 6.8) and several woodchip cars (*Trains*, 40(10), August, 1980: 17).

The main Seattle–Portland line, run by Burlington Northern (BN) was also affected by the Toutle River lahars. Early on May 18 the line was closed because of possible damage to the 57 m through-trussbridge over the Toutle River near Castle Rock by a lahar which raised the stream level 5 m threatening the upriver track. Trees and other debris careered into the Number 2, or easternmost, main: "A foot [30 cm] of mud was cleared off both lines, Main 2 was taken out of service, and BN had to contend with single track and a 10-mph [16 km h^{-1}] speed limit on one of the heaviest-density main lines in the West, a line shared with Union Pacific and also utilized by Amtrak" (*Trains,* 40(10), August, 1980: 17). Amtrak's *Coast Starlight* was cancelled between Portland and Seattle on May 18 and 19.

Tephra falls have also played their part in blocking railway lines and disrupting traffic. Following the 1906 eruption of Vesuvio the railway line from Naples to Salerno (see Fig. 3.3) was blocked by tephra fall for a few hours and the Circumvesuvian Railway was blocked for seven days (Anderson and Bonney, 1917: 652). In Bridgetown, Barbados, a fall of only 5 mm of tephra from the May, 1902, eruption of Soufrière, St Vincent, 160 km distant, was sufficient to run trams off the rails. Rail services in western Montana, where only 1–2 mm of tephra fell, were disrupted by the May 18, 1980, eruption of Mount St Helens: "Burlington Northern Railroad provides freight service for Missoula. Early Sunday night [May 18] the chief dispatcher received word of low visibility and breathing problems from a train crew in western Montana. Ten trains were shut down Sunday night through Monday afternoon. Crews were advised to wear masks for the next several days. Although speeds had been reduced to 35 mph [56 km h^{-1}] complaints were voiced by neighbours on Monday and Tuesday when trains stirred up dust along the track. One motor was ruined ($150 000 to repair) when a clogged filter was removed while the train was running. By Wednesday, the railroad had resumed normal operations" (Warrick *et al.,* 1981: 91).

In Ritzville, where up to 40 mm of tephra fell, problems with schedules remained longer. Railroad operations closed down on May 18. An engineer on a westbound freight train claimed that the wipers could not clear the ash fast enough. Restricted operations resumed about noon on May 19 but speeds were limited to 30 mph [48 km h^{-1}] and rolling stock was inspected every 32 km. These restrictions remained in force for 9 days. Full local service to Ritzville did not resume until May 29. Other problems for the railroads arose around Ritzville. "The phone system went out, and relays had to be replaced. Many employees could not get to work, and local personnel had to be summoned to work the section. Section maintenance crews had difficulty reaching inspection points because of highway closures and, therefore, could not do an adequate job. Finally, after rain, the ash, which turned out to be a good conductor of electricity, caused constant shorts which set off all the track signals" (Warrick *et al.,* 1981: 45-46).

Problems similar to those around Missoula and Spokane occurred throughout the tephra fall area. The first train into Yakima arrived on May 21. Trains continued to run around Spokane but crews were instructed to stop and inspect their trains every 16 km. Journal boxes were checked for unusual friction and air filters were cleaned and replaced. The Sheriff asked that trains not exceed 30 km h^{-1} in Spokane County to help keep the dust down. Burlington Northern restricted all their trains east of the Cascades to less than 50 km h^{-1} with inspections every 40 km but found that the tephra was more of an inconvenience than anything else. The first train over a line kicked up most of the dust, each succeeding train kicking up less and less. Mechanical problems were few. Air compressors which were usually changed every three months and which collapse and draw dust in if they become loaded, were changed every day as a precaution (*Trains* 40(10), August, 1980: 17; *Railfan and Railroad* 3(7), *November* 1980: 44; *Railway Age,* 181, June 9, 1980: 11).

Amtrak's *Empire Builder* (dubbed the "Ash Express") westward bound on May 22, picked up stranded motorists, making unscheduled stops at Cheney, Ritzville, Lind and Sprague. With standing room only, those taken aboard rode free to Seattle. Amtrak offered a further train at operating cost to pick up motorists stranded east of the Cascades but Washington State Governor Dixie Lee Ray turned the offer down (*Trains* 40(10): 17).

Although the eruption only caused relatively minor problems it was of more than passing interest to Burlington Northern (BN). Not only did BN own the square mile surrounding the summit of Mount St Helens and the land leased to Harry Truman for a resort at Spirit Lake but also 10 000 ha of timberland within a 19 km radius of the peak. Before May 18 the joke was around that BN owned "the hottest piece of property around" (*Railway Age,* 181, June 9, 1980: 11).

Although railroads were the least disrupted form of travel around Mount St Helens (Dueker *et al.,* 1980: 10), there is no reason to believe this to be true for all eruptions.

6.8 SHIPPING

Ships and ports can be damaged or schedules disrupted in a variety of ways by volcanic eruptions. Tsunami and other unusual waves have certainly damaged ships during a number of eruptions. Sea waves produced by the 1815 Tambora eruption swamped boats at Bima 75 km east of the volcano (Neumann van Padang, 1971: 59). Similarly, waves generated by the 1883 eruption of Augustine stranded the fishing fleet at English Harbor (Davidson, 1884). On Savai'i, during the Matavanu eruption, 3–4 m waves on October 6, 1907 wrecked a boat house and several boats (Anderson, 1910: 625). Waves created by the 1937 Rabaul eruption swamped two steamers and a schooner (Jaggar, 1945: 37). Many boats ended up high and dry (Fig. 6.9). A large tsunami generated by the Komagatake eruption of July 31, 1640 damaged many ships (Sapper, 1917: 122) while the Oshima collapse of 1741 reportedly destroyed 1500 ships (Siebert and Kienle, 1983).

Large sea waves produced during the 1883 eruption of Krakatau created problems for several ships in harbors. In most cases those ships *en route* through Sunda Strait noticed no unusual waves. The Dutch East Indies government steamer *Berouw,* anchored close to the pier at Telukbetung (see Fig. 2.24), was swept onto the beach by a wave about 7.30 a.m. on August 27. Shortly after 11 a.m. the largest wave of the eruption reached Telukbetung. "The wave lifted the gunboat *Berouw,* a vessel having a draft of six feet [2 m], carrying four guns, and having a complement of four Europeans and twenty-four natives [sic], from where she had been deposited in the Chinese quarter, carried her up the valley of the Koeripan River and deposited the vessel at a spot one and one-eighth miles [1.8 km] inland, leaving her stranded thirty feet [9 m] above sea level, behind a hill. All her crew were killed" (Furneaux, 1969: 109).

Three schooners and many small craft in the same port were also swept inland. The Dutch barque, *Marie,* a salt-trader of 570 tonnes, had taken the precaution of putting out a third anchor while at her moorings in the roadstead. At first she rode the large waves safely but eventually this ship was left high and dry on the beach (Judd in Symons, 1888: 15-16; Furneaux, 1969: 70-71, 109-110).

The tsunami also caused problems in Batavia: "The big dock with the *Augusta* at it, was in danger, because the heavy chains, with which the ship was moored, broke like glass, and it bumped into the *Siak,* a ship of the Navy, whose paddlebox broke and chimney toppled. However, the drydock did not float away, because some chains and cables held the anchors. The drydock of Amsterdam Island broke loose and floated to the South..." (Verbeek in Simkin and Fiske, 1983).

Vast fields of floating pumice, generated by the Krakatau eruption, also affected shipping. On the morning of August 28, the *Gouveneur General Loudon,* a steamship running down the Sumatran coast found its way completely blocked by pumice sheets. On forcing a way through the pumice, a layer piled up at the ship's bows indicating the sheet was 3 m thick (Furneaux, 1969: 125). Four days after the eruption the pumice was 9 m thick at Telukbetung and ships could not approach the shore.

In the first few weeks after the eruption, ships reported considerable difficulties negotiating the pumice fields in the Sunda Straits. On October 13, seven weeks after the eruption, the barque *Hottenburn* met tremendous fields of pumice which halted the vessel (Meldrum in Symons, 1888: 49) but Lampong and Semenka Bay which had been blocked by floating pumice, were largely free by the middle of December (*Nature,* 1884, 29: 437). In January 1884, patches of pumice off Jakarta reached several square miles [about 1500 ha] in extent and were thick enough to offer "considerable resistance" to steamships (*Nature,* 1884, 30: 288). "The *W.H. Besse* encountered wide sheets of pumice, on which lay the bodies of men and fish intermingled with green cocoanuts, six hundred miles [950 km] west of Java Head..." (Furneaux, 1969: 154-155). Three months after the eruption a passenger on one ship noticed that an iron bar thrown over the ship's side rested on the surface of the mass instead of sinking, and he saw trunks of trees and corpses embedded in the

Rabaul Harbour When the Worst Was Over

For a time the water was so thickly coated with pumice that navigation was almost impossible. The channel between Vulcan Island and the mainland had disappeared, and it was feared that submarine disturbances had altered the sea bed to such an extent that no vessel with any draught would be safe to move, but within a few days the pumice drifted out to sea with the tides and left the harbour clear. Earlier, however, as the bottom picture shows, a violent disturbance played havoc with many of the smaller craft, some of them being lifted high and dry on to the foreshores. Reporting on the situation last Sunday, the Administrator of the Mandated Territory (Brigadier-General McNicoll) said there was then no reason why shipping should not proceed to the wharves in the usual manner. H.M.A.S. Moresby took soundings within the harbour on Saturday and found plenty of depth in the main channel.

FIG. 6.9 (top) The *S.S. Durour,* 800 tonnes, on a slipway being repaired at the time of the 1937 Rabaul eruption, was stranded by tephra deposition on the shore. The Dutch volcanologist Dr C. Stehn is in the foreground (Photo: Bureau of Mineral Resources, Australia); (bottom) Two schooners stranded on the foreshore of Simpson Harbour, Rabaul, by a tsunami during the 1937 Rabaul eruption (Photo: Bureau of Mineral Resources, Australia).

pumice. Even more than a year after the eruption, sheets of pumice were reported at various locations across the Indian Ocean (Fig. 6.10).

Fields of floating pumice have also caused a hazard to shipping after other eruptions. The 1779 eruption of Sakura-jima filled the Seto channel (closed by a lava flow in 1914) with a floating sheet of pumice 1.5 m thick (Kotô, 1916: 32). A week after the 1815 eruption of Tambora floating rafts of pumice up to 5 km long were reported in the Flores Sea. Four years later these islands were still a nuisance to navigation (Neumann van Padang, 1971: 58). Following the Katmai-Novarupta eruption of 1912 small boats could not move where the tide had drifted the pumice (Erskine, 1962: 208). In the early 1960s pumice rafts in the South Sandwich Islands (58°S latitude) were reported to be up to 5 m thick as engine intakes at that depth sucked in pumice (Gass *et al.*, 1963).

Volcanic activity has proved hazardous to shipping in several other isolated instances. The most unusual was the disappearance of the Japanese Hydrographic Office vessel *Kaiyo-maru No.5* on September 24, 1952 while investigating, at close range, a submarine eruption at Bayonnaise Rocks about 430 km due south of Tokyo. Evidently the ship was directly over the vent when an eruption occurred, killing the crew of 22, and seven scientists. Wreckage from the ship was found later with numerous pieces of pumice trapped in it (Bullard, 1976: 395-397). At Deception Island in 1930 part of the sea floor dropped 4.5 m during an earthquake destroying part of the jetty and rocking ships at anchor. Earlier, in 1921, subsidence of the shore occurred near the whaling station and the water boiled, stripping paint from the hulls of anchored ships (Baker, 1969: 117).

Lava flows have also affected ports in a few instances. In 1706 lava from Tenerife buried the port town and harbor of Garachico, at the time the finest and most frequented harbor on the island. The port itself was filled by lava and converted into a promontory (Hulbert, 1827: 64). The port town of Heimaey was luckier in 1973. Figure 5.7 shows the changes in harbor configuration resulting from the eruption. At first it was feared that the fishing fleet would no longer be able to use the port but the lava flows actually improved the harbor, providing better shelter from Atlantic gales.

Tephra falls and electrical discharges have also provided dangers. During the climactic phase of the 1883 Krakatau eruption, the *Gouveneur General Loudon* was in Lampong Bay perhaps 80 km from the volcano. The captain reported that lightning struck the mainmast conductor five or six times and that "the mud-rain which covered the masts, rigging, and decks, was phosphorescent, and on the rigging presented the appearance of St Elmo's fire. The natives [sic] engaged themselves busily in putting this phosphorescent light out with their hands, and were so intent on this occupation that the stokers left the engine-rooms for the purpose, so that the European engineers were left to drive the machinery for themselves. The natives pleaded that if this phosphorescent light, or any portion of it, found its way below, a hole would burst in the ship, not that they feared the ship taking fire, but they thought the light was the work of evil spirits, and if the ill-omened light found its way below, the evil spirits would triumph in their design to scuttle the ship"

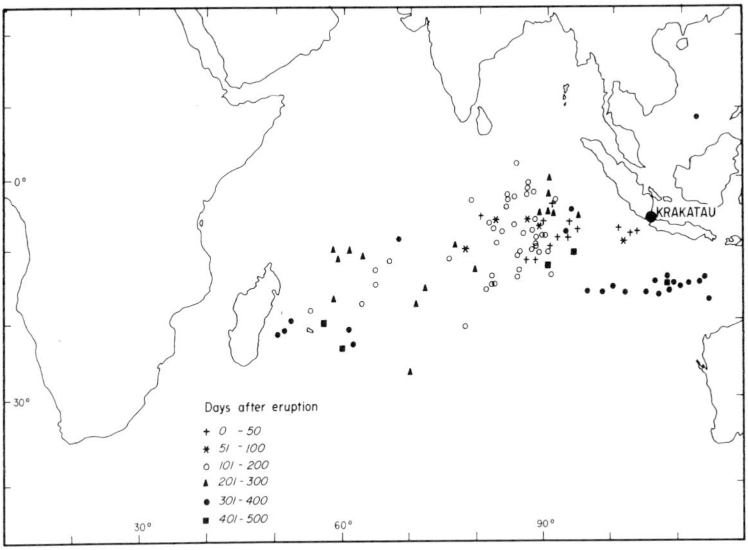

FIG. 6.10 Map to show locations and dates of pumice sightings in the Indian Ocean following the 1883 eruption of Krakatau (based on data in Symons, 1888).

(Judd, quoted in Symons, 1888: 21). Other ships in the vicinity experienced somewhat similar phenomena. All had to shift tonnes of dust and pumice from decks. The American cruiser *Philadelphia* (10 000 tonnes) experienced similar problems in 1944 when anchored only 19 km from Vesuvio. An Officer commented: "For two days we ate dust and slept with it... Then we got out in self-defense, and the crew shovelled her off... We were weeks eliminating the stuff" (*The New York Times*, November 7, 1944: 29).

Lahars have almost certainly affected shipping in other eruptions but the blocking of the Cowlitz River and the Columbia Shipping Canal on May 18, 1980 attracted attention because of the importance of the Columbia Canal to the economy of the Pacific Northwest. Figure 4.2 shows that the lahars which filled the Toutle also affected much of the Cowlitz River and part of the Columbia Canal. In the Cowlitz about 31.5 Mm^3 of sediment was deposited, clogging the channel to a depth of 4–5 m throughout and reducing the channel's capacity to carry water by about 85% (Collier, 1980: 14). Small boats along the river and its tributaries were stranded by the deposits (*The Columbian*, May 26, 1980: 10).

The mudflows that came down the Cowlitz reached the Columbia Canal on the incoming tide (Lawyer, 1980: 27; but see also Dudley, 1980: 31) with more material being carried upstream than downstream. The sediment altered the channel depth from 12 m to 4 m, the change being most pronounced at the mouth of the Cowlitz (Fig. 6.11). Changes occurred along more than 14 km of the Columbia. Estimates of the total volume of sediment vary from Lawyer's (1980: 27) early figure of 17 Mm^3 to Schuster's (1981: 711) 34 Mm^3.

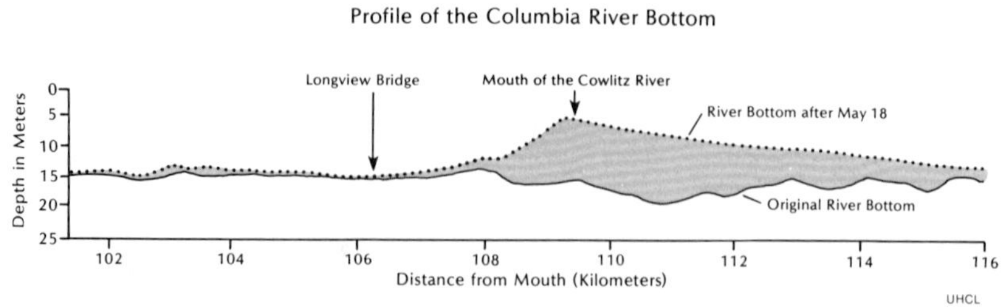

FIG. 6.11 Longitudinal profile of part of the Columbia River showing sedimentation from the Cowlitz River following May 18, 1980, lahars from Mount St Helens. Most of the sediment moved upstream (after Schuster, 1981: 714).

Early on the afternoon of May 18 the Coast Guard had issued a *Notices to Mariners* warning of the expectation that hundreds, if not thousands, of logs would empty into the Columbia from the Cowlitz and recommended that vessels not negotiate the river at night. The logs, and other flotsam caused numerous outages of downriver aids to navigation but it was only when the Norwegian ship *m.v. Hoegh Mascot,* drawing less than 6 m, grounded in mid-channel at 5 a.m. on May 19 just downriver of the Columbia mouth that it was realised that the navigation channel had filled with sediment (Dudley, 1980: 31).

As almost a dozen ocean-going freighters used the Columbia channel each day to reach the Port of Portland (Schuster, 1981: 711) the immediate closure of the Columbia in the Longview area (see Fig. 4.2) to all ships drawing more than about 3 m was quite serious. Eighteen ships were trapped in the Portland/Vancouver and Kalama River mouth areas and at least 22 vessels were prevented from coming up river.

Two hopper dredges began work on the channel on May 19. By the end of the first week four dredges were at work. Four additional dredges were on site by June 23. By June 12, 1.6 Mm^3 had been dredged from the channel; 11.5 Mm^3 had been moved before the end of 1980. The cost of excavating the new channel to 2 m deep and 180 m wide and maintaining it through 1983 was estimated at US $45 million (Schuster, 1981: 711).

While dredging proceeded, ships were able to traverse the channel. Dredged depths in the channel were computed daily and combined with tide data and information about water releases from upstream dams to set a daily window or time span when vessels of certain draft could safely pass up or downriver. Although the original safety zone allowed only vessels of less than 3 m draft to pass, this had been increased to 8.3 m by May 27 (*Feedstuffs,* 52(22), June 2, 1980: 11). By June 12 a draft of 10.6 m was permitted (Dudley, 1980: 33) but it was believed that complete restoration of the channel would take four to five months (Rainier National Bank, Economic Report No 2, June 11, 1980: 1). Between May 18 and June 12,

1980, only 65 upbound and 55 downbound vessels negotiated the blocked channel area, approximately 60% less than the shipping traffic for the same period in the previous year (Dudley, 1980: 33). In early June, some grain ships, including those trapped by the debris, were partially loaded at the grain terminals upriver of the blockage and then filled with barged cargo near Longview (*Feedstuffs*, 52(22): 11). However, it was expected at that time that large container ships would not be able to go upriver until channel restoration was complete (Rainier National Bank, Economic Report No. 2, June 11, 1980: 1).

It is interesting to note that the Port of Vancouver, Washington, also suffered side effects. Port officials attributed a downturn in shipping activity "largely to the interruption of Port activities from the Mt St Helens eruptions, but also cite misinformation and misunderstanding of current port capabilities. National publicity about port closures due to volcanic eruptions has discouraged business, port officials believe. This situation has prompted an advertising effort to dispel the psychological effects of the volcano. The campaign is being launched jointly by the Ports of Portland and Vancouver [Washington] and will include advertisements advising the shipping industry of current service conditions at both ports" (Rainier National Bank, Economic Report No. 2, June 11, 1980: 2).

6.9 ROADS

Roads can be affected by ground deformation, lava flows, lahars, drifting steam, tsunami, torrential rains accompanying eruptions, and tephra falls. As the most detailed information stems from the area around Mount St Helens affected by tephra falls these data are, rather arbitrarily, included in Chapter 6, Section 10. Chapter 6, Section 11, dealing with the effects on vehicles, is also largely about tephra fall and Mount St Helens.

Around Usu in 1944 ground deformation had quite spectacular effects on roads. Cracks developed rendering some roads impassable. By the end of October ground movements had shifted one road 60 m sideways (Minakami *et al.*, 1952: 268). Ground cracking across roads became quite a problem during the 1968 eruptions at Kilauea. Figure 6.12a illustrates the extent of cracks that opened during the August eruption. The maximum crack opening was only 51 mm with about 250 mm total dilation on the Chain of Craters Road. Small amounts of right lateral displacement were measured at some crack sets. Maximum vertical offsets were 54 mm (Jackson *et al.*, 1975: 8-9) These cracks disrupted traffic to only a minor extent. Urgent road repairs were required only to those cracks on Chain of Craters Road nearest to Ainahou Road on August 22, the day the eruption started. Lava emerged along those cracks marked "L" on the Escape Road (Fig. 6.12a). Ground cracking during the 1960 Kapoho eruption was more serious. Cracks across the road near Kapoho Village were only 20 mm wide on the morning of January 13. By early afternoon these cracks were 1 m wide with a 0.3 m throw (Macdonald, 1962: 263). Similar damage to roads occurred in 1924 (Fig. 6.12b).

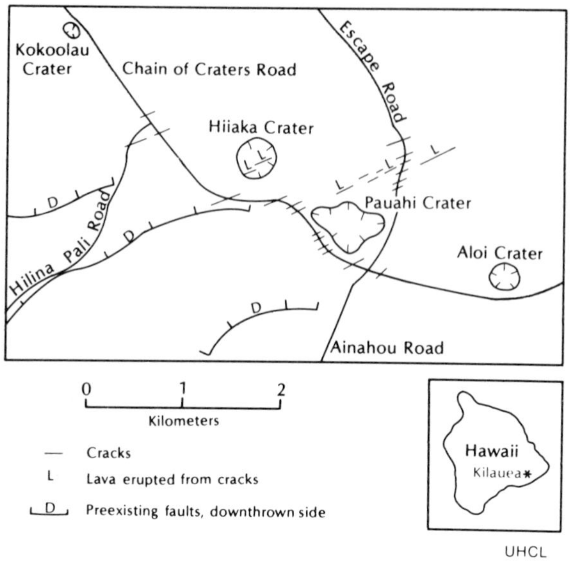

FIG. 6.12 (top) Crack patterns across roads around Kilauea, 1968 (after Jackson, *et al.*, 1975: 8); (bottom) A large crack and subsidence in road surface in Puna area in 1924 (Photo: Lyman House Memorial Museum, Neg. No. 1425 #1).

Lava flows from Kilauea have damaged roads on numerous occasions. In 1971 and 1982 the Crater Rim Road was cut. In 1969, 1973, 1974 and 1979 the Chain of Craters Road was buried in various places. In 1969–1974 Mauna Ulu eruptions covered about 20 km of the Chain of Craters Road (Fig. 6.13). The 1955 eruption buried 10 km of public road. Many earlier eruptions also buried roads (Macdonald and Eaton, 1964: 132; Hazlett, 1981; Carr and Greeley, 1980: 56-57 — Fig. 6.14).

Fluid lava flows have also buried roads around other volcanoes. On April 8, 1977, lava flows from Piton de la Fournaise cut Reunion Road number 57 and destroyed a bridge and submerged part of National Road number 2. Five days later road number 57 was cut again (*BVE*, 1979, 17: 15-16). In the same year a very fluid and fast-moving flow from Nyiragongo cut the main road linking north and south Kivu over a length of 10 km. The average thickness of these flows was one meter or less (Tazieff, 1977). Roads on the Fornazzo side of Etna have also been damaged in recent years. In 1971 (May 11 and 13) the main highway from Fornazzo to Cabancitelli was cut by a flow (*SISCLP,* 33-71: 1208). In 1979, one of the main roads out of Fornazzo was buried by a lava flow 90 m wide (Guest *et al.*, 1980). During the 1914 Sakura-jima eruption, the lava flow which blocked the Seto channel (see Fig. 3.19) rendered the coastal road impassable about 4m above sea level. A new road was made 36 m higher up the slope but this too was closed by the end of September, 1914 (Omori, 1916b: 253). The 1905–1907 eruption of Matavanu buried about 15 km of road along the north coast of Savai'i between Saleaula and Sataputu as shown on Figure 5.2 (Anderson, 1910: 631).

The Matavanu eruption also produced tsunami 2–3 m in height which flooded the coast road but did little damage to it (Anderson, 1910: 625). Along the shores of Sunda Strait in 1883, roads were destroyed along with everything else by Krakatau tsunami. Tenison-Woods (1884) noted: "The post road between Arigu and Tjaringen [Anyer and Tyringin, Fig. 2.24] ran close beside the sea. It was a very good road, broad, well-metalled, and fringed with trees. After the flood it was scarcely possible to discern a trace of it; the trees were, of course, laid level, and of the road a little patch of metal here and there showed where it had been" (quoted in Simkin and Fiske, 1983).

Lahars have been a major cause of damage to roads in a number of eruptions. Numerous bridges were destroyed by lahars during the 1877 Cotopaxi eruption, some of them tens of kilometers from the volcano (Wolff, 1978: 136-138). Bridges, in particular, are vulnerable to lahars both during and after eruptions. Around Agung in 1963 lahars following heavy rains washed out a number of bridges and roads (Fig 6.15). The suspension bridge over the River Klung Klung, the only bridge connecting the eastern and western parts of the island, was reinforced using concrete and anchor plates to survive lahars and flood toward the end of 1963. At the time of reinforcement the bridge had already been damaged three times by lahars. On February 3, 1964, this 70 m long bridge was swept away.

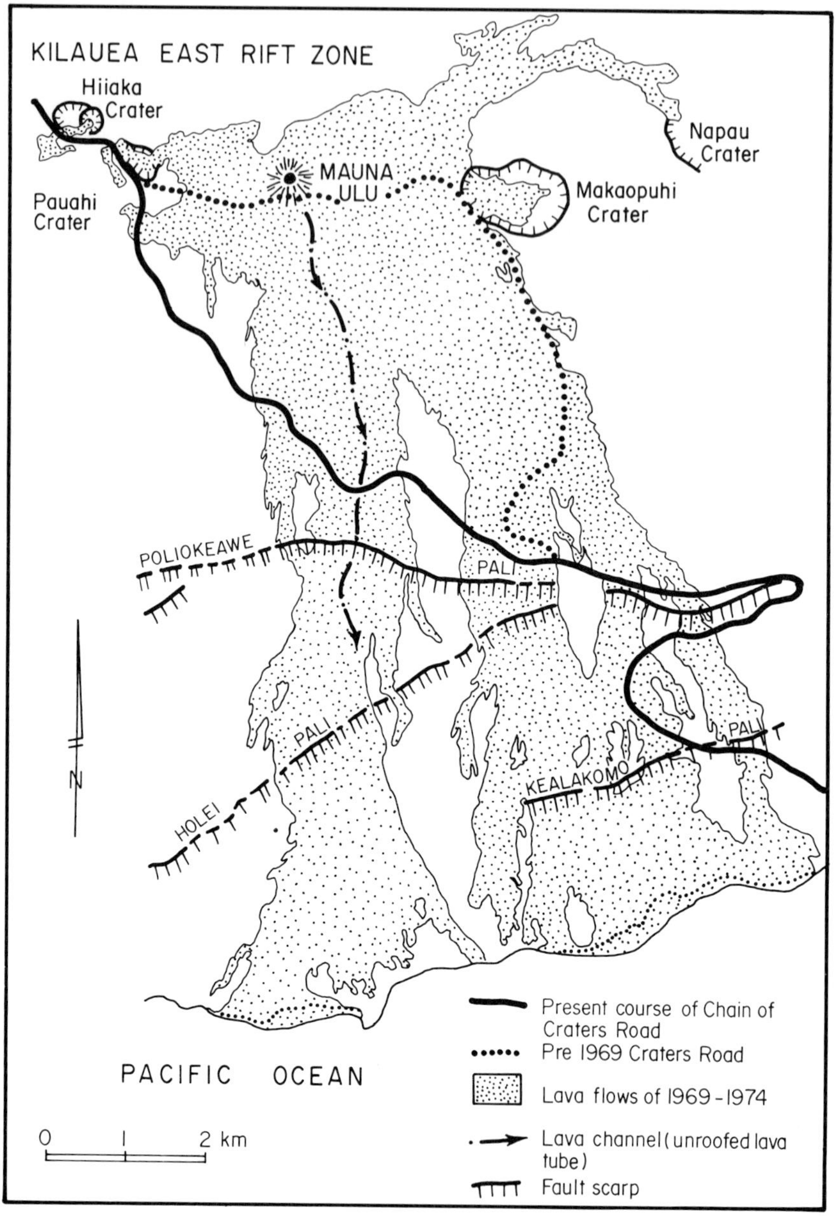

FIG. 6.13 Lava flows from Mauna Ulu and vicinity, 1969–1974. These flows covered about 20 km of the Chain of Craters Road opened in 1965 (modified after Hazlett, 1981: 33-34).

FIG. 6.14 1926 lava flow from the southwest rift of Mauna Loa crossing Kau Road (Photo: Lyman House Memorial Museum, Neg. No. 122 #1).

An emergency bridge for pedestrians only and constructed of bamboo was in use two days later. A new bridge with a five tonne limit was opened on June 17, 1964. The iron bridge over the River Bug Bug was destroyed during the eruption; a 10 m long emergency bridge built by the army was swept away in November, 1963. The same fate met the public works bridge in January 1964. A ford system was then built requiring constant maintenance. Similar systems devised for the River Krekuk near Subagan and for the River Djangga near Karangasem (see Fig. 3.18), were swept away (Jennings, 1969: 27, 144-145). Tens, if not hundreds, of other bridges were destroyed during and after the Agung eruption.

Similar problems occurred around Lamington after the 1951 eruption although there were few bridges at the river crossings. The eruption wiped out most of the jeep roads. In April, 1951, *Pacific Islands Monthly* reported that the rivers were still subject to heavy flooding in the lower Amboga River area and at Buna. The original roads to Sangara and Higaturu (see Fig. 2.15) were closed by washouts on the Amboga River, by ash deposition and by erosion of the upper crossing at Higaturu. River crossings that had been a few meters across widened to 200 m (*Pacific Islands Monthly*, April 1951: 35; May, 1951: 15). Torrential rains accompanying the 1937 Rabaul eruption severely damaged the roads around the town. Figure 6.16 illustrates the damage to one storm water channel.

FIG. 6.15 A bridge on the road near Rendang (see Fig. 3.18) washed out by a lahar during the 1963 eruption of Agung (Photo: National Geographic Society, Robert F. Sisson).

FIG. 6.16 Damage to the storm water drain alongside a road caused by torrential rain accompanying the 1937 Rabaul eruption. Figures standing on the bed of the channel provide scale (Photo: Bureau of Mineral Resources, Australia).

In 1971 lahars from Villarrica (Chile) destroyed concrete bridges over the Turbio, Carmelito, Correntoso and Chaillupen rivers and damaged the associated highways (*BVE*, 11, 1973: 42). A lahar from Calbuco (Chile) in 1961 covered the main highway south of Ensenada to a depth of 1.5-2.0 m. However, two days later it had hardened enough for motor vehicles to drive over it (Klohn, 1963). Lahars formed during the 1975 eruption of Ruapehu from melting snow and the ejection of crater lake water produced flood levels in the Whakapapaiti River up to 10 m above the bed and washed away swing bridges (pedestrian traffic only) or undermined their supports (Scott, quoted in Nairn, 1975). Around Irazú in 1964 massive lahars flowed down rivers, undercutting banks and depositing volcanic debris (Fig. 2.18). The undercut banks created landslides, sometimes with dramatic results for bridges (Fig. 6.17).

Lahars and pyroclastic flows also caused damage in 1980 to roads and bridges in the valleys draining Mount St Helens (Fig. 6.18). The extent of lahar and other deposits is shown on Figure 4.2. Along the North Fork of the Toutle River, 32 km of State Highway 504, many kilometers of private logging roads and several kilometers of Cowlitz County roads were destroyed or buried (Fig. 6.19a) under as much as 2 m of deposit (Schuster, 1981: 703-705). The lahars also destroyed: "or badly damaged 27 bridges in the valley [Fig. 6.18] including 5 of the 7 bridges lost

FIG. 6.17 The toe of a landslide, undercut by a lahar in the Rio Reventado, caused the failure of the bridge abutment (Photo: H.H. Waldron, US Geological Survey). See also Fig. 2.18.

on State Highway 504 ..., 15 highway... bridges privately owned and used to transport logs to downvalley lumber mills [Fig. 6.19b], and 2 highway bridges owned by Cowlitz County. Most of these were modern concrete and (or) steel structures, the largest being the 1670 m... steel-girder Coal Bank bridge that spanned the Toutle River on State Highway 504 near the town of Toutle. This bridge was destroyed by a mudflow that carried timber debris... With the exception of one log-stringer bridge, the private logging highway bridges on both the North Fork and South Fork Toutle River were modern concrete structures built between 1972 and 1979... Only four bridges spanning the Toutle River... were not destroyed or seriously damaged by the mudflows. Three of these were high enough that they received only minor, nonstructural damage from buffeting by timber debris...: the fourth... had sufficient clearance that it was not touched ... Interstate 5 traffic was detoured for 12 hr while the highway bridge was in immediate danger from the mudflow, flooding and debris..." (Schuster, 1981: 705-706). Had this bridge, and the adjacent Burlington Northern railroad bridge, been destroyed the effects on Washington's transport system would have been serious.

Lahars also occurred in tributaries of the Lewis River (Figs. 4.2 and 6.18). Sixteen bridges and several kilometers of United States Forest Service recreational and logging roads were destroyed or buried in this catchment. The bridges were concrete, steel and log-stringer structures ranging up to 104 m in length (Schuster, 1981: 709).

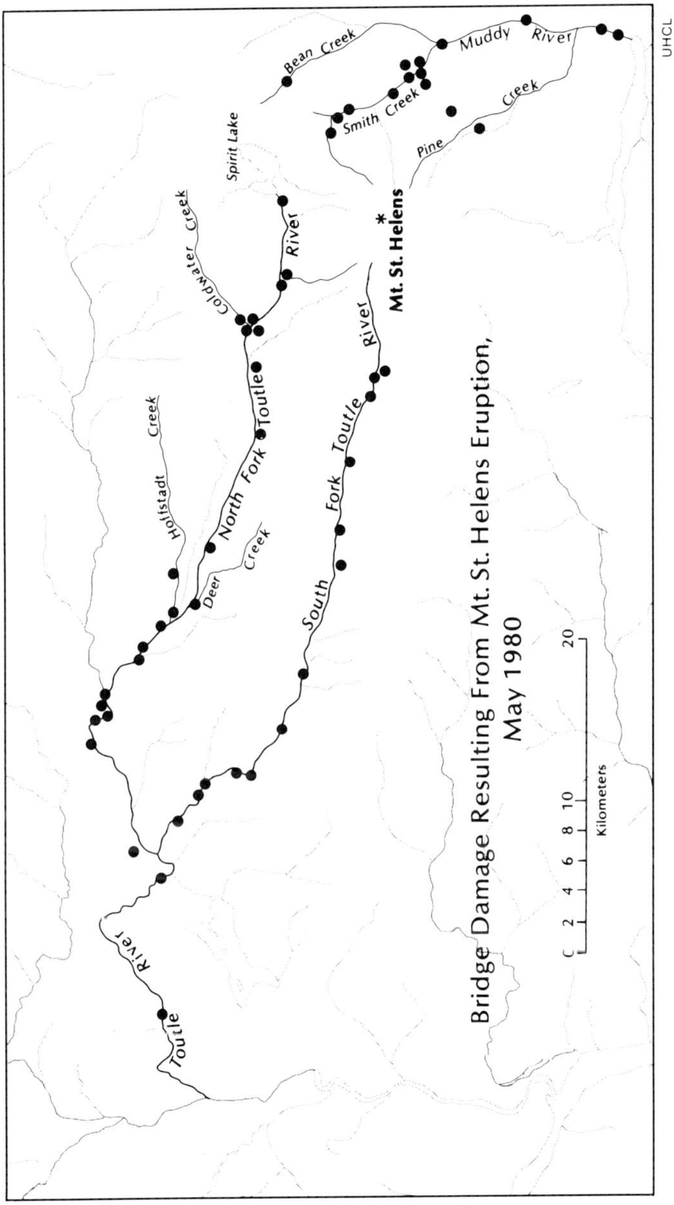

FIG. 6.18 Map of ocations of bridges damaged by lahars, Mount St Helens, 1980 (after Schuster, 1981: 794).

FIG. 6.19 (top) Lahar covering State Highway near Toutle to a depth of 2 m. In the photo is Dr D.R. Crandell; (bottom) Remains of bridge on private logging road, destroyed by a lahar down the North Fork of the Toutle River, Mount St Helens, May 1980. The thickness of the lahar can be gauged from the height of the marks on the trees in the middle distance (Photos: R.L. Schuster, US Geological Survey).

In all, the pyroclastic flows and lahars from Mount St Helens on May 18, 1980, destroyed or extensively damaged 300 km of road and 48 road bridges. Many kilometers of road were quickly restored to service by road crews with front-end loaders and similar equipment to move the sediment. By the end of the 1980 (northern) summer most highways and roads providing access for homeowners and timberworkers were useable and most highway bridges had been replaced with temporary structures, sometimes in nearby locations requiring shorter spans (Schuster, 1981).

In the Toutle Valley, parts of State Highway 504 were located as little as 2 m above river level and the destroyed bridges had clearances of only 5–6 m (except for a surviving bridge which had a clearance of 12 m). As prehistoric mudflows had thicknesses of about 20 m, even had it been considered, it would not have been feasible to design structures above lahar level (Schuster, 1981: 706-707). In such cases, few protective measures can be taken; little can be done except bear the loss. However, where bridges and roads are damaged by secondary lahars, as at Lamington 1951, Agung 1963 and Irazú 1963–1964, it may be possible to install warning systems, construct alternative escape routes, and erect diversionary structures and/or retention basins (see also Tazieff, 1965: 30). Roadside drains also need to be cleaned out and maintained to protect the road surface from excess runoff and scouring.

6.10 THE EFFECTS OF MOUNT ST HELENS TEPHRA ON ROADS

The Mount St Helens tephra fall of May 18, 1980, and later dates had widespread but short-lived effects on the vehicle-and-travel-oriented population of the United States Pacific Northwest. The May 18 eruption, in particular, not only reduced visibility on highways, caused accidents, altered traffic volumes and speed restrictions, stranded travelers and affected food stocks and the tourist and automotive repair industries but also led to diverse methods of clean up for the thousands of kilometers of tephra-coated roads. This section provides examples of each of these phenomena.

Visibility on roads was poor during and after tephra falls. On Sunday, May 18, most people stayed at home and the volume of traffic was relatively small. Poor visibility was probably a factor causing a fatal accident in Yakima (see Chapter 3, Section 12.). The May 25 eruption which sent tephra mainly west and southwest of the volcano (see Fig. 3.8) also reduced visibility especially along the Interstate-5 [I-5] freeway. Ash billowing up behind fast-moving vehicles seriously reduced visibility and probably was responsible for hundreds of accidents throughout the affected area. Intermittent whiteouts occurred on the Interstate-5 between Kelso and Olympia on May 27. "Brake lights could be faintly seen through the dust which rose 30 feet [9 m] above the freeway. Often, though, lights couldn't be seen, and all that could be heard was the tinkle of broken glass and the crunch of crumpled metal as cars from the rear headed into the ash clouds and rammed vehicles already hidden in the clouds" (*Seattle Post-Intelligencer,* May 28: A12).

Seven accidents occurred on one stretch of hill north of Kelso. Even as far away as northern Idaho and as late as June 10 billowing ash was reportedly the cause of a 10-car pileup (*Civil Engineering-ASCE,* August, 1980, 50(8): 74).

Both official and unofficial actions were taken to cut down accidents and other problems caused by the drifting tephra. Speed restrictions were imposed by most communities at some stage after the tephra fall. In Missoula and Cheney, speeds on city streets were restricted to 10 mph [16 km h^{-1}], while 15 mph [24 km h^{-1}] was allowed in Ritzville, Yakima, Ellensburg and towns west of the mountain. On rural roads around Ellensburg, a limit of 20 mph [32 km h^{-1}] was imposed (Warrick *et al,* 1981; *The Columbian,* May 26, 1980: 12; *Civil Engineering-ASCE,* August, 1980: 74). In Yakima, residents on some suburban streets constructed speed bumps from the tephra to slow down speeding motorists. On Interstate-5 freeway on May 27 truckers used CB (citizen's band) radio to block the highway with truck convoys spaced about a mile [1.6 km] apart, forcing motorists to keep speeds down to about 8 km h^{-1}. Some trucks in each convoy drove along the shoulder to prevent impatient motorists overtaking. Although the truckers' actions were presumably quite illegal, police credited them with preventing accidents and noted that where the convoys were not organised, it was almost suicidal. While some motorists phoned the patrol office to complain, others rang to compliment the truckers (*Seattle Post-Intelligencer,* May 28: A12).

Visibility and related problems also led to road closures. In Ritzville, although city streets remained open with speed restrictions, the Interstate-90 (I-90) freeway (Seattle–Spokane) was closed from May 18 to May 23. The I-90 and US 97 were also closed at Ellensburg, but by May 20, an emerging food shortage was apparent in the town so traffic was allowed westbound on I-90 and northbound on US 97 with a maximum speed of 40 km h^{-1} and at a rate of one vehicle every five minutes (Warrick *et al.,* 1981: 21, 45).

In Yakima and Spokane, motorists were advised to travel only when really necessary and the downtown areas of both cities were closed to traffic for several days to facilitate cleanup (Novak and Zais, 1981: 13). The Cheney administration responded to the tephra fall by closing down the city on May 19: "All non-essential traffic was banned. For the public, only emergency trips — for health or food — were permitted. Roadblocks were put in place and no cars were allowed to enter the city. Traffic from Spokane was turned back. Within the city, car keys were confiscated if drivers could not justify adequately their presence on city streets. Speed limits were reduced to 10 mph [16 km h^{-1}]. Strict enforcement lasted for five days. Restrictions were lifted after ten days on main streets, but speed limits of 15 mph [24 km h^{-1}] were recommended for streets with an ash residue" (Warrick *et al.,* 1981: 68).

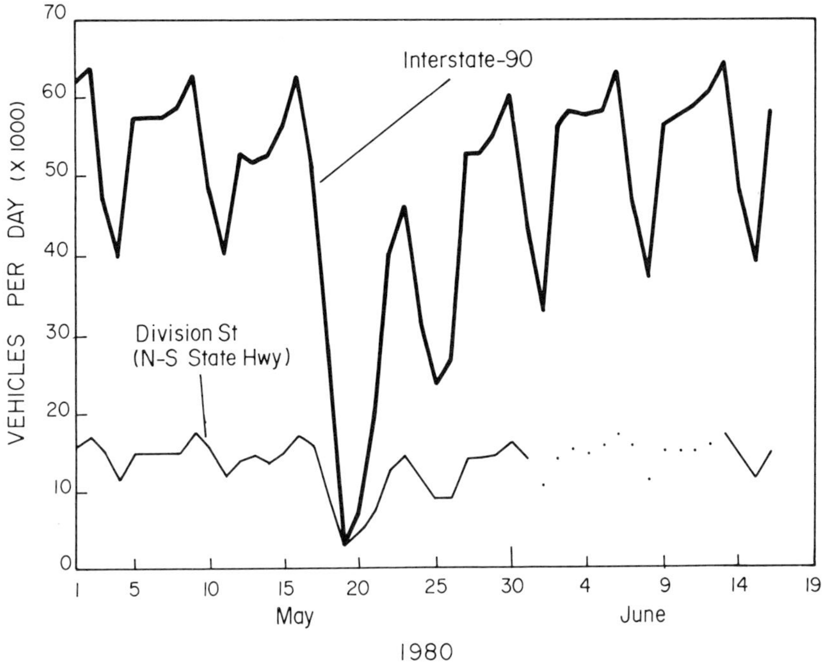

FIG. 6.20 Effects of Mount St Helens tephra fall on traffic volumes on two highways in Spokane, May 1980 (after Dueker *et al.,* 1980: 42).

As these examples suggest traffic volumes were much below average throughout most of the tephra fall area during the week following May 18. On that Sunday, most people stayed indoors or were stranded by poor visibility, road blocks, vehicle damage or public transport closures. In the following few days, many shops and businesses remained closed while citizens were involved in cleaning up the tephra around their homes. Citizens were urged to make only emergency trips. In Ellensburg, compliance with this request was good on May 19, but deteriorated a little as the cleanup began. In Ritzville, the intercity bus service was closed until May 24 and not fully operational until May 26. In Missoula, the traffic volume was only 10% of normal on May 19, 15% on May 20, 25% on May 21, and still only 60% on May 22 (Warrick, *et al.*, 1981: 20-21,43,89). In Yakima, May 18 traffic was only 30% of normal. In Spokane (Fig. 6.20) the traffic volume was less than 10% of normal on the I-90 (Dueker *et al.*, 1980: 42). Despite the multiple car accidents mentioned earlier, the total number of traffic accidents in the area was probably much below normal. In Yakima, only 50 accidents occurred in the period May 18 to 31 (of which 16 occurred on May 30 and 31) compared with 89 accidents for the same period in 1979, and 103 during May 1 to 18, 1980 (*Yakima Valley Sun,* June 12, 1980).

The widespread fallout of tephra meant that there was a huge cleanup problem so that visibility problems on roads would be eliminated or at least reduced. In the forest area around Mount St Helens, the United States Forest Service suffered fallout "damage" to about 2650 km of mostly gravel roads while another estimate suggested 30 000 km of county and city roads required a cleanup. Washington State Transportation Department estimated that it would cost US$50 million just to clear the tephra from the roads within 160 km of the volcano (*Engineering News Record,* July 31, 1980, 205(5): 14; EH & WS, 1980). Yakima County estimated the average cleanup cost at US$380 per km of road (County Road Report, 180: 6). Ritzville dumped 115 000 m^3 of tephra; Moses Lake dumped 230 000 m^3. Portland cleared 5300 m^3 from 2400 km of roads after the May 25 fall. The East Columbia Basin Irrigation District moved 350 000 m^3 of tephra from 2250 km of road. The Washington State Highway Department shifted 540 000 t and was still moving the ash from median strips, ditches and gutters along I-90 in the eastern part of the State five months after the eruption (Schuster, 1981: 715-716). Whatever the accuracy of these estimates, the cleanup was certainly a problem.

Each community tackled the cleanup of city streets in its own way using the available road cleaning equipment and, sometimes, the help of communities not affected by the May 18 eruption. Because the type of available equipment varied little, only minor variations in approach occurred. As the experiences of Yakima City have been well documented, the Yakima examples are reported here in some detail.

Yakima City departmental heads first met to discuss the likely problems resulting from the tephra fall on the afternoon of Sunday, May 18. As far as the cleanup was concerned, a high priority was given to removing tephra from the airport and the downtown area in case medicines, food and other supplies were needed (Larry Wittenberg, Assistant City Manager, personal communication, May 1981). Yakima tried to get assistance from the military, but found that a specific order of the President was required. Although the National Guard was called out by the Governor to help, most of Yakima's assistance came from Seattle and Portland, King and Multnomah Counties, and private contractors (Novak and Zais, 1981: 13). King County, for example, provided 38 pieces of equipment (mobile street sweepers, flush trucks, front-end loaders, graders, and large vacuum truck units) and 47 personnel, including mechanics (*In the Works*-King County Department of Public Works Employees' News, June 1980, 2(5): 3). During the cleanup campaign, the Central Business District of 16 blocks was cordoned off from Friday, May 23, to Monday, May 26, in order to minimise traffic problems. Streets leading into the area were signposted, barricaded and patrolled by the police. Unauthorised vehicles in the area were impounded. Businesses were urged to have the areas in front of their locations cleared by May 22 with the tephra piled near the kerb. During this period only essential personnel in governmental, media, utility and critical business activities were permitted access by law enforcement officers (City of Yakima Press Release, May 22, 1980).

The cleanup crews worked two 12 hour daily shifts. When the downtown area was reopened on May 26, the crews moved onto arterial roads and then to side streets (*Civil Engineering-ASCE,* August 1980, 50(8): 76). A number of different methods were tried to pick up the tephra: "It was quickly apparent that before municipal type street sweepers could begin work, the bulk material had to be removed by road graders and snow plows. A wide variety of front end loaders, supported by a massive fleet of haul trucks, were employed [sic] to remove the ash mounds created by snow plows and graders. However, belt loading equipment definitely had the advantage over bucket loaders where ash was windrowed and sufficient numbers of haul trucks were available. Street sweepers could then handle the remaining ⅛ inch [3 mm] of material left on the street by the previous operation. During all operations, large volumes of water were necessary to make the ash workable and minimize dust" (Markesino, 1981: 52).

The citizens of Yakima were organised into neighborhoods on a block-by-block basis. Individuals and neighborhoods swept and watered down roofs and allotments, piling the dampened tephra in windrows on the streets. When the street was ready to have the tephra removed, the block captain phoned in.

In general, the system worked extremely well. By June 13, the cleanup was essentially complete. Some minor hassles developed. Many of the larger stores were concerned when the Central Business District was cordoned off as the profitable Memorial Day Sales were about to begin — however the area was reopened in good time. Because of cross-connections with the sewage system (see Chapter 6, Section 3), efforts had to be made to keep the tephra out of storm drains. The tephra was hand swept 0.5 m outward from the gutter. Anyone found dumping tephra in a storm drain was threatened with a US$200 fine and the printing of his/her name on the front page of the newspaper — it was cheaper to keep the tephra out of the drains than to remove it later (*Civil Engineering-ASCE;* August 1980, 50(8): 74). Although police with bullhorns proceeded the cleanup teams requesting citizens to move their cars off the streets, the city had to spend about US$2400 removing legally parked cars (City of Yakima files, August 8, 1980). Minor hostility developed toward citizens who did not clean down their roofs or lots, and toward speeding drivers or those who drove back and forth over speed bumps to break them down. There was also a suggestion that well-organised neighborhoods got a higher priority with local government assistance than less organised areas or those with mixed land use (Kartez and Kelley, 1980: 9). Certainly, a number of people rang to complain that their back alleys had not been cleaned and to ask when was the city going to get on with it (L. Wittenberg, personal communication, May 1981).

The tephra removed from the streets of Yakima was stockpiled on three major and several smaller sites throughout the city. The largest site, on the north-west edge of town, covers 1.5–2 ha to a depth of 4.5 m and holds about 70 000 m^3 of tephra. An underground sprinkling system was installed, the tephra covered with 50 mm of topsoil and seeded to grass for a total cost of more than US$33 000

(McLucas, 180: 7; City of Yakima files, October 6, 1980). This site is now a city park. Two other dump sites contain 26 000 m^3 and 13 000 m^3 of tephra.

The cost of the cleanup in Yakima's 400 km of streets has been variously estimated from US$1.1 million (January 1981 City of Yakima estimate) to many millions of dollars. During the peak of the cleanup period when city crews and those from outside jurisdictions with their equipment were working two 12 hour shifts per day, the total direct cost was estimated at US$163 000 per day (City of Yakima files, May 26, 1980). Further discussion of these costs occurs in Chapter 8, Section 2.

The efficiency and success of the cleanup of Yakima's streets and buildings can be attributed to a number of things including the development of an excellent neighborhood spirit and the assistance of personnel and equipment from other local governments. The cleanup was also facilitated by the fact that the tephra was relatively coarse. Other communities were not so lucky.

The June 12 eruption of Mount St Helens produced cleanup problems for Portland and the northern part of Oregon. More than 1.5 mm of tephra covered Portland including the 560 km of arterial/collector roads and the 1700 km of residential streets. Although the tephra was rather thin, it was also fine-grained and it fell only two days before the Grand Floral Parade of Portland's annual Rose Festival. Luckily, Portland had gained some expertise in tephra removal as a result of assistance provided to Yakima the previous month and from a light dusting of tephra on May 25. The cleanup was also assisted by a timely rainfall of about 25 mm on June 13. The parade route was cleared using fire hoses and street flushers to move the tephra to catchment areas where storm drains had been protected with sand bags. By Monday, June 16, more than 400 people were at work on the street cleaning operation. Experience showed that a crew of two street sweepers, two power flushers and a supporting hand crew could clean about 8 km of road in a 12 hour shift. Other crews posted "No Parking" signs, towed vehicles and removed tephra piles placed in streets by residents and business people (on June 23 a parking lot owner was fined US$500 for not cleaning his allotment [*Civil Engineering-ASCE,* August 1980: 74]). Other maintenance functions such as street repairs and sewer cleaning were deferred. A first pass of the entire city was completed by July 11 and a second pass by the end of August: "Due to the small particle size of the ash, the street sweeping equipment was only about 50 per cent effective on the first pass. The street cleaning section began a second pass operation which lasted another 6½ weeks. However, even the second cycle failed to totally remove the ash. The actions of traffic and runoff continued to cleanse the pores of the pavement and deposit the ash along the curb. Given time, the ash would mix with normal debris loadings and eventually be removed by future sweeping passes" (Markesino, 1981: 54).

By the end of August 1980, Portland City had spent more than US $1.1 million on cleaning up more than 5300 m^3 of tephra. During the first pass operation, daily expenditures exceeded US$40 000 per day.

Most jurisdictions affected by the Mount St Helens tephra fall of May 18 initially employed the same strategies used to deal with heavy snowfalls. Ellensburg, for example, established roadblocks to stop traffic entering the area, mobilised road clearance personnel, imposed reduced speed limits and found emergency housing for stranded motorists (Warrick, *et al.*, 1981: 19).

However, all communities found that the tephra was far more difficult to deal with than snow, primarily because it didn't just melt away! Small communities like Ritzville (population 1900), where about 40 mm of fine tephra fell on the 300 ha jurisdiction, had enormous problems as the dust blew easily in the wind, and the town, although helped by Adams County, had limited resources available. The streets were bladed and broomed and the ash hauled to two dump sites. The larger site, an abandoned quarry, now contains about 67 000 m^3 of tephra. The estimated cleanup cost of US$0.5 million should be compared with the total annual municipal operating budget of about US$1.2 million (McLucas, 1980: 3; Rainier National Bank Economic Report No. 5, August 12, 1980: 3).

In some ways, even greater problems were presented to counties than to municipal jurisdictions because many of the roads were gravel — they could not easily be bladed and broomed, yet they caused similar dust and visibility problems. In Adams County (county seat at Ritzville) about 40 mm of tephra was bladed into adjacent ditches or onto fallow fields to be plowed under at the next cultivation (Lucas, 1980: 2-3). The blading operation resulted in the loss of about 40 mm of gravel from the road surface. About 300 000 m^3 of gravel was required as a replacement, equivalent to about 4.5 times that used in a normal year (McLucas, 1980). In Adams County the total cost of tephra cleanup from roads and other areas up to September, 1980, was US$1.5 million whereas the total county road maintenance budget for the year was only US$1.3 million (County Road Report, 1980: 3). Similar problems with tephra disposal, gravel replacement and costs occurred in Grant, Spokane, and Whitman Counties.

Various tephra stabilisation techniques were also investigated. Spokane County evaluated the use of a 32% calcium chloride solution as a dust palliative on a 1.6 km test strip. This solution penetrated the tephra, stabilised it and stood up to blading. Forty kilometers of roadside was treated with rock salt by Grant County The rock salt absorbed moisture from the air, moistened the tephra and created an ionic bond with it, thus decreasing the tephra's propensity to blow in the wind. Adams County and Whitman County sprayed some roads and roadsides with liquid sulphonate, a by-product of pulp manufacture. However, this solution cost US$100 per tonne, proved somewhat water soluble and required two applications (McLucas, 1980: 2,5,6). In Oregon, the Department of Transportation used 90 000 L of Coherex dust retardant, an emulsion of petroleum resins, on both streets and roadsides. The retardant nucleated the fine tephra particles, forming a non-toxic mat that adhered to both asphalt and Portland cement concrete surfaces (*Public Works*, December 1980: 37). While Coherex and other chemicals may suppress the dust problem for short or long periods, removal is still required from road surfaces. For roadside margins, vegetation will prove to be the only long term dust palliative.

As mentioned earlier, poor visibility, road closures, curtailment of public transport and fear of damage to vehicles drastically cut the number of vehicles on roads during the various Mount St Helens tephra falls, but particularly on May 18. Citizens who were at home when the ashfall began stayed there. In general, those some distance away from their prospective destinations (perhaps as many as 11 000 people — Anderson, 1982: 182) were stranded. Because mobility and automobile use were severely reduced, it is interesting to examine briefly some of the indirect effects produced by the road closures.

In Missoula, between 1000 and 1500 travelers were stranded, including several hundred truckers and a number of cyclists. Motels filled up and 40 people were housed by the Red Cross (Warrick *et al.,* 1981: 89). In Ellensburg more than 600 motorists were stranded, mostly on the Interstate-90 4 km from the downtown area. Two Greyhound buses were commandeered by the Sheriff's office to ferry motorists to town, but Greyhound headquarters in Los Angeles reacted quickly to news of the tephra fall by ordering all of its buses off the road. School buses were then used to continue the ferrying operation. The 600 motorists were offered accommodation in the university, a high school and the Presbyterian Church and in the Public Safety Building. In the next few days, food stocks in the town dwindled and some motorists were allowed to leave on highways that were still closed. Considerable confusion about highway closures resulted from erroneous reports by CB (citizens' band) and ham radio operators (Warrick *et al.,* 1981: 21).

In Ritzville, which has a population of only 1900, about 2500 motorists were stranded by closure of the I-90. Visibility was so poor flares had to be used to guide motorists to the town. Schools and churches provided accommodation and food (Warrick *et al.,* 1981: 45). Later, some travelers defied the travel ban but about 12 out of 75 cars were forced to return to the city because of poor road conditions (*The Columbian,* May 26 1980: 9).

In Yakima, traffic volumes on May 18 and 19 were only about 30% of normal. Those who did drive reported travel times two to three times as long as normal. Those who were stranded out of town reported delays in returning of three to eight days, mainly because of highway closures (Dueker *et al.,* 1980: 31). At least 350 travelers were accommodated in emergency shelters in Yakima high schools. All but 82 left on May 20 and the number of high schools used dropped from five to three (*Yakima Herald-Republic,* May 21, 1980). Citizens who reported making local journeys during the tephra fall went to secure their businesses, to open stores, to supply goods to emergency workers, or to provide moral support to relatives. Later, journeys were made only to go to work or to go shopping; discretionary trips were curtailed. Some "carpooling" occurred for shopping. Where more than one vehicle was available, it was the oldest or an employer's that was used. In many cases, the telephone was used as a substitute for travel (see Chapter 6, Section 5). Dueker *et al.* (1981: 33) also reported: "It appears that the group in Yakima whose travel was *least* affected by the ashfall was the elderly. Many of those interviewed normally depend heavily on friends, relatives, and social service programs for their travel needs and these same people (friends, relatives, and program personnel) went out of their way to assure that the shopping, medical and social needs of the elderly were met".

In more than 130 jurisdictions, schools were closed for at least one day, many remaining closed for the rest of the school year (Anderson, 1982: 183). Even in six Montana counties, schools remained closed for at least four days under emergency orders. In Idaho, school districts postponed trustee elections after the governor declared a state of emergency (*The Columbian,* May 26, 1980: 9). Many schools were dismissed up to two weeks early, largely because of transportation problems (Rainier National Bank, Economic Report No. 1, June 1, 1980). In Pullman, only a few miles from the Idaho border, 3358 students from Washington State University were granted emergency leave because of the fallout. Although the university said it was encouraging students to finish the semester, more than 2000 had left by May 28 (*Seattle Post-Intelligencer,* May 29, 1980: A18).

Undoubtedly, many other effects can be adduced to the closure of roads following tephra fallout. Undoubtedly, road closures have occurred following tephra falls elsewhere and cleanup problems have been overcome (for example, Irazú, 1963–1965 — see Clark and Lee, 1965), but they have not been as well documented as those in the Pacific Northwest. Furthermore, it is unlikely that the effects elsewhere would be as far reaching as they were following the Mount St Helens eruption as few societies are as mobile or as automobile dependent.

6.11 VEHICLES

Vehicles have undoubtedly been damaged by all volcanic agents but good descriptions are available only for damage by pyroclastic flows, lahars, projectiles and tephra falls.

Pyroclastic flows proved very damaging to vehicles around Mount St Helens. Vehicles, including heavy logging equipment, were battered and scorched by the May 18 pyroclastic flows which moved northwards from the mountain (see Fig. 3.17). Within the innermost two-thirds of the area where timber was felled, vehicles were moved and overturned (Fig. 6.21). Angular boulders up to 0.3 m in diameter penetrated vehicles up to 15 km north of the mountain. Similar effects occurred up to 5 km west of the mountain. A truck was overturned by the blast 15 km north-west of the volcano. In general, vehicles within 15 km were battered, while those 20 km away were only sandblasted (Waitt, 1981: 454).

The vehicles found in the downed-timber zone also showed the effects of heat: "... burned or blistered exterior paint, burned upholstery, melted exterior plastic, delaminated window glass. Even vehicles near the outer edge of the down-timber zone have melted plastic and delaminated windows in the west-north-west and north-east sectors — the same sectors in which fires in downed timber had been most concentrated" (Waitt, 1981: 454). Analysis of temperature effects on plastic

FIG. 6.21 Car battered by pyroclastic flow, Mount St Helens, May 18, 1980 (Photo: P. Francis).

indicate that these vehicles experienced temperatures of about 300°C in the northeast sector and 200°C–250°C in the west-north-west sector. These peak temperatures lasted only one to 10 minutes with the temperature then dropping rapidly to less than 150°C: "This short period of peak temperature is consistent with eyewitness reports, with the survival of paints, color film, cans of beer, polystyrene insulation, and, in the northern and western sectors of the devastated area, with the lack of charring of newsprint (charring temperature of about 250°C)" (Banks and Hoblitt, 1981: 303).

While plastic objects on the downflow side of vehicles and in protected sites behind seats and under other objects were generally unaffected, temperatures near clasts imbedded in plastic objects appear to have been about 100°C higher than that reached nearby (Banks and Hoblitt, 1981: 303).

Similar blast and temperature effects on vehicles were experienced during pyroclastic flows from Lamington in 1951. At Higaturu a battered jeep was stranded on top of two shattered tree trunks (Fig. 6.22). The temperature in this area was estimated at 200°C for 90 seconds (Taylor, 1958: 43,46) but the details of effects on the vehicles are not available.

Although vehicles have probably been damaged in lahars during many eruptions, few descriptions have been found. Schuster (1981: 713) includes a photograph of a Weyerhaeuser Company bus rather battered and partially buried by the

FIG. 6.22 Two jeeps severely damaged by pyroclastic flow from Lamington in 1951 (Photo: G.A.M. Taylor, Bureau of Mineral Resources, Australia).

North Fork Toutle lahar of May 18 (Fig. 6.23a). The bus was located at Camp Baker some 25 km from the volcano. About 40 km downvalley, observers noted on the afternoon of May 18, a fully loaded logging truck floating downstream submerged only to the level of the lowermost tier of logs. The lahar was described as having the consistency of fresh mortar. At 10.30 p.m. the lahar arrived at the mouth of the Toutle River, more than 80 km from the volcano, carrying amidst logs and buildings, a pickup truck (Cummans, 1981b: 482, 484). Figure 6.23b shows a Weyerhaeuser truck after its downstream journey in a lahar.

Vehicles have also proved susceptible to damage by volcanic projectile impacts. Projectiles ejected during the June 1, 1973, eruptions of Sakura-jima broke the windshields of 56 cars. On November 28 a further 15 windshields were broken by falling cinders (*BVE*, 1975, 13: 41; *SICSLP*, 1974: 125). Similarly, in 1976, 1982, and 1983, windshields on some cars parked near the foot of the volcano were broken by falling lapilli (*BVE*, 1978, 16: 40; *SEAN Bulletin*, 7(11), 1982: 11; *SEAN Bulletin*, 8(2): 14). The Bocca Nova explosion on Etna on September 12, 1979, scattered non-juvenile lithic blocks up to 0.3 diameter, weighing up to 6 kg, and still too hot to handle half an hour later, across the nearby parking lot. All the vehicles in the lot were damaged. A caravan used for emergency accommodation apparently received a direct hit which caused a gas cylinder to explode, totally destroying the caravan and burning several nearby tourists (Guest *et al.*, 1980).

A

B

FIG. 6.23 (top) Weyerhaeuser bus trapped by lahar, North Fork of Toutle Valley near Camp Baker. (bottom) Bogie of Weyerhaeuser logging truck swept downstream by lahar near Camp Baker (Photo: R.L. Schuster, US Geological Survey).

FIG. 6.24 An abandoned car near Rabaul in 1937. The hood has collapsed under the weight of tephra (Photo: Bureau of Mineral Resources, Australia).

Tephra falls can also damage vehicles. Around Rabaul in 1937, cars choked with dust (Fig. 6.24) were abandoned in the mud (Jaggar, 1945: 37). During the 1924 Kilauea eruption, a car driven the short distance from the Volcano Observatory to Halemaumau was so sand-blasted by the falling pumice, dust and fume that it had to be repainted and have its windshield replaced (Macdonald and Abbott, 1970: 82). The May 18, 1980, tephra fall from Mount St Helens reportedly raised blisters on the paint of cars in Wyoming (Holocaust, 1980: 42). Closer to the volcano, driving cars during the tephra fall and in the ensuing dusty conditions proved hard on the vehicles. An official of the Washington State Patrol reported: "We brought in one car that had more than 450 pounds [204 kg] of dust on it . . . and I mean IN it, like behind the glass on dashboard instruments gauges, through the brake seals and cylinder and IN the brake fluid . . . everywhere. And once you've run a car in those conditions for any amount of time, it's shot (*Seattle Post-Intelligencer,* May 28, 1980: A12).

In fact, police vehicles all other Washington State were pronounced "shot". Patrol Captain Mike Feldhausen was reported as saying that at least 165 out of 236 State Patrol cars in eastern Washington were disabled of which at least 50 had been "totalled". The initial damage estimate was put at $1.6 million (*Seattle Post-Intelligencer,* May 28, 1980: A12). Elsewhere on the same page of the newspaper 19 State Patrol cars were described as totally disabled but a later report (EH & WS,

1980) said that by May 25 the dust had ruined 60 out of 280 State patrol cars stationed in the eastern part of Washington. In Moses Lake four police cars and three pickup trucks were "lost to the dust" (*Seattle Post-Intelligencer,* May 28, 1980: A12). An article in the *Yakima Valley Sun* (June 12, 1980) reported that Washington State would ask the Law Enforcement Assistance Administration for US$5.8 million to cover equipment damage and overtime costs, US$1.6 million for the State Patrol, the remainder to be allocated to local police agencies. Similarly, *The Columbian* (May 26, 1980: 8) reported that the abrasive dust laden air had worn out hundreds of car and bus engines as air filters became clogged with the grit.

Some doubt is cast on the veracity of such reports, however, by the following example. An official of the Federal Highway Administration was reported in the August 1980 issue of *Civil-Engineering-ASCE* (50 (8): 76) as saying that: "in Yakima, police cars were dispatched to help citizens in trouble during the period of ashfall, and in just two or three hours, 28 police vehicles were put out of service". However, in May, 1981, Yakima Police Chief LaRue reported (personal communication) that all 30 of his vehicles had been in use 24 hours a day at the time, that all had been kept on the road and that there had been no down time other than service maintenance. Oil changes had been made five to seven times as frequently as normal. Later, the engines and transmission on a couple of cars were dismantled but no problems were found. Similarly, a report dated September 26, 1980, comments that no vehicle engine damage was noted by intercity bus and local transit operators (Mosebar, 1980). Yakima City also experienced little damage to equipment used in the cleanup. Oil was changed frequently and the equipment was not pushed hard. A year later there was no evidence of shortened vehicle life (L. Wittenberg, Asst. City Manager, personal communication, May, 1981). A spokesman for the insurance industry said that ash damage was covered by so-called "comprehensive" parts of most automotive insurance policies but there had been no substantial increase in damage claims. However, people who ruined their car engines just sightseeing in the dust could void their insurance cover (*Seattle Post-Intelligencer,* May 29, 1980: A12).

Perhaps damage to engines did occur further east where the tephra fall was finer and, in places like Moses Lake, thicker. Certainly, engines overheated as filters became clogged, leaving hundreds of motorists stranded (see Chapter 6, Section 10). Certainly, thousands left their cars at home, or stayed at home, fearful of the effects and warned by the publicity about "totalled" and disabled police vehicles.

There were also numerous reports on how to make engine air filters more effective. Panty hose over the carburetor inlet seemed to help filter out the dust. Others tried foam rubber in or around air intakes, or even toilet paper. A Ford Motor Company spokesperson echoed the drive slowly advice and suggested that motorists seal oil and transmission dipstick caps and ensure that all engine compartment hoses and gaskets were firmly sealed. The spokesperson noted that outside ventilation and air conditioning systems should be switched off and that auxiliary air filter systems were available (*Seattle Post Intelligencer,* May 28, 1980: A12). Washington State Department of General Administration issued information on the use of State Motor Pool Vehicles on May 30. Later, General Motors Corporation provided similar instructions in newspaper advertisements in tephra fallout areas (Fig. 6.25). Some owners protected their vehicles with plastic sheets and other covers.

The strong emphasis on preventive maintenance probably minimised actual damage. The "advertising" campaign was certainly effective. The May 25 tephra fall, mainly to the south-west of the volcano, triggered a stampede for car air filters (*The Columbian,* May 26, 1980: 12). The July 1980 issue of *National Petroleum News* (72(7): 7) reported: "The fallout of ash from Mt St. Helens in the state of Washington has been a boon to quick-lube facilities in Seattle. Spurred by recommendations from American Automobile Assn., police, insurance companies, and others, motorists are having their cars' oil, oil filter, and air filters changed every two or three weeks, or 300 miles [480 km]. Several quick-lube marketers reported they've done 500 changes in two days". And Seattle wasn't even in the zone of tephra fallout!

6.12 OTHER MECHANICAL AND ELECTRICAL EQUIPMENT

The effects of volcanic eruptions on railways, aircraft, water treatment plants, telecommunications and road vehicles have already been described. The present concern is with other items of equipment including electronic calculators, air conditioning systems and agricultural machinery. While such items have undoubtedly been damaged or destroyed by pyroclastic flows, lava flows, lahars and other volcanic agents the available literature has been concerned only with the effects of tephra on such equipment.

Virtually any item of equipment with moving parts is likely to be adversely affected by tephra. Wind abrasion by tephra ruined T.A. Jaggar's camera on a walk to the summit of Vesuvio following the 1906 eruption (Jaggar, 1956: 66). The continual drift of tephra onto San Jose, capital of Costa Rica, during the 1963–1965 eruption of Irazú made typewriters stick, linotype machines clog and doors jam. The acid aerosols attached to the tephra also corroded metals (*Time,* January 17, 1964: 31; *Newsweek,* October 7, 1964: 65; *US News and World Report,* June 29, 1964: 71-73). Tephra from the 1971 eruption of Cerro Negro, deposited on cotton balls, caused damaged to harvesting equipment (*UNDRO,* 1976: 11).

The list of damage to moving parts of equipment by the tephra falls around Mount St Helens in 1980 is even greater. Air filtration and conditioning systems were amongst those equipment items to suffer. At Lewiston in Idaho, Potlatch Corporation's sawmills were closed temporarily when tephra clogged machinery (*Pulp and Paper,* July 1980, 54(7): 17). In Yakima, shortly after the May 18 eruption, it

An important message to all drivers in areas suffering from volcanic ash fallout:

Because of severe vehicle operating conditions caused by highly abrasive ash from Mount St. Helens, special maintenance is required to help minimize damage to your car or truck. Here are some important steps you can take to help protect your means of transportation during this critical time.

Special Operating Recommendations

- Operate your vehicle only as required. A moving vehicle, or one standing with the engine running, "draws" the ash into various operating assemblies, fuel and lubrication systems, brakes, etcetera.
- If you must drive on surfaces covered with ash, operate your vehicle cautiously and at reduced speeds.
- If possible, install a vehicle-to-earth flexible "ground" strap to the underside of the vehicle. This helps discharge static electricity, which can cause ash to cling to your vehicle. Such straps are generally available from automotive supply stores.
- Do not operate windshield wipers to remove ash. Flood windshield with clear water, and avoid rubbing or scouring.
- In areas where water is readily available, wash vehicle exterior daily with clean water. Rinse thoroughly, especially horizontal surfaces and chrome parts. Avoid rubbing. Also, do not use high water pressure or brushes.
- Be certain air flow through the radiator and air conditioner condenser fins is not restricted.

Special Maintenance Recommendations

- Do not operate vehicles without air filters. It can cause severe engine damage.
- Do not place any material such as nylon mesh or cheesecloth between the carburetor and the air cleaner assembly. It could be drawn into the engine.
- If available, install a new, heavy-duty (H.D.) air cleaner element. Or you can take a regular filter and cover it with open-cell polyurethane wrap (plastic foam), which is generally available through AC Delco distributors and GM dealers who service trucks. However, be sure to soak the polyurethane in SAE 30-weight oil and squeeze out excess before installing. Wipe dust from inside air cleaner with a damp cloth before installing filter element.
- Clean your air filter(s) often and examine for holes or cracks. If damaged, replace. Do not knock ash from element. Also check rubber seal area of element to make certain there are no dents or damage to the seal. If element is equipped with a polyurethane wrap, clean & re-oil before reinstallation. Use compressed air at 30 psi to remove ash from the filter element, blowing from inside out. CAUTION: When using compressed air to remove ash, be sure to wear a respirator to avoid inhalation of dust and ash. In addition, eye protection is also recommended.
- Service the Positive Crankcase Ventilation (PCV) valve along with the engine air cleaner.
- In areas of heavy concentration of ash, change your oil and oil filter every 50 to 100 miles of operation. In areas of light concentration, change every 500 to 1000 miles of operation.
- Get a chassis lubrication every 30 days or 1,000 miles.

General Maintenance Information

- Install a filtering material such as open-cell polyurethane (plastic foam) over the outside ventilation system air intake grille. This is just forward of the windshield on most GM cars. Also install oil-treated filter material over the transmission and rear axle vents and intake to the PCV valve.
- Blow ash from all open electrical equipment such as the alternator, cranking motor and windshield wiper motor every 500 to 1000 miles.
- Under heavy dust conditions, clean front and rear brake assemblies with compressed air at 50 to 100 mile intervals.
- Change the Evaporative Control System (ECS) cannister filter—if so equipped—when the air cleaner is serviced. Keep fuel system vents clean.
- Be on the alert for unusual noises when operating your vehicle. Volcanic ash can cause abnormal wear and premature failure of many vehicle parts.
- Owners should consult their vehicle Maintenance Schedule for further information. There are many additional recommendations, including safety checks, which should be performed by a qualified technician rather than the vehicle operator. Depending on the operating environment of the vehicle, it may be necessary to shorten the recommended service interval.

GM

A PUBLIC SERVICE ANNOUNCEMENT FROM GENERAL MOTORS CORPORATION

FIG. 6.25 An important message to all drivers in areas suffering from volcanic ash fallout. A General Motors Corporation message that appeared in many Pacific Northwest newspapers in June, 1980 (courtesy General Motors Corporation).

was believed that air conditioning and refrigeration fan motor bearings and compressors would be adversely affected by the tephra. There and elsewhere filters were changed and condensers cleaned frequently in preventive maintenance programs (*Air Conditioning, Heating and Refrigeration News,* June 2, 1980, 150(5): 47). Later it was suggested that in order to help reduce breakdowns in air filtration systems building superintendents should: "inspect and, if necessary, clean cooling and heating coils. Replace the filters if pressure drop is excessive. Upgrade filtering efficiency in certain locations for critical applications. Temporarily reduce outside air intake to the minimum permissible level to reduce the load on the filters" (*Air Conditioning and Refrigeration News,* July 14, 1980, 150(11): 11). Preventive maintenance seemed to work well. Some Yakima dealers had plenty of service business, cleaning condensing units. While some clients lost condensers, fan motors or exposed electrical relays, one serviceman noted that most losses were the result of lackadaisical maintenance. He believed that those who kept intake vents closed and filters clean had no problems (*Air Conditioning, Heating and Refrigeration News,* January 26, 1981: 50).

In general, electrical problems were the most common sources of failure in equipment in urban areas. The Yakima Clean Air Authority's high volume sampling machine was put out of action by dust between the points in a switch. Dust concentrations had reached more than 35 000 $\mu g\ m^{-3}$ when the equipment failed. Equipment to measure carbon monoxide concentrations and an anemometer also failed (*Yakima Valley Sun,* June 12, 1980).

Banks in Yakima and elsewhere had problems with electronic cash machines and other equipment. In Portland, the First National Bank of Oregon had 40 out of 43 cash machines out of service following the minor tephra fall of May 25. The electrical conductivity of the ash seemed to be the cause of the problem (*Seattle Post-Intelligencer,* May 29, 1980: A11). In Yakima, although the Rainer National Bank's cash machines were covered, they were still serviced before being put back in operation two weeks after the May 18 tephra fall. This bank also had seven drive-up lanes in Yakima, the tellers in a kiosk sending notes and checks to the bank itself via five pneumatic tubes. Although these worked on May 19 they were then closed for a week for cleaning (Rainer National Bank spokesperson, personal communication, June, 1980).

Computer equipment was also regarded as highly susceptible to damage by the tephra. A California technical service manager pointed out "The volcanic ash is almost 100% pumic-like abrasive material. It will literally polish away the magnetic costing on data media". It was also suggested that "the liner material with which floppy disks are packaged will retain volcanic ash to continue its abrasive damage" (*Data Channels,* July 1980, 7(7): 4). Users were advised to protectively package material and to have a back-up copy of all valuable data. The Washington State Library computer system did report problems caused by the tephra fall. The first problems resulted from air conditioning failures or potential failures; the second from tephra entry to terminals (*Bookmark,* September 1980, 33(1): 3).

Although it was initially expected that companies engaged in manufacturing electronic equipment and instruments would have problems as a result of the tephra fall in the Pacific Northwest, Mosebar (1980: 14) reported that there was no significant loss of production. Many companies installed temporary additional filtration equipment, restricted access to areas where the tephra ingress could be a problem and utilised shoe and clothing vacuuming equipment.

Despite the general expectation that mechanical and electrical equipment would suffer grievous harm from the tephra fall, Yakima office equipment companies reported only a marginal increase (less than 5%) in business. As the industry across the rest of the country showed a real downtown in the first quarter of 1981, this marginal increase may have been quite significant (two Yakima service companies, personal communication, May, 1981). The work was mainly general cleaning, some of the work coming from offices where windows had been left open on May 18, or from office areas where sealing was poor. Calculators seemed to be more susceptible than other types of office machinery. The problem seemed to be shorting out of chips, but it proved difficult to pin this down specifically to tephra ingress. However, tephra did cause office machinery to become sluggish with printer problems showing up six to eight months after the tephra fall, when bushings became scraped, evidently where tephra particles had dried up the oil. This problem also occurred in typewriter motors. Other problems may not show up for years — these will prove even more difficult to relate to the tephra fall. The slight increase in business, however, suggests that only a small proportion of business machines were adversely affected.

In the agricultural areas of the Pacific Northwest, more problems were experienced with farm machinery. Irrigation pumps certainly had problems; where pumps were not sheltered from the tephra fall, arcing occurred in some cases, particularly when the tephra was wet. Sprinkler heads also had maintenance problems especially in those areas where irrigation water was contaminated with tephra. Warrick *et al.* reported on the Ellensburg area: "Irrigation systems were subjected to a variety of malfunctions from the ash. Electrical panels shorted out because of the high electrical conductivity of wet ash. Bushings, brushes and commutators on pump and drive motors suffered increased wear. Thrust bushings on rainbird-type sprinkler heads and pumps wore down faster. Generally, responses consisted of cleaning switch panels and trying to keep ash out of motors in any way possible. Blowing the ash off with air compressors was one common procedure. Farmers were advised to try to run water destined for pumping irrigation systems through settling basins before running it through the pumps" (1981: 18).

The County Extension Service of the United States Department of Agriculture made other suggestions after receiving advice from an irrigation dealer in Grant County (County seat, Ephrata) that he had repaired 70 pumps in the week following May 18. The service recommended that bearings and packings on water-lubricated pumps should be regularly inspected but the packing glands should not be overtightened, a slow drip being desirable to lubricate and to wash particles on through and packing. They also suggested that overheating could result from dust buildup in the pump ventilation system. The dust problem around the pump could be alleviated by sprinkling, incorporating the tephra into the soil, or by working straw into the nearby ground (*Ash and Agriculture,* Washington State University and United States Department of Agriculture, June 17, 1980).

Other problems with farm machinery resulted from the abrasiveness of the tephra: "Some estimates suggest that wear to gears, bearings, cutting sickles, and other moving parts of equipment during the growing season of 1980 will equal the wear expected in three or four years before the ashfall" (Cook *et al.*, 1981: 18). One dairyfarmer only 30 km northeast of Mount St Helens was impressed by the abrasiveness of the tephra when he chopped a field of alfalfa and grass for silage three weeks after the fall: "I grind about an inch [2 cm] off my flail chopper blades with five passes around the field" (*Dairy Herd Management,* August 1980, 17(9): 10). Reports of blades wearing out rapidly also came from Lewis County (*Ash and Agriculture,* August 12, 1980). In the same county, Christmas tree growers maintained that tree shearing knives were dulled more quickly than before the tephra fall (*Ash and Agriculture,* July 1, 1980).

Chain-driven equipment also had problems. Weyerhaeuser shut down its logging operation at Vail tree farm because tephra was clogging chain saws (*Chemical Week,* May 28, 1980: 19). Chains wore out quickly if they were not lubricated and tephra accelerated the wearing process. However, tephra with grease formed an abrasive compound. Either way, chain wear was accelerated. It was suggested that silicon or graphite based lubricants would provide some lubrication without attracting the abrasive tephra (*Ash and Agriculture,* August 12, 1980).

Farm machinery, like road vehicles, also had some problems with air cleaning systems. However, similar precautions, better filters, more frequent change of oil and filters, all meant that farm machinery damage was not as severe as anticipated. With some machinery (tractors for example) air cleaners could be extended high above the vehicle and away from the recirculated tephra picked up by wheels or tracks.

In most cases, many of the problems with mechanical and electrical equipment can be overcome by improved filtration of air and/or by better or more frequent cleaning. Improved maintenance, especially preventive maintenance, eliminates most problems with machinery and office equipment. Good filtration systems remove particles larger than 20 μm. As the oil film between moving parts in most engines if 25 μm thick, particles smaller than 20μm should not prove damaging (Cook *et al.*, 1981: 18). This analysis suggests then that it is the abrasive action of larger particles on moving parts such as chain drives and chopping blades that is the least preventable.

The evidence presented in this chapter clearly shows that volcanic eruptions can seriously impact a wide variety of elements in a country's economic infrastructure.

Although many of the examples presented here stem from the 1980 eruption of Mount St Helens, similar effects are probably quite common. Communication systems and public utilities can be destroyed or damaged. Individual and corporate property may be similarly affected. The damage bill can be enormous or it may comprise only the cost of increased maintenance. Certainly, the specific remedies adopted around Mount St Helens have a general applicability wherever equipment and effects are similar. Frequently, the effects are not as bad as they at first seem. Either way, the cost of a volcanic eruption may well be profit to someone else. As we will see in the next two chapters, volcanic eruptions have positive as well as negative effects on agriculture and economic activity.

7

THE EFFECTS ON AGRICULTURE

7.1 INTRODUCTION

Agriculture has been severely affected in many eruptions, sometimes with the deaths of farm animals or the destruction of crops. In turn, human deaths may result not from an eruption *per se* but from starvation (or from disease and malnutrition); for example, starvation (or disease and malnutrition) killed more than 9000 people in Iceland as a result of the Laki eruption of 1783 and about 80 000 people died following the 1815 eruption of Tambora (see Chapter 3, Section 3.12). Crop destruction by tephra fall from the seventeenth century eruption of Long Island caused the Urii people of Papua New Guinea to (briefly) become cannibals (Professor Peter Lawrence, University of Sydney, personal communication, 1980).

The present chapter is concerned with the effects of volcanic eruptions on crops, livestock, aquatic life and other aspects of agricultural production. While agriculture is the main focus, the effects of non-agricultural plants and animals are discussed where it is necessary to illustrate specific points. Inevitably, tephra falls are the volcanic agents causing the most common impacts on agriculture, because tephra falls can be very widespread (see Figs. 2.11 and 2.12).

7.2 CROPS

Agricultural crops have, no doubt, been adversely affected by all of the volcanic hazards discussed in Chapter 2 but tephra falls, pyroclastic flows, lahars and acid rains and gases are the most destructive agents. Table 7.1 provides some brief examples, but including all the information readily available, of the effects of a few

Table 7.1 Effects of some Indonesian eruptions on croplands

Eruption	Volcanic hazard	Effects	Reference
Galunggung, 1822	pyroclastic flow	1 668 000 coffee trees destroyed	*CIRV*, 1979: 794
Raung, 1870	?	250 000 coconut palms destroyed	*CIRV*, 1979: 796
Krakatau, 1883	tsunami	66 km^2 arable land destroyed	*CIRV*, 1979: 796
Kelut, 1919	lahar	135 km^2 arable land destroyed	*CIRV*, 1979: 798
Semeru, 1946	tephra and lahars	~150 ha rice fields destroyed	van Bemmelen, 1949: 26
Merapi, 1954	?	1 ha rice, 675 ha corn, 263 ha vegetables, 250 ha forest, 2130 ha other crops destroyed	*CIRV*, 1979: 800
Batur, 1963-64	?	1600 ha arable land destroyed	*CIRV*, 1979: 801
Agung, 1964	tephra, pyroclastic flows and lahars	55 417 ha rice, 54 289 ha arable land, 11 745 ha forest destroyed	*CIRV*, 1979: 800
Iya, 1969	?	1600 ha arable lands, 30 000 coconut palms destroyed	*CIRV*, 1979: 801
Dempo, 1974	?	~350 ha of trees destroyed; >20 t of 1974 tea smelled of sulfur	*BVE*, 1976, 14: 29

Indonesian eruptions on croplands. Many more examples with similar "precise" figures are presented in *CIRV* (1979).

Although few details are available, lava flows must have destroyed plantation and other crops on numerous occasions. As the lethal temperatures for hydrated tissues of plants is about 50°–60°C (but also dependent on the log of duration of exposure — Baker, 1950; Gill, 1976), crops some distance beyond the lava flow front will also be killed or damaged. Certainly, the 1944 Vesuvio, the 1811 and 1971 Etna, and the 1808 San Jorge lava flows destroyed significant areas of vineyards, olive groves and/or arable land (Anon, 1811; *BVE*, 1979, 11: 9; Clapperton, 1972: 166; Dabney, 1809: 343-344). The 1938 Nyamuragira lava flows set hundreds of square kilometers of forest on fire (Coleman, 1946: 168). The 1977 Nyiragongo lava flows rapidly covered 5 km^2 generating severe temperature gradients and strong winds which uprooted big trees and tore down banana groves. In some areas lavas less than 10 mm thick flowed over and around banana leaves but did not dry out the leaves or burn the trees down (Tazieff, 1977). In the same year, lava flows on Piton de la Fournaise destroyed 2 km^2 of arable land, mainly planted with sugar cane (*BVE*, 1979, 17: 15-16). The 1955 Kilauea eruption covered with lava (and tephra) 280 ha of sugar cane (Fig. 7.1), 140 ha of other crops and 1100 ha of forest-land. An additional 500 ha of cane and other crops were burnt (Macdonald and Eaton, 1964: 132). In 1960 Kilauea flows advanced through papaya plantations. The largest lava eruption of the millenium covered 14 farmsteads around Lakagigar in 1783 (Thorarinsson, 1969).

FIG. 7.1 A 1955 lava flow advancing through a sugar cane plantation, Hawaii (Photo: G.A. Macdonald, US Geological Survey).

Lightning strikes have also damaged tree crops. Seven kilometers from Soufrière, St Vincent, in 1979 coconut trees were decapitated (Rowles, 1979). At Mount St Helens lightning strikes from a pyroclastic flow or the tephra cloud started an estimated 50 to 100 forest fires most of which were doused by falling tephra.

Pyroclastic flows have also affected agriculture during numerous eruptions. Two examples are given in Table 7.1. Many plants including trees could be killed by the temperature effects on hydrated tissues mentioned above but exposure times in individual flows are generally only a few minutes. Blast effects, resulting from high flow velocities, are probably more significant (Figs. 7.2 and 7.3). Where the effects are most severe, such as in the inner zones of destruction at Mount St Helens and Lamington, trees and other vegetation have been completely removed. In some eruptions, for example, Taal 1965, pyroclastic flows of the base surge type "sandblasted" large areas of wood from the sides of trees facing toward the volcano (Fig. 7.4). Nevertheless, maps of the affected areas around Lamington 1951 and Mayon 1968 (see Figs. 2.15 and 2.17) indicate large areas of partial devastation where the vegetation has been only charred or singed.

At Mount St Helens some 300 km^2 of tree crops were damaged by pyroclastic flows. Some 3000 ha of the downed timber belonged to Weyerhaeuser Corporation (*Pulp and Paper,* July 1980, 54(7): 17). Perhaps as much as 80% of this timber was subsequently salvaged (Rainer National Bank, Economic Report, No 1, June 1980:

FIG. 7.2 Tropical rainforest destroyed by the March 8, 1979, pyroclastic flow on Karkar. The volcanologists Rob Cooke and Elias Ravian were killed near the site shown in this photograph.

FIG. 7.3 Trees felled by the pyroclastic flow on May 18, 1980, at Mount St Helens (Photo: US Geological Survey, P10-80-180c).

FIG. 7.4 "Sand-blasting" by a base surge eruption removed large amounts of wood from the volcano side of trees around Taal in 1965 (Photo: J.G. Moore, US Geological Survey, P10-65-70K).

2). Timber and related losses were later assessed at US$695 million but by no means all of this loss can be attributed to pyroclastic flows.

Other crops have also been destroyed by pyroclastic flows (Table 7.1). On the western side of Sakura-jima island in 1914 pyroclastic flows uprooted a large number of mandarin trees. Sugar cane at Hakamagoshi (see Fig. 3.19) was flattened "mostly outwards or away from the centre of the disturbance, although the cases of the contrary direction were also not wanting" (Omori, 1916b: 211).

Rees (1970: 17) has suggested four obvious causes for the death of crop plants and other vegetation caused by tephra fall: (i) complete burial of the plant; (ii) partial burying, restricting root access to oxygen; (iii) defoliation and prolonged absence of leaves; and (iv) tephra covering foliage surfaces, clogging stomata and blocking out sunlight. Limbs may also be removed by the tephra and/or ballistic projectiles (Fig. 7.5). In a few cases tephra may be at a sufficient temperature to damage or destroy crops.

Experiments by Miller and Lee (1966: 13-14) during the 1963–64 Irazú tephra falls indicate that the retention of tephra on foliage is a function of: (i) leaf surface area; (ii) foliage density per unit area of ground surface; (iii) the shape, size and orientation of individual leaves; (iv) leaf surface characteristics such as roughness, and the presence of fibres and oils; (v) the tephra particle size and volume; (vi) the velocity of air through the canopy; and (vii) the relative humidity especially if the humidity is above 90%. With constant wind speed and humidity, the degree of exposure or shielding of the leaf and the dry weight of the leaf relative to the exposed area were the most important factors influencing foliage contamination. Other leaf characteristics were of only secondary importance in tephra retention (Miller and Lee, 1966: 112). While these results should be used with caution as they seem to be influenced by the experimental method, Table 7.2 summarises the relative concentrations of leaf contaminations recorded by Miller and Lee (1966: 102). Geraniums and lettuce retained about six times as much tephra per unit area per unit dry weight of foliage as did gardenias. When the relative humidity was above 90% tephra concentrations retained on foliage were about twice those found with lower relative humidities. It should be noted the concentrations measured were relatively low (equivalent to less than 1 mm of tephra fall) and that there is no necessary direct relationship between concentration retained and effects on the plant.

Damage to crops by lapilli or cinders has been reported during several eruptions. Huggins (1902: 13) wrote, concerning the May 7, 1902, Soufrière, St Vincent, eruption: "The pebbles and globular drops of mud fell so quickly that clouds of leaves were struck from the trees and the larger foliage was perforated in a number of places. The writer counted 205 perforations in a breadnut leaf, taken haphazard, and 14 in a Liberian coffee nut but neither of these leaves had been touched by the larger size pebbles."

In the April, 1903, eruption cinder-like pieces of ejecta damaged cocoa trees. The leaves were riddled with holes and turned brown; flowers on the upper sides of

Table 7.2 Relative leaf contamination for a variety of plants measured during tephra falls around Irazú, 1963–1964.

Tephra retention	Plants
highest tephra retention on leaves	geranium, lettuce, potato, grass, beet, bean, agapanthus, tomato, orange, bougainvillea
lowest tephra retention on leaves	carrot, camellia, rose, cabbage, cypress, leak, corn, gardenia

From Miller and Lee, 1966: 102.

FIG. 7.5 (top) Branches stripped off *Casuarina* on Casuarina Road, Rabaul in 1937 (Photo: Australian Bureau of Mineral Resources, M2579/10); (bottom) Branches *(Larix kaempferi)* bowed down under a 90 mm load of tephra near Usu in 1977 (Photo: courtesy of K. Saito).

FIG. 7.6 Holes were punched in large tropical leaves by a fall of fine lapilli during the 1979 eruption of Karkar.

branches were knocked off while those on the underside were seldom injured (*HMSO* 1903: 65-67). Pond and Smith (1886: 364) observed the effects of tephra on the strong leaves of the *tupaki* [? tupakihi, Coriaria spp.] on a journey from Ohinemutu on the margin of the ashfall area toward Tarawera in 1886: "At first the leaves were only bespattered with mud, further on they were perforated by the small stones, still further on the fleshy portion had been beaten out, leaving only the midrib, while beyond this not a vestige of a leaf had been left on any of the bushes". Figure 7.6 illustrates the effects of fine lapilli on vegetation around Karkar in 1979. Hot projectiles caused forest fires on the upper flanks of Sakura-jima on numerous occasions in 1973, 1974, 1976 and 1977 (*BVE*, 1975 (13): 41; 1976 (14): 40; 1978 (16): 40; 1979 (17): 41). Around Ngauruhoe in 1974 the heat from bombs charred small circular patches of alpine vegetation but no fires started.

Table 7.3 summarises the reported effects of tephra falls of varying thickness on a variety of crops. Several points need to be kept in mind when interpreting this table. The crop plants reach a wide variety of heights above the ground surface. Plants in less temperate areas may be adapted to snow falls and may be less sensitive to tephra loads on branches and limbs. As the observations were made at varying (and generally unknown) times after the tephra fall, abrasion by tephra blowing in the wind may be an important factor influencing crop damage. Heavy rainfalls may have occurred soon after the tephra fall, washing ash from the foliage.

Table 7.3 Summary of reported effects of tephra falls on crops

Tephra fall thickness (mm)	Effects on crops*
>2000	All vegetation killed
1000-1500	Most vegetation killed; some recovery
200	Rice paddy destroyed
150	Coffee trees sevorely damaged
100	Some branches break under tephra load; palm fronds broken
50	Banana plants damaged; "forced ripening" of some crops
40	50% loss of lentils; 15-30% loss of wheat, spring barley, peas, hay
30	Some cherry damage; some plants unable to push through tephra crust
25	Sugar cane prostrated in patches; branches broken off young breadfruit trees; damage to sweet potatoes, yams, tannias
20	30% loss of lentils; 5-15% loss of winter wheat, barley, peas, grass seed; mandarin oranges, mulberrry tree crop and vegetables considerably damaged
15	Loss of first alfalfa cutting; grass hay windrowed before tephra fall not usable
10	Breadfruit, apples, cotton balls dropped; cotton stems broken down; 20-40% scarring on blueberries; banana, spring wheat, mulberry, tobacco and vegetable crops somewhat damaged; hay crop loss 25-30% because of prostration
<10	Strawberry plants flattened increasing berry rot; minor damage to corn and hay; some burned leaf and salt damage of blueberries

*Compiled from data on the following eruptions: Tambora, 1815; Cosiguina, 1835; Tarawera, 1886; Ambrym, 1894; Soufrière, St Vincent, 1902, 1903; Taal, 1911; Katmai-Novarupta, 1912; Boqueron, 1917; Rabaul, 1937; Parícutin, 1943–1952; Taal, 1965; Ulawun, 1967; Cerro Negro, 1968; Sakura-jima, 1974; and Mount St Helens, 1980.

The tephra falls reported in Table 7.3 were certainly of different particle sizes and composition; coarse pumiceous tephras are unlikely to hold much moisture available for plant growth (see Smathers and Mueller-Dombois, 1974: 8-9). Particle size will also influence susceptibility to wind and/or water transport. The reported tephra falls also occurred at different seasons; the stage of plant growth is an important factor influencing crop damage. Finally, tephra compaction occurs rapidly after deposition (Blong, 1982: 138) so that the reported thicknesses may not be strictly comparable, depending on when the observations were made.

Some of these points can be illustrated by reference to the damage to vegetation resulting from the eruptions of Katmai-Novarupta and Parícutin.

Rees (1970: 17) defined four zones of vegetation damage around Parícutin (Table 7.4). The "total kill zone" had a tephra thickness of greater than 1500 mm (see also Fig. 4.4). In the nearly "total kill zone" ash thickness averaged 1500 mm. Most individuals of all size classes of all species were eliminated plus an area of fir killed on Tancitaro where other species survived in relatively shallow ash. The "first zone of partial survival" averaged 500–1500 mm of tephra. There tree damage and a heavy kill of shrubs and herbs occurred. In the "second zone of partial survival" ash deposition of 150–500 mm resulted in slight tree damage and partial survival of shrubs and herbs.

Rees (1970: 13) also noted the following effects: where more than 500 mm of tephra fell, large tree branches were broken. Where more than 250 mm fell (at a distance of about 8 km) crops were killed and agriculture became impossible. The same thickness of tephra at a distance of less than 12 km defoliated fruit trees. Where the tephra thickness was much less than 100 mm fine ash entered avocado flowers preventing pollination.

The varying susceptibility of individual species to tephra was clearly indicated by Eggler's (1948) work around Parícutin 2.5 years after the eruption commenced. Oaks apparently survived in ash depths up to 1780 mm, elder in depths of 1500 mm, some pines in depths in 1240 mm, *Crataegus* in depths of 1020 mm, and trailing arbutus in depths up to 890 mm. The most hardy shrubs were *Cestrum* and *Smilax* which survived burial by as much as 1270 mm of tephra while the common *Baccaris* survived falls of only 560 mm. Two herbaceous plants, the prickly poppy and mimosa, survived tephra falls up to 740 mm while Bermuda grass remained in areas receiving less than 250 mm of tephra (Eggler, 1948, cited in Wilcox, 1959: 486).

Furthermore, Eggler's (1948; 1963) results showed that the effects of tephra varied with the basal diameter of the tree — at least for pine species around Paricutin (Rees, 1970: 17-18). Pine seedlings and small trees were killed as a result of excessive bending and burial, while large mature trees suffered from branch breakage under loads of ash. Pines with basal diameters of 100–300 mm survived best because their stems were strong enough to resist excessive bending yet sufficiently flexible to dump part of the ash load and thus avoid breakage.

Figure 7.7 indicates the area around the volcanoes Katmai and Novarupta.

Table 7.4 Zones of vegetation destruction versus tephra thickness around Parícutin, 1943–1952

Zones	Tephra thickness (mm)	Total Area	Area (hectares) Forest Area	Forest Proportion	Cleared* Area	Cleared* Proportion	Villages Area	Villages Proportion
Total kill zone	>1500	2 480	975	39.3%	1 374	55.4%	131	5.3%
Nearly total kill zone	~1500	4 375	3 435	78.5%	940	21.5%	0	0%
1st zone of partial survival	500-1500	5 245	3 866	73.7%	1 379	26.3%	0	0%
2nd zone of partial survival	150-500	26 110	17 363	66.5%	8 695	33.3%	52	0.2%

*After Rees, 1970: 17. Cleared land = lands under cultivation, in annual fallow, or long fallow shifting cultivation without timber.

Figure 7.8 shows the area affected by tephra fall. Where more than 500 mm of tephra fell, more than 60 km from the volcanoes, most trees were killed. A total area of about 19 000 km^2 was covered deeply enough to destroy small plants (Griggs, 1919: 209).

In contrast to the results reported by Eggler around Parícutin, Griggs (1919: 173) notes "the remarkable absence of specific effects of the Katmai eruption proper on different plants". Griggs believed "most of the flowering plants suffered alike; where any perished, most perished, and where any escaped, almost all escaped...The marked differences shown in the recoveries of different species are traceable in almost every instance to some readily observable difference in habitat or adaptation. Thus *Equisetem arvense* came through where nothing else could, not because of any greater toughness of constitution or greater capacity to endure burial, but because it could send out runners of greater length than other plants. *Salix alaxensis* recovered better than the poplars, not because more hardy, but because of its capacity of forming adventitious roots which the poplars lacked. Where here and there a few plants survived in the region of total destruction, it was for the most part due not to superior resistance but to a fortunate location" (1919: 193).

Griggs emphasises the point by comparing the condition of surviving plants (three years after the eruption) in Soluka Valley, where more than 1800 mm of ash fell and a pyroclastic flow occurred, with the area around Kodiak where 300 mm of tephra fell (Table 7.5; Fig 7.8). All the species mentioned were common in both localities before the eruption. Most of the survivors in the Soluka Valley were found in an upland bog from which the tephra fall had largely been eroded. The reason for the survival of the species mentioned in the Soluka Valley when they were practically exterminated around Kodiak "clearly lies not in any peculiarity of the species themselves, but in the accident of a favorable situation" (Griggs, 1919: 193).

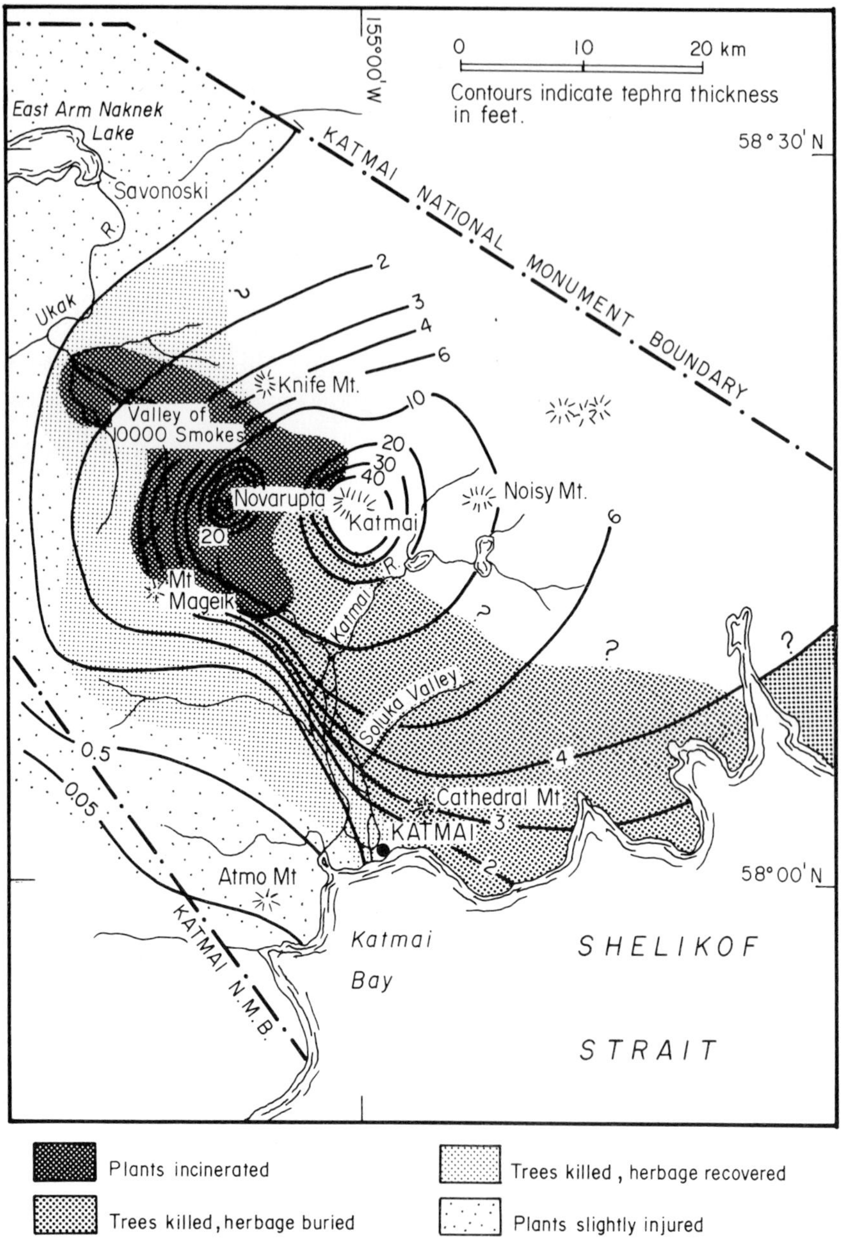

FIG. 7.7 Zones of vegetation damage around Katmai-Novarupta in 1912 (after Griggs, 1919).

Table 7.5 Effects of Katmai-Novarupta tephra fall on vegetation in Soluka Valley and at Kodiak

Species	Soluka Valley	Kodiak
Alnus sinuata	Only one seen alive	Not injured
Equisetum arvense	Scarce	The most abundant survivor
Athyrium cyclosorum	Thrifty	No survivors observed
Trientalis arctica	Flowering	No survivors observed
Ledum decumbens	Flowering	No survivors observed
Betula rotundifolia	Thrifty	Only occasional survivors
Empetrum nigrum	Thrifty	Very rarely survived, was ubiquitous
Vaccinium uliginosum	Fruiting	No survivors observed
Sanguisorba sitchensis	Flowering freely	Abundant
Cornus suecica	Thrifty	Rarely survived
Vitis-idoea vilis-idoea	Thrifty	Rarely survived

After Griggs, 1919.

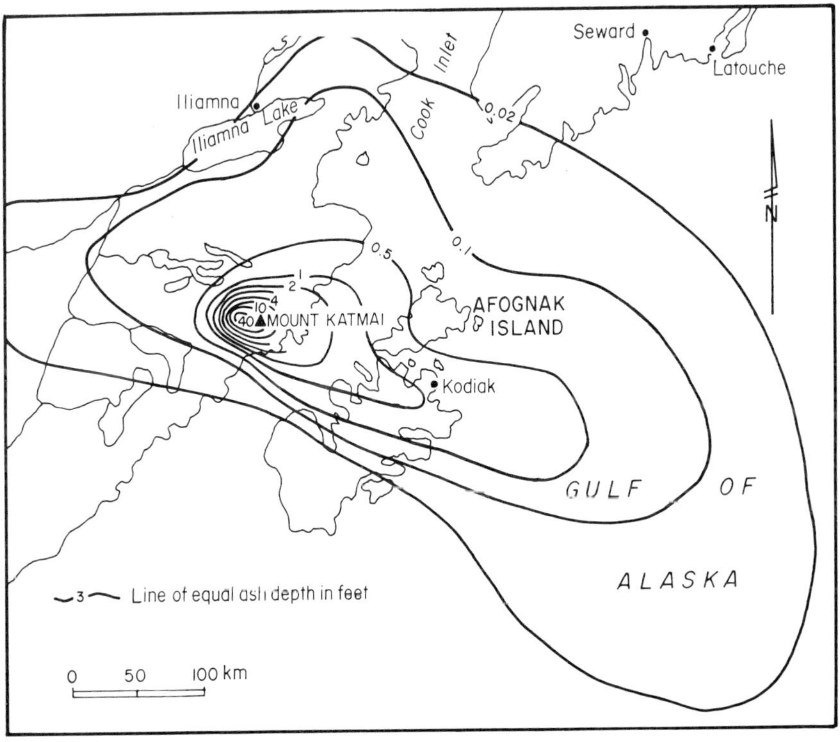

FIG. 7.8 Area affected by tephra fall from Katmai-Novarupta, 1912 (after Wilcox, 1959).

A number of other observations by Griggs are of interest:

1. In places the ash lay undisturbed for three years until a flood in July, 1915, removed much of it. Plants from old roots, which must have remained dormant for three years, then came up with all their original vigor.
2. Plants that could put out new root systems at the surface of the ash had a big advantage. Thus willows had a big advantage over poplars. The willows did not have to carry over-extended tops and roots of the old trees. Where Alaska willow (*Salix alaxensis*) grew among poplars in three years the willows were as thriving as ever, while the poplars were more dead than alive (Griggs, 1919: 200-201).
3. At the beginning of the fourth season (1915) after the eruption there were still practically no seedlings around Kodiak, most of the growth being directly traceable to old roots. The most striking example of this occurred in a field near Kodiak which had been ploughed before the eruption. Where the cultivation destroyed the weeds, no new vegetation appeared for five years; on adjacent cultivated land plants came up vigorously and completely covered the ground (Griggs, 1918: 31).
4. Where the tephra fall was 900 mm or more in thickness it was coarser than it was around Kodiak so that the ash did not crack. In these places conditions were not favorable for penetration of the ash by buried herbage (Griggs, 1919: 189). On the other hand, beach grass grew through cracks in 280 mm of tephra, within five weeks of the tephra fall (Fig. 7.8).

Thus, significant differences can occur in vegetation survival following tephra fall. In assessing these differences it should also be remembered that Parícutin produced tephra over a period of years whereas the Katmai-Novarupta eruption produced large amounts of tephra only on one day. Although the descriptions above contains some evidence of vegetation recovery following tephra fall, further discussion of this topic is contained in Chapter 7, Section 7; the important consideration for the moment is the interpretation of Table 7.3.

The various reports summarised in Table 7.3 span a wide range of vegetation responses and illustrate the point that, while a few millimeters of tephra may cause some damage to crops, other plants will survive tephra falls of up to one or two meters. Furthermore, some of the effects included in the table result only indirectly from tephra fall. For example, increased strawberry rot is reported as resulting from less than 10 mm of tephra fall. In fact, the tephra fall weighed down the plants pressing the fruit into the moist soil; strawberry infection and decay was caused by contact with the moist soil not by the tephra *per se*.

Some other points also require further consideration. In the Yakima Valley and elsewhere, increased apple drop was reported following the May 18 Mount St Helens tephra fall. Some growers indicated that the drop was more severe with red delicious than with golden delicious varieties. This increased drop possibly resulted from a reduced rate of photosynthesis where ash clung to the hairy (pubescent) leaves (Cook *et al.*, 1981: 20). On the other hand, it has also been suggested that pronounced apple drop only occurred where certain thinners had been sprayed a few days before the tephra fall. In either case, where the tephra was removed immediately, by shaking, use of wind devices (including hovering helicopters!) or sprinkler irrigation, the fruit drop was less. Furthermore, where fruit drop did occur, it was expected that some of the loss would be offset by the better size and grade of the remaining apples (*Ash and Agriculture,* July 22, 1980: 28; Cook *et al.*, 1981: 20). Table 7.3 refers particularly to the effects on crops rather than to the economic damage resulting from tephra fall. Strawberries, raspberries, peaches, apricots and other pubescent or rough-skinned fruits received lower market prices or could not be sold because adhering tephra could not easily be removed. On the other hand, clean-skinned (glabrous) fruit such as cherries and apples could be washed and suffered little in the market place. Of course, such distinctions are only relevant in developed economies; at the subsistence level the most important difference glabrous and pubescent skins make is in the amount of tephra ingested by the local population!

The stage of plant growth at the time of tephra fall is obviously important but is not taken into account in Table 7.3. For example, six weeks after the May 18 Mount St Helens tephra fall it was reported that spring wheat still looked worse than winter wheat in the area between Ellensburg and Ritzville (see Fig. 3.8). In the same area, the condition of peas depended on type; upright types were reported to be growing well whereas prostrate types were diseased. In some fields, peas were blooming but few pods were forming while in others pods were forming normally. It was not clear whether this was related to pea variety or maturity at the time of tephra fall (*Ash and Agriculture,* 1980, June 17: 8; June 24: 2; July 8: 2).

Another difficulty with Table 7.3 derives from the fact that natural hazards are likely to get the blame for all sorts of occurrences during the aftermath. For example, the 1979 Karkar eruption produced a tephra fall of about 10 mm (compacted) on the south-east side of the island. Two weeks later taro tubers were observed rotting in the ground. Local subsistence farmers insisted that the rot was caused by the ash whereas the local agricultural officer believed that taro blight, a quite common plant disease, had caused the decay. Where coconut palms had several fronds broken down, the agricultural officer thought rhinoceros beetle (*Scapanes australis; Oryctes rhinoceros*) had caused the damage but plantation owners blamed the tephra. Similarly, in the Yakima Valley after May 18, 1980, loss of about half the cherry crop due to heavy rains was blamed on the tephra particles in the atmosphere causing excessive rain (*Ash and Agriculture,* July 1, 1980: 3). In each of these cases it is not easy to substantiate one view or the other.

A further major difficulty in interpreting some of the data in Table 7.3 stems from the recognition that some tephra falls, particularly those of fine-grained ash, are accompanied by acid aerosols, or acid rains. Several examples of damage to crops by acid rains, aerosols and gases can be cited.

In the Mediterranean, a "great Smoak" from the 1707 eruption of Santorini burnt and destroyed ripe grapes on Thera "especially in such Vineyards as lay most exposed to the South" (Bourguignon, 1708: 204), while acid rains "ruined" hundreds of thousands of dollars worth of orchards and vineyards around Vesuvio in 1928 (*Literary Digest,* June 23, 1928).

Gases produced during the 1905 Matavanu eruption also damaged crops. Where the lava flows ran into the sea (see Fig. 5.2) the gases produced were considered even more poisonous than those discharged from the crater. Coconut palms 3 km from the lava flow were damaged; some died but most recovered. Some breadfruit trees also died while some recovered with the loss of upper branches (Anderson, 1910: 634). Acid rains and gases produced by the Katmai-Novarupta eruption killed vegetation 400 km to the northeast at Seward (Fig. 7.8). At Latouche, 480 km northeast, tender annual garden plants were completely destroyed and the leaves of many native perennials were so burnt that they dropped off (Griggs, 1922: 25). Nearer to the volcanoes, on Kenai Peninsula and around Prince William Sound, the "leaves of the currants, salmon berries and many other of the shrubs and herbs ... were blighted by the dust or acid rain which fell there. This effect, curiously enough did not occur in the district of thicker ash" (Martin, 1913: 180). Acid rains or aerosols produced during the Mayon, 1928, eruption wilted and browned, sometimes with black spots, the leaves of abacá plants 10–15 km from the volcano. Acacia, santal and pilli trees lost most of their leaves while cogon and other grasses, including bamboo, were burnt brown. In some cases only the portions of the plant facing into the wind were damaged (Faustino, 1929: 30). Following the 1938 eruption of Mayon, Galvez (1938: 767) noted that foliage first yellowed and then blackened. "In more serious cases, the leaves were scorched, with their edges shriveled ... It was also noted ... that plants having soft, tender, and hairy leaves suffered more than those with hard, thick and glossy leaves." However, it was reported that abacá and papaya were particularly sensitive whereas coconut, coffee, banana, gabi and rice appeared unaffected (Galvez, 1938: 766; Galvez *et al.,* 1939).

Miller produced some useful observations on the effects of acid rains on vegetation in a study of tephra retention by plants around Irazú in 1963–1964: "... the growth of many plants was adversely affected by occasional showers of ceniza-arena [tephra] particles that carried sufficient quantities of sulfuric acid or sulfate salts to burn the foliage. Single showers of such particles killed entire bean crops and burned the corn leaves to a uniform gray color. Large squash plants so affected died within a few days. Tomato leaves were easily burned at their edges and tips; the burned edges first dried and became brittle, but within a few days, under damp conditions, the dead tissue became soft and rotted. The tips of the onion foliage were burned, but otherwise the onion foliage did not retain enough particles to seriously affect the growth of the plants. (The acidic solution would run down the stems to the ground rather than concentrate in place on the foliage as the water evaporated, as was the case for the other leafy plants.) The oat and rye foliage was less sensitive to burn than was the wheat and barley (the latter was most sensitive) foliage; all the grains suffered leaf-tip burns to some degree. Cabbage was never affected by acid burn... The severest burn condition on the plants resulted when the acid-carrying particles landed on the foliage during the night or early morning hours when the foliage was damp with dew. The acid then dissolved into the dew and spread more or less uniformly over the whole top surface of each contaminated leaf. When the temperature increased after sunrise, the water rapidly evaporated; in the process, the acidic salts were concentrated on the surface of the leaves, and, as dryness approached, the acidic salts dehydrated and burned the leaves" (1966: 182).

Detailed studies of volcanic acid rain effects on plants were also made during the Kilauea 1970–1973 eruption when acid rains fell in the Kona area 75 km northwest of the volcano. Acid rain with pH 4.0 and appreciable quantities of Cl^- and SO_2^- caused blossom drop, poor fruit set, hollow, small or almost seedless fruits and a less luxuriant appearance in tomatoes. On the other hand tomatoes grown under plastic shelters grew normally (Kratky *et al.*, 1974).

These studies have allowed some more general observations to be made. Faustino (1929: 39) noted that hydrochloric acid gas has a strong affinity for water (see Chapter 2, Section 8) and is at once absorbed on the ash, thus causing the damage when the dust settled on the leaves. However, hydrochloric acid also evaporates quickly so that samples tested in the laboratory in Manila did not produce damage to leaves. He also noted that damage to the vegetation around Mayon was only temporary. Griggs (1919: 176) made the point that the occurrence of acid rains is sporadic, being dependent not only on the drift of fumes but also on the presence of the atmospheric conditions necessary to produce precipitation at the time when acid-laden clouds are passing overhead. However, this view is unlikely to be completely correct; as many acids have a strong affinity for water, plant damage can occur during periods of high humidity. The situation is further complicated by the observation of Kratky *et al* (1974) that the rainwater samples form a single heavy rainfall were less acidic than from cumulative samples of lighter rainfalls indicating that the first portion of a rainfall absorbs most of the pollutants from the air and is thus most damaging to plants. Wilcox (1959: 452) summarised information on the effects on plants of industrial gases similar to volcanic gases (Table 7.6). While threshold concentrations at which damage to plant tissues occurs can be cited, it "is necessary also [to note] that the threshold value may vary greatly, not only from one plant type to another, but also that the susceptibility of a given plant may increase drastically, for instance, during periods of higher humidity and more active growth". Investigation of the effects of industrial pollutants such as SO_2 and fluorides are of considerable relevance to considerations of crop damage caused by volcanogenic acid rains, gases and aerosols. The interested reader is referred to various chapters in Kozlowski and Mudd (1975).

Table 7.6 Effects of gases on crops

Gas	Threshold concentration (ppm)	Effect of overdose
Ammonia	10	"Cooked" green appearance becoming brown upon drying
Carbon dioxide		No marked effect
Carbon monoxide	500	Epinasty, chlorosis, shedding of leaves
Chlorine	1	Marginal and intravenal leaf lesions
Fluorine	0.1	Marginal leaf lesions, shedding of leaves
Hydrochloric acid	10	Marginal and intravenal leaf lesions
Hydrofluoric acid	Variable	Marginal leaf lesions, shedding of leaves
Sulfur dioxide	1	Chronic: Yellowing and bleaching of leaves Acute: Collapse of marginal and intercoastal areas, drying to ivory or brownish red.
Sulfuric acid (aerosol)	Variable	Scorching or charring depending on moisture content, size of aerosol droplets.

After Wilcox, 1959.

7.3 LIVESTOCK AND OTHER MAMMALS

The destruction or burial of vegetation by volcanic hazards, particularly tephra falls, will have widespread effects on livestock and other herbivores. Livestock may also be directly affected by lava flows, pyroclastic flows, tephra falls, atmospheric effects, gases and tsunami. In some cases carnivores will be affected by the fate of their prey.

Although the main focus of this section is the effects of eruptions on livestock, non-domesticated animals form part of the food resource of subsistence agriculturalists; thus some consideration is given to such animals and to household pets.

No doubt, livestock and other animals have occasionally been killed by lava flows but few such cases have been recorded. However, the 1938 lava flows from Nyamuragira set fire to large areas of forest, the fires trapping numerous antelope which were later found completely carbonised (Coleman, 1946: 168). Livestock have also been killed by lightning strikes during the 1631 eruption of Vesuvio (Hamilton, 1772: 164) and the 1902 Soufrière, St Vincent, eruption (Anderson and Flett, 1903). During the 1924 Kilauea eruption a cow that fell into a crack which opened during an earthquake was reportedly killed when the crack closed, crushing the animal. According to one account, a photograph showed the cow's legs protruding from the ground (Mullins, 1977: 23). Certainly, many cattle were killed by the tsunami produced during the 1883 Krakatau eruption (*CIRV,* 1979: 796) and tsunami resulting from the 1827 eruption of Unimak Island (Alaska) killed hogs on neighbouring Chernobour Island (Dall, 1870: 335). However, such volcanic agencies have been responsible for relatively few mammal deaths when compared with pyroclastic flows, lahars, tephra falls and gases.

Gases and aerosols have been responsible for a large number of animal deaths. Even before the May 8, 1902, eruption of Mt Pelée sulfur fumes were so strong in the streets of St Pierre that horses fell down dead (Heilprin, 1903: 39, 62). Accumulation of CO_2 and H_2S in Death Gulch at Yellowstone in 1897 killed hares, squirrels and eight bears (Jaggar, 1956: 11). After the 1947 Hekla eruption CO_2 accumulated to sufficient depth in depressions to kill 15 sheep, and a fox. These CO_2 "lakes" (mofettes) formed in calm weather on old lava flows, the gas issuing from ground water saturated with CO_2. About 100 mofettes formed; after they were fenced off only dead birds and insects were found (Kjartansson, 1957).

Of far more significance than gases, however, are fluorine aerosols attached to fine tephra (see Chapter 2, Section 8). Fluorine poisoning of sheep following the 1970 Hekla eruption occurred across a very wide area, even where only 1 mm of tephra fall was recorded (see Fig. 2.29). Poisoning evidently occurs where the fluorine content of dried grass exceeds 250 ppm. As contents up to 4000 ppm were recorded it is not surprising that 1500 ewes and 6000 lambs died. However, it should be noted that many of the sheep were in poor condition before the eruption (*SICSLP,* 1970: 55; Thorarinsson and Sigvaldason, 1972: 275). Before death, fluorine poisoning causes lesions in the nose and mouth and hair to fall out from around the mouth (Nielsen, 1937: 9). Examination of stock during and after the 1947 Hekla eruption showed that only sheep were affected although cattle, horses, dogs, cats and poultry lived in the same area. Lesions on legs were much less common than those on jaws, the appearance of nodules in the legs of the afflicted sheep evidently indicating advanced cases. Many sheep appeared lame but their condition changed from day to day and their recovery rate was high. Sheep fed 20–40 mg of fluorine daily for six months developed fluorosis, the experiments suggesting that a mild degree of chronic fluorosis could be produced by feeding sheep 1 mg of fluorine per kilogram body weight per day for that period (Sigurdsson and Palsson, 1957).

Fluorine poisoning has also been an important cause of deaths and injuries in livestock following other Icelandic eruptions. Livestock died, probably as a result of fluorine poisoning, after the 1934 Grimsvotn eruption (Nielsen, 1937: 9), and after the 1693, 1776 and 1845 Hekla eruptions (Thorarinsson, 1970: 15, 17). However, the greatest death toll was recorded after the 1783 Lakagigar eruption (see Fig. 2.4). Animal deaths resulted from fluorosis, other sicknesses and hunger, the latter the result of a stunted grass crop caused by tephra and atmospheric pollution including SO_2 and SO_3. In all, about 50% of the horses and cattle and 79% of the sheep in the whole of Iceland died (*Bulletin Volcanologique* Supplement, 1974, 38: 57). Herbet and Bardossi (1968: 56) give the numbers killed as 11 500 cattle, 28 000 horses and 190 000 sheep. Stephensen's (1785) account provides an exceptional view of the suffering of livestock as a result of the Laki eruption:

> All the quadrupeds of the island had thriven wonderfully, and gained strength, during the mild winter and beautiful spring of 1783, but this did not prevent them from dying off in considerable numbers, during the week or fortnight immediately subsequent to the eruption, with inflammatory diseases caused by the poisonous quality of food. Such was particularly the case with the sheep, of which, in the district of Skaptefield, it was remarked that, whereas in Iceland they generally walk facing the wind, they now regularly turned away from it; naturally anxious to avoid the strong sulphureous smell, which the infected breezes brought along with them. As the cold, too, at a distance from the fire, was unusually piercing, they instinctively approached the current of lava, by which many of them were overwhelmed and destroyed, in spite of all the exertions that were made to save them. Nor was the situation of the cows and horses much better; for, although the disease was to them not equally fatal, yet they became excessively lean, and, even in the best season of the year, the cows gave scarcely any milk... (229-230).
>
> In addition to the inflammatory disease just mentioned as so fatal to the sheep: so early as the commencement of autumn, 1783, when they were collected from the hills, several of them were found to be attacked with a distemper hitherto unknown to the natives. The poor animals could neither walk nor stand: their teeth were loose, so as to prevent them from chewing their food: their cheeks were full of swellings; and their joints were contracted. Towards Christmas the sickness began to shew itself in a still greater degree, even among the stall-fed sheep, and also among the horned cattle, which rendered it necessary for them to be slaughtered. Many, however, fell victims to the distemper much sooner than was expected, when the disease attacked them internally. Thus it was only found that the heart, liver, lungs, and kidnies of those miserable animals were covered on all sides with boils and ulcers: they were in some cases much swollen, in others quite destroyed and hollowed out: one of the kidnies was frequently considerably distended, while the other was porportionably shrivelled. The jaw-bones were perforated, as if they had been bored with an instrument, and the ribs were knit together in a most extraordinary manner. The bones were reduced to a substance resembling gristle, and even the hardest became at the joints so tender, that they might easily be separated from each other... (232-233).
>
> No striking external marks of the disorder were perceptible among the horses, out of the district of Skaptefield, but it had nevertheless prevailed there, if not as the sole cause, yet certainly in union with others, to produce a general destruction both among them and the horned cattle: many having died suddenly, when they had a plentiful supply of hay; others when in pastures where there was a sufficiency of grass, of which they were never deprived either by ice or snow. To our utter astonishment, we saw horses in the most miserable state of leanness, in the richest meadows, and even actually starved to death, having preferred eating substances the most injurious, such as the wood of houses, the hair from each other's coats, or whatever else was within their reach, rather than touch the grass of last year's crop, still remaining in the pastures. This appears to me to be a sufficient proof of the poisonous state of the herbage, during the year 1783; and, although the circumstance has not yet been investigated, I am full convinced that the entrails of the horses, have been equally, with those of other animals, infected with the distemper. The few inhabitants, who had still left them some of the old hay, of the year 1782, preserved their cattle in a healthy and good condition; but even here, when the new hay came into use, the disease began to appear among them.
>
> [Stephensen, 1785: 235-236]

Pyroclastic flows and lahars have also taken their toll on livestock. Table 7.7 presents some cattle death tolls for Indonesian pyroclastic flows and lahars. Generally, it is not possible to ascribe death to only one volcanic agency. Brief examples can also be cited from other areas. Around Legaspi during the 1897 eruption of Mayon, numerous cows, horses, pigs and carabao (water buffalo) were burnt to death in pyroclastic flows (Anon., 1897). Pyroclastic flows produced during the

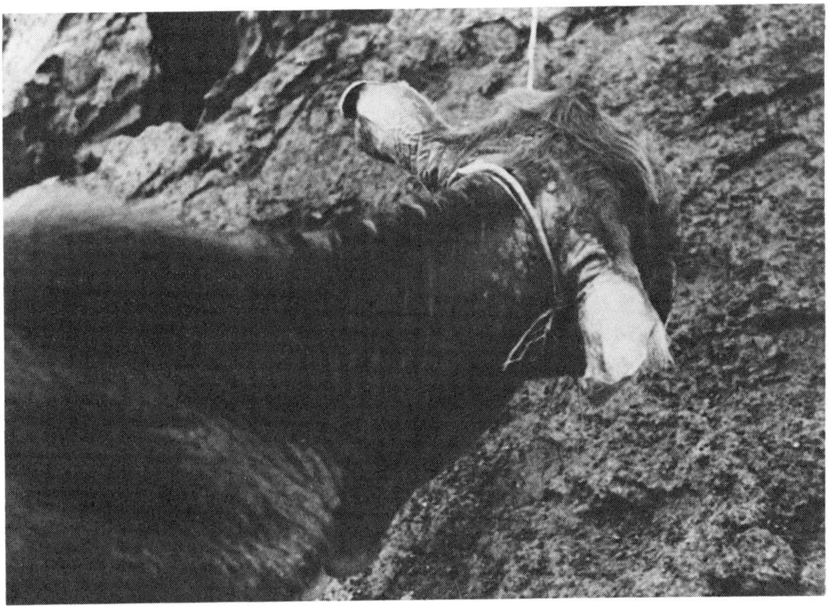

FIG. 7.9 One cow was found alive but with burnt and battered ears following the 1965 eruption of Taal (Photo: US Geological Survey P10-65-70L[2d]).

1914 Sakura-jima eruption killed more than 20 horses and pigs at Hakamagoshi on the western tip of the island (see Fig. 3.19) (Omori, 1916). The 1911 Taal eruption killed about 700 animals including pigs, goats, horses and cattle (Rivers, 1911: 14). In the 1965 eruption all the cattle on Taal island were killed except for one beast found alive in a semi-protected area but with burnt and battered ears (Figure 7.9). On the outskirts of Mambajao in 1951 following the Hibok-hibok pyroclastic flows, a *Time* reporter observed a carabao bathed in ash and dead but still standing (*Time*, December 17, 1951: 36). The Washington Department of Game estimated that casualties in pyroclastic flows around Mount St Helens included 11 000 hares, 6000 black-tailed deer, 5200 elk, 1400 coyotes, 300 bobcats, 200 black bears and 15 mountain lions (Del Moral, 1981: 58).

Table 7.7 Cattle deaths in some Indonesian eruptions

Eruption	Volcanic hazard	Death Toll	Reference
Galunggung, 1822	Pyroclastic flow	958	*CIRV*, 1979: 793
Merapi, 1930	Tephra fall, pyroclastic flows, lahars	2 140	*CIRV*, 1979: 798
Dieng, 1944	Tephra fall	332	*CIRV*, 1979: 798
Iliwerung, 1948	Pyroclastic flow	300	van Bemmelen, 1949: 27-28
Agung, 1964	Pyroclastic flows	18 801	*CIRV*, 1979: 800

Downstream, on the Cowlitz Valley bottomlands, five prize heifers floated 0.8 km down valley and became trapped in mud 1.5 m deep, unable to move, 200 m from the edge of the lahar. Three bulls and a heifer were also trapped in a barn on the lahar. All were eventually rescued (*The Columbian*, May 26, 1980: 11).

Household pets may have a fairly strong tendency to survive pyroclastic flows. One newspaper report said that the only creature found alive in the Mount St Helens pyroclastic flow zone was a dog, apparently with a family killed by the blast (*The Columbian*, May 26, 1980: 10; EH & WS, 1980). Similarly, Taylor (1958: 50) reported the survival of a dog on the margin of the zone of total devastation around Lamington in 1951. Cases were also noted in St Pierre in 1902, following the May 8 eruption of Mt Pelée, where cats and dogs were found alive in closed rooms where all humans had succumbed (Lacroix, 1904: 287 in Taylor, 1958: 50). According to Coleman (1946: 124), while all 11 sulfur workers were killed in the White Island lahar of 1914 which destroyed all the buildings, the pet cat survived.

Jökulhlaups have also killed large numbers of stocks in Iceland. In 1727 about 600 sheep and 150 horses were killed by jökulhlaups from Öraefajökull and tephra (including projectile) fall (Thorarinsson, 1958). Similarly, in 1973, hundreds of animals were reported buried by mud in the Huemules Valley following fumarolic activity on ice-clad Mt Hudson (*SICSLP*, 43-73:1603).

In contrast to the rather sketchy accounts of the effects of pyroclastic flows and lahars on mammals, the impacts of tephra falls are quite well documented.

By 1945, the eruption of Parícutin, which began in 1943, had killed 4500 cattle, 500 horses and an indeterminate number of sheep and goats as a result of breathing volcanic ash. Animals that were initially exposed to, but subsequently removed from, the thick tephra zone often died months later. Post-mortem examinations often revealed that the animals' lungs had an ash–mucous coating that interfered with respiration (Rees, 1970: 13). The mucous coating probably resulted from aerosols attached to the tephra rather than from the tephra itself (Dr M. Murphy, Yakima otorhinolaryngologist, May, 1981, personal communication).

The initial stages of the long-continued Irazú eruption of 1963–1964 killed about 2000 cattle (*Newsweek*, October 7, 1963: 65). About 6000 head of cattle grazing on the slopes of the volcano, had to be moved down to the plains because some 6700 ha of good pasture land had been ruined by tephra. This produced some crowding of animals on the plain: "Control programs for cattle diseases have been intensified to compensate for the increased dangers caused by the crowding. Some increase in slaughter in addition to the normal culling of the herds has been required owing to illness from the ash and also from the moving. Cattle illness due to the ash has been mainly intestinal irritation by swallowed ash with resulting diarrhea. No respiratory illness has been seen. Illnesses from moving are mainly increased abortion and nutritional disturbances, thought to be caused by bringing animals acclimatized to high altitudes down to the level of the plain. No evidence of injury to cattle from volcanic gases has been reported at any time . . . Milk production has decreased to slightly below current requirements of the community for fresh milk, but no shortage of milk is anticipated. Production will not return to the normal level for several years because of increased slaughter, abortions and deceased breeding" (Horton and McCaldin, 1964: 928).

Similar effects resulted from the Arenal 1968 eruption. About 800 cattle died in the eruption, 80 000 were ordered slaughtered for immediate sale, and a further 100 000 had to be moved to new pasture lands (*Time*, August 9, 1968: 27-28).

Apart from evacuation, respiratory and intestinal problems and some longer term diseases, cattle and other large mammals may also suffer from teeth problems following tephra fall. Around Parícutin, animals suffered from the grinding away of teeth surfaces as a result of eating tephra-laden vegetation (Bolt *et al.*, 1975: 79). Reindeer suffered the same problem following the eruption of Aniakchak Crater in 1931. Tephra covered the moss on which the reindeer fed. The adults moved on in the search for food leaving the fawns to die. The moss which the reindeer eventually did find to eat was so coated with gritty tephra as to grind their teeth down to the gums (Trowbridge, 1976: 73).

Reindeer were also seriously affected by the Katmai-Novarupta eruption of 1912 even in areas where as little as 25 mm of tephra fell (Martin, 1913: 180). Other animals were also affected. It was reported that some bears were blinded by the ash. Around Kodiak, where more than 300 mm of tephra fell, surviving bears killed a cow and a calf. Foxes also came to the edge of town searching for food two weeks after the eruption (Martin, 1913: 180; Erskine, 1962: 205). Cattle evidently survived the tephra fall at Kodiak quite well despite little water and no food. They obtained water from rain and ice, some food from the shoots that poked through on steep slopes, and salt from kelp beds on the shore. They suffered from severe nasal infections and running eyes, but they did survive. The Galloway herd at the Agriculture Experiment Station was sent elsewhere in the United States for about two years (Erskine, 1962: 151, 209; Griggs, 1922: 44). Small mammals did not, however, survive the Katmai-Novarupta tephra fall nearly as well. At Kodiak ermines and marmots almost all died (Erskine, 1962: 210). At Iliamna Lake (Fig. 7.8) where less than 100 mm fell, many rabbits died. Martin (1913: 180) also reports the death of squirrels, marmots and mice at Cold Bay where the tephra fall was only 6 mm.

Similar effects have been reported elsewhere following tephra falls. In eastern Java where about 230 mm of tephra fell from Tambora in 1815 (Fig. 2.13), 126 horses and 86 cattle died "chiefly from want of Forage during a month from the time of the Eruption" (Ross, 1816: 349). Four days after the 1812 eruption of Soufrière, St Vincent, cattle were reported to be starving for want of forage as not a blade of grass could be seen (Huggins, 1902: 28). Following the Tarawera eruption of 1886, even on the coast where only 80 mm of tephra fell, cattle moaned with hunger, their parched tongues and nostrils filled with mud, and their eyes bloodshot. At Te Wairoa, just across the lake from Tarawera, rats and mice "made bold by hunger" were seen hunting for food in the ruined buildings. A breeding sow reportedly dug its way out after three weeks of burial. A cat also survived burial for several days but others (Fig. 7.10) were not so lucky (Grayland and Grayland, 1971:

FIG. 7.10 A mummified cat found in 1901 at Wairoa buried in tephra by the 1886 Tarawera eruption (Photo: The Alexander Turnbull Library, 82159 ½).

101-103). Many cats also died during the 1973 Heimaey eruption, not as a result of burial but because the cinders were needle-sharp. Although most animals were evacuated with the townsfolk, other cats left behind could not be caught until they had bleeding paws and were half dead with hunger (Gunnarsson, 1973: 76).

Around Kodiak and Tarawera, with tephra falls of more than 300 mm and more than 80 mm respectively, the effects on cattle were quite minor; thus, we would expect the effects around Mount St Helens to be similar or of less significance. Although information is available, much of it is rather unsatisfactory. Table 7.8 summarises data from *Ash and Agriculture* for June and July 1980. Problems arise because: (i) data were presented by counties but tephra thicknesses varied markedly across individual counties; (ii) tephra thicknesses reported on the map accompanying *Ash and Agriculture* are grossly exaggerated (thicknesses given in Table 7.8 are uncompacted thicknesses abstracted from the maps of Sarna-Wojcicki *et al.*, 1981 — see Figure 3.8); and (iii) more than one tephra fall is represented in the table. The table perhaps tells us more about the reporters than the reported but the effects are in general agreement with those described following other tephra falls. It can also be noted that some of the problems are similar to those experienced by humans in the same area (see Chapter 3, Section 5.2). While livestock deaths were reported in some areas, most such reports are unsubstantiated. Specific causes of deaths have not been determined and may be unrelated to tephra fall. However, it is possible that some deaths resulted where animals ate poisonous plants which appeared more attractive than tephra-laden grass (McReynolds *et al.*, 1980: 5).

Perhaps the more important aspect of the Mount St Helens tephra fall is that several experiments were set up to test the effects of tephra on livestock. Young calves were fed a diet including 10% ash for three weeks but their stomachs and intestinal tracts showed no ash accumulation and no tissue damage. A five week trial on cows using an ash ingestion rate up to 1.5 kg per day had no observable

Table 7.8 Effects of Mt St Helens tephra fall on livestock

Tephra thickness (mm)	Reported effects
30-40	No livestock health problems
30	Some eye irritation in small calves; eye disease greater than normal; some respiratory problems, young and old animals reported as having most difficulty; some diarrhea in cattle; reduced milking giving lower calf weight
15	Livestock in general doing well No noticeable dairy health problems
4	Some cases of animal respiratory problems; eye irritation, minor diarrhea
1	A few early reports of cattle refusing pasture

effect on milk production or milk composition although feed intake was slightly suppressed at the highest tephra level. There was no visual damage to the digestive system. It also appeared that concentrations of heavy metals (As, Cd, Co, Cu, Mo) in the tephra were below levels toxic to livestock (Cook *et al.*, 1981: 20; *Ash and Agriculture* July 22, 1980: 5). Few problems were reported amongst domestic pets. A few dogs developd wheezes and coughs but the tephra was not their primary problem. Some cats developed diarrhea as a result of licking ash from their fur. In some cases, this led to prolapsed rectums (*DVM, the News Magazine of Veterinary Medicine*, 1980, July 11 (7): 40-41).

There are some difficulties in reconciling the various reported effects on large mammals discussed in this section, particularly as far as tephra falls are concerned. The higher mortality rates reported at Parícutin and Irazú when compared with Mount St Helens, Katmai-Novarupta and Tarawera may result from: (i) increased exposure resulting from the prolonged eruptions; (ii) greater accumulation of tephra; (ii) toxic components within or attached to the tephra; (iv) prior stress in the animals; or (v) any combination of these. Without further information we can only conclude that, in many cases, tephra falls of less than 50 mm are unlikely to unduly stress cattle, and that falls of 50–300 mm or more can be readily survived by the majority of animals. These limits are probably too high for close-grazing animals such as sheep and reindeer which may find it more difficult to obtain suitable food and which also suffer increased wear of teeth as a consequence of grazing tephra-mantled vegetation. These limits may also be incorrect for carnivores and omnivores where small animal prey have been selectively killed.

7.4 AQUATIC LIFE

Aquatic life is affected in a few direct ways by volcanic eruptions — by increased acidity, increased turbidity, changed temperature or changes in food supply. Frequently, more than one of these conditions is altered during an eruption but only occasionally are reports sufficiently detailed to allowed determination of the cause of death or injury.

Acidic waters may have a number of effects on fish as reported by Bortleson *et al.* (1977), following the release of acid waters from Mount Baker in 1975. When acidity can be measured at levels less than pH 5.0 it is lethal to the eggs and fry of salmon whereas waters with a pH of 6.0–6.5 are unlikely to be toxic. They also note that H_2S toxicity to fish is greatly increased in acidic conditions but that the ultimate effects of low pH on fish may also depend on factors such as temperature, dissolved oxygen, the concentration of various other dissolved constituents and on the prior acclimatisation of the fish to acid conditions. The timing of an infusion of acid water may also be important as a sudden drop in pH when fish are spawning or when eggs are hatching may cause malformation or death even at a pH higher than that normally harmful to adult fish (Bortleson *et al.*, 1977: 29).

Less detailed accounts suggest that increased acidity has been responsible for fish kills. In 1969 a sulfurous lahar flowed into the upper tributaries of the Wanganui River from Ruapehu. At least as far as 77 km downstream dead trout and eels more than 2 kg in weight were found along the banks of the river. Others were still alive but their skins peeled when they were picked up, suggesting that the water was quite acidic. Although some trout were found alive in small tributary streams these contained no appreciable food source; it was believed that re-establishing the trout would be difficult. Similar fish kills were reported after the 1945 Ruapehu and the 1983 Rincón de la Vieja eruptions (*SICSLP*, 1969: 46; *SEAN Bulletin*, 8 (3): 17). The January, 1979, eruption of Karkar killed two species of fish, freshwater prawns and eels in rivers draining south-east from the caldera. Although less than 10 mm of tephra fell in the area where hundreds of dead fish were found, tephra falls in the headwaters were certainly heavier; some small secondary lahars may even have occurred. Some fish, eels and prawns were still alive. All had a strong sulfurous smell but local subsistence farmers reported that they were edible.

Lava flows into water and even primary lahars may increase water temperatures sufficiently to kill fish and other marine life. At Santorini in 1707, dead fish found on the shore were believed killed by the elevated water temperature. However, live oysters, prised from the growing dome itself, were found to be delicious (Bourguignon, 1708: 202). Similarly many fish were believed killed by high water temperatures (and increased acidity) during the 1925 Santorini eruption (Washington, 1926: 359).

Around Bogoslof in 1884 no fish were found in the warm water or in nearby waters of normal temperatures even though repeated attempts were made to catch them. Among the sea lions that were observed near the volcano there were a number that had been scalded so that their hair had come off. Merriam (1902: 303, 316) believed that many sea lions had been killed. In 1925 Beebe (1926: 254-256) measured water temperatures near the Albemarle (Galapagos) lava front at 27°C–37°C compared with the surrounding sea at 24°C. Tunnies in a large school were observed swimming panic-stricken toward the cooler water while a few small fish drifted by on their backs and sea worms writhed and twisted. A full-grown sea lion was seen to suicide on the lava after getting into the hot water. At Matavanu in

1905–1906 sea water boiled as far as 100 m from the visible lava front. Fish that were killed by the boiling water were collected and eaten by the Samoans (*Nature*, 75: 351).

A variety of effects have been reported where tephra has fallen into rivers and lakes, presumably increasing turbidity. Around Tarawera in 1886, numerous small fish were washed to the lakes' shores dead or dying. Many were partially discolored and had a bruised appearance (Mair, 1886: 374). The Hekla 1693 eruption killed trout and salmon in the lakes and rivers while fishing was spoilt by the 1776 eruption where the tephra fall was 20–40 mm (Thorarinsson, 1970: 15). A 30 mm fall at Macassar from Tambora in 1815 killed many pond fish (Anon., 1816: 253). The 300 mm tephra fall at Kodiak in 1912 from Katmai-Novarupta proved lethal to spawning salmon; it took a number of years for the spawning grounds to become fertile again (Erskine, 1962: 153, 221). Martin (1913: 180) reported that some of the small fish in Alaskan creeks were killed and fish in lakes were driven offshore into deeper water.

In the seas around Katmai and Kodiak still other effects were noted. The halibut fishing grounds around Kodiak Island were evidently unaffected by the deposition of 300 mm of tephra but the floating mats of pumice frightened the fish away (Erskine, 1962: 204, 208). In Katmai Bay (see Fig 7.7) where more than 600 mm of tephra fell, barnacles and mussels as far down as the low tide mark were mostly dead and kelp appeared dead as far as the eastern end of Afognak Island where about 90 mm fell (see Fig. 7.8). Cod and halibut were reported to have died in great numbers in the shallow waters of Cook Inlet (Martin, 1913: 80).

Even though tephra falls over water from the 1947 Hekla eruption were relatively minor, cod seemed to move into deeper water for several days (Thorarinsson, 1954). On the other hand, the relatively minor tephra fall into the Bay of Naples from the 1906 Vesuvio eruption reportedly caused much suffering to both the floating fauna and some of the bottom life (Lacroix, 1906: 235). Similarly, the 1952 eruption of Barcena, which produced a great deal of pumice, had severe effects on marine life. While offshore fishes probably did not suffer, intertidal fishes and marine invertebrates appear to have been seriously depleted, presumably as a result of the abrasive action of pumice (Brattstrom, 1963:512-514).

Much more detailed studies have been made of the effects of the 1977 Usu eruption on marine life. Before-and-after collections indicated that there was little effect on the composition of the benthic invertebrate fauna in Usu Bay. However, the tephra fall was only a few millimeters (Yamada, 1978). Similarly, comparison of the physiological condition of cultured scallops in areas where tephra did and did not fall revealed no evident differences (Suzuki, 1978; Fuji and Nakao, 1978; Taniguchi *et al.*, 1978). Furthermore, while turbidity must have increased greatly as a result of tephra transport to Funka Bay by the Kimonbetsu River at Date city there seemed to have been no effect on estuarine fish communities in terms of the numbers netted or the biomass caught (Mishima *et al.*, 1978).

The popular fishing areas of the Toutle and Cowlitz Rivers were severely affected by the lahars produced by the May 18, 1980, Mount St Helens eruption. A wide range of effects were reported. Certainly, many fish were killed by the high temperature of the mudflows into the Cowlitz. On May 19 the temperature in the lower Cowlitz (below the Toutle confluence — see Fig. 4.2) was reportedly 32°C, too hot for fish to survive. Summer-run steelhead trout and spring chinook salmon repeatedly lept into the air or deliberately beached themselves. However, adult chinook salmon scheduled to travel up the Cowlitz to spawn evidently swam past the mouth of the overheated river and ascended the adjacent Kalama River. On the Toutle River one hatchery lost three million coho and chinook fingerlings. In all, more than 240 km of trout stream and 50 lakes were severely affected by the eruption; some will undoubtedly take years to recover. In Spirit Lake alone, it was estimated, 126 000 trout were killed, the total for 26 affected lakes reaching 11 million fish. Although hatchery fish were replaced fairly easily the wild stock fish could take years to recover. The eruption also ruined a good deal of fisheries research, particularly genetic studies and fish-tagging activities (EH & WS, 1980; *The Columbian*, May 26, 1980: 4, 10; Del Moral, 1981: 38). On the other hand, increased diatom abundance has been reported in Alaskan lakes after tephra fall, presumably because of the increased availability of readily soluble silica. It also has been suggested that lake fertilisation by tephra fall increased the abundance of red salmon (Barsdate and Dugdale, 1972).

As with much of the material in this volume, there is enough data to provide examples but not enough hard evidence on which to base conclusions. Certainly, tephra falls of 30 mm or less have caused large fish kills but the deaths may have resulted more from changes in pH than from turbidity changes. Although the Usu example suggests that a few millimeters of tephra will not adversely affect benthic invertebrates or economically important shellfish such as scallops we cannot be certain that this specific case represents a general rule. Changes in tephra particle size, heavy metal constitutents or attached aerosols could all influence the extent and nature of the effects. Furthermore, the effects may often be more subtle than indicated here; only one link in a food chain needs to be affected in order to create a detrimental (or favorable) effect on an economically significant population.

7.5 BIRDLIFE

An amazing variety of eruption effects on birds have been reported, although rarely in any detail. Bird migration, roosting, flying ability and feeding activity, have all been influenced by various aspects of volcanic eruptions. Although few of these effects relate to bird species of direct economic importance they do reflect possible effects on domesticated species. Furthermore, many bird species are important in that they are game birds, prey on crop pests, or are themselves crop pests.

Tephra falls from the Katmai-Novarupta eruption in 1912 and the Tarawera eruption of 1886 provide good examples of the range of effects on birdlife. At Kodiak, during the tephra fall (300 mm), small birds fell out of the air apparently gassed while large birds flew around in panic until they hit something. Afterwards, dead gulls, snipe, ptarmigan and ducks were found everywhere — at sea, buried in the ash and tangled in trees. Sea birds that survived the tephra fall could not find open water to land on, or take off from, because of the massive sheets of floating pumice. Many subsequently died or drowned (Erskine, 1962: 65, 208, 210). Even at Cold Bay where only 6 mm of tephra fell, and at Iliamna Lake where the fall was 25–100 mm, many of the small birds died (Martin, 1913: 180). The Tarawera eruption killed pigeons, ducks and sparrows in large numbers. Surviving sparrows were blinded, at least temporarily, with their eyelids gummed together by the falling mud. Pheasants had so much fine wet tephra in their feathers that they were unable to fly and pheasants and quails came almost to house doors searching for food (Mair, 1886: 374; Grayland and Grayland, 1971: 102-103).

Unfortunately, it is not possible to relate the Tarawera examples to specific thicknesses of tephra. Such is the case with most observations. Large numbers of birds are reported to have died around Etna during the 1329 eruption (Rodwell, 1878: 86) and around the Soufrière, St Vincent, eruption in 1812 (Huggins, 1902: 28). Ptarmigan died in large numbers as a result of the Hekla eruption of 1693 (Thorarinsson, 1970: 15). Numerous dead swallows were observed in areas of heaviest tephra falls (about 40 mm) at Mount St Helens in 1980 (EH & WS, 1980). Around Vesuvio in 1906 birds with wide open beaks, as though asphyxiated, lay dead where the lapilli fell thickest (de Lorenzo, 1906: 479). At Macassar, 400 km from Tambora, a tephra fall of 30 mm was sufficient to kill small birds in 1815 (Anon., 1816: 253). Tephra falls of up to 600-900 mm killed many birds which refused to, or were unable to, leave the ground or nests during the 1952 eruption of Barcena volcano on San Benedicto Island (Mexico). Brattstrom (1963: 505, 516) reported that the skeletons of birds, some on nests with eggs, were revealed by erosion of the tephra in late 1953. Bird populations on the island may have been reduced to one per cent of pre-eruption levels.

The frequency of such examples suggests that tephra falls of as little as a few millimeters can kill at least some of the bird population and that small birds are the more vulnerable. However, sulfurous fumes and other gases may have been responsible for bird deaths in some of the examples cited above. For example, Hamilton (1772: 107) makes reference to the deaths of birds at the Phlegrean fields and during various eruptions of Vesuvio. Jaggar (n.d.: 53) reports that numerous dead birds were found around the Mt Pelée vent several days before the August 5, 1851, eruption. Numerous wheatear, meadow pit-pit, redwing, ravens, great black-backed gulls and mallards were killed by CO_2 accumulation in depressions (mofettes) around Hekla following the 1947 eruption (Kjartansson, 1957: 8, 16).

Bird deaths may also have resulted indirectly from tephra falls when food sources have been destroyed. Tephra fall from Mount St Helens decimated many insect populations (see Chapter 7, Section 6). As the eruption occurred in the spring this was expected to have a deleterious effect on the populations of young chukar partridges, pheasants, quail and songbirds (EH & WS, 1980). At Heimaey in 1973 attempts were made to feed snowbunting where they converged in large flocks. However, the feed soon became covered in tephra and the birds gradually became fewer in number (Gunnarsson, 1973: 76).

On the other hand, food sources for some bird species may be enhanced during some volcanic eruptions. Beebe (1926: 255) reported that five or six species of birds were attracted by floating fish and other organisms, the noises, and the clouds of steam, to the warm water adjacent to the lava at Albemarle in 1925. Frigate birds and shearwater, in particular, dived into the near boiling water. Some, however, were killed.

Other birds have also taken advantage of warm rocks near volcanoes as nesting sites. Ground-nesting species in New Britain and on Niuafo'ou deliberately seek out geothermally-warmed nesting sites, utilising the heat as an aid to incubation of eggs. At Bogoslof Island in the Aleutians, courting sea birds nested wherever it was cool enough. Murres may have relied on the warm rock to aid incubation. One observer reported that it was remarkable that the birds should have voluntarily taken up quarters in places where hot steam and fumes were almost suffocating. Some evidently died as "such as flew into the cloud of steam and smoke of the belching volcano, as many did, immediately perished" (Merriam, 1902: 330).

Other eruptions have disturbed roosting behavior. The 1967 eruption on Deception Island covered east coast penguin rookeries apparently retarding the breeding cycle. At the Bailey Head rookery tephra covered the small stones used for lining the nests, creating intense competition for the few stones available. The less successful birds had to fetch stones from the beach 100 m below the nesting site. On the west coast of the island, where no tephra fell, the breeding cycle of the rookeries proceeded normally (Baker, 1969: 119). The 1925 eruption of Santorini affected the migration pattern of quail *en route* to North Africa. The quail habitually use the island as a resting place but the noise, smoke and tephra from the eruption caused the quail to abandon the island (Washington, 1926: 365).

While nearly all of these examples concern species that are not directly of economic significance a few reports provide evidence of effects on domestic fowl. During the 1973 evacuation of Heimaey, many domestic chickens were evacuated to the mainland along with their owners. However, of the 1100 chickens transported only 129 survived (Gunnarsson, 1973: 68,72). At Kodiak in 1912 nearly all the domestic chickens died, either gassed, or smothered by the 300 mm tephra fall (Erskine, 1962: 210). After the May 1980 Mount St Helens eruption concern developed about the effects of tephra on domestic fowls. A seven week trial in which tephra made up to 30% of the rations of broilers produced no toxic effects. However, day-old chicks, highly sensitive to diet, showed a 6% decrease in growth rate for each 10% increase in the tephra content of their food. This can be compared with a 4% reduction in growth rate for each 10% increase in sand in their diet (Cook *et al.*, 1981: 20; *Ash and Agriculture*, July 22, 1980: 5). The most bizarre observation concerning domestic fowls comes from a reporter surveying the damage on the outskirts of Mambajao caused by pyroclastic flows during the 1951 Hibok-hibok eruption: "As I left a chicken crossed my path. Its right side was grey and seared, the other side was untouched. It was the only living thing I saw there" (*Time*, December 17, 1951: 36).

7.6 INSECTS AND OTHERS

As with most aspects of the effects of volcanic eruptions on agriculture, tephra falls are the most important hazard affecting insects, primarily because tephra can cover such a wide area. As some of the examples presented here indicate, there may be as many benefits as costs depending on which insects are most severely affected. It is also apparent that the balance of beneficial and deleterious effects on insects and, hence, on crops, is very dependent on the timing of the eruption in relation to stages of insect (and crop) growth.

Cook *et al.*, 1981, provide an excellent summary of unpublished work on the effects of tephra fall on insects around Mount St Helens in 1980: "The ash particles were especially destructive to insects, largely due to abrasion to the epicuticular wax layer, which caused rapid desiccation and death. This wax layer... is only 0.25 μm thick on the average and is the principal means of protecting the insect against desiccation. The problem is compounded by the insects' large surface area relative to their size. Adult house flies (*Musca domestica*), brown-banded cockroach nymphs (*Supella longipalpa*), and adult orchard, mason bees (*Osmia lignaria*) lost 28 to 66 percent of their body moisture and 50 to 100 percent died within 4 to 8 hours after exposure to ash at 24.5°C and 31 percent relative humidity. In contrast, great wax moth larvae (*Galleria mellonella*) and mealworm larvae (Tenebris molitor) were only slightly affected by exposure to the ash. Yellowjackets (*Hymenoptera*) exposed to the ash on 18 May, when brought to the laboratory and given access to water, still died within 3 to 6 hours. The ash on insect bodies also affected their ability to fly and is a continuing problem for their survival in the area" (1981:18).

Crop pests killed by the Mount St Helens tephra fall included the Colorado potato beetle (*Leptinotarsa decemlineata*) and grasshoppers (*Melanoplus spp*). Adult Colorado potato beetles did not recover after 24 hours' exposure to the tephra except where overhead irrigation occurred within a day. Some growth stages were more sensitive, and some more resistant, to the tephra than the adult stage (Cook *et al.*, 1981: 18). Grasshoppers in Whitman, Adams and Lincoln Counties, where tephra falls ranged from less than 5 to 40 mm (uncompacted thickness), were reported almost completely killed by mid-June, their demise possibly assisted by wet cold weather in some areas. A pea leaf weevil infestation in Whitman County was also reported much reduced (*Ash and Agriculture*, June 10, 1980: 2, 4; June 17: 3). However, other crop pests including aphids and leafrollers, were in less vulnerable stages of their life cycles or were protected by plant leaves, thus escaping serious damage. Grain beetles were reported to be relatively unaffected by the tephra fall, possibly because they are already adapted to dusty environments (Cook *et al.*, 1980: 18). Around Yakima severe mite infestations on ornamental plants and some vegetables were believed to be due, in part, to the dustier conditions (*Ash and Agriculture*, August 26, 1980: 2).

Beneficial species of insects were also severely affected by Mount St Helens tephra. These included honey and pollinator bees, predatory and parasitic wasps and yellowjackets. These insects suffered abrasion of the surface wax layer, trapping of the dust in dense coverings of body hair and plugging of tracheal tubes. Yellowjackets and bumble bee populations were also severely reduced because queens were killed before their annual nests were established (Johansen, 1980: 500; Cook *et al.*, 1981: 18). It was estimated that 12 000 out of the 15 000 honey bee colonies in the worst-hit tephra fall areas were destroyed (Johansen, 1980: 501) but some recovery in populations did occur in June and early July, 1980. While pollination of fruit trees in the Yakima Valley had been completed before the tephra fall, some loss of alfalfa seed production was expected as pollination had not started before May 18 (Cook *et al.*, 1981: 18).

The "mixed blessing" evident in the selective destruction of insect populations by the Mount St Helens tephra fall was earlier noted around Parícutin. Twenty kilometers west of the volcano (see Fig. 4.4) mangos and guavas were able to grow more successfully for several years due to the temporary elimination of a destructive fruit fly. On the other hand, in the same area, the tephra fall evidently killed a beneficial insect which preyed on a sugar cane borer. A plague of cane borers caused a widespread loss of cane harvests in 1944 and early 1945. These effects occurred in an area that received considerably less than 100 mm of tephra (Segerstrom, 1950: 22; Wilcox, 1959: 454; Rees, 1970: 13; Bolt *et al.*, 1975: 78).

Another interesting contrast in effects is provided by considering the fate of mosquitoes. Martin (1913: 180) reported that around Katmai in 1912 mosquitoes were entirely exterminated throughout the greater part of the district in which tephra fell. On the other hand, a Guatemalan plantation owner, A.H. Gehrke, reported to Tempest Anderson concerning the aftermath of the 1902 Santa Maria eruption: "One of our greatest troubles was that of sickness, owing to the balance of Nature having been upset by the eruption, which, having killed all the birds for some hundreds of miles, enabled the flies, mosquitoes, and rats to multiply to such an extent that life to man became nearly unbearable. The immediate consequences was an epidemic of malaria, which cost more lives than the eruption itself — many times more. It has passed, happily; the birds having come again, the breeding of these pests is checked, and the district again enjoys the excellent reputation for health that it deservedly had before" (1908:488). Similarly, the 1937 Rabaul eruption, which caused minor areas of subsidence and numerous waterlogged bomb craters, developed ideal grounds for mosquito breeding. The Administration worked hard to avert a malaria epidemic (Cilento, 1937: 48-49).

Other reports concerning the effects of eruptions on insects and other organisms are of minor significance but, nonetheless, of some interest. At Matavanu in 1905–1906 the surface of the cone was covered with the countless bodies of moths, attracted by the glow of the crater and killed by vapors rising from cracks in the surface of the lava flow (*Nature*, 1907, 75: 351). At Kilauea in 1924 a tephra fall of about 12 mm around the observatory on the night of May 16 killed hundreds of earthworms, probably, Finch (1924: 219) believed, because the coating of dust blocked pores rather than resulting from any poisonous constituent in the dust. Similar observations were made around Vesuvio in 1906: "where the ash was light and uncompacted numerous and varied trails were to be seen made by lizards, small snakes and coleoptera. At the end of the trails made by the last-named, the animal itself could often be found still living or entombed in the ash" (Lacroix, 1906: 235). However, the most interesting story concerning this eruption was told by Frank Perret (1950: 125) who described how countless living caterpillars fell with the tephra at the volcano observatory.

7.7 RECOVERY

Agricultural areas overrun by lava flows, lahars, pyroclastic flows and thick deposits of tephra will obviously take a long time to recover. Revegetation of lava flows can be expected to take decades or even hundreds of years even where environmental conditions are favorable. However, such areas are a very minor proportion of the total area affected by the deposition of volcanic material. Similarly, while pyroclastic flows and lahars are generally much more extensive than lava flows (see Chapter 2) and the apparent intensity of destruction just as great, many such flow deposits are less than 1 meter thick. Furthermore, the areas covered by lava flows and pyroclastic flows are frequently steep sparsely utilised terrain adjacent to the vent. In short, flow deposits which destroy agricultural land are relatively few in number and limited in extent by comparison with both the extent and the frequency of impact of airfall tephra deposits.

Before considering the recovery of agriculture it is necessary to make three important points:

1. Tephra compacts very rapidly, within one or two weeks, to about half its uncompacted thickness. The bulk density of the tephra changes from perhaps 0.7 g cm^{-3} to 1.5 g cm^{-3} (Blong, 1982: 138).
2. Some tephra develop a surface crust which: (i) decreases wind erosion of the tephra surface; and (ii) inhibits infiltration and encourages runoff. The reasons for the surface crust formation are not known precisely but are presumably related to particle size and particle surface chemistry.
3. Erosion of tephra deposits, encouraged by changes in bulk density and the development of a surface crust, occurs quite rapidly (Fig. 7.11). Waldron (1967:

FIG. 7.11 Vast talus slopes of Katmai-Novarupta, 1912, tephra accumulated in the upper Katmai Valley as the ash eroded from the higher slopes. This photograph was probably taken in July, 1915 (Photo: National Geographic Society, R.F. Griggs).

11) for example, indicated that one-half to one-third of the freshly fallen tephra on the upper slopes of Irazú was removed within a year. Similarly, Anderson and Flett (1903: 453) believed that more than half the new deposits around Soufrière, St Vincent, were removed to the sea within six months of the 1902 eruption.

These three changes, probably characteristic of pyroclastic flow deposits as well as tephra mantles, have important implications for crop growth. While increased bulk density and surface crusting make it more difficult for plants to push through the deposit, erosion means that some areas have been stripped of the new material or that the thickness had been much reduced. Despite the different viewpoints of Griggs (1919) and Eggler (1963) concerning plant survival around Katmai-Novarupta and Parícutin, both emphasize the importance of erosion for revegetation. Of course, erosion also means that deposition increases the thickness of the new deposit somewhere else.

Around Mount St Helens after May 1980 one of the main problems with the tephra fall was that it prevented infiltration of rain and irrigation water. Experiments indicated that even if the ash was not totally incorporated into the soil, small volumes of tephra interspersed with the soil helped to increase infiltration. Mixing the soil was usually achieved by disc ploughing or using a spring tooth harrow. One orchardist, in an area where about 12 mm of tephra fell, reported that after mixing, the soil did not "bake" as badly as it had done before the tephra fall; he also believed that mixing had improved the water intake of the soil. Another researcher

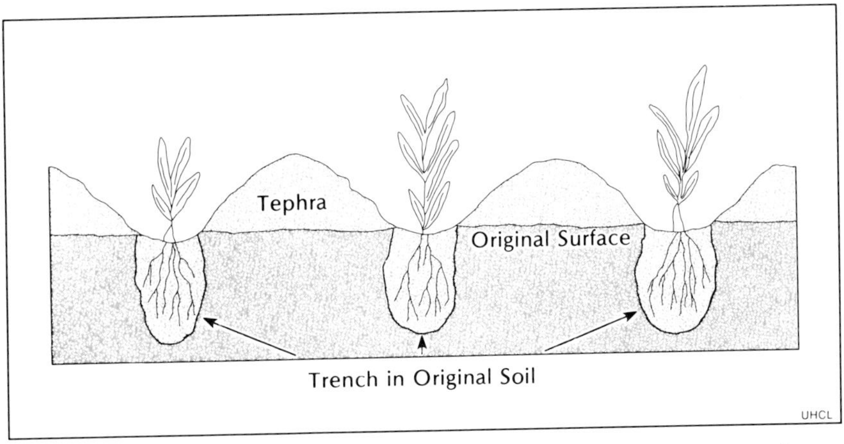

FIG. 7.12 A technique used around Soufrière, St Vincent, following the 1902 eruption to allow the successful planting of sugar cane, arrow root and other crops (after HMSO, 1903: 156).

found that a crop stubble on the field surface helped water infiltration (*Ash and Agriculture*, 1980, June 17: 8; July 8: 2; August 26: 3).

These discoveries were not new. Experiments on St Vincent after the 1902 Soufrière eruption showed that the freshly fallen tephra alone did not contain sufficient nutrients to support plant life but that fair crops could be grown providing the tephra and the underlying soil were mixed (Sands, 1912: 28-31). Another technique suggested was to pile the ash in ridges (Fig. 7.12) and to plant the sugar cane, arrowroot or other crop in trenches dug in the original soil (*HMSO*, 1903: 156). At Kodiak, following the 1912 Katmai-Novarupta tephra fall, agriculturists believed that mixing the tephra with the underlying soil improved the physical condition of the old soil which was inclined to be heavy, mucky and poorly drained (Griggs, 1918: 6-7). Pot trials conducted by Espino (1940a,b) showed that the stunted and chlorotic growth of radishes and sweet corn in a mixture of Lipa clay and 1938 Mayon tephra was not due to soil compaction or low pH and could be easily corrected by the addition of sulfate of ammonia. At Paricutin, replanting commenced in individual holes, made to break up the tephra crust and to reach the buried soil (Wilcox, 1959: 455). Other techniques, to encourage erosion of the tephra, were much less successful (Fig. 7.13). Around Agung in 1964 farmers dug holes in the tephra to mix the soil and tephra before planting sweet potato and cassava (Jennings, 1969: 114, 118).

Although revegetation in totally devastated and deeply buried areas, such as the Krakatau group of islands after 1883, probably always results from recolonisation from outside, revegetation where deposits are thin or rapidly eroded may generally be from old root stock. Such regeneration may occur even where all surface parts of plants have been totally destroyed by hot and fast-moving pyroclastic flows. It is

FIG. 7.13 Furrows were cut by the men of Zirosto down an old field in an unsuccessful attempt to accelerate erosion of the 1.4 m thick Parícutin tephra mantle (Photo: K. Segerstrom, US Geological Survey).

not intended here to enter the regeneration versus recolonisation debate (see Docters van Leeuwen, 1936; Eggler, 1959; Smathers and Mueller-Dombois, 1974), but it is worth while pointing out some examples of recovery of vegetation including crops.

The first two examples concern the recovery of subsistence crops following the passage of pyroclastic flows. At Lamington the area described is close to Higaturu (see Fig. 2.15) in the totally devastated zone where the flow velocity was estimated at 120 km h^{-1} and the temperature at 200°C for 1.5 minutes (Taylor, 1958: 44, 46): "Five days after the eruption, taro shoots appeared through the ash near the Coffee Mill and a fortnight later similar plants had penetrated the nine inches [23 cm] of ash over the gardens north of Hupo village above Higaturu [Fig. 7.14]... Within two months the monotonous grey of the devastated areas was broken by many patches of bright green where garden plants such as taro, yams, sweet potato and bananas were bursting into luxuriant growth... " (Taylor, 1958: 51). Similarly, the totally devasted Taal island was partly revegetated within a few years of the 1911 eruption. By early 1914 and in the absence of human habitation on the island, sugar cane, bananas, papaya, sweet potato, tomatoes, peanuts, cassava and a single rice plant were re-established. Some clumps of bamboo had grown to nearly their normal height and a few plants were even growing in the 1911 crater (Gates, 1914: 401-404, 421-422; Brown *et al.*, 1917).

These two examples indicate that recovery may be reasonably rapid even where the eruption left the terrain looking more like a lunar landscape than a tropical

FIG. 7.14 These taro plants (foreground) penetrated 230 mm of tephra within 17 days of the 1951 pyroclastic flow from Mount Lamington (background). (Photo: G.A.M. Taylor, Bureau of Mineral Resources, Australia, M/1759-11).

agriculturalist's garden.* It is instructive, therefore, to look at recovery after tephra deposition in an area with another climatic regime.

Thorarinsson (1971: 228) has provided a useful summary of the effects of tephra fall on agriculture in Iceland using historical records spanning almost a thousand years. He concluded that falls of 200 mm of uncompacted tephra in highland areas of Iceland rather marginal to agriculture caused abandonment of rural settlements for decades. In the more fertile lowland areas 300–500 mm of uncompacted tephra also caused lengthy abandonment. Where farms were abandoned for one to five years 150 mm or more had fallen. Absences from farms for less than a year had been associated with falls of at least 100 mm.

The temptation to generalise the applicability of Thorarinsson's data to the rest of the world must be resisted. Some of the data refer to the period of the Little Ice Age when Iceland's climate was marginal for agriculture. Furthermore, where environments are more temperate, we could anticipate that greater tephra thicknesses would be required to cause the abandonment of essentially subsistence agriculture.

*Several detailed studies of biological activity and vegetation recovery in the pyroclastic flow area around Mount St Helens were presented in a symposium "Ecological recovery at Mount St Helens" at the Northwest Scientific Association 56th Annual Meeting, Evergreen State Collge, Olympia, Washington, March 24-26, 1983.

Economic factors would have a large influence on decisions to abandon or continue other forms of agriculture.

Although the agriculturist's initial reaction to tephra fall is almost universally despair, the longer term view is frequently much less pessimistic. Several examples illustrate the viewpoint that the effects of tephra fall on agriculture are not all negative.

In referring to the tephra fall from the May 1812 eruption of Soufrière, St Vincent, Huggins wrote: "This dust was found to be a powerful fertiliser, and our neighbours in Barbados treasured it under the name 'May dust' for many years utilizing every little deposit as it was discovered and carefully applying it to the cultivation. All vegetables having been destroyed in the Carib county here [on St Vincent], the planters had great difficulty in feeding their slaves, but in the course of a few months the provisions planted grew and yielded so luxuriantly that the negroes [sic] revelled in the excess of plenty, and the whole county felt the great benefit of the fertilising properties of the fallen dust; doubtless, therefore, the ejecta of the late eruption [1902] must possess the same qualities and even now at this early period the benefit is to be seen in the rich vivid green and the rapid upward growth of the vegetation" (1902: 14).

A similar report stems from the aftermath of the 1902 Santa Maria eruption. A.H. Gehrke owned the Helvetia plantation 10 km south of Santa Maria and in an area where 2.1–3.7 m of tephra fall. After five years the tephra had eroded and compacted down to 0.9–1.5 m thick and Gehrke reported that he was anxious not to lose it as it was "useful in keeping down the growth of weeds, besides other advantages". Immediately after the tephra fall the coffee trees, 4–5 m high, had only a few leafless twigs remaining above the tephra surface. Four months after the eruption a few leaves appeared on the coffee. After spending a lot of money and effort "we have succeeded in bringing the plantation to a state of perfection that it never had before, and the crops are, as a rule, much heavier and of better quality than previously" (Gehrke, quoted in Anderson, 1908: 487).

After the 1906 Vesuvio eruption, Lacroix (1906: 235) noted that the fine ash seemed to destroy everything, but the buds on the vines were still closed and the damage to vegetation was less than expected. By the beginning of May the countryside was again luxuriantly green. Similar appraisals were made by Delmé-Radcliffe (1906), by Hamilton, (1772: 5) following the 1764 eruption and by Varenius (1683: 51), referring to an eruption of Vesuvio in the time of Vespasian (the AD 79 eruption): "But then the Conflagration ceasing in the Superficies of the Mountain here and there was great fertility of Wine". In areas around Parícutin where 30 mm of tephra fell the wheat and barley crops were reported to be excellent (Rees, 1970: 13). Some Papua New Guinea Highlanders regard tephra fall as so beneficial to crop growth and other aspects of life that they perform elaborate rituals to bring about a recurrence (Blong, 1982: 130-131; Glasse, 1965).

Such reports are not limited only to warmer climes. Icelandic farmers noted improved grass growth on low-lying lands following the 1918 Katla and the 1947

FIG. 7.15 Enhanced annual increments of growth in a small tree which experienced about 300 mm of tephra fall near Kodiak, Alaska, in 1912. The pre-eruption diameter of the tree was only 16 mm. The 1912 growth ring is scarcely perceptible in the photograph. The 1913 ring indicates recovery. In 1914 the tree produced five times as much new wood as in any previous year. (Photo: National Geographic Society, E. Wisherd).

Hekla eruptions (Thorarinsson, 1971). Luxuriant vegetation also returned after earlier Hekla eruptions (Stephensen, 1785, 225). Concerning the 1912 Katmai-Novarupta tephra fall, Griggs (1918: 6) wrote that three years after the eruption everyone agreed that the tephra fall was "the best thing that ever happened to Kodiak... Never was such grass before, so high or as early. No one ever believed that the country could grow so many berries, nor so large, before the ash". Three years earlier serious consideration had been given by the townsfolk to abandoning the area!

Even though the fertility of volcanic soils is renowned, some actual evidence is necessary to support these claims of enhanced plant growth. Figure 7.15 indicates increased growth increments in a small tree following the Katmai-Novarupta eruption. Although Oswalt (1957) demonstrates decreased white spruce (*Picea glauca*) growth in 1783 along the Alaska arc, the two lines of evidence are probably not in conflict. Oswalt's data came from the area affected by decrease in solar radiation resulting from decreased transparency of the atmosphere following the 1783 Laki and Asama eruptions. On the other hand, Wisherd's photo comes from the area of tephra fall where the surviving plants suffered from less competition. In support of this argument, Eggler (1967) also found that the diameter of survival trees in tephra fallout areas around Parícutin had increased greatly (see also Smathers and Mueller-Dombois, 1974: 6).

These examples indicate that recovery of agricultural production following a tephra fall may not be long delayed and that previous productivity expectations may even be exceeded. As noted earlier, insect predators on crops may be decimated or beneficial insects favoured; diatom and fish growth in lakes may also be enhanced. Despite the emphasis on destruction of agricultural production cataloged here, Huggins' (1902: 14) comment on the aftermath of the 1812 and 1902 Soufrière eruptions may often be correct: "Thus the wind is tempered to the shorn lamb and what in themselves are great misfortunes are turned into benefits for afflicted humanity".

8

THE EFFECTS ON ECONOMIC ACTIVITY

8.1 THE RANGE OF EFFECTS

Volcanic eruptions have some obvious effects on economic activity. Some people become unemployed because their workplaces are destroyed or damaged; others have new employment opportunities because their skills as builders, mechanics, morticians or social workers are in demand. Those who are employed by emergency service organisations are likely to work overtime. Manufacturers of window glass, respirators and/or engine filters may receive new orders.

Of course, the economic effects of volcanic eruptions are much more all-encompassing than these examples suggest. Three areas – agriculture, tourism and commerce – are considered in some detail here but first a number of very brief examples are given to illustrate the diversity of economic effects and to emphasise that such impacts can result from each of the volcanic hazards recognised in Chapter 2. In each case the economic effects are implied rather than explicit. Some of the examples appear trivial but the economic (and social) consequences almost certainly were not. This section also serves to summarise some of the material in preceding chapters although a number of the examples have not been mentioned previously.

8.1.1 Lava flows

March, 1973, lava flows on Heimaey covered 60 houses, severed a 30 000 volt submarine cable, one of two freshwater pipelines, and damaged three fish processing plants and the power house (Williams and Moore, 1973: 15-16).

Kapoho village was almost completely destroyed by lava flows from Kilauea rift in January–February, 1960 (Lachman and Bonk, 1960: 1095).

In the Sakura-jima 1914 eruption the 500 m wide Seto channel (see Fig. 3.19) was filled with lava; Sakura-jima island became a peninsula.

No money was used on Heimaey during the first days after the eruption in 1973 (Gunnarsson, 1973: 75).

8.1.2 Tephra falls and projectile impacts

About 100 000 people left Naples during the five days of the April, 1906, Vesuvio eruption (Perret, 1924: 27).

From early 1944 all able-bodied men from the village around Parícutin could enlist as contract laborers in the United States whereas quotas were imposed for most other Mexican communities. These contract laborers experienced life in a different cultural setting and accumulated capital before returning to their villages (Nolan, 1979: 308).

Following the 1937 eruption the volcanologists Stehn and Woolnough recommended the abandonment of the Rabaul town site (*Pacific Islands Monthly*, December 1937: 62-64).

The burial of Pompeii by tephra fall inspired Lord Lytton (Edward George Earle Bulwer) to write *The Last Days of Pompeii* in 1834. Hundreds of other books have also been inspired by Pompeii's burial and re-excavation (see Leppmann, 1966).

8.1.3 Pyroclastic flows

Following the 1902 Soufrière, St Vincent, eruption, the government withdrew labor from the plantations in the devastated area and resettled the people in the south of the island (Huggins, 1902: 23).

The Kambara pyroclastic flow swept away 1265 houses during the 1783 eruption of Asama (Aramaki, 1956: 198).

Fifty five thousand people evacuated homes around Taal in 1965 (*Newsweek*, October 11, 1965).

Five hundred men took advantage of the government's offer of relief work on roads following the 1911 eruption of Taal (Rivers, 1911: 17-18).

Although villages were able to use the land they occupied before the 1951 Lamington eruption, the administration centre, the hospital and the mission were all relocated. A new network of roads developed and the new villages were located in relation to these (Schwimmer, 1977: 304).

8.1.4 Lahars

Levees with a volume of 1.5 Mm^3 and other works were constructed to protect Cartago from lahars produced on the slopes of Irazú in 1963–1965 (Waldron, 1967: 33).

Secondary lahars from the slopes of Kelut in 1966 cut railway bridges and isolated the town of Blitar (Jennings, 1969: 69).

Around Parícutin, eroded tephra silted up canals and destroyed irrigation systems (Rees, 1970: 13).

8.1.5 Seismic activity and ground deformation

Earthquakes and ground deformation around Usu (1944) damaged water pipes and plastered walls at Toyako Spa, dammed irrigation canals, moved the road and railway laterally, and destroyed terraces in grain fields (Minakami *et al.*, 1952: 264-265, 268; *Time*, July 4, 1949: 45; Minakami, 1974: 327).

As a result of the decision to evacuate 73 000 people from around Soufrière de Gaudelope in 1976, livestock left behind were in danger of starvation and vegetable prices increased (*Time*, August 30, 1976: 7).

Subsidence of the shore near the Deception Island whaling station in 1930 destroyed the jetty (Baker, 1969: 117).

8.1.6 Tsunami and pumice sheets

Tsunami produced during the Matavanu eruption of 1905–1907 destroyed a boat house, a buggy and several boats and lifted an 1800 L water tank off its foundations (Anderson, 1910:625).

Four years after the 1815 Tambora eruption large floating sheets of pumice were still a nuisance to navigation (Neumann van Padang, 1971: 58).

The government steamer *Berouw* was swept 1800 m inland and stranded 9 m above sea level by tsunami during the 1883 eruption of Krakatau (Furneaux, 1965: 109).

8.1.7 Atmospheric effects

Shock waves produced by Stromboli in 1907 broke nearly every window on the island (Perrett, 1912: 286).

The radio could not be used for four days at Kodiak during and following the Katmai-Novarupta 1912 eruption (Erskine, 1962: 59, 151).

Lightning struck animals dead and damaged buildings at Fancy (see Fig. 3.12) and trees on all estates during May 7, 1902, Soufrière St Vincent eruption (Anderson and Flett, 1903).

Continued noise and shock waves kept people awake on Thera and Therasia during the first months of the 1925 Santorini eruption (Washington, 1926: 365).

8.1.8 Gases and acid rains

At Seward a sulfuric rain from the Katmai-Novarupta 1912 eruption corroded the metal work of buildings (Erskine, 1962: 72; Griggs, 1922: 25).

Fluorine-rich aerosols attached to the fine tephra killed 1500 ewes and 6000 lambs around Hekla following the 1970 eruption (*SICSLP,* 1970: 55).

Carbon dioxide gas produced during the Heimaey 1973 eruption was such a problem that part of the already-evacuated town had to be closed to maintenance traffic (Gunnarsson, 1973).

Plantations downwind of Masaya have frequently been damaged by sulfur gas (Bolt *et al.,* 1975: 112-114).

8.2 SOME GENERAL CONSIDERATIONS

The examples above leave much to the imagination; no firm idea is conveyed of the actual costs involved. We are left only with impressions of the total costs and, perhaps, of how such costs were distributed. Unfortunately, even where seemingly more precise estimates are available, there are still numerous difficulties (see also Dacy and Kunreuther, 1969: 5,140). For example, Table 8.1 presents rough estimates of the cost of the Mount St Helens eruptions of 1980. Data from a number of sources have ben used as shown in the footnotes. The report to the United States House of Representatives Committee on Ways and Means arrived at a cost of US$1.213 billion but inexplicably left out the cost of dredging the Cowlitz River, damage to water treatment facilities and many costs to the tourist industry, private businesses and individuals (Rainier National Bank, Economic Report, no. 6, October 14, 1980). Moreover, the estimated cost to agriculture in that report, put at US$192 million, is nearly twice as high as the later figure given by Cook *et al.* (1981: 16) and used in Table 8.1. The figures in this table do not include costs to individuals who lost jobs or could not go to work because of closed roads, losses by private businesses who were forced to close during the cleanup period or who suffered a decline in customers or sales, or costs to government and other agencies who lost income such as that derived from the sale of hunting and fishing licences for streams draining the slopes of the volcano.

A summary of costs and losses compiled by the Washington State Department of Commerce and Economic Development (Hunt and MacCready, 1980) in November, 1980, put the total bill at only US$859.4 million (Table 8.2). Although the two tables are not strictly comparable, the big discrepancy between this estimate and that provided in Table 8.1 results largely from totals for forestry of US$449.9 million and agriculture of US$39.1 million. Comparison of the two tables suggests that some other items have also been assessed at lower costs. However, Hunt and MacCready (1980: 39) do not include the costs of dredging and flood control in the Cowlitz-Toutle system, estimated in Table 8.1 at US$300 million for the period through 1985 (Schuster, 1981: 709). The differences between the two tables serve to emphasise the point that costs are rarely known with any accuracy.

Table 8.1 Losses and costs estimates — Mount St Helens eruptions of 1980–(I)

Losses	Costs (US$ millions)
Timber	695[a]
Agriculture	100[b]
Fisheries	95[a]
Damage to public roads and bridges	112
Dredging Columbia River	45[c]
Dredging Cowlitz River	300[c]
Destruction/damage to 200 Toutle River houses	10[d]
Damage to sewerage treatment plants	5[e]
Cleaning tephra from roads	75[a,f]
Individual homes cleanup	35[g]
Individual business premise cleanup	10[h]
Total	about $1.5 billion

a *The economic effects of the eruptions of Mt St Helens*, a report to U.S. House of Representatives Committee on Ways and Means, cited in Rainier National Bank, Economic Report No. 6, October 14, 1980.
b *Cook et al.*, 1981: 16.
c Schuster, 1981: 709, 711.
d Number of houses from Schuster, 1981; estimated $50 000 per house.
e Guesstimate based on Chapter 6, Section 3.
f Probably a gross over-estimate as community cleanup (mainly road cleaning expenses for 40 Eastern Washington communities totaled only $6.24 million based on FEMA figures in Warrick *et al.*, 1981: 142.
g Estimated 100 000 houses in fallout area of May 18 eruption (330 000 people in 40 Washington communities considered by Warrick *et al.*, 1981: 142); average insurance payment for home cleanup in Yakima c. $350 (see Chapter 8, Section 5).
h Guesstimate.

Some costs are likely to remain incalculable. For example, it is likely that the downturn in the sale of hunting licences in one area is compensated, to some extent, by an upswing in other areas. Many of the outlets that pumped less gas than usual in the week following May 18 because motorists stayed at home probably experienced increased sales of air filters and engine oil.

We might also note that Tables 8.1 and 8.2 fail to include an estimate of the economic cost of about 60 deaths and an indeterminate number of injuries. In the mid-1970's the cost of an individual human life in the United States was placed at US$300 000 and the average cost of an injury at US$10 000 (White and Haas, 1975: 77); however, allowing US$30 million for such costs would add less than 2% to the US$1.5 billion estimate.

As a general rule, and as indicated by Tables 8.1 and 8.2, cost estimates decline with time following a volcanic eruption (Dacy and Kunreuther, 1969: 9-11). This probably applies particularly to eruptions where tephra falls appear to cause widespread destruction but, in fact, damage little of the infrastructure. Recovery, especially in the agricultural sector, is more rapid than at first imagined (see Chapter 7, Section 7). There is probably also a natural tendency on the part of the media to focus on spectacular examples of destruction and on the part of politicians and administrators to overestimate the losses in order to maximise outside sympathy and support.* With these comments in mind, we may note that early estimates by

Table 8.2 Losses and costs estimates — Mount St Helens eruptions of 1980 — (II)

Sector	Costs (US$ millions)	
Clean-up		274.2
Local (includes ports)	50.5	
State	5.0	
Federal	209.0	
Private	9.7	
Agriculture		39.1
Hay	29.4	
Dairy	5.6	
Beef — Calf	2.2	
Feeder	0.3	
Bees	1.6	
Income		8.9
Wages & salary	3.6	
Business revenue	5.3	
Transport losses		2.1
Ports	1.4	
Airports	0.1	
Property (including roads and bridges)		85.3
Private	39.2	
State	2.5	
Federal	43.6	
Forestry		449.9
Timber-Private	218.1	
Timber-State	63.7	
Timber-Federal	168.0	
Total		859.4

After Hunt and MacCready, 1980: 4.

the Washington State government placed the damage bill at US$2.7 billion. As a result of estimates by 12 Federal agencies, Congress appropriated US$945.88 million in response to the eruption. By October, 1981, only US$385.8 million (41%) of these funds had been spent on Mount St Helens disaster recovery work. Small Business Administration spent only 15% of the allocated US$430 million because its initial estimates of needs were based on the assumption that the ash would destroy all farm crops and the land would not be usable for a considerable length of time (GAO, 1982: 17-21). It is also interesting to illustrate this "principle" by reference to the Yakima experience following the May 18, 1980, tephra fall.

Table 8.3 is based on projected economic losses for Yakima County resulting from a tephra fall of 8–10 mm. The data on which the table is based were compiled by an Economic Assistance Task Force of the Greater Yakima Chamber of Commerce. The time periods and loss assessment criteria in the left hand column formed part of the basis for the May 24, 1980, estimates. Losses to agriculture and government have not been included. The May 24 figures were compiled for a congressional committee; work began on the estimates two days after the tephra fall at

*GAO (1982) is recommended to the interested reader. This fascinating document details many of the fiscal problems from a Federal point of view.

Table 8.3 Projected economic losses, Yakima County

Damage	Losses (US$million)		
	May 24 1980	May 30 1980	Early July 1980
Equipment damage (90 days)	100	10	2.5
Automotive maintenance and repair, $450 per vehicle (90 days)	90	32	2.0
Retail sales (60% loss for 30 days; 40% loss 30-60 days; 20% loss, 60-90 days)	89.4	14	7.5
Manufacturing (45 days)	76	5	
Housing cleanup (a1000 per house including equipment damage)	50	10	4
Tourism	15.3	8	11
Professional services (30 days)	4	2.5	2.5
Real estate (90 days)	3.5	2.0	2.0
Other services	1.0	0.5	0.25
Miscellaneous (includes business cleanup)	20.0	10	5.0
Total	448.2	95	41.75

Table 8.4 City expenditures on clean-up of May 18 tephra fall

Community	Initial estimate — May-June 1980		Final estimate submitted to FEMA*	
	Total($)	Per capita ($)	Total ($)	Per capita ($)
Ellensburg	58 300	4.90	71 538	6.08
Cheney	37 000	4.86	89 118	11.71
Ritzville	227 000	126.18	483 154	268.57
Missoula	230 000	3.70	100 000	1.60

*FEMA = Federal Emergency Management Agency
After Warrick *et al.*, 1981: 116.

a time when the clean-up campaign had hardly begun. By the time the second set of figures were compiled (May 30) the clean-up of much of the city was complete, the Central Business District had reopened (see Chapter 6, Section 10), and it had rained. Thus, the May 30 figures reflect some optimism. The July figures reflect even more the revival of spirits as by then it was apparent that damage to irrigated orchards in the Yakima Valley was neither severe or permanent, and tourists were returning to the area. Although we have no way of knowing how accurate the early July total is, it is certainly likely to be a better assessment of the cost of the eruption to Yakima County (excluding government and agriculture) than are the May 24 figures. Thus, the estimated costs dropped by more than an order of magnitude in six weeks (see also Dacy and Kunreuther, 1969: 9-11, 138).

Although Table 8.3 illustrates the general principle that loss estimates generally fall as the costs are reassessed, three of the four examples in Table 8.4 provide exceptions to the rule. The important general point about Tables 8.1, 8.2, 8.3 and 8.4 is that we rarely know when loss estimates were made, or on what basis they were made. These tables emphasise the dangers inherent in uncritically accepting isolated loss estimates.

A second "general principle" which may be considered in relation to Table 8.1 can be stated: "one person's loss is, very frequently, someone else's gain". As intimated earlier, repairs are made, overtime is worked, new business opportunities are developed and clean-up equipment is in demand. It may well be fair to state that natural disasters result more in the redistribution of incomes and expenditures than they do in losses. An extreme example where the gains from disaster impact exceeded the losses is represented by Hurricane Iwa which struck the Hawaiian islands of Kauai and Oahu in 1982: "Hurricane Iwa will result in gains to the Hawaiian economy that are greater than the losses it caused, an ad hoc committee established by Gov. George Ariyoshi to assess the economic impact of the Nov. 23 storm has calculated..." (*Honolulu Star Bulletin,* January 28, 1983: 193).

A group of United States Pacific Northwest economists came to almost similar conclusions about the Mount St Helens eruptions of 1980: "Although the tally of direct damages caused by the eruptions is not final, it is expected to be relatively small compared to other natural disasters, with some relief coming from the state and federal governments in the form of disaster relief funds. Whether this relief will provide a net positive income to the region is not known at this time" (Mosebar, 1980: 2). The possibility of net gains from a disaster results from the fact that the capital required for repairs to damaged infrastructure and private property comes from federal aid and insurance companies located outside the region.

Of course the losses and gains are not shared equally amongst the various sectors of the economy as is apparent in Table 8.1. Nor are the losses and gains shared equally through time. Table 8.5 attempts to illustrate the effects on various sectors of the 1980 Mount St Helens eruptions. Given the quality of the data in Table 8.1 and 8.2, we can regard Table 8.5 as no more than impressionistic. However, for agriculture and fishing at least the downturn in the three to six month period was of the order of 5%.

Table 8.5 attempts to put economic effects in the context of the whole state. Thus, while the timber industry in the area around Mount St Helens was dealt a savage blow by the pyroclastic flows and lahars of May 18, that area represents a very small part of the timber industry of the Pacific Northwest. Furthermore, while a great deal of infrastructure in the form of roads, bridges and logging equipment was destroyed much of the timber in the blow-down area (see Fig. 3.17) proved salvageable. It is also necessary to recognise that some local governments will take years to recover from the effects of the tephra fall, having spent many times their annual maintenance budgets on the cleanup (see Chapter 8, Section 6 below).

Hidden within the gross generalisations and oversimplifications characteristic of presentations such as Table 8.5 is the certainty that losses and gains are not evenly distributed. Nonetheless, loss sharing does occur, a process which tends to even out some of the costs. Figure 8.1 suggests a model of loss sharing in a United States disaster impact, such as the eruption of Mount St Helens. Clearly, the dead and their surviving families and the injured bear the greatest costs. Deaths are assessed at

Table 8.5 The effects of Mount St Helens eruptions in 1980 on various economic sectors in Washington State

	Effects		
Economic sector	Immediate 0-10 days	Short term 3-6 months	Long term up to 5 years
Agriculture	↓↓	↓	↑
Fishing	↓↓	↓	–
Timber	↓↓↓	↓↓	–
Transportation	↓	–	–
Public utilities	↓	–	–
Manufacturing	↓	–	–
Construction	↓	↑	↑
Tourism	↓	↑	↑↑
Financial	↓	–	–
State and local governments	↓	↓	–

↓ downturn, ↑ upswing, – effect insignificant.
The extent of the effect is roughly proportional to the number of symbols.

US$300 000 each and injuries at US$10 000 per person injured (White and Haas, 1975: 77). The third category shown on Figure 8.1 includes the socially dislocated: "The high cost of social dislocation includes the irreplaceable loss of employment, residence, community, treasured possessions, and familiar surroundings. It includes also the loss of replaceable property that is not replaced, or compensable earnings that go uncompensated because of the pride or ignorance of the victim or the inequity and inefficiency of the welfare system of disaster sharing . . . even in situations where there is a great generosity in relief and compensation, the elderly are forced to move, older workers search futilely for another job, the ethnic community is dispersed, and the fiercely independent are left in want. All these suffer a special loss that has never been measured adequately either in monetary or mental-health terms but is nonetheless real and poignant" (Burton *et al.*, 1978: 120-121). The "damaged" include those who suffer some property loss and/or some loss of employment. The "disturbed" include those living nearby whose lives are disrupted to some degree, probably including some loss of employment. Tens of millions of "donors" give to the American Red Cross and other charities or give directly to victims. Finally, some costs are passed on to all the taxpayers of the United States (White and Haas, 1975: 76-77; Burton *et al.*, 1978: 119-121).

The model is clearly very Western in orientation. Nonetheless, it has some relevance to most cultures particularly if the ordinate on Figure 8.1 is read as "units of social cost". The model can then be interpreted as indicating, in a general sense, loss sharing between victims, relatives, communities and, perhaps, the government and the international disaster relief agencies.

These three "principles" — (i) costs are rarely known with any accuracy; (ii) one person's loss is frequently another's gain; and (iii) loss sharing occurs — will rarely be addressed specifically in the sections that follow, but they do need to be kept in mind.

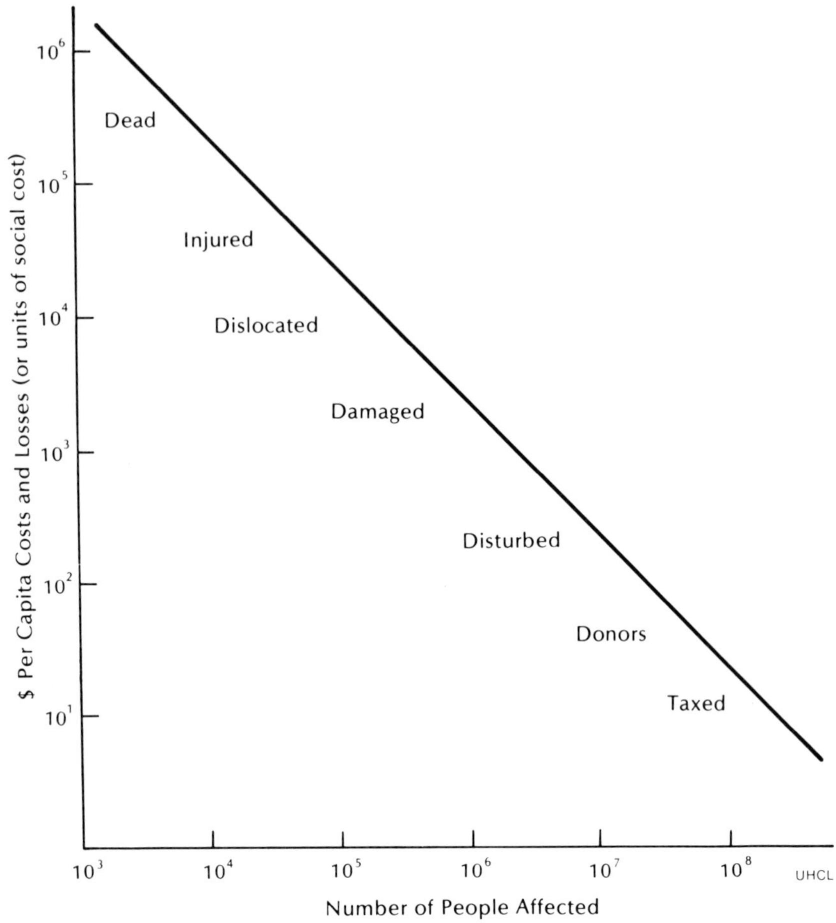

FIG. 8.1 A model of loss sharing in natural hazard impacts indicating the distribution of effects through the population of the afflicted country – in this case the United States (after White and Haas, 1975: 76).

8.3 AGRICULTURE

The range and diversity of economic impacts on agriculture have already been outlined in Chapter 7 but can be illustrated further here using the eruptions of the Soufrière, St Vincent, in 1902 and Mount St Helens in 1980 as examples which are indicative rather than complete assessments.

The 1902 Soufrière eruption "destroyed" the large estates (Fig. 8.2), such as Rabaka and Wallibu on the northern end of St Vincent (see Fig. 3.12). Twenty

FIG. 8.2 Damage to an estate near Georgetown, St Vincent, resulting from the 1902 eruption of Soufrière (Photo: National Geographic Society, Israel C. Russell).

estate owners made claims to the government (of England) for compensation. The average claim was for 4000 English pounds but ranged in value from 75 to 19 378 pounds (Anderson and Flett, 1903: 467). The government paid on average about 35% of the value claimed, with individual claimants receiving as little as 15% or as much as 76% of their reckoning. An administrator assessed the total damage caused by the several eruptions of 1902 at 78 850 English pounds, divided as shown in Table 8.6. A further breakdown of assessed losses is given in Table 8.7 for Wallibu Estate. The administrator's assessment of destruction places the damage to land second to damage to produce but Wallibu's owner made the damage to his land 71% of his claim. It is likely that this differing perception of damage results from the fact that the estate owner made his assessment some months before the administrator compiled his figures. As mentioned in Chapter 7, Section 7, much of the land returned rapidly to productivity.

While these two tables provide some idea of the economic impact on agriculture of the eruption, the figures need to be considered alongside those in Table 4.9 which sets out disaster relief expenditure for various periods after May 7, 1902 — the economic effects on agriculture were certainly felt more widely than is suggested by Tables 8.6 and 8.7 alone.

Table 8.6 Damage assessed on Soufrière, St Vincent, 1902

Item	Value (£)	Proportion of total
Land	19 350	24.5%
Houses	10 000	12.7%
Furniture and clothing	2 000	2.5%
Produce	20 000	25.4%
Stock	7 500	9.5%
Loss of earnings	13 500	17.1%
Miscellaneous	6 500	8.2%
Total	78 850	99.9%

After *HMSO*, 1903.

Table 8.7 Claim made for Wallibu Estate

Item	Value (£)
House	120
Furniture (incl. piano)	73
Estate (265 ha)	3 000
Livestock	
5 horses	58
1 mule	17
3 asses	10
78 cattle	325
50 pigs	51
40 goats	17
27 sheep	8
Produce	
11 bags cocoa	46
5 bags coffee	25
383 barrels sugar	319
40 puncheons molasses	83
4 carts	58
80 empty barrels	4
Total claimed	4 214

HMSO, 1903: 17.

The total value of agricultural production in Washington State in 1979 was about \$2.43 billion. The cost to agriculture of the Mount St Helens eruptions has been variously estimated at \$39.1, \$100, and \$192 million (Hunt and MacCready, 1980; Cook *et al.*, 1981: 16; Rainier National Bank, Economic Report No. 6, October 14, 1980) indicating a downtown of 1.6%, 4% or 8% compared with the previous year. The lowest figures (Hunt and MacCready, 1980) are most likely to be correct. Clearly, losses were very unevenly distributed. A very large proportion of the state received no tephra fallout; some producers in these areas may have received superior prices. Farmers in the lower Toutle Valley were more or less wiped out, some with nearly all their land buried under lahars. Some strawberry and blueberry producers lost 20%–40% of their crop, because of salts in the tephra, rotting on the ground or, in the case of strawberries, the difficulties of cleaning the pubescent skins (see Chapter 7, Section 2). Some milk producers lost production when tankers failed to make collections during difficult periods immediately after various tephra fallouts, but milk production was not reduced. Honey producers were severely affected as bees proved very susceptible to the tephra (see Chapter 7, Section

6). Costs to the industry have been put at only $1.56 million but in relative terms the bee industry was the most severely affected in the agricultural sector (Hunt and MacCready, 1980: 29). Winter wheat was also affected but the loss was less than 10% and this was offset by heavy planting and favorable weather. Although some individual growers experienced problems with excessive apple drop, the annual crop was a record high (Cook *et al.*, 1981; Mosebar, 1980; Rainier National Bank, Economic Reports, 1980). On balance, no losses can be attributed to the eruption (Hunt and MacCready, 1980: 22).

The most significant agricultural impact occurred in the hay industry, which experienced a 10% fall in the first alfalfa cutting. The worst-case scenario of Hunt and MacCready (1980: 24) indicates a loss of US$29.4 million in Grant, Adams and Franklin counties (Fig. 8.3) where 14 000 ha of alfalfa were severely affected. This loss represents 75% of the total losses to agriculture calculated by Hunt and MacCready (1980: 25) who also note: "To provide a perspective, winter-kill reduced hay output by about 20 percent in the '78–'79 season. In addition, another 20–25 percent of value in first-cut hay was lost to untimely rains this year in some areas largely unaffected by ash". Concern by dairy farmers about the effects of the tephra fall on hay quality and availability pushed the price of first-quality hay upwards, with average prices paid for a few weeks after the eruption reaching $5–$15 per tonne above those anticipated.

As wheat, milk and apples are by far the most important agricultural products, in terms of value of production, in the state, the effects of the 1980 eruptions were clearly minor when considered *in toto*. This conclusion does not deny, however, that some individual producers were severely affected.

A number of points need to be made about these examples. First, they illustrate the three "principles" propounded earlier. In neither example, are the costs known with any certainty. It is not possible to determine the proportion of excessive "apple drop", experienced by a few growers, due to Mount St Helens tephra. What proportion of strawberry rot would have occurred even if tephra had not fallen? What proportion of machinery wear is due to the tephra? Some producers benefited because of others' losses. Hay producers outside the tephra fall area received above average prices for quality feed. Presumably, sugar, rum and molasses producers around the Caribbean received better prices in 1902 because St Vincent production fell substantially. Loss sharing also occurs in the agricultural sector. St Vincent plantation owners received some compensation for their crop and stock losses. Some Washington producers had crop insurance, and Small Business Administration loans were available. State funds were also made available on a cost-sharing basis to incorporate the tephra into fields and/or to remove crop debris that could not be salvaged.

The Soufrière, St Vincent, and Mount St Helens examples should not be taken as anything more than that. It is doubtful that these examples of economic effects on agriculture can even be regarded as typical of those produced by moderate-sized Plinian or Pelean eruptions as a great deal depends on the timing of the eruption in

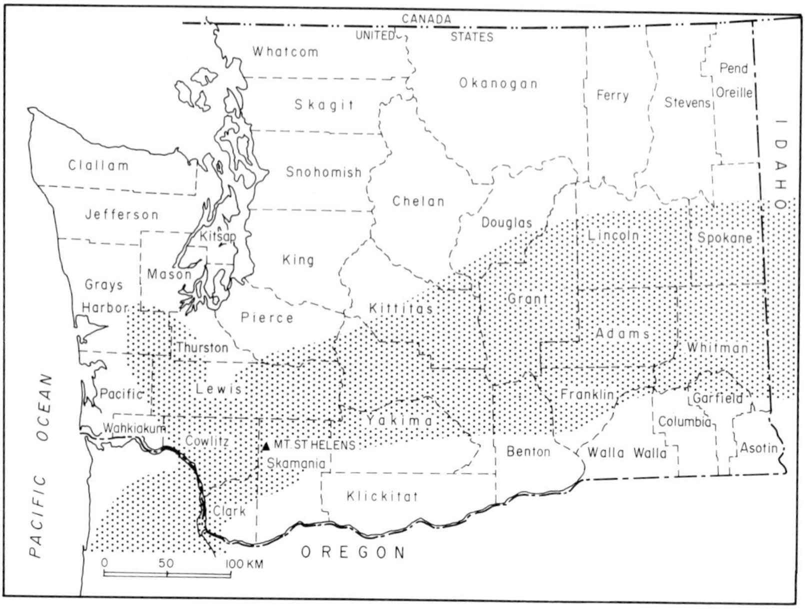

FIG. 8.3 Counties in Washington State: the shaded zone indicates areas receiving more than 1mm of tephra fall during 1980 eruptions of Mount St Helens. A detailed isopach map is presented in Fig. 3.8.

relation to crop growth, weather and other factors after the impact, and market conditions.

8.4 TOURISM

Volcanic activity usually increases tourist activity although there may be a substantial decline in the number of tourists while the eruption is in progress. However, this depends on the type of eruption; frequently eruptions attract sightseers.

A few examples indicate that tephra falls drive tourists away. Tephra fall in Naples from the 1906 Vesuvio eruption caused a great exodus of tourists from the city. Additional trains were provided to cope with the departures early on in the eruption. So many tourists left that some of the grand hotels dismissed their servants (Hobbs, 1906: 640). The tourist industry was also hard hit by the Usu 1977 eruption. Cancellations poured into the resorts in the hot springs areas which were isolated anyway by tephra-blocked railroads and highways (*Time*, August 22, 1977: 18). Cancellations also seriously affected the tourist industry of Washington State when Mount St Helens erupted in 1980 (see below). Similar effects were reported in the summer of 1982 in Mammoth Lakes (Long Valley, California) when newspapers and radio broadcasts reported the possibility of an eruption (*Discover*, 1982, October: 20).

On the other hand, tourism is likely to be enhanced or even created by volcanic activity. In Japan, thousands visit Aso in Kyushu every year to pay homage to the diety that the volçano represents (Anderson, 1907: 30). Scenic monuments, such as Crater Lake in Oregon, Yellowstone Park in Wyoming and Katmai National Monument in Alaska are all the product of volcanic activity. It is even doubtful that Katmai would attract much tourist attention if it had not been created a national monument in 1919. The area around Lassen, last active in 1914–1917, draws 400 000 visitors a year, while Hawaii Volcanoes National Park attracts 2.5 million tourists annually (Mosebar, 1980: 5). Kilauea attracted numerous tourists as early as the mid-nineteenth century. In 1866 Mark Twain wrote of the newly-opened Volcano House: "The surprise of finding a good hotel in such an outlandish spot startled me considerably more than the volcano did. The house is new-built three or four months ago – and the table is good" (Mullins, 1977: 27).

In Hawaii an eruption generates additional tourist traffic. Tourists either make non-stop over-the-eruption flights from Honolulu, fly from Honolulu to Hilo and then travel by road to the eruption and/or make scenic flights from Hilo. During the January, 1983, eruptions, in part of the park inaccessible by road, small charter operators in Hilo flew sightseeing flights from dawn to 10.30 p.m. for several days. (*Honolulu Star-Bulletin,* January 19, 1983: A20). In 1968 the traffic volume of Hawaiian Airlines increased 22% over 1967 figures, largely as a result of the eruption which ended in June 1968. The airline even had a list of people who wanted to see each new eruption from special flights. Once news of a new eruption was received a series of marketing plans were initiated (*Aviation Week and Space Technology,* September 9, 1968: 39).

Some Etna eruptions provide a similar boom with tourists, volcanologists and the media flocking to the volcano. The 1971 eruption had a carnival-like atmosphere with tourists everywhere enjoying one of the most exciting spectacles for years. It was suggested that the lava flow be diverted by bombing (see Chapter 5, Section 2) but this was not well-received by local businessmen who hoped that Etna would continue in eruption, that the tourists would keep arriving and that the summer would be a bountiful one for all (*Newsweek,* May 17, 1971: 92-93; *Time,* May 31, 1971: 44).

It is not only effusive eruptions such as those of Kilauea, Etna and Parícutin but also more explosive eruptions such as those of Mayon and Irazú which have increased tourist traffic. Commercial airlines in Manila were fully booked on southbound flights when Mayon erupted in 1938 and the railroad put on extra coaches to cater for the sightseeing traffic (Miller, 1942: 41). The 1963–1965 eruption of Irazú also proved to be a major tourist attraction. Spectators were able to drive to the summit of Irazú, park near the lip of the crater and peer into the vent. However, after a mid-April eruption killed two and injured 10 people on the summit, the road was closed (*Time,* April 24, 1964: 46-47).

The most bizarre example of the effects of volcanic activity on tourist activity stems from Oshima in Tokyo Bay. Before World War II, 80 or 90 Japanese, many of them lovers, threw themselves down the vent. According to *Time* (June 6, 1965) the steamship company that served the island proved profitable, carrying tourists to see the volcano.

The 1980 eruptions of Mount St Helens also had important effects on tourism. Between March 20 and May 18 sightseers flocked to the area around the mountain, venturing as close as they were allowed. When the merchants of Cougar complained that a roadblock was cutting off their share of the tourist business, the barrier was conveniently moved east of the town (*Volcano,* 1980: 20). However, after the May 18 eruption things were different. Potential visitors, impressed no doubt by media reports, believed the whole state of Washington was knee-deep in ash that caused silicosis. Chehalis did not get any of the May 18 tephra but weekend visitors dropped from a usual 40 000–60 000 to about 10 000 (*Seattle Post-Intelligencer,* June 12, 1980). Seattle also missed out on the tephra but experienced a "significant visitor drop", with revenues off about 10% and business at the Seattle Space Needle restaurant down 19% in June (*Business Week,* July 21, 1980: 9).

In the ash fall areas tourist business was even worse. Spokane had experienced a visitor boom in early May but after May 18 it lost 7 conventions, with some hotels reporting business down by 50% (*Business Week,* July 21, 1980: 9). In Yakima a projected loss to the tourist industry of US$15.3 million was eventually changed to US$11 million (Table 8.3). About 55% of this loss was calculated as a drop in room revenues and 33% as a loss on food, beverages and banquets. In the first week after May 18 the Visitors Bureau was notified of the cancellation of numerous events including seven major state-wide athletic tournaments. Although the city's motels usually experienced occupancy rates of 70%–85% during the Memorial Day Weekend the rates were only 0%–13% on that weekend in 1980. Although mid-June rates had climbed back to about 50%, some motel owners still had only 10% of their rooms let (Gary Webster, Yakima Chamber of Commerce, personal communication, June 1980; May 1981).

Places on the Olympic Peninsula which received just a touch of tephra on May 25 (see Fig. 3.8) in the middle of the normally booming Memorial Day Weekend also suffered massive cancellations when guests cut short their vacations (EH & WS, 1980). Portland lost an estimated US$1 million in revenue when the American Federation of Television and Radio Artists switched its 500-delegate convention to Michigan. Even in July, hotel occupancy in Portland was reported down 15% (*Business Week,* July 21, 1980: 9).

Although impossible to estimate with any accuracy, annual gross revenues from tourism in Washington were expected to be down US$200–$300 million for 1980, a fall of about 10% (Rainier National Bank, Report No. 3, June 27, 1980: 3). The United States Forest Service expected to lose US$4.8 million in revenues from Gifford Pinchot National Park, the area surrounding the volcano (*Seattle Post-Intelligencer,* June 12, 1980). However, Hunt and MacCready (1980: 32) note that the nationwide recession and climbing gasoline prices in 1980 meant that decreased tourism would have been predicted even without the eruption.

The decline in tourist visits led to several advertising campaigns. Portland used the theme: "The only Portland ash left is in California gift shops" but the campaign turned sour on June 12, the day it was launched, when the city received 3 mm of tephra (*Business Week,* July 21, 1980: 9). Yakima began a "clean and green" campaign, some motel owners offering two rooms for the price of one. Although many conventions were cancelled, some as much as a year ahead, the city managed to keep the National High School Rodeo, worth an estimated US$3 million in revenue. Two board members were flown to Yakima to convince them that the city really was "clean and green." Holding this event gave a big fillip to commercial morale (*Business Week,* July 21, 1980: 9; Gary Webster, personal communication, May 1981). The Seattle–King County Convention and Visitors Bureau also developed an effective media campaign designed to convince people that Mount St Helens was not across the street from the Space Needle but 190 km away. United Air Lines flew in 80 journalists from around the United States to demonstrate that most of the Pacific Northwest was unaffected by the eruption (*Business Week,* July 21, 1980: 9; Rainier National Bank Report No. 3, June 27, 1980: 2).

Other publicity came in journals and trade magazines. *Business Week* (June 2, 1980) pointed out good vantage points along the Interstate-5 (I-5) freeway and some of the best buildings in Portland (80 km away!) from which to get a view. It also announced that the Oregon Museum of Science and Industry was running a 20 minute videotape on the volcano while the Pacific Science Center in Seattle had a seismogram hookup and a 20 minute movie on the 1973 Heimaey eruption. Charter operators put up signs along highways with directions to their airfields — Hillsboro, Troutdale, Kelso, Toledo, Portland International and Seattle Boeing. The small plane operators did well even before the May 18 eruption and business continued to boom. One company in Kelso bought five more planes in the winter of 1980–1981, but still doubted that it would be able to meet demand. On clear days there were so many planes around the volcano it was surprising there were no accidents (see Chapter 6, Section 6.2; *The Columbian,* May 26, 1980: 2; *The Wall Street Journal,* May 18, 1981).

On the ground, the United States Forest Service opened two visitor information centers early in July, 1980. One center close to the I-5 was reportedly averaging 6000 visitors a day; the other, 6 km from the freeway, 800 visitors a day. In 1981 the Forest Service expected one million visitors through these hastily built centers (Mosebar, 1980: 5; *The Wall Street Journal,* May 18, 1981). By early June 1981, 726 286 visitors had passed through the centers (*Forest Service News,* Gifford Pinchot National Forest, June 2, 1981). Perhaps President Carter was right when he said, after viewing the devastation, that the area could become "a tourist attraction that will equal the Grand Canyon "(*The Columbian,* May 26, 1980: 9).

Of course, the tourist influx to the area west of Mount St Helens led to all sorts of souvenirs. The first Mount St Helens T-shirt appeared on May 28, 1980, only eight days after the initial seismic signs of an impending eruption (*Holocaust,* 1980: 12).

Table 8.8 gives an idea of the range of Mount St Helens souvenirs available in May 1981. As *The Wall Street Journal* (May 18, 1981) noted, the thriving souvenir trade "supports the theory that nobody ever went broke underestimating the taste of American tourists". Horowitz (1983) argued, however, that most tourist businesses have lacked customers.

Table 8.8 A selection of souvenirs on sale around Mount St Helens, May, 1981

Souvenirs
T-shirts with about 25 messages
Hats with a variety of messages
Jewellery — necklace, ear rings, rings (embossed metal or ash ceramic)
Crockery with ash glaze
Pen holder, ash-filled ballpoint
Embossed ashtrays *inter alia*
Hundreds of wall posters, photos, postcards
8 mm films, cassettes, slides
Harry Truman record "I don't want to go Mr Weyerhauser" *et al.*
Model volcano with removable top and side
Erupting model volcano — baking soda and vinegar
Pet volcano
Ground ash for pet rock food
Baby pumice with eyes
Knitted Mount St Helens cushion covers
Stamped metal plates
More than 11 books
Souvenir newspapers (12 +)
Volcano card game, board game
Mount St Helens card deck
Bumperstickers, pennants
Liqueur glasses, beer can coolers, beer mats
Soap (with ash)
Frisbees, tote bags, desk memo pads, rulers
Travel mugs, spoons, miniature shovels, 3 purses
Key rings, howdee hands, trays, lapel buttons
Mirrors, more than five jigsaws, belt buckles, eye shades
Mount St Helens passports
Labelled empty bottles to put your own ash in!

8.5 COMMERCIAL ACTIVITY

From various comments made earlier we might expect volcanic eruptions to have both positive and negative effects on commercial activity. Generally, shops and other commercial outlets will be closed during a nearby explosive eruption or tephra fall; the aftermath may also represent a poor trading period for many outlets, but some items will be in demand. On the other hand, as suggested in the previous section, effusive eruptions may stimulate tourism and service industries.

In St Pierre, even five days before the 1902 eruption of Mount Pelée that obliterated the town, the local newspaper recorded the closing of more than 20 businesses (Heilprin, 1903: 64), presumably because the proprieters left the town despite the forthcoming election and the pressure exerted by the governor. During the 1906 eruption of Vesuvio the tephra fall in Naples caused many shops to close early (Hobbs, 1906: 640).

Tephra falls around Mount St Helens generally caused a low period for retail outlets. Some businesses reported slumps of as much as 80% in turnover. At White Pass, in the Cascades north of the volcano, the Rimrock Grocery and Resort was down US$10 000 compared with usual Memorial Day Weekend trading, even though the area was on the margin of the tephra fall. However, vacationers did appear in the northern Cascades, well beyond the area of tephra fallout. Merchants there reported above average turnovers (*Seattle Post-Intelligencer,* June 12, 1980).

In Yakima the Central Business District was closed for four days to facilitate the clean-up (see Chapter 6, Section 10) so that many retail outlets lost trade. The enforced closures meant that the downtown shops cancelled television advertisements. The television channels offered commercials at half the usual rate. One enterprising suburban family, owners of a hobby shop unaffected by the enforced closures, ordered a saturation advertising campaign. Parents and children from surrounding suburbs, bored by the clean-up, long unbroken periods at home and admonishments not to drive, flocked to the hobby shop, forcing it to stay open extra hours. This resulted in the best trading period ever for the hobby shop, far exceeding even the usual pre-Christmas boom.

The May 25 eruption produced similar variations in retail fortunes in the area south-west of the volcano. In Vancouver (Washington) the Vancouver Mall was closed on Sunday, May 25, but the Jantzen Beach shopping centre was open with just a few stores closed. Groceries experienced a brisk trade as many shoppers, fearing the worst, stocked up on bread, milk, eggs and other staples. One Battle Ground store reported record business for a Sunday. Air filters for cars were also in demand. Pharmacies experienced a rush for refill prescriptions, especially those relating to respiratory ailments, and numerous requests for face masks. Some restaurants in the Vancouver area reported the day as normal, but others found it slow and closed early (*The Columbian,* May 26, 1980: 12).

During the recovery period some items were in demand and retail sales boomed. In Yakima and, no doubt, everywhere else that experienced tephra fall, brooms, shovels and wheelbarrows were soon out of stock. The 3M plant in St Pauls worked at full capacity manufacturing face masks but could not keep up with the demand. At the end of May 150 000 masks a day were being shipped into the Pacific Northwest (*Seattle Post-Intelligencer,* June 28, 1980). Later, quick-lube facilities, chain saw manufacturers, the makers of the souvenirs listed in Table 8.8 and hundreds of other merchants benefited from the volcanic activity.

Advertisers also had a lot of fun with Mount St Helens. Apart from the big campaigns mounted by United Airlines, Yakima, Portland, Seattle and other cities to encourage tourists back into the north-west (Chapter 8, Section 4), Mount St Helens appeared in a lot of advertisements. In Yakima, advertising slogans for real estate and retail sales were headed “Volcanic sales”, “St Helens Blow Out Sale”,

"Big Blow up Sale", and "Mount St Helens Ash Sundae". A local nutritional food store ran an advertisement:

Be Prepared When... DISASTER STRIKES
Mount St Helens Stopped Transportation
Cut Food Supplies and More!
Are you Prepared? Neo-life offers a nutritionally complete line of food storage items and vitamin supplements
• Finest Quality •
• Designed for Long Term Storage •
• Satisfaction Guaranteed •
[*Penny Press*, Yakima, June 10, 1980]

Kentucky Fried Chicken shipped 20 tonnes of tephra from its five franchises in the Yakima Valley to outlets in Northern California and Nevada. Customers buying chicken were able to get 100 g of ash free; those who only wanted the souvenir had to pay 50 cents. A Palo Alto franchise reported that it had passed out its supply of 1000 samples in a couple of days. On a national level, Olympus Cameras used a superb photo of the May 18 Plinian eruption with the caption: "When the world around sou is exploding, there's one camera you can trust not to blow a shot".

Of course, the volcano, both before and after the May 18 eruption, attracted photographers. Those who witnessed, photographed and survived the Sunday morning eruption did very nicely. Gary Rosenquist reportedly made $25 000 in sales to *Time, Life* and other publications. Vern Hodgson's pictures went from the *Everett Herald* to its parent company *The Washington Post* and then to its subsidiary, *Newsweek*. Keith Ronnholm sold superb sets of slides and posters of the early morning eruption sequence. Sunday hikers who had taken photographs of the eruption made prints for posters, postcards and even soap wrappers. Fotomat outlets gave away prints of the eruption at their drive in kiosks (*American Photographer*, November, 1980: 70-71). However, all these entrepreneurial efforts need to be put in an additional perspective — Reid Blackburn, a photographer for *The Columbian* and on assignment to *National Geographic* died in that Sunday morning eruption and Keith Ronnholm gave a proportion of his photographic earnings to endow a scholarship in the name of his friend David A. Johnston, the US Geological Survey scientist also killed in the eruption.

The mass of disaster photos and the eruption itself created a publication boom. As early as March 27 *The Columbian* published a special section headed "Mt St Helens Diary" which sold nearly 225 000 copies (*Editor and Publisher*, May 10, 1980: 26). Stephen Harris's book, *Fire and Ice*, which is about the Cascade volcanoes and includes a chapter on Mount St Helens but nothing on the 1980 eruptions, sold 10 000 copies in late April and early May 1980 (*Seattle Post-Intelligencer*, May 29, 1980: 1). On May 19, the day after the tephra fall, the *Yakima Herald-Republic* printed and sold an additional 12 500 souvenir copies (*Yakima Herald-Republic*, May 21, 1980). *The Longview Daily News* devoted its entire staff and reporting effort to Mount St Helens earning a Pulitzer prize and accolades from volcanologists and others for its accuracy.

The real publication rush began straight after the May 18 eruption. The combined staff of the *Longview Daily News* and the *Bellevue Journal-American* produced the 96 page *Volcano* in 11 days with copies on sale by the end of June. The initial print run was 120 000 copies, and a third printing was being considered by November, 1980 (*American Photographer,* November, 1980: 71). At least 10 similar books were on the market by the end of the year — some produced by media outlets, some by individuals. Stephen Harris's *Fire and Ice* also appeared in a revised 1980 edition. Almost every weekly and monthly in the United States (and many in other countries) carried items on Mount St Helens — testament to the fact that the eruption touched every facet of life in the Pacific Northwest. A bibliography of Mount St Helens, published in August, 1981, and by no means complete, contains 291 references to publications on the 1980 eruptions (Harnly and Tyckoson, 1981). Other bibliographies were maintained by the United States Geological Survey in Reston, Virginia and by a University of Washington scientist.

The insurance industry was one of the many facets of commercial activity affected by the eruption. Some companies give victims of the eruption two months to pay life insurance premiums due in the May–June 1980 period (*Seattle Post-Intelligencer* May 29, 1980: A10). For most companies the eruption meant a big task in assessing claims. In Yakima, for example, about 70 companies were involved in assessing claims for damage to real estate (see Chapter 5, Section 3.2 for descriptions of the types of damage). One company with 2500 claims flew in 26 loss adjusters from other parts of the United States for periods of two to three weeks. Average claims were about US$350 according to several companies, with much of the damage actually being caused by the clean-up process. Homeowners were paid by most companies at the rate of US$5 per hour for period ranging 40 to 100 hours for undertaking the clean-up. There were big differences in payment policies from one company to another, the final decisions on payment being made in head offices in Chicago, New York or wherever. Some companies took four months to make up their minds whether or not they were going to pay; in the end about seven out of the 70 companies did not pay out on the clean-up (Vince Trapp, Hal Baker, personal communication, Yakima, May 1981).

Although some concern was expressed initially that the eruption would result in higher insurance rates, this turned out to be not necessarily the case. Very likely, insurance rates did go up in 1981 but as a result of normal fluctuations rather than because of the eruption or other 1980 disasters (Hurricane Agnes, the summer heatwave, and tornadoes). A disaster such as Mount St Helens should not effect property insurance rates as the probability that disasters will occur is already built into the rate (*Business Insurance,* August 25, 1980: 15, 19). Later, many insurance companies added "volcanic action peril" to homeowner's policies in the western United States at no extra charge (*USA Today,* April 4, 1983).

The Mount St Helens eruptions clearly had effects on both unemployment and employment opportunities. Closures of shops and businesses and the difficulties of getting to work caused by the tephra fall meant temporary unemployment for thousands of people. The pyroclastic flows and lahars affecting the Toutle–Cowlitz–Columbia River systems also had short-term effects on all kinds of employment. Sawmills and pulp and paper mills in Oregon, Washington, Montana and Idaho were closed for periods ranging one to four days while Weyerhaeuser's large operations in the Toutle Valley were affected for much longer periods. Even in Idaho temporary closure of Potlach Corporation's operation affected more than 1000 people (*Pulp and Paper,* July, 1980: 17; *Chemical Week,* May 28, 1980: 19). Flour mills and bakers were closed across Washington, generally by staff transportation problems (*Milling and Baking News,* May 20, 1980: 8). Silver mines and lead and zinc operations in Idaho and copper facilities in Montana were all closed for one to two days by the ash fall, the main problem again being worker transportation (*American Metal Market,* May 22, 1980:1,20). Presumably some workers lost wages as a result of the eruption; some of those who spent the time cleaning their houses got some wages back from their insurance companies. In terms of lost employment, the effects of the tephra fall were similar to those expected with a heavy snowfall (Mosebar, 1980: 2).

An estimated 2500 Chicano and Mexican farm workers engaged in the Yakima Valley and Central Basin areas were more seriously affected by the May 18 eruption. These workers were engaged in harvesting asparagus, thinning apples, training hops, cultivating potatoes and weeding onions. Innova (1980) estimated the average acute economic loss per wage earner as US$546 of which nearly 50% was added expenses and the remainder loss of income. This study also estimated average chronic loss of income due to tephra-related illness at US$10.17 per day. On the plus-side the eruption created a lot of overtime and new temporary jobs. In Yakima, at the height of the clean-up campaign, the city estimated labor costs at $86 000 per day. Of this 26% can be regarded as the normal wage bill, 14% overtime payments to those employees, 10% as wages for temporary employees and 9% as payments to private contractors. The remaining 42% was payments to employees of other government agencies. In all 625 people were employed on the cleanup (City of Yakima cost estimate, May 26, 1980). As the 100 employees from outside Yakima required room and board, some additional employment was generated in the hotel industry — or more correctly, the fall off in trade was not as bad for some owners as for others. Despite the creation of some jobs such as those resulting from the cleanup, the Employment Security Commissioner estimated the loss of 1000 jobs in the Yakima Valley in the short term. On the other hand, the Commissioner noted, anticipated federal assistance of US$600 million for construction and related services to replace roads, bridges and buildings and create 20 000–25 000 jobs of more than one year's duration for the state (*Yakima Valley Sun,* June 12, 1980). According to the Executive Summary Report of the Governor's Recovery Assistance Team, Mount St Helens Volcano Disaster (December, 1980) more employment was created than lost by the eruptions. However, Horowitz (1983) believed that once salvage operations were finished, jobs would be lost and the resource base would be smaller than it was prior to the eruption.

8.6 GOVERNMENT

If the United States Government paid all the costs of the Mount St Helens eruptions of 1980 (Table 8.1) the expenditure would work out at US$4–$7 per citizen. Clearly, the situation is much more complicated, and possibly more expensive, than that calculation suggests. Clearly, the costs of the emergency services provided around the mountain, the efforts of the Federal Emergency Management Agency (FEMA), and the costs of the new David A. Johnston Cascades Volcano Observatory in Vancouver, Wash., are not included. Nor are the costs of Small Business Administration and Farmers Home Administration Loans. On the other hand, if a federal grant of $600 million for reconstruction does create 20 000–25 000 long term jobs, the federal government should receive back about one quarter of its outlay in taxes.

Although the governmental role is complex and multifaceted, (see GAO, 1982) in many ways the major role of a federal government is in loss-sharing. In the United States, once the President declared the area around Mount St Helens a disaster area, FEMA was able to assist with clean-up costs incurred by local governments. Individuals, businessmen and farmers were eligible for low interest loans for refinancing and reconstruction. Maximum loan amounts were generally US$0.5 million with interest rates varying 3%–14% depending on the purpose of the loan and the recipients ability to repay and/or obtain bank finance (*Congressional Report,* Mike McCormack, 1980). Disaster-related losses could also, in most cases, be claimed as tax deductions. Given the plethora of assistance sources available and the complicated details of each of them it was not surprising that additional offices and phone lines were provided to assist with enquiries.

Table 8.2 provides some idea of the distribution of costs of the Mount St Helens eruption between local, state and federal governments. On a per capita basis the burden of the costs clearly falls on local governments. The wide range of costs experienced by local governments is shown in Table 8.9 which provides data on a county basis but includes costs to municipal governments. County locations are shown on Figure 8.3. Adams, Cowlitz and Grant Counties were the most severely affected, more than 50% of the total local government costs being attributable to the three counties. Adams, Lincoln, Grant and Whitman Counties, the areas with the heaviest tephra falls (see Fig. 3.8) and relatively low population densities, experienced the highest per capita costs. However, these estimates should not be regarded as final as higher maintenance costs may appear and flooding may continue in Cowlitz County for some years.

Figure 8.4 indicates the wide range of cleanup costs experienced by 40 municipal governments in central and eastern Washington State. The costs seem to be more related to tephra thickness than to community size; Hallagan *et al.,* (1981) suggested that street kilometers (actually miles) x tephra depth (ash volume) provided a good first estimate of clean-up costs. They noted, however, that it is more than twice as expensive to clean up a street with 20mm of tephra than it was to clean the

Table 8.9 Expenditures by counties (including municipal governments) in Washington State, after Mount St Helens eruptions of 1980

County	1980 population	Expenditures $ (000's)	Per capita expenditures $
Adams	13 300	7 178	540
Benton	108 700	22	0.2
Clark	192 100	840	4.4
Cowlitz	79 500	6 906	87
Franklin	34 600	557	16
Grant	48 000	8 316	173
Grays Harbour	66 400	81	1.2
Jefferson	15 900	3	0.2
Kittitas	26 900	248	10
Lewis	55 450	1 246	22
Lincoln	9 600	3 164	330
Pacific	17 200	37	2.2
Pend Oreille	8 600	1	0.1
Pierce	482 700	16	0.03
Skamania	7 900	34	4.3
Spokane	341 060	4 817	14
Thurston	124 200	34	0.3
Whitman	40 300	4 766	118
Yakima	170 800	3 012	18
Total		$41 278	average $22.4

After MacCready, 1981:222.

same street with only 10 mm of ash cover. Hunt and MacCready (1980: 43) suggested that the financial resources of the community and experience in intergovernmental relations also influenced the degree of financial strain experienced.

The tephra falls and the associated cleanup costs also produced some significant longer term financial effects for local governments as pointed out by MacCready: "Following the eruptions, there were indications that some political subdivisions in the state would have difficulties marketing their bonds. An immediate increase in the interest rates necessary to sell bonds in certain areas caused concern in local government within the state. Municipal bonds are very important to local governments because they are the primary method of financing construction and major maintenance projects such as schools, water supplies, public transit, etc. Since bonds for small communities are generally financed within the state, there has been little difficulty in selling these because the local financial community remains confident of the viability of the local communities... Larger communities which finance through national money markets, however, face some degree of uncertainty. At first, the national financial community indicated an aversion for municipal bonds of Washington communities. This aversion took the form of rejection of certain municipal bonds and a demand for higher interest rates on those found acceptable. The reaction of the national financial community appears to have tempered over time... The long-term reaction of the national financial community to local government bonds in Washington State is unclear. Some predictions can be made on the basis of reaction to other major events. The negative reaction is likely to

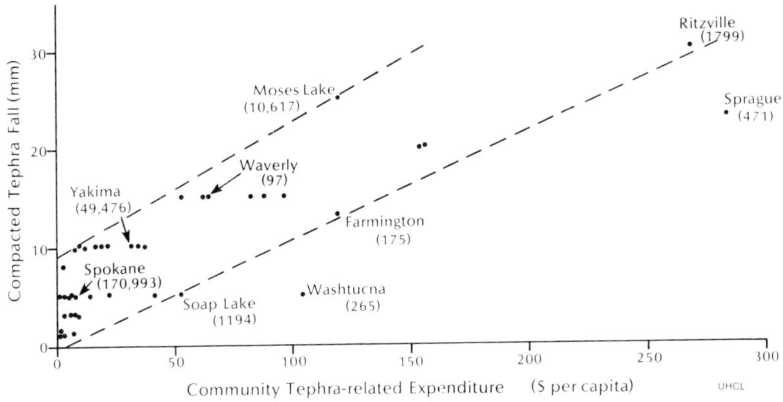

FIG. 8.4 Per capita cleanup costs of tephra fall in 40 eastern Washington communities following the May 18 eruption of Mount St Helens. The numbers below city names provide population estimates (data from Warrick *et al.,* 1981).

dissipate within one to two years. After the financial community has adjusted to the events, a return to normal expectations can be anticipated" (1981: 223).

It is not intended here to analyse the role of government at every level following a volcano-induced disaster, but it is interesting to consider further the role of FEMA following the Mount St Helens's eruptions of 1980. Following the Presidential disaster declaration on May 22, FEMA manned Disaster Assistance Centers in several areas (for example, Kelso, Vancouver [Wash.], Moses Lake, Yakima) providing information and assistance. As an example of the range of assistance programs available, about 100 families visited the FEMA office in Kelso before closing time on May 24, 1980; 37 enquirers asked for temporary housing, 26 requested grants, 51 were interested in Small Business Administration loans, 57 wanted to know about tax rules concerning lost or damaged property, and 57 were interested in disaster unemployment benefits (*The Columbian,* May 26, 1980: 11). During the next few months FEMA also published 33 Mount St Helens Technical Information Bulletins, mainly concerned with health care and maintenance procedures for equipment.

FEMA also played an important role in liaising with local governments concerning the costs of cleaning up the tephra fall. Federal funds, through FEMA, could be obtained for many of the clean-up costs, but not for all. For example, benefits for temporary city employees, supervisory costs, and interest on borrowed interim funds were not eligible (Rainier National Bank, Economic Report, No. 5, August 12, 1980). Unbelievably, a broom was an eligible cost but a shovel was not, apparently because a broom was worn out in cleaning up one tephra fall, whereas a shovel was not. As Yakima City Treasurer Don Horan said: "I don't know what the lifespan of a shovel is — all I know is the city didn't need 60 new shovels" (*Yakima Herald-Republic,* August 14, 1980). Similarly, FEMA paid for topsoil to cover an ash dump but not the grass seed to stabilise it even though wind-blown topsoil seemed no more acceptable than wind blown ash to residents.

But the story goes on: "Horan also has been forced to work with how FEMA determines what is an eligible cost. He said instead of taking an employee and simply figuring out how many hours he worked, FEMA uses equipment as the basis for their payback . . . The agency wants every piece of equipment documented for each day it was used. They want to know who used it, how they used it, the mobilization cost to get the equipment here. I'm just giving every broom a number. Then I say John Doe used broom No. 1 for so many hours doing this job" (*Yakima Herald-Republic,* August 14, 1980).

The Rainier National Bank (Economic Report no. 5, August 12, 1980) estimated that about 80% of the costs incurred by local governments were eligible as far as FEMA is concerned. However, FEMA paid only 75% of eligible costs. Of that amount, 75% could be paid right away with the remainder held until an audit was held, perhaps a year or more later. This means, that in the short term a local administration could expect to receive back only 45% of its clean-up costs, at the same time creating a paper-trail which kept full-time city employees from their regular work. In the longterm, Yakima believed it would be lucky to get back 59% of its costs. Spokane thought it would probably get less than 50%. Some of the smaller administrations, less able to document the FEMA paper-trail could well get back even less. Even though Yakima was determined to appeal every disallowed cost, resolution of such claims could take six to seven years. The basic problem with FEMA, according to at least some administrators, was that its rules were made for flood disasters not volcanic disasters (*Yakima Herald-Republic,* August 14, 1980; Larry Wittenburg, Assistant City Manager, Yakima, personal communication, May, 1981).

By January 9, 1981, Yakima had received US$785 145 or 37.5% from FEMA out of an anticipated total cost of US$2.093 million. Adding in amounts received from other agencies, amounts due from all agencies and subtracting normal salary expenses during the clean-up period left the net cost to Yakima at US$402 465 or about US$8 per citizen (City of Yakima memo, January 9, 1981). As Warrick *et al.* (1981: 142) appear to have used FEMA data for per capita clean-up expenditures, Yakima administrators believe the clean-up cost to have been US$42.3 per capita and not the US$31.7 per capita shown in Figure 8.4. This significant discrepancy only serves to emphasise a point made early on in this chapter — the costs of volcanic eruptions are rarely known with any accuracy.

But the FEMA story does not end with the clean-up costs. Late in 1980 every Yakima homeowner received through the post box a brochure from FEMA summarising for the householder what to do in the event of a volcanic eruption. This added insult to injury. The citizens of Yakima had eaten tephra, slept with it, swept, slogged, sweated and carried the stuff to the dump. The citizens had discovered how to deal with the stuff, not FEMA. And now FEMA was wasting tax dollars telling the experts how to manage ash! My informants, still annoyed six months after the brochure arrived, suggested that it might be some time before FEMA regained much credibility in central and eastern Washington State.*

8.7 SOME PERSPECTIVES

So much of this chapter has concerned the economic impact of Mount St Helens in 1980 that it is necessary to provide some perspectives.

Certainly, some individuals suffered enormous losses as a result of the eruptions, losing relatives, friends, jobs and/or property. As a group, the taxpayers of several counties have incurred significant costs (Table 8.9). Although some hay producers lost almost their entire crops the effect on the industry within the state has been assessed as no more than the effects of "untimely rains" or "winter-kill" (Chapter 8, Section 3). Although the costs to forestry total half of the estimated costs presented in Table 8.1 and 8.2, the downed timber represents less than one per cent of the available timber land in Washington State (Hunt and MacCready, 1980: 7). Although many individuals, particularly in Cowlitz County lost jobs, creating all sorts of unrecorded social and economic costs, more employment was created than lost by the eruption (Chapter 8, Section 5). Finally, the economic consequences were relatively small in relation to the size of the Washington economy — Hunt and MacCready (1980: 3) note: "the overall economic effect was less than the Boeing downturn of a decade ago".

The Mount St Helens example provides an excellent case study of the myriad economic effects of a moderately large volcanic eruption on a developed western economy. We can probably assume that effects of a similar range and severity have occurred (or will occur) in Japan, Italy, Iceland, New Zealand and California.

On the other hand, detailed information is not available concerning the economic effects of eruptions on underdeveloped or subsistence economies. Nonetheless, two more general examples provide some perspectives, suggesting that the effects may be much more severe than in the case of a western economy and reminding the reader that economic costs cannot be divorced from social costs.

As Rees (1970, 1979) and Nolan (1979) have indicated, the Parícutin eruption of 1943–1952 produced a wide variety of economic effects. The removal of oxen from the tephra fall area (and in some cases, their deaths) meant that farmers could not plow their lands so they turned the most poorly paid traditional forest activity of collecting firewood. As more farmers turned to this activity, despite the shortage of animals for transport, oversupply occurred and the price of wood and charcoal

*It is only fair to add that FEMA seems to be highly regarded in areas west of Mount St Helens.

fell. Another economic activity, the collection of pine resin, was also prevented because the distilleries would not accept resin with tephra admixed. For some, migration was the only solution (see Chapter 4, Section 4). On the positive side, the arrival of tourists provided employment opportunities as guides for at least a few years, and many young men went to the United States for employment bringing money and ideas back to the villages. Outside aid brought electricity, water supply and poultry-raising projects and numerous cottage industries. However, many of these enterprises failed because of insufficient technical training and poor product quality.

Stephensen (1785) provides a more dramatic picture of the economic and social costs of the 1783 Laki fissure eruption:

> The volcanic eruption having thus been productive of devastation and sickness, both among man and beast, a great famine and unexampled misery throughout the country, naturally ensued. The peasant, who, with the loss of his cattle, was likewise deprived of his sole means of subsistence, and of the best and most valuable part of his property, had nothing else (after having eaten the animals that died by famine and sickness) wherewith to satisfy the painful cravings of hunger, but skins and old hides, which he then boiled and devoured. Many, driven to the last extremity, have killed the few healthy cattle and sheep that still remained, and afterwards, when these were consumed, wandered with their whole families down to the sea-side, where they have become an intolerable burthen and source of improverishment to the inhabitants of the coast. At the same time, too, that the uplands are become desolate, the condition of the inhabitant of the coasts is so much the more pitiable; as he can no longer continue his laborious toil through storms and frosts, with vigor and energy, unable as he is, to obtain the smallest quantity of butter or other strengthening articles of food to add to his present wretched fare; and being reduced to water, too as his only drink; since whey, which was his usual beverage, is denied him. All this, as is known by long and sad experience in Iceland, renders the fishermen weak and disspirited, and unfits them for ordinary occupations: thus, each hanging on each, the misery that began with one runs through all. The want of skins for sea-clothing will likewise for some years be a great obstacle to the carrying on of thefisheries with advantage; for although, since the mortality among the cattle, there is so great a quantity of hides in the country that they are considered as scarcely of any value, yet it is a well-known fact that those of animals which have died of hunger are in general unfit for use, and these, therefore, will neither answer the purpose of making coats or even of being manufactured into the shoes in use in the country.
>
> The loss of the horned cattle and sheep was very severely felt by the Icelanders, but that of the horses was equally so, especially by the inhabitants of the interior of the country, who thus found themselves deprived of their last resource, the means of having provisions and other necessaries conveyed from the coast, through long and tedious roads. Nay, many who are totally destitute of horses are under the necessity of carrying every load of hay into the outhouses upon their own backs, and frequently from a very considerable distance. Nor is there any prospect of these invaluable animals being soon replaced.
>
> [Stephensen, 1785: 245-248]

9

SUMMARY, CONCLUSIONS, IMPLICATIONS

9.1 A SUMMARY AND SOME CONCLUSIONS

While the size of volcanic eruptions can be estimated, for example, in terms of volume of products, distance of spread, or rate of energy expenditure, the relationship to volcanic hazard is by no means straightforward as the properties which cause human deaths can be different to those which damage buildings, lifelines, machinery or crops.

Table 1.4 summarises, *inter alia,* the relative frequency of occurrence of volcanic hazards. The data presented in the table and in Chapter 2 indicate that tephra falls and atmospheric effects such as shock waves and lightning discharges occur most frequently during volcanic eruptions, with pyroclastic flows, lahars and tsunami occurring occasionally. These sections also indicate, in a rough fashion, the magnitude and extent of death, damage or destruction for each volcanic hazard, given that the particular hazard occurs. Thus, damaging tephra falls, acid rains and gases and tsunami may sometimes be experienced at distances of more than 1000 km from a volcanic source. At the other extreme, lava flows rarely travel, and volcanogenic earthquakes are rarely experienced, more than 30 km from a volcano. Estimates of mean and maximum distances and areas affected and the likely duration of warning periods are presented in Table 2.14.

However, the damage done or the death toll is not necessarily related to the frequency of occurrence or the area affected by a volcanic hazard. Data summarised in Figure 1.7 indicate that the Halmahera, Calabrian and New Zealand volcanic belts have been the most active since records began. While two-thirds of the known deaths in volcanic eruptions since AD 1600 have occurred in the Indonesian region, the New Zealand and Calabrian death tolls are insignificant in the worldwide scheme. Furthermore, far fewer than 20% of the Indonesian deaths in that period have occurred in the Halmahera arc.

Identification of the most significant volcanic hazards as far as human deaths are concerned varies depending on the time period considered and whether or not deaths resulting from disease and starvation are included (Tables 3.2 and 3.3). The relative importance of some volcanic hazards as agents of death could be altered by one major eruption but, given the timely arrival of disaster relief, pyroclastic flows, tsunami and lahars are likely to remain the greatest killers. On the other hand, lava flows, volcanogenic earthquakes, atmospheric effects and gases and acid rains are likely to cause, relatively, very few deaths. Survival rates (Table 3.15) provide a useful measure of personal risk from volcanic hazards but few accurate statistics are available.

As indicated in Chapter 3, various personal protection measures can be taken to mitigate the effects of volcanic hazards. Such protection might include goggles, particle and/or gas mask, and hard hat. While such items offer considerable protection from tephra falls, ballistic missiles and acid rains and gases, the protection afforded in a pyroclastic flow or lahar may be no more than momentary. Thus, issuing such equipment to all populations at risk, even were that feasible, would have little effect on death tolls in volcanic eruptions. If reduction in human death tolls is the main aim of volcanic hazards research (rather than the reduction of injury tolls, the mitigation of building and lifeline damage, or the amelioration of agricultural devastation), then the central issue involves human vulnerability to pyroclastic flows. Detailed medical and technological research is required. We do not know the precise causes of death in pyroclastic flow victims. Would small particle/gas masks prevent ash asphyxiation? What types of everyday and/or protective clothing reduce the risk of life-threatening burns?

However, the hazard offered by volcanoes cannot be measured solely by death tolls. For example, from the point of view of disaster relief, morbidity may be more important than mortality, or survival rates more significant than the body count. Information presented in Chapter 4, section 3 suggests that the psychological effects of an eruption will be influenced both by the casualty and survival rates and by the rapidity and scope of relief efforts. Both the death toll and the amount and type of disaster relief will be influenced by the perception of officials and citizens, and by the risks that both groups perceive and/or are prepared to accept. Perception depends on experience, beliefs, attitudes and, at least in western societies, media reports. Evacuation is a function of perception or, in some cases political or official will, in itself also a function of perception. Similarly, political perceptions (in some societies) influence funds for volcano surveillance. While Chapter 3 demonstrates that personal risk is a function of eruption style, Chapter 4 shows that the effects of volcanic hazards are imbedded in a complex social matrix, the more profound social effects being experienced in less developed areas (Chapter 4, Section 9). Unfortunately, we know far less about the social effects of hazard impacts in these areas than we do about the consequences of natural disasters in western societies.

Table 5.1 summarises the effects of volcanic hazards on buildings — a wide variety of effects is noted, ranging from complete demolition and/or ignition to relatively minor damage. The most destructive volcanic hazards are lava flows, tsunami, pyroclastic flows and lahars. Gases and acid rains and atmospheric effects are rarely totally destructive of buildings though shock waves and lightning strikes may severely damage or ignite structures. A number of ways of reducing building damage may be successful. Increasing seismic resistance is likely to also make buildings more resistant to most other volcanic hazards. Sealing buildings against the ingress of floodwaters would probably increase protection from pyroclastic flows, decreasing both interior damage and human vulnerability. Diversionary or protective barriers may save buildings from, or ameliorate the effects of lava flows, lahars and tsunami. Hastily erected diversion barriers, or the application of massive quantities of water to lava flows, have been successful in saving buildings. The impact of lahars can be reduced by draining water from crater lakes and the construction of containment structures.

The effects of volcanic hazards on community lifelines may be quite different to those on individual buildings as lifelines frequently need to fail at only one point in order to cause failure of the entire system. Consideration of the examples presented in Chapter 6 indicates that the hazards most destructive of utility systems and machinery are pyroclastic flows, lahars and tsunami. Tephra falls of less than 200 mm pose only relatively minor problems for most lifeline systems; however, the disruptive potential of wet (and magnetic) tephra for electrical and communication systems and machinery should not be underestimated. Although adequate data are not available, it seems likely that lahars offer by far the greatest threat to lifelines. It should also be noted that the absence of lifelines such as electricity and gas from a community may reduce the risk to buildings although the absence of telecommunications and a variety of transportation links are likely to decrease the number of warning mechanisms and delay the arrival of relief.

Although other volcanic hazards damage agricultural systems more intensively, the very widespread effects created by tephra falls suggest that volcanic ash poses the greatest threat to crops and animals. However, the adverse effects resulting from tephra falls of less than a few hundred millimeters are likely to be short-lived unless plants (and animals) are at a particularly sensitive growth stage. The same tephra fall may preferentially inhibit or promote the proliferation of crop pests and/or animal parasites. The examples presented in Chapter 7 suggest, however, that for a mixed agricultural economy the effects will not be all bad and recovery is likely to be more rapid than believed possible in the immediate aftermath (Chapter 7, Section 7).

All types of volcanic hazards create a range of economic impacts, simply as a result of deaths and injuries and the effects on buildings, utilities, communication and transport networks, machinery and agriculture. Although it is generally assumed that such impacts are negative the 'benefits' may outweigh the losses. Certainly, some individuals, communities, industries and other commercial groups gain financially and in other ways. The role of governments, apart from the obvious task of organising disaster relief, can often be seen as loss-sharing. In less-developed societies financial losses and gains may be smaller but the impacts and aftermath are more life-threatening and/or society threatening.

This selective summary emphasises the complexity of determining volcanic risk. Clearly, the risk presented to people by a volcanic hazard need not be the same as the risk to various elements of community infrastructure, animals or machinery. Equally clearly, the risk varies within a group of humans, and within a group of buildings. Tables 9.1 and 9.2 attempts to summarise such variation.

Table 9.1 Degree of risk with volcanic hazards

High risk	Intermediate risk	Low risk
Pyroclastic flow	Tephra fall	Lava flow
Lahar	Gases and acid rains	Seismic activity
Tsunami		Atmospheric effects

Table 9.2 A summary of factors which increase or decrease risk in volcanic eruptions

High or increased risk	Risk factor	Low or decreased risk
A. Volcanic and topographic factors		
VOLCANIC FACTORS		
Subduction zone	*Location*	Rift zone/mantle plume
Stratovolcano	*Morphology*	Shield volcano
High	*Silica content*	Low
High	*Gas content*	Low
Long	*Repose period*	Short
None	*Surveillance*	monitored
Short/unknown	*Eruption history*	long/known
TOPOGRAPHIC FACTORS		
Near	*Distance*	Far
Downwind	*Direction*	Upwind
Valley floor (within 'flood' zone)	*Altitude*	Ridge crest (above 'flood' level)
Proximal	*Slope aspect*	Distal
B. Shelter and Lifeline factors		
SHELTER		
Inadequately braced, flimsy, flammable	*Construction*	Reinforced, massive, fire-, wind-, seismic-resistant
Weak, no connections	*Foundation*	Solid, connected
Impact forces maximised	*Orientation*	Impact minimised
Within 'flood' zone	*Elevation*	Above 'flood' level
Large windows, doors, etc.	*Wall openings*	Few, small openings
Absent	*Resistant compartment*	Present
Susceptible to differential load	*Roof*	Adequately braced against static and live loads
Low	*Roof and wall cladding penetration resistance*	High
Gas, electricity	*Utilities connected*	None (water only?)
Low	*Corrosion resistance*	High

Table 9.2 cont'd

LIFELINES AND COMMUNICATIONS		
High with numerous electromagnetic components	*System complexity*	Low
Single channels	*Network integration*	Integrated
Above ground, numerous river crossings	*Location*	Below ground, on high ground.
In open	*Housing*	Enclosed
Inadequate	*Filtration system*	Adequate
Infrequent careless	*Maintenance*	Regular, frequent, careful
C. Agricultural factor		
Subsistence	*System*	Capital intensive (plantation, truck, etc.)
Reliance on one staple	*Season of eruption*	Several staples
Early in growth season		During dormant season
Crop grows close to ground	*Crops*	Tree crops, root crops
Small, young	*Animals*	Large, mature
Beneficial species wiped out; pests multiply	*Salient insects, birds, etc.*	Pests wiped out
Short growing season	*Recovery*	Long growing season
D. Human factors		
CULTURAL		
Traditional subsistence	*Culture*	Post-industrial
Fatalistic	*Religion*	Christian?
Weak links	*Kinship*	Strong links
Human dominated	*Nature*	Dominates humans
Transient	*Residence*	Resident
Marginalised	*Economy*	Self-sufficient
PHYSICAL		
Very old, very young	*Age*	Late teens, early adult
Sedentary/handicapped/ immobile	*Fitness*	Agile/mobile
Malnourished/ undernourished/ overnourished	*Diet*	Adequate
Cardiac condition, asthmatic/bronchial condition	*Health*	No problems
PSYCHOLOGICAL		
None	*Hazard experience*	Recent
Youthful	*Age*	Old, very young
Human dominated	*Attitude to nature*	Dominates humans
Gambler	*Risk taking*	Doesn't believe in gambling
Short	*Horizons*	Long
Unaware of range	*Awareness of adjustments*	Aware of range
Worries about remote danger; imminent dangers produce monstrous fantasies and lack of preparation; impact seen as punishment; overcome with terror; cannot bear revival of experience in memory; continued tension produced by unassimilated trauma	*Adaptiveness*	Does worry about remote dangers; imminent dangers taken seriously, preparations made; develops scenarios and precautions; becomes inured to emotional impact in imagination; at impact preserves confidence in own survival; can tolerate revival of experience in memory; assimilates event so can look back on it with equanimity

Five important group of factors are recognised. Table 9.1 generalises the risk presented by each of the hazards considered in this volume. Table 9.2 is presented in four parts: part A summarises volcanic and topographic factors which contribute to risk, and owes much to the unpublished work of R. W. Decker (Hawaiian Volcano Observatory); part B considers the risk associated with the various shelter and lifeline factors; part C lists a number of factors which increase or decrease the risk to agricultural systems; and part D recognises that a range of cultural, physical and psychological factors all contribute to the risk offered to humans.

In broad terms the four parts of Table 9.2 are arranged in order both in terms of risk and from the general to the specific. Each of the four parts assumes the information in earlier sections — for example, whether or not a volcano is under scientific surveillance does not appear as a risk factor in part D, important as that is to human vulnerability, because it is already listed in part A. Within each group, factors are also arranged in the general order of importance. It should not be surprising that there are some seeming contradictions. Thus, for example, old people are more at risk because they are less agile and sometimes in poorer health (and, frequently, less willing to evacuate under threat) but have decreased risk because of hazard experience resulting simply from longevity. Every reader will, of course, know exceptions to such generalisations.

Tables 9.1 and 9.2 purport to be nothing more than a very general summary. The important point about the tables is that they recognise again that risk or vulnerability is not only a function of volcano characteristics but also an expression of a variety of cultural and human factors.

In the Preface, the influence of geophysical, public official and citizen subsystems on volcanic hazard impacts in the areas around Kilauea and Mount St Helens was emphasised. It was recognised that differences resulted not only from eruption character but also from the topography and landuse and from the attitudes and actions of both public officials and citizens. Despite the differences, the Kilauea and Mount St Helens events had similarities — the same highly trained scientists and technological back-up, the same legal, political, financial systems, and the same law enforcement and emergency services. How much greater might the differences have been if all these aspects of the cultural infrastructure had also varied?

Tables 9.1 and 9.2 can also be regarded as a basic procedure for determining vulnerability in relation to individual volcanoes. Each risk factor could be assigned a numerical value reflecting its importance in the specific situation, *vis à vis* the other relevant factors. Such a procedure suggests that four of five maps could be compiled to define risk in the area around the one volcano. One map would delimit the areas of greatest volcanic hazard, relying on risk factors such as those in Table 9.1 and part A of Table 9.2. In effect such maps are quite common — an example for Galunggung is presented in Figure 4.7. However, other types of vulnerability maps, using bases similar to those outlined in parts B, C and D of Table 9.2 are rarely constructed (or even considered) — Figure 6.1a and Table 6.1 based on Cullen's work on lifeline systems vulnerability in the Puyallup valley present a rare exception. Although it can readily be argued that such maps (or data bases) are of only secondary importance they seem essential for detailed counter-disaster planning. While the general maps based on factors of the type in Table 9.1 and part A of Table 9.2 provide general answers to questions of vulnerability, the more specific questions of "vulnerability of what or whom" require the more specific data based on factors of the types listed in parts B, C and D of Table 9.2.

9.2 HAZARD PERCEPTION

Given the high human vulnerability and the relatively low resistance offered by most buildings to the high risk volcanic hazards (pyroclastic flows, lahars and tsunami), evacuation in the face of the hazard seems the most sensible course of action. Such evacuation implies that: (i) adequate warning systems exist; and (ii) the hazard is perceived as such by public officials and citizens alike.

Consideration of the nature and adequacy of volcano monitoring is beyond the scope of this work (see, for example, Decker, 1980; Shimozuru, 1972) but several relevant points must be made: (i) only a small proportion of the world's volcanoes is monitored, and only an insignificant proportion is adequately monitored, mostly in Japan, Italy and the United States; (ii) the best monitored volcanoes are those that are most approachable, and those that erupt most frequently for example Kilauea, and Etna; (iii) in general, only those volcanoes are monitored that have a recent history of eruption; and (iv) some evidence suggests that the explosivity of an eruption increases with the interval between eruptions (Chapter 1, Section 3). The conclusion from these points is that, on the world scene, monitoring and surveillance of volcanoes is totally inadequate. Furthermore, the most dangerous volcanic hazards are the most difficult to predict.

It is likely to be a long time before the scientific understanding of volcanic processes is at a level where pyroclastic flows can be predicted, not in the sense of a statistical prediction, but by an actual warning system. Such systems are sometimes practical for lahars, especially at some distance from the vent. Many lahars follow well-defined stream channels, move only moderately rapidly, and frequently occur as secondary lahars. Their passage along stream channels means that many can be monitored using techniques akin to hydrological techniques for flood warnings (Chapter 5, Section 5). While a tsunami warning system is in operation in the Pacific Basin, it should be recalled that tsunami propagation rates range up to 800 km h^{-1} (Chapter 2, Section 6), and that there can be virtually no warning close to the source as shorelines within 100 km of the source volcano will receive the first seismic sea waves within 10 minutes of their generation.

That hazard perception by public officials as well as by the populace is important is easily demonstrated. Official attempts to keep the citizens in St Pierre for the following election certainly contributed to the high death toll in 1902 on Martinique. The work of Sorenson and others around Mount St Helens before the May 18 eruption indicated the general inability of the citizens and officials interviewed to visualise or verbalise the hazard character. Certainly, the long-continued media attention to the May 18, 1980, eruption and the production and sale of an enormous variety of posters, postcards, color slides and other memorabilia has altered Pacific Northwest perceptions of the volcano hazard. Whilst it is appealing to suggest that hazard perception in the area will remain at a high level this is unlikely to be the case. Hazard awareness declines with time elapsed since impact. The Rabaul, 1937, eruption occurred less than half a century ago but less than 5% of the present population experienced the event. Various techniques for tephra clean-up were developed in Anchorage in 1951 and in San Jose in 1963–1964; these seem to have been unused or reinvented after Mount St Helens erupted. The earlier experiences were just unknown. Other evidence presented in Chapter 4, Section 2, indicates that human hazard perception, influenced as it is by race, age, religion and experience (in essence, by the factors in part D of Table 9.2), declines with time elapsed since the last experience.

Unless community hazard perception is maintained at a high level neither authorities nor individuals are likely to take measures to reduce vulnerability through counter-disaster planning, protective measures, evacuation or even loss-sharing.

Although numerous cases of evacuation before, during and after volcanic eruptions have been presented in Chapter 4, Section 3, it is rarely known whether these evacuations were officially or individually motivated. Many are probably an amalgam of both. In either case some individuals are likely to remain behind whatever the pressures to evacuate. Some, like Harry Truman at Spirit Lake, may eventually be placed in the position where they cannot evacuate even if they wanted to without considerable loss of face. Some may be killed or injured by the evacuation process. Public perceptions of threat and attitudes toward evacuation may be colored by almost inevitable false alarms such as that on Guadeloupe in 1976. Inevitably, a community will reflect a whole range of attitudes toward evacuation, as expressed by the villagers of Parícutin in 1943.

While evacuation remains the most practical method of reducing vulnerability in the face of the most dangerous volcanic hazards there are real problems in raising and maintaining public and official awareness of the threat in the absence of volcanic activity, or possibly, with the presence of false alarms.

9.3 THE FUTURE

Much of the information presented in Chapters 5, 6 and 8 implies a developed community infrastructure and economy. Reconsideration of Figure 1.1 suggests that many of the examples in these chapters have limited relevance for at-risk areas around many of the world's volcanoes. Where such infrastructure does exist the monetary cost of eruption damage is likely to be much higher than it is in areas without technological lifelines. Similarly, in the case of damage to agricultural production (Chapter 7), the dollar cost of the loss of truck or plantation crops will almost always be higher than the dollar cost of lost subsistence crops. This point is made implicity in Chapter 8 — but it is made in a negative sense as consideration of the effects of volcanic eruptions on economic activity is almost entirely concerned with eruptions in developed countries — in fact, almost entirely with Mount St Helens.

It is perhaps a truism to note that volcanic hazard impacts produce mainly economic costs in developed societies but largely social costs in underdeveloped communities. Undoubtedly, many people lost possessions or income during and after the May eruption of Mount St Helens. Some lost their livelihood and their way of life. Some suffered psychological impairment. In all cases, however, some form of loss sharing is available. Federal disaster assistance provided monies to rebuild community infrastructure. Farmers Home and Small Business administrations and several other bodies noted in Chapter 8 provided insurance and/or low interest loans. State and federally funded programs were available to assist those with mental health problems engendered by the eruption. Federal, state and private organisations provided assistance with emergency food and shelter and advice on the tax deductibility of losses. No one starved.

Such a plethora of assistance sources, all providing some means of sharing personal, community and state losses, is a phenomena characteristic of the most developed countries. At the other end of the spectrum lie the subsistence farmers typical of most underdeveloped countries. Such people have few opportunities to share their losses beyond their relatives who almost invariably provide food and shelter, at least in the short term. But with subsistence crops perhaps destroyed, the need is for longer term assistance. Local money lenders may provide loans at usurious rates. State or international disaster relief may, or may not, arrive, but the level of recovery reached will rarely equal that expected in developed communities. The Parícutin example of economic destruction presented at the end of Chapter 8 typifies the effects of disaster impact in underdeveloped countries. Increased economic marginalisation as a result of hazard impact is the norm rather than the exception. In many cases, the inhabitants of volcanically hazardous areas become more disaster prone in the aftermath of an eruption.

Sadly then, much of Chapter 8 concerns Mount St Helens with almost no information about the effects of volcanic eruptions on economic activity in underdeveloped countries. Sadly, because on the world scale of concern about the effects of volcanic hazards, Mount St Helens, indeed, volcanic hazards in developed countries, are largely irrelevant. Another glance at Figure 1.1 indicates that volcanoes are concentrated in low latitude areas. If deaths are regarded as the single most important indicator of volcanic risk, then Table 3.1 clearly suggests that tropical countries have the greatest volcanic hazards. In fact, a brief profile of a volcanic hazard-prone country suggests a relatively underdeveloped land, tropical, excolonial, with a relatively low gross national product, a poor communication network, limited infrastructure development, and a heavy reliance by the populace on subsistence agriculture.

It is in these areas where the need for the type of data base presented in Table 9.1 is greatest. In such cases, the various vulnerability maps need to be incorporated into country profiles which allow adequate assessment of likely international disaster needs (Chapter 4, Section 6).

While, on a world-scale, volcanic hazard impacts such as the 1980 Mount St Helens eruption, may be largely irrelevant, the information gained and the experience are invaluable. As a result of the eruption we are in a better position to estimate the effects of tephra falls of various thicknesses on a variety of crops, for example (Table 7.3), and to protect vehicle engines (Figure 6.25) and aeroplanes (Chapter 6, Section 6.2). While Chapters 5, 6 and 8 focus on the varied effects of the Mount St Helens eruptions, the examples given should be viewed as evidence of the range of effects that have occurred in many eruptions, but have not been adequately recorded. Although the communication networks and machinery around Merapi, Mayon, Masaya and El Misti may be less sophisticated than those affected near Mount St Helens, the effects and the protective mechanisms may well be the same. At least such possible effects need to be considered during the development of contingency plans.

The various chapters of this volume indicate that we now know a considerable amount about the effects of volcanic hazard impacts. But the information is unevenly distributed. We need to know more about clinical causes of death and mental health during the aftermath, more about effective evacuation measures and protective design of buildings, and more about methods of reducing the vulnerability of lifelines and ameliorating the effects on agriculture. But most of all, we need to know much more about the effects of volcanic hazard impacts on underdeveloped communities and on the social and economic fabric of subsistence agriculturalists.

REFERENCES

LIST OF PUBLICATIONS REFERRED TO IN THE TEXT

Much use has been made of reports and information contained in the following list of publications. Where information is referred to in the text, a specific reference is given in parentheses. As the publications below have been consulted frequently and at widely varying dates of publication only the names and, where appropriate, locations of the publications are given.

Air Conditioning, Heating and Refrigeration News; Agricultural Engineering; Akron Beacon Journal; American Metal Market; American Photographer; American-Scandinavian Review; Anchorage Daily Times; APA Ψ Monitor (American Psychological Association)*; Ash and Agriculture; Asian Building and Construction; Atlantic and Iceland Review; The Australian Physicist; Aviation Week and Space Technology; Baltimore Sun; Barbados Agricultural Reporter; Bookmark; Business Insurance; Business Week; Chemical Week; The Christian Century; Civil Engineering – ASCE; The Columbian;* (US) *Congressional Report; Dairy Herd Management; Data Channels; Discover; DVM, the Newsmagazine of Veterinary Medicine; Editor and Publisher; Engineering News Record; FAA General Aviation News; Feedstuffs; Forest Service News; Honolulu Star Bulletin; In the Works* (King County Department of Public Works Employees' News)*; Literary Digest; Keesing's Contemporary Archives; Kobe Herald; Milling and Baking News; Morbidity and Mortality Weekly Report* (Centers for Disease Control)*; National Petroleum News; Nature; Newsweek; The New York Times; Pacific Islands Monthly* (Sydney)*; Paradise Magazine* (Honolulu)*; Penny Press* (Yakima)*; PNG Post Courier* (Port Moresby)*, Public Works; Pulp and Paper; Railfan and Railroad; Railway Age; Rainier National Bank Economic Reports; SEAN Bulletin* (Smithsonian Institution SEAN (Scientific Event Alert Network) Bulletin)*; Seattle Post-Intelligencer; SICSLP* (Smithsonian Insitution Center for Short Lived Phenomena – card catalog)*; The Spokesman Review* (Spokane)*; Time; Trains; USA Today; US News and World Report; Volcano Letter* (A weekly news leaflet of the Hawaiian Volcano Research Association)*; The Wall Street Journal; Washington Cattleman; Washington Geologic Newsletter; County Road Report* (US County Road Administration Board)*; Yakima Herald-Republic; Yakima Valley Sun.*

Acord, J.E. 1980, State perspective and activities: Montana, *Forum on the effects of the Mount St. Helens eruption on water resources,* Pacific Northwest River Basins Commission, Vancouver, June 12, 1980, 71-72.

Adams, R.D. and Dibble, R.R. 1967, Seismological studies of the Raoul island eruption, 1964, *New Zealand Journal of Geology and Geophysics,* 19, 1948-1961.

Alaska's volcanoes: Northern link in the ring of fire, 1976, Alaska Geographic Society [Henning, R.A., Rosenthal, C.H., Olds, B., and Reading, E. (eds.)], Anchorage.

Alfano, G.B. and Friedlaender, I. 1929. *Die Gesichte des Vesuv,* Dietrich Reimer, Berlin

Algue, J.S.J. 1911, Phenomena accompanying the eruption of Taal volcano, January 30, 1911. *London Quarterly Journal of the Royal Meteorological Society,* 37, 273-274.

Allen, J.E. 1980, Final report on the operations of the Mount St. Helens research and educational coordinating committee, Gifford Pinchot National Forest Office, Vancouver, WA.

Allen, J.E. 1982, Scientific access to Mount St. Helens, 1980 (abstract) in S.A.C. Keller (ed.), *Mount St. Helens: One year later,* Eastern Washington University Press, Cheney.

Anderson, J. 1982, Mount St. Helens May 18 ashfall: The human ecology of an unanticipated natural hazard, in S.A.C. Keller, (ed.) *Mount St. Helens: One year later,* Eastern Washington University Press, Cheney, 181-189.

Anderson, R. 1907, The great Japanese volcano Aso. *Popular Science Monthly,* June, 29-39.

Anderson, R., Björnsson, S., Blanchard, D.C., Gathman, S., Hughes, J., Jónasson, S., Moore, C.B., Survilas, H.J. and Vonnegut, B. 1965, Electricity in volcanic clouds. *Science,* 148, 1179-1189.

Anderson, T. 1908, The volcanoes of Guatemala. *Royal Geographical Society,* 473-489.

Anderson, T. 1910, The volcano of Matavanu in Savaii. *Quarterly Journal of the Geological Society,* 66, 621-639.

Anderson, T. and Bonney, T.G. 1917 *Volcanic studies in many lands,* vol. 2, John Murray, London.

Anderson, T. and Flett, J.S. 1903, Report on the eruptions of the Soufrière in St. Vincent, and on a visit to Montagne Pelée in Martinique. *Philosophical Transactions of the Royal Society of London, A,* 200, 353-553.

Anderson, T. and Johnston-Lavis, H.J. 1888, Notes on the late eruption in the island of Vulcano. *British Association for the Advancement of Science,* Section C, 2, 15-17.

Anon. 1717, Extract of a letter of Mr Edw. Berkeley from Naples giving several curious observations and remarks on the eruptions of fire and smoak from Mount Vesuvio. *Philosophical Transactions,* 30, 708-13.

Anon. 1771, An account of the eruption of Vesuvius, in 1767. *Transactions of the American Philosophical Society, held at Philadelphia,* 1, 281-285.

Anon. 1810, Eruption of Mount Aetna. *American Register* (Philadelphia), 6, 343-345.

Anon. 1811, Eruption of Etna. *American Register* (Philadelphia), 7, 216-220.

Anon. 1813, Description of the eruption of the Souffrier mountain, on Thursday night, the 30th of April, 1812, in the Island of St. Vincent. *New England Journal of Medicine and Surgery,* 2, 95-100.

Anon. 1815, Description of volcano and earthquake St. Vincent, 1812. *Literary and Philosophical Society of New York* 1, 315-323.

Anon. 1816, Miscellaneous observations on the volcanic eruptions at the islands of Java and Sumbawa, with a particular account of the mud volcano at Grobogar. *Quarterly Journal of Science and the Arts, London,* 1, 245-58.

Anon, 1824, The last eruption of Vesuvius. *The Port Folio* (Philadelphia), 17(5s), 52-57.

Anon. 1832, Eruption of Mount Galoungoun. *The monthly repository and library of entertaining knowledge,* 2, 413-415.

Anon. 1847, Volcanic dust of Hekla. *American Journal of Science,* 3(2), 272-273.

Anon. 1888, The tsunami of 13 March, 1888, from *Nachrichten uber Kaiser Wilhelms-Land und den Bismarck-Archipel Herausgegeben von den Neu Guinea Compagnie zu Berlin fur 1888* (trans. G. Stange, 1979), 147-149.

Anon. 1897, *Baretá dapit can volcan Mayong sa Albay can Junio de 1897* [News about Mayon volcano in Albay in June, 1897], [written by M.P. (full author's name unknown) and translated by Chris Newhall], Nueva Caceres, La Sagrada Familia.

Anon. 1902, Volcanic disturbances in Guatemala. *National Geographic,* 13(12), 461-463.

Anon. 1965, Taal casualty reports: Erratic and confusing. *Press forum – monthly publication of the Philippine Press Institute,* 11(2), 1,2,8.

Anon. 1968, Volcanic eruption compels evacuation of three bases on Deception Island. *Antarctic,* March, 23-26.

Anon. 1977, La Soufrière, volcanology and forecasting, *Nature,* 269, 96-97.

Anon. 1983, Bureaucratic and political obstacles to disaster relief, *Disasters,* 7(2), 79-80.

Aramaki, S. 1956, The 1783 activity of Asama volcano. Part I. *Japanese Journal of Geology and Geography,* 27, 189-229.

Aramaki, S. 1957, The 1783 activity of Asama volcano. Part II. *Japanese Journal of Geology and Geography*, 28(1-3), 11-33.

Aramaki, S., Shimozuru, D., and Ossaka, J. Asama volcano, Part 2 in *Field excursion guide to Fuji, Asama, Kusatsu-shirane and Nantai volcanoes, Symposium on arc volcanism,* Tokyo and Hakane, Volcanological Society of Japan and the International Association for Volcanology and Chemistry of the Earth's Interior, Japan and IAVCEI.

Arthur, T., Nelson, A.S. and Hearon, H. (eds) 1981, *A mental health training manual for disaster relief workers: Project Ashlift.* Mental Health Division, Washington State Department of Social and Health Services.

Aspinall, W.P., Sigurdsson, H. and Shepherd, J.B. 1973, Eruption of Soufrière volcano on St. Vincent Island, 1971-1972. *Science,* 181, 117-124.

Axelrod, D.I. 1981, Role of volcanism in climate and evolution. *Geological Society of America Special Paper,* 185.

Baker, P.E. 1969, A volcano erupts beneath the Antarctic ice. *Geographical Magazine,* November, 115-126.

Baker, P.E. 1979, Geological aspects of volcano prediction. *Journal of the Geological Society of London,* 136, 341-345.

Baksi, A.K. and Watkins, N.D. 1973. Volcanic production rates: Comparison of oceanic ridges, islands, and the Columbia Plateau Basalts. *Science,* 180, 493-496.

Ball, R.S. 1906, *In starry realms,* Pitman, London.

Banks, N.G. and Hoblitt, R.P. 1981. Summary of temperature studies of 1980 deposits, in P.W. Lipman and D.R. Mullineaux (eds), The 1980 eruptions of Mount St. Helens, Washington, *US Geological Survey Professional Paper,* 1250, 295-313,.

Banks, N.G., Koyanagi, R.Y. and Honma, K.T. 1981, Preliminary reports by U.S. Geological Survey Team, Pagan volcano – May 1981. *Hawaiian volcano observatory memorandum,* June 9, 1981.

Barberi, F. and Gasparini, P. 1976, Volcanic hazards. *Bulletin of the International Association of Engineering Geology,* 14, 217-232.

Bardintzeff, J.-M. and Boudon, G. 1979, Nuées ardentes de l'éruption d'avril-mai 1979 de la Soufrière de Saint-Vincent (Antilles). *Comptes rendus Academie Science Paris,* 289(D), 891-894.

Bardsdate, R.J. and Dugdale, R.C. 1972, Effects of volcanic ash falls on chemical and sediment characteristics of two Alaskan lakes, *Journal of the Fisheries Research Board of Canada,* 29(3), 229-236.

Bardswell, N. 1901, Volcanic action as a cause of outbreaks of epidemic disease. *Edinburgh Medical Journal,* 321-335.

Baross, J.A., Dahm, C.N., Ward, A.K., Lilley, M.D. and Sedell, J.R. 1982, Initial microbiological response in lakes to the Mt St Helens eruption, *Nature,* 296, 49-52.

Baxter, P.J., Ing, R., Falk, H., French, J., Stein, G.F., Bernstein, R.S., Merchant, J.A. and Allard, J. 1981, Mount St Helens eruptions, May 18 to June 12, 1980: An overview of the acute health impact. *Journal of the American Medical Association.,* 246(22), 2585-2589.

Baxter, P.J., Bernstein, R.S., Falk, H., French, J. and Ing. R. 1982, Medical aspects of volcanic disasters: An outline of the hazards and emergency response measures, *Disasters,* 6(4), 268-276.

Baxter, P.J., Stoiber, R.F. and Williams, S.N. 1982, Volcanic gases and health: Masaya Volcano, Nicaragua. *Lancet,* July 17, 150-151.

Beavers, J.E. (ed.) 1981, *Earthquakes and earthquake engineering: The eastern United States,* Proceedings of 1981 Knoxville conference, Ann Arbor Science Publications, Ann Arbor, 2 vols.

Beck, A.C. 1950, Volcanic activity at Mt Ruapehu from August to December, 1945. *New Zealand Journal of Science and Technology* 31B(5), 1-13.

Beebe, W. 1926, The birth of a volcano, in Chalmers Roberts (ed.) *The World Today,* 47, 243-257, Worlds Work, 1913.

Belshaw, C.S. 1951, Social consequences of Mt. Lamington eruption. *Oceania,* 21(4), 241-52.

Bernstein, R.S., 1983, Assessment of respiratory hazards associated with air pollutants from volcanic eruptions, Kilauea volcano, Hawaii, Public Health Service, Centers for Disease Control, Atlanta.

Bernstein, R.S., McCawley, M., Attfield, M., Green, F., Olenchock, S.A., Dollberg, D. and Engelberg, A.L. 1980, International woodworkers of America, Region III, Gladstone, OR and the Weyerhaeuser Company, Tacoma, WA. *HHE Interim report* no. GHE-80-112.

Bernstein, R.S., McCawley, M.A., Attfield, M.D., Green, F., Dollberg, D.D., Baxter, P.J., and Merchant, J.A. 1982, Epidemiologic assessment of the risk for adverse pulmonary effects from persistent occupational exposures to Mount St. Helens' volcanic ash (tephra) in S.A.C. Keller (ed.) *Mount St. Helens: One year later,* Eastern Washington University Press, Cheney, 207-213.

Block, J.W. 1975, Sodom and Gemorrah: A volcanic disaster. *Journal of Geological Education,* 26(3), 74-77.

Blong, R.J. 1981, Some effects of tephra falls on buildings, in S. Self and R.S.J. Sparks (eds), Tephra Studies, Proceedings NATO Advanced Studies Institute, Laugarvatn and Reykjavik, June 18-29, 1980, Reidel, Dordrecht, Series C, 75, 405-420.

Blong, R.J. 1982, *The time of darkness, local legends and volcanic reality in Papua New Guinea.* University of Washington Press, Seattle/Australian National University Press, Canberra.

Bolt, B.A. 1976, *Nuclear explosions and earthquakes: The parted veil,* Freeman, San Francisco.

Bolt, B.A., 1978, *Earthquakes, a primer.* Freeman, San Francisco.

Bolt, B.A., Horn, W.L., Macdonald, G.A., and Scott, R.F. 1975. *Geological Hazards.* Springer-Verlag, New York.

Bonis, S. and Salazar, O. 1973, The 1971 and 1973 eruptions of volcano Fuego, Guatemala, and some socio-economic considerations for the volcanologist, *Bulletin Volcanologique,* 37(1), 394-400.

Booth, B. 1974, Persistent Etna, *Geographical Magazine,* 46(7), 315-317.

Booth, B. 1979, Assessing volcanic risk. *Journal of the Geological Society of London,* 136, 331-340.

Bortleson, G.C., Wilson, R.T. and Foxworthy, B.L. 1977, Water-quality effects on Baker Lake of recent volcanic activity at Mt Baker, Washington. *US Geological Survey Professional Paper,* 1022-B.

Bostok, D. 1978, A deontological code for volcanologists? *Journal of Volcanology and Geothermal Research,* 4 (1), 1.

Bourguignon. 1708, A relation of the new Island thrown up near the Island of Santerini. *Philosophical Transactions,* 26, 200-208.

Brattstrom, B.H. 1963, Bárcena volcano, 1952 – its effect on the fauna and flora of San Benedicto Island, Mexico, *Proceedings, 10th Pacific Science Congress,* 1961, Bishop Museum Press, 499-524.

Brigham, W.T., 1909, The volcanoes of Kilauea and Mauna Loa, *Memoirs of the B.P. Bishop Museum,* 2(4).

Brown, W.H., Merrill, E.D., and Yates, H.S. 1917, The revegetation of Volcano Island, Luzon, Philippine Islands, since the eruptions of Taal Volcano in 1911. *Philippine Journal of Science,* 12C(4), 177-248.

Bruner, M.T., 1974, The Westmann Islanders come home to Heimaey, Geography Department, University of Washington.

Bryson, R.A. and Murray, T.J. 1977, *Climates of hunger: Mankind and the world's changing weather,* Australian National University Press, Canberra.

Bullard, F.M. 1976, *Volcanoes of the earth.* University of Texas, Austin.

Bultitude, R.J. 1976, Eruptive history of Bagana Volcano, Papua New Guinea, between 1882 and 1975, in R.W. Johnson (ed.), *Volcanism in Australasia, a collection of papers in honour of the late G.A.M. Taylor,* Elsevier, Amsterdam, 317-336.

Burdecki, F. 1964, Meteorological phenomena after volcanic eruptions. *Weather,* 19, 113-114.

Burlison, J. 1981, Relief work after the south Italy earthquake, November 1980, *Disasters,* 5(4), 349-354.

Burton, I., Kates, R.W. and White, G.F. 1978, *The environment as hazard,* Oxford University Press, New York.

BVE, *Bulletin of Volcanic Eruptions,* Volcanological Society of Japan, International Association of Volcanology and Chemistry of the Earth's Interior and the International Union of Geology and Geophysics, in *Bulletin Volcanologique.*

Cadle, R.D. and Blifford, I.H. 1971, Hekla eruption clouds. *Nature,* 230, 573-574.

Cadle, R.D. and Frank, E.R. 1968, Particles in the fume from the 1967 Kilauea eruption. *Journal of Geophysical Research,* 73(14), 4780-4783.

Cadle, R.D., Wartburg, A.F., Pollock, W.H., Gandrud, B.W. and Shedlovsky, J.P. 1973, Trace constituents emitted to the atmosphere by Hawaiian volcanoes. *Chemosphere,* 6, 231-234.

Cadle, R.D., Fernald, F.G. and Frush, C.L. 1979, Combined use of lidar and numerical diffusion models to estimate the quantity and dispersion of volcanic eruption clouds in the stratosphere: Vulcán Fuego, 1974 and Augustine, 1976. *Journal of Geophysical Research,* 82(12), 1783-1786.

Caldcleugh, A., 1836, Some account of the volcanic eruption of Coseguina, in the Bay of Fonseca, on the western coast of Central America, *Philosophical Transactions,* 1, 27-30.

Capaldi, G. and 12 others, 1978, Stromboli and its 1975 eruption. *Bulletin Volcanologique,* 41(3), 259-285.

CAVW, 1951, *Catalogue of the active volcanoes of the world including solfatara fields,* International Association of Volcanology, Rome, 22 vols.

Carr, M.H. and Greeley, R. 1980, *Volcanic features of Hawaii, a basis for comparison with Mars,* NASA Scientific and Technical Information Branch, Washington.

Carroll, J.J. and Parco, S.A. 1966, *Social organization in a crisis situation: The Taal disaster.* Philippine Sociological Society, Manila.

Centers for Disease Control, Atlanta, Georgia. Department of Health and Human Services, Public Health Service. Mount St. Helens volcano health reports #1-23 (June 2, 1980-Feb 23, 1981).

Christiansen, R.L. and Peterson, D.W., 1981, Chronology of the 1980 eruptive activity, in Lipman, P. W. and Mullineaux, D.R. (eds), The 1980 eruptions of Mount St. Helens, Washington, *US Geological Survey Professional Paper,* 1250, 17-30.

Cilento, R. 1937, The volcanic eruption in Blanche Bay, Territory of New Guinea, May, 1937. *Historical Society of Queensland Journal,* 3, 37-49.

CIRV: Kusumadimata K. (ed.), 1979, *Data Dasar Gunungapi Indonesia.* Catalogue of references on Indonesian volcanoes with eruptions in historical time, Republic Indonesia, Direktorat Vulkanologi.

Clapperton, C.M. 1972, Patterns of physical and human activity on Mount Etna, *Scottish Geographical Magazine,* 88(3), 160-167.

Clapperton, C.M. 1973a, Thrice threatened Heimaey, *Geographical Magazine,* 45(7), 495-500.

Clapperton, C.M. 1973b, Back home to work on a volcanic island, *Geographical Magazine,* 46(2), 83-90.

Clark, D.E. and Lee, H. 1965, *Ceniza-arena cleanup in San Jose, Costa Rica: Operational aspects as related to nuclear weapon fallout decontamination,* Stanford Research Institute, Project MU-5069,.

Coleman, S.N. 1946, *Volcanoes, new and old.* John Day, New York.

Collier, C.R. 1980, Hydrologic effects. *Forum on the effects of the Mount St. Helens eruption on water resources,* Pacific Northwest River Basin Commission, Vancouver, WA, June 12, 1980, 13-16.

COMVOL 1975, Commission on Volcanology. *Mayon Volcano.*

Connor, J.J. 1979, Geochemistry of *ohia* and soil lichen, Puhimau thermal area, Hawaii. *Science of the Total Environment* 12, 241-250.

Cook, R.J., Barron, J.C., Papendick, R.I., Williams, G.J., 1981, Impact on agriculture of the Mount St. Helens eruptions. *Science,* 211, 16-22.

Cooke, R.J.S. 1981, Eruptive history of the volcano at Ritter Island, in R.W. Johnson (ed.) Cooke-Ravian volume of volcanological papers, *Geological Survey of Papua New Guinea Memoir,* 10, 115-123.

Cordero, C. 1964, Costa Rican volcano in eruption since March 1963. *Americas,* 16, 22-27.

Corliss, W.R. 1977, *Handbook of unusual natural phenomena,* Source Book Project, Glen Arm, MD.

Crandell, D.R., 1971, Postglacial lahars from Mount Rainier volcano, Washington. *US Geological Survey Professional Paper,* 677.

Crandell D.R. 1973, Potential hazards from future eruptions of Mount Rainier. *US Geological Survey Miscellaneous Geologic Investigations,* Map I-836.

Crandell, D.R. and Mullineaux, D.R., 1978, Potential hazards from future euptions of Mount St. Helens volcano, Washington. *US Geological Survey Bulletin* 1383-C.

Cullen, J.M. 1978, Impacts of a major eruption of Mount Rainer on public service delivery systems in the Puyallup Valley, Washington, Departments of Geography and Urban Planning, University of Washington, in Hodge, *et al.* 1978.

Cummans, J. 1981a, Mudflows resulting from the May 18, 1980, eruption of Mount St. Helens, Washington. *US Geological Survey Circular,* 850-B.

Cummans, J. 1981b, Chronology of mudflows in the South Fork and North Fork Toutle River following the May 18 eruption, in P. W. Lipman and D.R. Mullineaux (eds), The 1980 eruptions of Mount St. Helens, Washington, *US Geological Survey Professional Paper,* 1250, 479-486.

Curtis, G.H., 1968, The stratigraphy of the ejecta from the 1912 eruption of Mount Katmai and Novarupta, Alaska. *Geological Society of America Memoir,* 116, 153-210.

Dacey, D.C. and Kunreuther, H. 1969, *The economics of natural disasters,* The Free Press, New York.

Dall, W.H., 1870, *Alaska and its resources,* Lee and Shephard, Boston.

Dana, J.D. 1850, Historical account of the eruptions on Hawaii. *American Journal of Science,* 9(2), 347-364.

Dabney, J.B. 1809, Account of a volcanic eruption at St. Georges, one of the Western Islands. *The Medical Repository* (New York) 12, 254-256.

Davidson, G. 1884. Notes on the volcanic eruption of Mount St. Augustin, Alaska, Oct. 6, 1883, *Science,* 3(54), 186-189.

Davison, C. 1912, The eruptions of the Asama-yama (Japan) in 1909-11. *Nature,* 89, 487-488.

Day, A.L., and Allen, E.T. 1925, The volcanic activity and hot springs of Lassen Peak. *Carnegie Institute of Washington Publication,* No. 360.

Day, T.G. and Fisher, J.E. 1980, Mt. St. Helens: How a wastewater plant coped with its aftermath. *Water Pollution Control Federation Journal,* 52(8), 2082-2089.

Decker, R.W. 1980, State of the art in volcano forecasting, *International Union of Geodesy and Geophysics Chronicle,* 145, 505-517.

Deiss, J.J. 1966, *Herculaneum, Italy's buried treasure.* Crowell, New York.

Delmé-Radcliffe, C. 1906, *Report by Lt.-Colonel Delmé-Radcliffe on conditions following the eruption of Vesuvius in 1906,* [14404 No. 1] H.M.S.O., London.

Del Moral, R. 1981, Life returns to Mount St. Helens, *Natural History,* 90(5), 36-46.

De Lorenzo, G. 1906, The eruption of Vesuvius in April 1906. *Quarterly Journal of the Geological Society of London,* 62, 476-83.

Dilke, O.A.W. and Dilke, M.S. 1979, Vesuvius: August 79. *Geographical Magazine,* 51(11), 743-749.

Dillman, J.J. and Roberts, M.L. 1982, The impact of the May 18 Mount St. Helens ashfall: Eastern Washington residents report on housing-related damage and cleanup, in S.A.C. Keller (ed.), *Mount St. Helens: One year later,* Eastern Washington University Press, Cheney, 191-198.

Dilmy, A. 1965, Pioneer plants found one year after the 1963 eruption of Agung in Bali. *Pacific Science,* 19, 498-50l.

Dion, N.P. and Embrey, S.S. 1981, Effects of Mount St. Helens eruption on selected lakes in Washington, *US Geological Survey Circular,* 850-G, 1-25.

Docters van Leeuwen, W.M. 1936, Krakatau 1883-1933 *Annales du Jardin Botanique de Buitenzorg,* vols. 46-47, E.J. Brill, Leiden.

Dudley, H. 1980, Marine activity. *Forum on the effects of the Mount St. Helens eruption on water resources,* Pacific Northwest River Basins Commission, Vancouver, June 12, 1980, 31-33.

Dueker, K.J., Pendleton, P., Robinson, L., Gihring, T., and Rabiega, B. 1980, *Impacts of volcanic ashfall on travel behavior,* Center for Urban Studies, School of Urban Affairs, Portland State University, Oregon.

Dynes, R.R. and Quarantelli, E.L. 1976, The family and community context of individual reactions to disaster in Parad, H.J., Resnik, H.L.P. and Parad, L.G. (eds), *Emergency and Disaster Management: A mental health sourcebook,* Charles Press, Bowie, MD, 231-244.

Eggler, W.A. 1948, Plant communities in the vicinity of the volcano El Parícutin, Mexico, after two and a half years of eruption. *Ecology,* 29(4), 415-436.

Eggler, W.A. 1959, Manner of invasion of volcanic deposits by plants, with further evidence from Parícutin and Jorullo, *Ecological Monographs,* 29(3), 267-284.

Eggler, W.A. 1963, Plant life of Parícutin volcano, Mexico, eight years after activity ceased, *The American Midland Naturalist,* 69(1), 38-68.

Eggler, W.A. 1967, Influence of volcanic eruptions on xylem growth patterns. *Ecology,* 48(4), 644-47.

Eisele, J.W., O'Halloran, R.L., Reay, D.T., Lindholm, G.R., Lewman, L.V. and Brady, W.J. 1981, Deaths during the May 18, 1980, eruption of Mount St. Helens, *The New England Journal of Medicine,* 305(16), 931-936.

Endo, E.T., Malone, S.D., Noson, L.L. and Weaver, C.S. 1981, Locations, magnitudes, and statistics of the March 20-May 18 earthquake sequence, in P.W. Lipman, and D.R. Mullineaux (eds) The 1980 eruptions of Mount St. Helens, Washington, *US Geological Survey Professional Paper,* 1250, 93-107.

Erskine, H.D. 1940, Katmai's black-out. *Alaska Sportsman,* 16-17, 22.

Erskine, W.F. 1962, *Katmai,* Abelard-Schuman, London.

Espino, R.B. 1940, Studies of the fertilizing value of Mayon volcano ash: 1. Effects upon plants of liberal application of the volcanic ash. *Philippine Agriculturist,* 28, 133-142.

Espino, R.B. 1940, Studies of the fertilizing value of Mayon volcano ash: 2. Beneficial effects of adding ammonium sulfate. *Philippine Agriculturist,* 28, 260-270.

EH&WS, Everett Herald and Western Sun, 1980, *Mount St. Helen erupts.* Daily Herald Co., Everett, WA.

FAA, 1980, Federal Aviation Administration, *General aviation airworthiness alerts: Volcanic ash hazard.* U.S. Dept of Transportation AC No. 43-16, May 21, 1980.
Faustino, L.A. 1929, Mayon volcano and its eruptions. *Philippine Journal of Science,* 40(1), 1-43.
Finch, R.H. 1924, Seismic sequences of the explosive eruption of Kilauea in May, 1924. *Bulletin of the Seismological Society of America,* 14(4), 216-222.
Fisher, B.F. 1902, Vesuvius, destroyer of cities. *The Cosmopolitan,* 32(6), 573-584.
Fisher, N.H. 1939, Geology and vulcanology of Blanche Bay and the surrounding area, New Britain, *Territory of New Guinea Geological Bulletin,* 1, 68.
Fisher, R.V. 1961, Proposed classification of volcaniclastic sediments and rocks. *Geological Society of America Bulletin,* 72, 1409-1414.
Fisher, R.V., 1964, Maximum size, median diameter, and sorting of tephra. *Journal of Geophysical Research,* 69(2), 341-355.
Fisher, R.V., and Schmincke, H.-V. 1984, *Pyroclastic Rocks,* Springer-Verlag, Heidelberg.
Fisher, R.V., Smith, A.L. and Roobol, M.J. 1980, Destruction of St. Pierre, Martinique, by ash-cloud surges, May 8 and 20, 1902. *Geology,* 8, 472-476.
Fiske, R.S. 1979, A deontological code for volcanologists? – a response to Derek Bostok's editorial. *Journal of Volcanology and Geothermal Research,* 5, 211-212.
Fiske, R.S. 1980, Volcano hazards: Lessons learned in the eastern Caribbean, *Earthquake Information Bulletin,* 12(4), 150-153.
Fiske, R.S. 1981, Scientists and the news media: Contrasting interactions during two volcanic crises in the eastern Caribbean. *International Association for Volcanology and Chemistry of the Earth's Interior Conference Abstracts,* Tokyo, 93.
Flood Loss Reduction Associates, 1981. *Cooperative flood loss reduction – a technical manual for communities and industry,* SEDA-Council of Governments and U.S. Water Resources Council.
Foster, Ted. 1976, *Bushfire – history, prevention, control,* A.H. and A.W. Reed, Sydney.
Francis, P. 1976, *Volcanoes.* Penguin, Harmondsworth.
Francis, P. 1979, Vesuvius inquest: August 1979. *Geographical Magazine,* 51(11), 750-755.
Francis P.W. and Baker, M.C.W. 1977, Mobility of pyroclastic flows, *Nature,* 270, 164-165.
Franklin, B. 1784, Meteorological imaginations and conjectures, in *Eos,* August 10 1982, 601.
Fritz, W.J. 1980, Stumps transported and deposited upright by Mount St. Helens mud flows. *Geology,* 8, 586-588.
Fudali, R.F. and Melson, W.G. 1972, Ejecta velocities, magma chamber pressure and kinetic energy associated with the 1968 eruption of Arenal volcano. *Bulletin Volcanologique,* 35(2), 383-401.
Fuji, A. and Nakao, S. 1978, Preliminary observations for the effect of volcanic ashes on scallop culture in Uchiura Bay, *Usu Eruption and its Impact on Environment,* Hokkaido University, Dec 1978, 233-237 (trans. US Geological Survey).
Furneaux, R., 1965, *Krakatoa,* Secker and Warburg, London.
Galunggung, 1982, Lions Club of Bandung, Alumni Publishers.
Galvez, N.L. 1938, The chemical and physical composition of the fine ejecta of Mayon volcano, *Philippine Agriculturist,* 27, 765-794.
Galvez, N.L., Aquino D.I. and Mamisao, J.P. 1939, Agricultural value of fine ejecta of Mayon volcano. *Philippine Agriculturist,* 27, 844-849.
Gamroth, M. 1980, In Washington State everything's coming up ashes, *Dairy Herd Management,* 17(9), 8-10,14.
GAO, 1982, *Federal involvement in the Mount St Helens disaster: Past expenditures and future needs,* Report to the Congress of the United States, U.S. General Accounting office, AO/RCED-83-16.
Gass, I.G., Harris, P.G. and Holdgate, M.W. 1963, Pumice Eruption in the area of the South Sandwich Islands. *Geological Magazine* 100(4), 321-330.
Gates, F.C. 1914, The pioneer vegetation of Taal volcano, *Philippine Journal of Science,* 9C(5), 391-434.
Gemmellaro, M. 1825, An account of the eruption of Mount Etna, on the 27th May, 1819. *The Carolina Journal of Medicine, Science and Agriculture,* 1, 344-349.
Gerlach, T.M. 1980, Evaluation of volcanic gas analyses from Surtsey Volcano, Iceland, 1964-1967. *Journal of Volcanology and Geothermal Research,* 8, 191-198.
Glasse, R.M. 1965, The Huli of the southern Highlands, in P. Lawrence and M.J. Megitt (eds) *Gods, ghosts and men in Melanesia,* Oxford University Press, Melbourne, 27-49.
Goldberry, J.P. 1916, Eruption of Bulusan volcano. *Philippine Journal of Science,* A 11, 251-256.

Golan, N., 1978, *Treatment in crisis situations,* Free Press, New York.

Gorshkov, G.S. 1959, Gigantic eruption of the volcano Bezymianny. *Bulletin Volcanologique,* 20, 76-109.

Gorshkov, G.S. 1963, Directed volcanic blasts. *Bulletin Volcanologique,* 26, 83-88.

Gorshkov, G.S. 1972, General introduction, in UNESCO, *The surveillance and prediction of volcanic activity,* 9-13.

Gorshkov, G.S. and Dubik, Y.M., 1970, Gigantic directed blast at Shiveluch volcano (Kamchatka), *Bulletin Volcanologique,* 24, 261-288.

Granville, A.B. 1817, A report on a memoir by Signor Monticelli, perpetual secretary of the Royal Academy of Sciences of Naples, entitled "Descrizione dell'Eruzione de Vesuvio avvenutane'giorni 25 e 26 Dicembre, 1813", *The Quarterly Journal of Science and the Arts,* London, V.2, 25-34.

Grayland, E. and Grayland, V. 1971, *Tarawera,* Hodder and Stoughton, Auckland.

Green, F.H.Y., Vallyathan, V., Mentnech, M.S., Tucker, J.H., Merchant, J.A., Kiessling, P.J., Antonius, J.A. and Parshley, P. 1981, Is volcanic ash a pneumoconiosis risk. *Nature,* 293, 216-217.

Green, F.H.Y. and 19 others. 1982, Health implications of the Mount St. Helens' eruption: Laboratory investigation. *Annals of Occupational Hygiene,* 26(1-4), 921-933.

Greene, M.R., Perry, R.W., Lindell, M.K. 1980, *The March, 1980 eruptions of Mt. St. Helens: Citizen perceptions of volcano hazard,* Batelle Human Affairs Research Centers, Seattle.

Greene, M., Perry, R. and Lindell, M. 1981, The March 1980 eruptions of Mt. St. Helens: Citizen perceptions of volcano threat. *Disasters,* 5(1), 49-66.

Griggs, R.F. 1918. Scientific results of the Katmai expedition of the National Geographic Society: 1. The recovery of vegetation at Kodiak. *Ohio Journal of Science,* 19(1), 1-57.

Griggs, R.F. 1919, Scientific results of the Katmai expedition of the National Geographic Society: 4. The character of the eruption as indicated by its effects on nearby vegetation. *Ohio Journal of Science,* 19(3), 173-209.

Griggs, R.F. 1922, *The valley of ten thousand smokes.* National Geographic Society, Washington, D.C.

Guest, J.E. and Murray, J.B. 1979, An analysis of hazard from Mount Etna volcano. *Journal of the Geological Society of London,* 136, 347-354.

Guest, J., Murray, J., Kilburn, C. and Lopes, R. 1980, Eruptions on Mount Etna during 1979. *Earthquake Information Bulletin,* 12(4), 154-160.

Gunnarsson, F. 1973, *Volcano, ordeal by fire in Iceland's Westmann Islands,* Iceland Review Books, Reykjavik.

Hadikisumo, D. 1974, The rise and drop of Mt. Kelut crater bottom after paroxysmal eruptions. *Tectonophysics,* 23, 341-347.

Hallagan, W., Joerding, W. and Koepke, D. 1981, Cost function estimates of ash cleanup after Mt. St. Helens, *Washington State University Economics Department,* Working Paper 181-1.

Hallauer, W.G. 1980, State perspective and activities: Washington, *Forum on the effects of the Mount St. Helens eruption on water resources,* Pacific Northwest River Basins Commission, Vancouver, June 12, 1980, 61-64.

Hamilton, W. 1772, *Observations on Mt. Vesuvius, Mt. Etna and other volcanoes,* Cadell, LONDON.

Hammer, C.U., Clausen, H.B. and Dansgaard, W. 1980, Greenland ice sheet evidence of post-glacial volcanism and its climatic impact. *Nature,* 288, 230-235.

Harnly, C.D. and Tyckoson, D.A. 1981, *Mount St. Helens, the 1980 eruptions: A bibliography,* Vance Bibliographies, Public Administration Series, P-786, Monticello, Illinois.

Harris, S.L. 1976, *Fire and ice: The Cascade volcanoes,* The Mountaineers, Pacific Search Books, Seattle.

Harrison, P., 1980, Disaster relief: relief or disaster, *New Society,* July 10, 1980, 66-67.

Hatori, T. 1964, A study of the damage to houses due to a tsunami, *Bulletin of the Earthquake Research Institute,* 42, 181-191.

Hayakawa, Y., Aramaki, S., Shimozuru, D., Ossaka, J., 1981, Kusatsu-shirane volcano, Part 3 in *Field excursion guide to Fuji, Asama, Kusatsu-shirane and Nantai volcanoes, Symposium on arc volcanism,* Tokyo and Hakone, Volcanological Society of Japan and the International Association for Volcanology and Chemistry of the Earth's Interior.

Hazlett, R.W. 1981, *Guide to points of geologic interest in Hawaii Volcanoes National Park Hawaii,* Circum-Pacific Council for Energy and Mineral Resources, Honolulu.

Heathcote, R.L. 1979, The threat from natural hazards in Australia, in Heathcote, R.L. and Thom, B.G. (eds) *Natural hazards in Australia,* Australia Academy of Science, Canberra, 3-12.

Hédervári, P. 1971, Energetical calculations concerning the Minoan eruption of Santorini, *Acta 1st International Scientific Congress on the volcano of Thera,* Greece, 15-23 Sept. 1969, Archaeological Services of Greece, Athens, 257-276.

Heilprin, A. 1903, *Mont Pelée and the tragedy of Martinique.* J.B. Lippincott, Philadelphia.

Heiser, V. 1936, *An American doctor's odyssey,* Norton & Co., New York.

Herbet, D. and Bardossi, F. 1968, *Kilauea: Case history of a volcano,* Harper & Row, New York.

Higashi, S., Araya, T. and Onodera, H. 1978, Studies on the debris-flow and erosion control works at Mt. Usu. *Usu eruption and its impact on environment,* Hokkaido University, 87-99 (trans. by U.S. Geological Survey).

Hill, M.R. 1970, Mount Lassen is in eruption and there is no mistake about that. *Mineral Information Service, Publication of California Division of Mines and Geology,* 23(11), 211-224.

Hill, R.T. 1902a, Report by Robert T. Hill on the volcanic disturbances in the West Indies. *National Geographic,* 13(7), 223-267.

Hill, R.T. 1902b, A study of Pelée: Impressions and conclusions of a trip to Martinique. *Century Magazine,* 64, 764-785.

Hirose, H. 1979, Volcanic eruption and local politics in Japan: A case study. *Mass emergencies,* 4, 53-62.

Hirschboeck, K.K. 1980, A new worldwide chronology of volcanic eruptions. *Palaeogeography, Palaeoclimatology, Palaeoecology,* 29(3-4), 223-241.

Hitchcock, C.H. 1909, *Hawaii and its volcanoes,* Hawaiian Gazette Co., Honolulu.

HMSO, 1903, His Majesty's Stationery Office. Further correspondence relating to the volcanic eruptions in St. Vincent and Martinique in 1902 and 1903.

Hobbs, W.H. 1906, The grand eruption of Vesuvius in 1906. *Journal of Geology,* 14, 636-655.

Hoblitt, R.P., Miller, C.D. and Vallance, J.W. 1981, Origin and stratigraphy of the deposit produced by the May 18 directed blast, in Lipman P. W. and Mullineaux D.R. (eds). The 1980 eruptions of Mount St. Helens, Washington, *US Geological Survey Professional Paper,* 1250, 401-419.

Hodge, D.C., Sharp, V.L., Marts, M.E., Sheridan, F.E., MacGregor, J.M. and Cullen, J.E., 1978, *Social implications of volcano hazard — case studies in the Washington Cascades and Hawaii,* National Science Foundation — RANN Program, Project No. ENV-76-20735.

Hoffer, W., 1982, *Volcano, the search for Vesuvius,* Summit, New York.

Holocaust, 1980, *Mt. St. Helens. Holocaust, a diary of destruction,* Boone Publication, Texas.

Holway, R.S. 1915, The volcanic activity of Lassen Peak, California. *Popular Science Monthly,* 86(3), 290-305.

Horoguchi, K. and Kinoshita, S. 1978, On the infiltration of an alkaline element in volcanic ash into the ground. *Usu eruption and its impact on the environment,* Hokkaido University, 141-145 (English abstract).

Horowitz, H. 1983, Mt. St. Helens: An economic and cultural explosion, Association of American Geographers Annual Conference, Denver, 24-27 April 1983 (extended abstract).

Horton, R.J.M. and McCaldin, R.O. 1964, Observations on air pollution aspects of Irazú Volcano, Costa Rica. *Public Health Reports, Washington,* 79(10), 925-929.

Hovey, E.O. 1902, The eruptions of La Soufrière, St. Vincent, in May, 1902. *National Geographic,* 13(12), 444-459.

Howard, Lord, 1931, The study of natural catastrophes. *Geographical Journal,* 78, 457-60.

Huggins, P. F. 1902, *Report on the Soufrière Saint Vincent, B.W.I.,* (no publisher given).

Hulbert, C., 1827, *Volcanic wonders, and scenes of astonishment: being historic and scientific descriptions of the volcanoes of the Azores, and a general view of burning mountains, in various parts of the globe,* Langman & Co., London.

Hunt, C.E. and MacCready, J.S., 1980, *The short-term economic consequences of the Mount St. Helens volcanic eruptions in May and June, 1980,* Washington State Department of Commerce and Economic Development, Research Division.

Hyde, W. W. 1916, The volcanic history of Etna. *The Geographical Review,* 1(6), 401-418.

ICAO. 1983, The effects of volcanic ash on aircraft operations. *International Civil Aviation Organization, Working paper.*

Iida, K. and Ishimoto, M. 1978, Study on the influence of the eruption of Usu volcano and safety countermeasure in Tourist Town, Toyako Spa Town, *Usu eruption and its impact on environment,* Hokkaido University, 465-485 (trans. US Geological Survey).

Ikeya, H., 1979, *Introduction to Sabō works — The preservation of the land against sediment disaster,* The Japan Sabō Association, Tokyo.

Imbò, G. 1979, Seismic prediction at Pozzuoli. *Nature,* 277, 511.

Ingleby, I. 1966, Mount Lamington fifteen years later. *Australian Territories,* 6, 28-34.

Innova, 1980, Executive Summary, *An assessment of the economic and health impacts of volcanic ashfall in region X (Idaho, Oregon and Washington only) on migrant and seasonal farmworkers,* Innova Corporation, Seattle.

Jackson, D.B., Swanson, D.A., Koyanagi, R.Y. and Wright, T.L. 1975, The August and October 1968 East Rift eruptions of Kilauea Volcano, Hawaii. *US Geological Survey Professional Paper,* 890.

Jackson, E.L. 1982, The Laki eruption of 1783: impacts on population and settlement in Iceland. *Geography,* 67(1), 42-50.

Jaggar, T.A., 1902, Field notes of a geologist in Martinique and St. Vincent. *Popular Science Monthly,* 352-368.

Jaggar, T.A. 1904, The eruption of Pelée, July 9, 1902. *Popular Science Monthly,* January, 219-231.

Jaggar, T.A. 1906, The volcano Vesuvius in 1906. *Technology Quarterly and Proceedings,* 19(2), 104-115.

Jaggar, T.A. 1924, Sakura-jima, Japan's greatest volcanic eruption. *National Geographic,* 45, 441-470.

Jaggar, T.A. 1945, *Volcanoes declare war; Logistics and strategy of Pacific Volcano Science,* Paradise of the Pacific Ltd, Honolulu.

Jaggar, T.A. 1956, *My experiments with volcanoes,* Hawaiian Volcano Research Assn, Honolulu.

Jaggar, T.A. n.d., The eruption of Mt. Pelée, 1851. trans. of the French article by Le Prieur, Peyraud and Rufz, Harvard University, 51-73.

Janda, R.J., Scott, K.M., Nolan, K.M. and Martinson, H.A. 1981, Lahar movement, effects, and deposits, in Lipman P.W. and Mullineaux D.R., The 1980 eruptions of Mount St. Helens, Washington, *US Geological Survey Professional Paper,* 1250, 461-478.

Jashemski, W.F. 1979, Pompeii and Mount Vesuvius, A.D. 79, in Sheets, P.D. and Grayson D.K. (eds) *Volcanic activity and human ecology,* Academic Press, New York, 587-622.

Jennings, P., 1969, Disruptions of the environmental balance: The eruptions of Mt. Agung and Mt. Kelut, Indonesia, M.A. Geography thesis, University of Hawaii.

Jensen, H.I. 1906, The geology of Samoa, and the eruptions in Savaii. *Proceedings of the Linnean Society of New South Wales,* 23, 641-672.

Johansen, C., 1980, Mount Saint Helens blows the season for many Washington beekeepers, *American Bee Journal,* 120(7), 500-502.

Johnson, R.W., Davies, R.A. and White, A.J.R. 1972, Ulawun Volcano, New Britain, *Bureau of Mineral Resources Geology and Geophysics, Australia, Bulletin* 142 (Bull PNG5).

Johnston-Lavis, H.J. 1914, *On the Effects of Volcanic Action in the Production of Epidemic Diseases in the Animal and in the Vegetable Creation, and in the Production of Hurricanes and Abnormal Atmospherical Vicissitudes,* John Bale, Sons and Danielsson, London.

Judd, J.W. 1888, On the volcanic phenomena of the eruption, and on the nature and distribution of the ejected materials, in Symons, G. (ed.) 1888, 1-46.

Kadomura, H., Yamomoto, H., Suzuki, R. and Suzuki, K. 1978, Some aspects of erosional form and processes on tephra-covered hillslopes produced by the 1977 eruption of Usu volcano, Hokkaido, Japan: August 1977 — early August 1978. *Usu eruption and its impact on environment,* Hokkaido University, 121-139 (trans. US Geological Survey).

Kadomura, H., Takahasi, H., Yamamoto, H., Suzuki, R., Arai, K., Inagawa, T. and Tanabe, T. 1979, Behaviour of mudflows on valley flats and alluvial fans at Usu Volcano, *Investigation of Usu Volcano mud-flow disasters on October 24, 1978,* 41-76 (in Japanese with English abstract).

Kadomura, H., Yamamoto, H., Imagawa, T., Takahashi, H., Suzuki, R., Tanabe, T., Arai, K., Suzuki, K., and Tajime, J. 1980, Erusion and mudflows at Usu Volcano after the 1977-1978 eruption: August 1977-December 1979, *Environmental Science,* Hokkaido, 3(2), 155-184.

Kaneta, H. 1978, The dynamics of the disaster behaviours by the eruption of Mt Usu, 1977, *Usu eruption and its impact on environment,* Hokkaido University, 365-388, (trans. US Geological Survey).

Kasahara, M. and Suzuki, S. 1978, Seismic activity immediately before and in the early stage of the 1977 eruption of Usu volcano, Hokkaido, Japan, *Usu eruption and its impact on environment,* Hokkaido Univiversity, 13-24, (trans. US Geological Survey).

Kataoka, T., Umeda, Y., Sakurada, J. and Nagasawa, T. 1978, The restoration works of the damage on agricultural land by the 1977 eruption of Usu volcano, *Usu eruption and its impact on environment,* Hokkaido University, 357-363, (trans. US Geological Survey).

Katsui, Y. and 21 others, 1978, Preliminary report on the 1977 eruption of Usu volcano. *Journal of the Faculty of Science Hokkaido University,* Series IV, 18(3), 385-408.

Katsui, Y., Yokoyama, I., Watanabe, H., Murozumi, M., 1981, Usu Volcano, Part I in Katsui, Y. (ed.) *Field excursion guide to Usu and Tarumai volcanoes and Noboribetsu Spa,* Volcanological Society of Japan and the International Association for Volcanology and Chemistry of the Earth's Interior.

Keenan, G. 1902, *The tragedy of Pelée: A narrative of personal experience and observation in Martinique,* Outlook, New York.

Keesing, F. 1952, The Papua Orokaiva versus Mt. Lamington: Cultural shock and its aftermath. *Human Organisation,* 11(1), 16-22.

Kellogg, W.W., 1983, The causes of climatic variations, Abstr. *Volcanoes and climate conference,* Boulder, CO, 18 March.

Kieffer, S.W. 1981, Fluid dynamics of the May 18 blast at Mount St. Helens, in Lipman P. W. and Mullineaux D.R. (eds) The 1980 eruptions of Mount St. Helens, Washington, *US Geological Survey Professional Paper,* 1250, 379-400.

Kienle, J. and Forbes, R.B. 1976, Augustine — evolution of a volcano, *Annual Report,* Geophysical Institute, University of Alaska, 26-48.

Kienle, J. and Shaw, G.E. 1979, Plume dynamics, thermal energy and long-distance transport of vulcanian eruption clouds from Augustine volcano, Alaska, *Journal of Volcanology and Geothermal Research,* 6, 139-164.

Kikuchi, K., Endoh, T., Konno, T. and Murakami, M. 1978, Atmospheric electrical observations of the 1977 eruption of the Mt. Usu volcano. *Usu eruption and its impact on environment,* Hokkaido University 75-85 (trans. US Geological Survey).

Kilijanek, T.S. 1981, *There she blows: The search and rescue response to the Mount St. Helens volcano,* Monograph 11, Department of Sociology, University of Denver.

King, A.R. 1962, *The efficiency of rural fire fighters,* CSIRO, Melbourne, Chemical Research Laboratory Technical Paper No. 4.

King, S.S. 1964, Eastern Bali digs out of year old volcanic ash, *New York Times,* March 30, 1964.

Kish, T. 1980, Mt. St. Helens' effects on Washington's treatment plants, *Water Pollution Control Federation Journal,* 52(9), September 1980, 2310-2313.

Kjartansson, G. 1951, The eruption of Hekla 1947-48 II, 4. Water flood and mudflows, *Societes Scientiarum Islandica,* Reykjavik.

Kjartansson, G. 1957, The eruption of Hekla 1947-1948 III, 1. Some secondary effects of the Hekla eruption, *Societes Scientiarum Islandica,* Reykjavik.

Klohn, E. 1963, The February, 1961 eruption of Calbuco volcano, *Seismological Society of America Bulletin,* 53, 1435-36.

Kobe Herald correspondents, 1914, *The eruption of Sakurajima; condition of the island as viewed by Kobe Herald correspondents,* Kobe Herald, Kobe.

Koster, A.L. 1956, City of the dead. *Natural History,* 65, 412-415.

Koto, B. 1916, The great eruption of Sakura-jima in 1914. *Journal of the College of Science Imperial University of Tokyo,* 38(3), 229.

Kozlowski, T.T. and Mudd, J.B. 1975, *Responses of plants to air pollution,* Academic Press, New York, 383p.

Kratky, B.A., Fukunaga, E.T., Hylin, J.W. and Nakano, R.T. 1974, Volcanic air pollution: Deleterious effects on tomatoes. *Journal of Environmental Quality,* 3(2), 138-40.

Krauskopf, K.B. 1948, Lava movement at Parícutin volcano, Mexico. *Geological Society of America Bulletin,* 59, 1267-1283.

Kudo, N., Yoshii, G., Fujimoto, Y., Yanagawa, R., Namioka, S., Kanno, T. and Mikami, T 1978, Comparative medical studies on influences of eruption of Mt. Usu in wild animals, *Usu Eruption and its Impact on Environment,* Hokkaido University, 155-168 (trans. US Geological Survey).

Kuenzi, W.D., Horst, O.H., McGehee, R.V. 1979, Effect of volcanic activity on fluvial-deltaic sedimentation in a modern arc-trench gap, southwestern Guatemala, *Geological Society of America Bulletin* 90, 827-838.

Kutschera, M. 1903, The volcanic eruption of Torishima. *Geographical Journal,* 21, 436-439.

Lachman, R. and Bonk, W.J. 1960, Behaviour and beliefs during the recent volcanic eruption at Kapoho, Hawaii. *Science,* 131, 1095-1096.

Lacroix, A. 1904, *La Montagne Pelée et ses eruption,* Masson et cie, Paris.

Lacroix, A. 1906, The eruption of Vesuvius in April, 1906. *Smithsonian Institution Annual Report,* 223-248.

Lamb, H.H. 1970, Volcanic dust in the atmosphere; with a chronology and assessment of its meteorological significance, *Philosophical Transactions of the Royal Society of London,* 266A, 425-533.

Lamb, H.H. 1971, Volcanic activity and climate, *Palaeogeography, Paleoclimatology, Paleoecology,* 10(2-3), 203-230.

Latter, J.H. 1981, Tsunamis of volcanic origin: Summary of causes, with particular reference to Krakatoa, 1883, *Bulletin Volcanologique,* 44(3), 467-490.

Lawyer, D. 1980, Navigation and flooding, *Forum on the effects of the Mount St. Helens eruption on water resources,* Pacific Northwest River Basins Commission, Vancouver, June 12, 1980, 27-29.

Legowo, D. 1980, Volcanic debris control applied in Indonesia, *Journal of Hydrology* (New Zealand), 20(1), 71-79.

Le Guern, F., Tazieff, H. and Faivre Pierret, R. 1982, An example of health hazard, people killed by gas during a phreatic eruption: Dieng Plateau (Java, Indonesia), February 20 1979, *Bulletin Volcanologique,* 45(2), 153-156.

Leik, R.K., Leik, S.A., Ekker, K. and Gifford, S. 1982, Under the threat of Mt. St. Helens: a study of chronic family stress, FEMA contract No FEMA/EM W — C — 9454, Work Unit No 2234F.

Leppmann, W. 1966, *Pompeii in fact and fiction,* Elek, London.

Leus, X., Kintanar, C. and Bowman, V., 1981, Asthmatic bronchitis associated with a volcanic eruption in St. Vincent, West Indies. *Disasters,* 5(1), 67-69.

Lewis, H.E., Roberts, D.F. and Edwards, A.W.F. 1972, Biological problems, and opportunities, of isolation among the islanders of Tristan da Cunha, in Glass, D.V. and Revelle, R., (eds). *Population and social change,* Arnold, London, 383-417.

Lewis, J. 1979, Volcano in Tonga. *Journal of Administration Overseas,* 18(2), 116-121.

Lirer, L., Pescatore, T., Booth, B. and Walker, G.P.L. 1973, Two Plinian pumice-fall deposits from Somma-Vesuvius, Italy, *Geological Society of America Bulletin,* 84, 759-772.

Lockwood, J.P. and Torgerson, F.A. 1980, Diversion of lava flows by aerial bombing — lessons from Mauna Loa Volcano, Hawaii, *Bulletin Volcanologique,* 43(4), 727-741.

Lombard, R.E., Miles, M.B., Nelson, L.M., Kresch, D.L. and Carpenter, P.J. 1981, Channel conditions in the Lower Toutle and Cowlitz Rivers resulting from the mudflows of May 18, 1980. *US Geological Survey Circular,* 850-C.

Loomis, B.F. 1926 (1971), *Pictorial history of the Lassen volcano,* Loomis Museum Association, California.

Lowdermilk, W.C. and Bailey, R.W. 1946, A look at Parícutin. *Soil Conservation,* 11(12), 280-288.

Lowenstein, P.L. 1982, Problems of volcanic hazards in Papua New Guinea. *Geological Survey of Papua New Guinea,* Report 82/7.

MacCready, J.S. 1982, Some economic consequences of the eruptions, in Keller, S.A.C. (ed.) *Mount St. Helens: One year later,* Eastern Washington University Press, Cheney, 215-224.

Macdonald, G.A. 1962, The 1959 and 1960 eruptions of Kilauea volcano, Hawaii, and the construction of walls to restrict the spread of the lava flows. *Bulletin Volcanologique,* 24, 249-294.

Macdonald, G.A. 1972, *Volcanoes,* Prentice Hall, Englewood Cliffs, NJ.

Macdonald, G.A. and Alcaraz, A. 1956, Nuées ardentes of the 1948-1953 eruption of Hibok — hibok. *Bulletin Volcanologique,* 18, 169-178.

Macdonald, G.A. and Eaton, J.P. 1964, Hawaiian volcanoes during 1955. *US Geological Survey,* Bulletin 1171.

Macdonald, G.A. and Abbott, A.T. 1970, *Volcanoes in the sea — The geology of Hawaii,* University of Hawaii Press, Honolulu, 441.

Macdonald, G.A. and Hubbard, D.H. 1982, *Volcanoes of the National Parks in Hawaii,* Hawaii Natural History Assn, Honolulu (revised R.W. Hazlett).

Machado, F., Parsons, W.H., Richards, A.F. and Mulford, J.W. 1962, Capelinhos eruption of Fayal volcano, Azores, 1957-1958. *Journal of Geophysical Research,* 67(9), 3519-3529.

Mair, W.G. 1886, Notes on the eruption of Tarawera Mountain and Rotomahana, 10th June, 1886, as seen from Taheke, Lake Rotoiti, *New Zealand Institute Transactions and Proceedings,* 19, 372-374.

Maiuri, A. 1961, Last moments of the Pompeians. National Geographic, 120(5), 651-669.

Malin, M.C. 1980, Lengths of Hawaiian lava flows, *Geology, 8,* 306-308.

Marca de la Torre, D.J. 1753, Description of a torrent of fire, that issued from Mount Vesuvius in the year 1751, in *Observations upon the antiquities of the town of Herculaneum,* J.L. Bellicard, London, 206-236.

Marinatos, S. 1939, The volcanic destruction of Minoan Crete, *Antiquity,* 13, 425-439.

Marinos, G. and Melidonis, N. 1971, On the strength of seaquakes (tsunamis) during the prehistoric eruptions of Santorin, *Acta 1st International Scientific Congress on the Volcano of Thera,* Greece, 15-23 Sept. 1969, Archaeological Services of Greece, Athens, 277-282.

Markesino, J. 1981, Mount St. Helens ash cleanup. *Public Works,* January 1981, 52-55.

Martin, C. 1911, Observations on the recent eruption of Taal volcano. *Philippine Journal of Science,* 6A(2), 87-90.

Martin, G.C. 1913, The recent eruption of Katmai volcano in Alaska. *National Geographic,* 24(2), 131-81.

Masó, M.S. 1904, Volcanoes and seismic centres of the Philippine Archipelago, *US Census Bureau Bulletin,* 3.

Masó, M.S., 1911, *The eruption of Taal volcano, January 30, 1911,* Department of the Interior, Weather Bureau, Manila.

Mathews, A. 1965, *Night of Purnama,* Jonathan Cape, London.

McIntosh, D.H. 1963, *Meteorological Glossary,* HMSO, London.

McKee, C.O., Wallace, D.A., Almond, R.A. and Talai, B. 1981, Fatal hydro-eruption of Karkar volcano in 1979: Development of a maar-like crater, in Johnson R.W. (ed.), Cooke-Ravian Volume of volcanological papers, *Geological Survey of Papua New Guinea Memoir 10,* 63-84.

McKee, C.O., Johnson, R.W., Lowenstein, P.L., Riley, S.J., Blong, R.J., de Saint Ours, P. and Talai, B. in press. Rabaul caldera Papua New Guinea: Volcanic hazards and eruption contingency planning. *Journal of Volcanology and Geothermal Research.*

McLucas, G.B. 1980, Cleanup and disposal of Mount St. Helens ash in eastern Washington. *Washington Geologic Newsletter,* 8(4), 1-7.

McReynolds, B., Preston, R., Males, J. and Gates, N. 1980, Influence of the volcanic eruption on animal production, *Washington Cattleman,* June 1980, 4-5.

Melson, W.G. and Saenz, R.R. 1968, The 1968 eruption of Volcan Arenal, Costa Rica: Preliminary summary of field and laboratory studies. *Smithsonian Institution Center for Short-lived Phenomena.*

Merriam, C.H. 1902, Bogoslof, our newest volcano, in *Alaska: History, Geography, Resources.* vol. 2. Harman Alaska Expedition, Doubleday, Page & Co., New York, 291-336.

Mészáros, S. 1978, Some words on the Minoan tsunami of Santorini, in Doumas, C. (ed.), *Thera and the Aegean World,* I, Thera and the Aegean World, London, 257-262.

Metzger, E. 1884, Gleanings from the reports concerning the eruption of Krakatoa. *Nature,* Jan 10, 240-244.

Miller, C.D., Mullineaux, D.R. and Hall, M.L. 1978, Reconnaissance map of potential volcanic hazards from Cotopaxi volcano, Ecuador, *U.S. Geological Survey Miscellaneous Investigation Series,* Map I-1072.

Miller, C.D., Mullineaux, D.R. and Crandell, D.R. 1981, Hazards assessments at Mount St. Helens, in Lipman, P.W. and Mullineaux, D.R. (eds), The 1980 eruptions of Mount St. Helens, Washington, *US Geological Survey Professional Paper,* 1250, 789-802.

Miller, C.F. 1966, Operation Ceniza-Arena: The retention of fallout particles from Volcan Irazu (Costa Rica) by plants and people Part 2, Stanford Research Institute, SRI Project No MU-4890.

Miller, C.F. and Lee, H. 1966, Operation Ceniza-Arena: The retention of fallout particles from Volcan Irazu (Costa Rica) by plants and people, Part I, Stanford Research Institute, SRI Project No MU-4890.

Miller, H.H. 1939, Mayon volcano in eruption. *Asia,* 39, 474-476.

Miller, H. 1942, Restless giant of the Philippines, *Natural History,* 49, 40-49.

Miller, T.P. 1976, Augustine volcano, in Alaska's volcanoes northern link in the ring of fire, *Alaska Geographic,* 4(1), 16-29.

Miller, T.P. and Smith, R.L. 1977, Spectacular mobility of ash flows around Aniakchak and Fisher calderas, Alaska, *Geology,* 5, 173-176.

Minakami, T. 1942, On the distribution of volcanic ejecta, I, The distribution of volcanic bombs ejected by the recent explosions of Asama, *Bulletin of the Earthquake Research Institute,* 20(1), 65-92.

Minakami, T., Ishikawa, T. and Yagi, K. 1952, The 1944 eruption of Usu, in Hokkaido, Japan. *Smithsonian Institution Annual Report,* 1952, 261-272.

Minakami, T., Utibori, S. and Hiraga, S. 1969 The 1968 eruption of Volcano Arenal, Costa Rica. *Bulletin of the Earthquake Research Institute,* 47, 783-802.

Minakami, T. 1972, Introduction to physical methods in UNESCO *The surveillance and prediction of volcanic activity,* UNESCO, Paris, 17-18.

Minakami, T. 1974, Prediction of volcanic eruptions, in Civetta, L., Gasparini, P., Luongo, G., Rapolla, A. (eds) *Physical volcanology,* Developments in Solid Earth Geophysics, 6, Elsevier, Amsterdam, 313-333.

Mishima, S., Shimazaki, K., Ogi, H. and Nishime, F. 1978, On the influence of volcanic ejecta from the Usu volcano upon the marine fish community, *Usu eruption and its impact on environment,* Hokkaido University, 243-249 (trans. US Geological Survey).

Moen, W.S. and McLucas, G.B. 1981, *Mount St. Helens ash — properties and possible uses.* State of Washington, Department of Natural Resources, Division of Geology and Earth Resources, Report of Investigations, 24.

Moore, J.G. 1967, Base surge in recent volcanic eruptions. *Bulletin Volcanologique,* 30, 337-363.

Moore, J.G. and Koyanagi, R.Y. 1969, The October 1963 eruption of Kilauea Volcano, Hawaii. *US Geological Survey Professional Paper,* 614-C.

Moore, J. . G., Nakamura, K. and Alcaraz, A. 1966, The 1965 eruption of Taal volcano. *Science,* 151, 955-960.

Moore, J.G. and Melson, W.G. 1969, Nuées ardentes of the 1968 eruption of Mayon Volcano, Philippines. *Bulletin Volcanologique,* 33(2), 600-620.

Moore, J.G. and Sisson, T.W. 1981, Deposits and effects of the May 18 pyroclastic surge, in Lipman, P.W. and Mullineaux, D.R. (eds), The 1980 eruptions of Mount St. Helens, Washington. *US Geological Survey Professional Paper,* 1250, 421-438.

Mosebar, J.C. (ed.), 1980, Economic impact of Mt. St. Helens, Washington State Department of Commerce, Olympia.

Mullins, J. 1977, *The goddess Pele,* Aloha Graphics and Sales, Honolulu.

Munro, S. 1980, Electric power, *Forum on the Mount St. Helens eruption on water resources,* Pacific Northwest River Basin Commission, Vancouver, Washington, June 12, 1980, 35-37.

960, On the mud-flows of the 1926 eruption of volcano Tokachi-dakè, Central Hokkaido, Japan. *Bulletin of the Earthquake Research Institute,* 38, 55-70.

Murai, I. and Hosoya, Y. 1964, The eruptive activity of Mt. Asama from 1958 to 1961 and the associated minor pyroclastic flows. *Bulletin of the Earthquake Research Institute,* 42, 203-236.

Murase, T. and McBirney, A.R. 1973, Properties of some common igneous rocks and their melts at high temperatures. *Geological Society of America Bulletin,* 84, 3563-3592.

Murata, K.J. 1966, The 1959-60 eruption of Kilauea volcano, Hawaii; an acid fumarolic gas from Kilauea Iki *US. Geological Survey Professional Paper* 537-C, 1-6.

Murata, K.J., Dondoli, C. and Saenz, R. 1966, The 1963-65 eruptions of Irazú volcano, Costa Rica (The period of March 1963 to October 1964). *Bulletin Volcanologique,* 29, 765-793.

Murton, B.J. and Shimabukuro, S. 1974, Human adjustment to volcanic hazard in Puna District, Hawaii, in White G.F. (ed.), *Natural hazards — local, national, global,* Oxford University Press, New York, 151-159.

Nairn, I.A. 1975a, Immediate report, field investigations of 1975 April 24 and 27 Ruapehu eruptions. *New Zealand Geological Survey Preliminary Report,* (with 2 appendices by B.J. Scott and others).

Nairn I.A. 1975b, Summary of Ngauruahoe eruption observations 1975 February 19-20. *New Zealand Geological Survey Report* (unpublished).

Nairn, I.A., Hewson, C.A.Y., Latter, J.H. and Wood, C.P. 1976, Pyroclastic eruptions of Ngauruhoe volcano, central North Island, New Zealand, 1974 January and March, in Johnson, R.W. (ed.) *Volcanism in Australia, a collection of papers in honour of the late G.A.M. Taylor,* Elsevier, Amsterdam, 385-405.

Nairn, I.A. and Self, S. 1978, Explosive eruptions and pyroclastic avalanches from Ngauruhoe in February 1975. *Journal of Volcanological and Geothermal Research,* 3, 39-60.

Nagayama, M. and Tamura, H. 1978. Effect of the volcanic ash fall from Mt Usu on the quality of water in Lake Toya, *Usu eruption and its impact on environment,* Hokkaido University, 289-292 (trans. by US Geological Survey).

Nakano, T. 1974, Natural hazards: Report from Japan, in White, G.F. (ed.) *Natural hazards — local, national global,* Oxford University Press, London, 231-243.

National Academy of Sciences, Committee on International Disaster Assistance. 1979 *Assessing international disaster needs,* National Research Council, Washington.

National Academy of Sciences, Committee on Disasters and the Mass Media. 1980, *Disasters and the Mass Media,* Proceedings of the Committee on Disasters and the Mass Media Workshop, February, 1979, National Research Council, Washington D.C.

National Academy of Sciences, Committee on Earthquake Engineering Research. 1982, *Earthquake Engineering Research 1982, overview and recommendations,* National Research Council, USA, National Academy Press, Washington.

Neall, V.E. 1976, Lahars as major geological hazards. *Bulletin of the International Association of Engineering Geologists,* 14, 233-240.

Neumann van Padang, M. 1951, *Catalogue of the active volcanoes of the world, including solfatara fields,* I, *Indonesia.*

Neumann van Padang, M. 1960, Measures taken by the authorities of the Volcanic Survey to safeguard the population from the consequences of volcanic outbursts. *Bulletin Volcanologique,* 23, 181-192.

Neumann van Padang, M. 1971, Two catastrophic eruptions in Indonesia, comparable with the Plinian outburst of the volcano of Thera (Santorini) in Minoan time. *Acta 1st International Scientific Congress on the Volcano of Thera,* Greece, 15-23 September, 1969, Archaeological Services of Greece, Athens, 51-63.

Newhall, C.G. 1982, A method for estimating intermediate- and long-term risks from volcanic activity, with an example from Mount St. Helens, Washington. *US Geological Survey Open-file Report* 82-396.

Newhall, C.G. and Self, S. 1982, The volcanic explosivity index (VEI): An estimate of explosive magnitude for historical volcanism. *Journal of Geophysical Research,* 87(C2), 1231-1238.

Nielsen, N. 1937, A volcano under an ice-cap, Vatnajokull, Iceland, 1934-36. *Geographical Journal,* 90(1), 6-23.

Nielson, L. 1980, Water quality and municipal water supply. *Forum on the effects of the Mount St. Helens eruption on water resources,* Pacific Northwest River Basins Commission, Vancouver, June 12, 1980, 25-26.

NIMH, 1978, National Institute of Mental Health, *Training manual for human service workers in major disasters,* Faberow, N.L. and Frederick, C.J. (project directors) US Dept of Health, Education and Welfare, Maryland, (DHEW Pub. No. (ADM) 77-538).

Nolan, M.L. 1979, Impact of Parícutin on five communities, in Sheets, P.D. and Grayson, D.K. (eds), *Volcanic activity and human ecology,* Academic Press, New York, 293-338.

Novak, T. and Zais, R. 1981, How to manage your ash, *Public Management,* Jan-Feb 1981, 12-15.

Official Handbook of Territory of New Guinea, 1'37, Commonwealth Government Printer, Canberra. (Supplement to 1943 reprint).

O'Flaherty, C.A. 1974, *Highways, vol. 2 — Highway engineering,* Arnold, London.

Okada, S. and Suzuki, K. 1978, The fluoride concentration in the local drinking water near the Mt Usu after its eruption. *Usu eruption and its impact on environment,* Hokkaido University, 293-302 (trans. US Geological Survey).

Oliver-Smith, A. 1979, The Yungay avalanche of 1970: Anthropological perspectives on disaster and social change. *Disasters,* 3(1), 95-101.

Omori, F. 1894, The eruption of Azuma-san. *Seismological Journal of Japan,* 19, 1-23.

Omori, F. 1914, The Sakura-Jima eruptions and earthquakes, I. (General Account). *Bulletin of the Imperial Earthquake Investigation Committee,* 8(1), 1-34.

Omori, F. 1916a, The Sakura-Jima eruptions and earthquakes, II. *Bulletin of the Imperial Earthquake Investigation Committee,* 8(2), 35-179.

Omori F, 1916b, The Sakura-Jima eruptions and earthquakes, III. *Bulletin of the Imperial Earthquake Investigation Committee,* 8(3), 181-321.

Omori, F. 1920, The Sakura-Jima eruptions and earthquakes, V. *Bulletin of the Imperial Earthquake Investigation Committee,* 8(5), 353-466.

Omori F. 1922, The Sakura-Jima eruptions and earthquakes, VI. *Bulletin of the Imperial Earthquake Investigation Committee,* 8(6), 467-525.

Óskarsson, N. 1980, The interaction between volcanic gases and tephra: Fluorine adhering to tephra of the 1970 Hekla eruption. *Journal of Volcanology and Geothermal Research,* 8, 251-266.

Oswalt, W.H. 1957, Volcanic activity and Alaskan spruce growth in A.D. 1783, *Science,* 126, 928-929.

PAHO, 1981, Emergency vector control after natural disaster. *Pan American Health Organisation Scientific Publication,* No. 419.

Palfreyman, W.D. and Cooke, R.J.S. 1976, Eruptive history of Manam volcano, Papua New Guinea, in Johnson, R.W. (ed.) *Volcanism in Australasia,* Elsevier, Amsterdam, 117-131.

Palmieri, L. 1873, The eruption of Vesuvius in 1872, Asher & Co., London (trans. with introduction by R. Mallet.)

Parad, H.J., Resnik, H.L.P. and Parad, L.G. (eds) 1976, *Emergency and disaster management: A mental health sourcebook,* Charles Press, Bowie, MD.

Parkin, J. 1841-53, *On the remote cause of epidemic diseases; or the influence of volcanic action in the production of general pestilences,* 2 parts, London.

Parkin, J. 1877, *The volcanic origin of epidemics,* Sampson Lowe & Co., London.

Parsons, W.H. and Mulford, J.W., 1958, Capelhinos Volcano, Fayal Island, Azores. *Cranbook Institute of Sciences News Letter,* 28(2), 10-21.

Pennebaker, J.W. and Newtson, D. 1983, Observation of a unique event: Psychological impact of Mt. St. Helen's volcano in Reis H. (ed.), *New directions for methodology of behavioral science,* Josey-Bass, San Francisco.

Perret, F.A. 1906, The stages of Vesuvius's eruption. *World's Work,* 12(6), 8091-8098.

Perret, F.A. 1912, Report on the recent eruption of the volcano 'Stromboli'. *Smithsonian Annual Report* 285-289.

Perret, F.A. 1912, The flashing arcs: A volcanic phenomenon. *American Journal of Science,* 4, 329-335.

Perret, F.A. 1916, The lava eruption of Stromboli, summer-autumn, 1915. *American Journal of Science,* 42, 443-463.

Perret, F.A. 1924, The Vesuvius eruption of 1906; study of a volcanic cycle. *Carnegie Institute of Washington,* Publication 339.

Perret, F.A. 1937, Eruption of Mt. Pelée 1929-32. *Carnegie Institute of Washington,* Publication 458.

Perret, F.A. 1950, Volcanological observations. *Carnegie Institute of Washington,* Publication 549.

Perry, R.W. 1979, Incentives for evacuation in natural disaster, *American Planning Association Journal,* October, 449-447.

Perry, R.W., Green, M.R. and Lindell, M.K. 1980, *Human response to volcanic eruption: Mt. St. Helens, May 18, 1980,* Batelle Human Affairs Research Center, Seattle.

Petroeschevsky, W.A. 1949, A contribution to the knowledge of the Gunung Tambora (Sumbawa). *Tijdschrift van het Koninklijk Nederlandsch Aardrijkskundig Genootschap,* ser 2, 66, 688-702.

Pichler, H. W. 1970, Volcanism in eastern Sicily and the Aeolian Islands in Alvarez, W. and Gohrbandt, K.H.A. (eds) *Geology and History of Sicily,* Petroleum Exploration Society of Libya, 261-281 (trans. K.H.A. Gohrbandt).

Pond, J.A. and Smith, S.P. 1886, Observations on the eruption of Mount Tarawera, Bay of Plenty, New Zealand, 10th June, 1886. *New Zealand Institute, Transactions and Proceedings,* 19, 342-371.

Porter, S.C. 1981, Recent glacier variations and volcanic eruptions. *Nature,* 291, 139-142.

Powers, S. 1916, The eruption of Yake-dake, Japan, in 1915, *The Geographical Review,* 1, 359-362.

Prasad, A.S. 1978, *Trace elements and iron in human metabolism,* Plenum Medical Book Company, New York.

Pratt, W.E. 1911, The eruption of Taal volcano, January 30, 1911. *Philippine Journal of Science,* 6A(2), 63-83.

Preusser, H. 1973, Der Vulkanausbruch auf Heimaey/Vestmannaeyjar und seine Auswirkungen. *Geographische Rundschau,* 25(9), 337-350.

Priest, M.J. and Blong, R.J. in prep., The effects of volcanic eruptions on radio communications. *Journal of Electrical and Electronics Engineering, Australia.*

Prinz, M. 1974, Fiery Vesuvius. *Natural History,* 88(4), 41-48.

Project Ashlift, 1981, A Mount Saint Helens Disaster Assistance and Crisis Counseling Program, Final Report, Nelson A.S., Hearon H., Arthur T., Barnett B, Heid K and Schwindt F. (eds) Mental Health Division, Washington State Department of Social and Health Services.

Purey-Cust, H.E., 1895, *Report on the eruption of Ambrym, New Hebrides, October and November 1894.* HMSO London.

Quarantelli, E.L. and Dynes, R.R. 1972, When disaster strikes it isn't much like what you've heard (and read about). *Psychology today,* 5(9), 67-70.

Quarantelli E.L. and Dynes, R.R. 1977, Response to social crisis and disaster. *Annual Review of Sociology,* 3, 23-49.

Raffles, T.S. 1830, *The history of Java,* 2 vols., John Murray, London.

Raible, J.A.B. 1980, Human needs and Mt. St. Helens. *Report to Washington State Legislature* (R -3178) July 14, 1980.

Ramsay, C.S. and Plumb, C.P. c.1938, *Tin Can Island, A story of Tonga and the swimming mail man of the South Seas,* Hurst and Blackett, London.

Rees, J.D. 1970, Parícutin revisited: A review of man's attempts to adapt to ecological changes resulting from volcanic catastrophe. *Geoforum,* 4, 7-25

Rees, J.D. 1979, Effects of the eruption of Parícutin volcano on landforms, vegetation, and human occupancy, in Sheets P.D. and Grayson D.K. (eds), *Volcanic activity and human ecology,* Academic Press, New York, 249-292.

Richards, A.F. 1958, Transpacific distribution of floating pumice from Ilsa San Benedicto, Mexico. *Deep Sea Research,* 5, 29-35.

Richards, A.F. 1965, Geology of the Isla San Benedicto 1952-61 following the birth of the Volcan Barcena. *Bulletin Volcanologique,* 28, 381-403.

Richardson, D. 1968, Glacier outburst floods in the Pacific Northwest. *US Geological Survey Professional Paper,* 600-D, 79-86.

Rittman, A. 1962, *Volcanoes and their activity,* Interscience Publications, New York (trans. E.A. Vincent).

Rivers, J.P.W. and Brown, G.A. 1979, Physiological aspects of shelter deprivation. *Disasters,* 3(1), 20-23.

Rivers, W.C. 1911, Taal volcano eruption of January 30, 1911. *American Red Cross Bulletin,* 6(4), 8-18.

Robock, A. 1981, The Mount St. Helens volcanic eruption of 18 May 1980: Minimal climatic effect. *Science,* 212, 1383-1384.

Rodwell, G.F. 1878, *Etna, A history of the mountain and of its eruptions,* Kegan Paul, London.

Rogers, G. 1981, The evacuation of Niuafo'ou, an outlier in the Kingdom of Tonga. *Journal of Pacific History,* 16(3), 149-163.

Rose, W.I. 1972, Notes on the 1902 eruption of Santa Maria volcano, Guatemala. *Bulletin Volcanologique,* 36, 29-45.

Rose, W.I. 1977, Scavenging of volcanic aerosol by ash: atmospheric and volcanologic implications, *Geology,* 5, 621-624.

Rosenblum, M. 1981, *Coups and earthquakes; reporting the world to America,* Harper Colophon, New York.

Ross, J.T. 1816, Narrative of the effects of the eruption from the Tomboro Mountain in the island of Sumbawa on the 11th and 12th of April 1815. *Lembaga Kebudajaan Indonesia Verhandelingen,* 8, 343-360.

Rowley, K. 1979, Soufrière: A volcano in the Caribbean environment, *Trinidad Naturalist,* 2(9), 19-27.

Russell, I.C., 1902, Volcanic eruptions on Martinique and St Vincent. *National Geographic,* 13(12), 415-436.

SICSLP, Smithsonian Institution Center for Short-Lived phenomena, Annual Reports.

Saito, K., Taniguchi, N., Kumashiro, M., Sasaki, T., Sato, Y., Takakuwa, E., Murao, M., Osaki, J. and Abe, S. 1978, Effect of the 1977 eruption of Usu volcano on human living environment and health. *Usu Eruption and its Impact on Environment,* Hokkaido University, 169-206 (trans. US Geological Survey).

Saito, K. and Seki, K. 1983, Effect of the 1977 eruption of Mt. Usu volcano on human living environment and health matters, *1st International Symposium on Public Health in Asia and the Pacific Basin,* Honolulu, March 7-9.

Sapper, K. 1917, *Katalog der geschichtlichen Vulcanausbruche,* Tribner, Strasburg.

Sarna-Wojcicki, A.M., Shipley, S., Waitt, R.B., Dzurisin, D., and Wood, S.H. 1981, Areal distribution, thickness, mass, volume and grain size of air-fall ash from the six major eruptions of 1980, *US Geological Survey Professional Paper,* 1250, 577-600.

Sax, N.I. (ed.) 1963. *Dangerous properties of industrial materials,* 2nd ed, Reinhold, New York.

Sax, N.I. (ed.), 1974, *Industrial pollution,* Van Nostrand Reinhold, New York.

Scheidegger, A.E. 1973, On the prediction of the reach and velocity of catastrophic landslides. *Rock Mechanics,* 5, 231-236.

Schuster, R.L. 1981, Effects of the eruptions on civil works and operations in the Pacific Northwest. *US Geological Survey Professional Paper,* 1250, 701-718.

Schuster, R.L. 1983, Engineering aspects of the 1980 Mount St. Helens eruptions. *Bulletin of the Association of Engineering Geologists,* 20(1), 125-143.

Schwimmer, E.G. 1969, Cultural consequences of a volcanic eruption experienced by the Mt. Lamington Orokaiva. *University of Oregon, Department Anthropology Report,* 9.

Schwimmer, E.G. 1977, What did the eruption mean? Leiber, M.D. (ed.) *Exiles and Migrants in Oceania,* University of Hawaii Press, Hononulu, 296-341.

Searle, E.J. 1964, Volcanic risk in the Auckland Metropolitan District, *New Zealand Journal of Geology and Geophysics,* 7(1), 94-100.

Searle, .J. and Mayhill, R.D. 1981, *City of volcanoes, a geology of Auckland,* Longman Paul, Auckland.

Seed, H.B. 1975, Earthquake effects on soil-foundation systems, in Winterkorn, H.F. and Fang, H-Y. (eds) *Foundation Engineering Handbook,* Van Nostrand Reinhold, New York, 700-732.

Segerstrom, K. 1950, Erosion studies at Parícutin, State of Michoacan, Mexico. *US Geological Survey Bulletin,* 965-A.

Segerstrom, K. 1960, Erosion and related phenomena at Parícutin in 1957, *US Geological Survey Bulletin,* 1104, 1-18.

Seki, K. Yamamura, E. and Kagaya, S. 1978, A system analysis on the structure of the social damages. *Usu eruption and its impact on environment,* Hokkaido University, 429-436 (trans. US Geological Survey).

Sekiya, S. and Kikuchi, Y. 1889, The eruption of Bandai-san. *Journal of the College of Science, Imperial University, Tokyo,* 3(2), 91-172.

Self, S., Sparks, R.S.J., Booth, B. and Walker, G.P.L. 1974, The 1973 Heimaey Strombolian scoria deposit, Iceland. *Geological Magazine,* 111(6), 539-548.

Self, S. and Rampino, M.R. 1981, The 1883 eruption of Krakatau. *Nature,* 294, 699-704.

Self, S. and Rampino, M.R. 1982, Comments on "A geophysical interpretation of the 1883 Krakatau eruption" by I. Yokoyama. *Journal of Volcanology and Geothermal Research,* 13, 379-386.

Self, S., Rampino, M.R. and Barbera, J.J. 1981, The possible effects of large 19th and 20th century volcanic eruptions on zonal and hemispheric surface temperatures. *Journal of Volcanology and Geothermal Research,* 11, 41-60.

Sheets, P.D. and Grayson, D.K. (eds) 1979, *Volcanic activity and human ecology,* Academic Press, New York.

Shepherd, J.B., Aspinall, W.P., Rowley, K.C., Pereira, J., Sigurdsson, H., Fiske, R.S. and Tomblin, J.F. 1979, The eruption of Soufrière volcano, St Vincent, April – June 1979. *Nature,* 282, 24-28.

Sheridan, M.F., Barberi, F., Rosi, M. and Santacroce, R. 1981, A model for Plinian eruptions of Vesuvio. *Nature,* 289, 282-285.

Sheridan, M.F. and Wohletz, K.H. 1981, Hydrovolcanic explosions: The systematics of water-pyroclast equilibration. *Science,* 212, 1387-1389.

Shibata, T., Joh, O., Arai, Y., Magara, S., Ohno, K., Okano, M. and Nakata, M. 1978, Damages to building structures in Toyako and Sobetsu Spa-Districts caused by the eruption of Usu Volcano, 1977. *Usu eruption and its impact on environment,* Hokkaido University, 407-427 (trans. US Geological Survey).

Shimozuru, D. 1972, A seismological approach to the prediction of volcanic eruptions, in *The surveillance and prediction of volcanic activity,* UNESCO, Paris, 19-45.

Siebert, L. and Kienle, J. 1983, Volcanogenic tsunamis produced by large volcanic debris avalanches. *International Union of Geodesy and Geophysics, 18th General Assembly, Hamburg, 15-27 August 1983, Hamburg, Programme and Abstracts, Vol. 1,* Hamburg, 314.

Sigit, S. 1965, The rehabilitation of the Kelut tunnel works. *Bulletin of the Geological Survey of Indonesia,* 2(1), July, 1965, 3-11.

Sigurdsson, H., 1982, Volcanic pollution and climate: The 1783 Laki eruption. *Eos,* 10 August, 601-602.

Sigurdsson, H., Cashdollar, S. and Sparks, R.S.J. 1982, The eruption of Vesuvius in AD 79: Reconstruction from historical and volcanological evidence. *American Journal of Archaeology,* 86, 39-51.

Sigurdsson, B. and Pálsson, P.A., 1957, Fluorosis of farm animals during the Hekla eruption of 1947-1948, *Societas Scientiarum Islandica,* 3(3), 1-12.

Sigvaldasson, G.E. 1978, Reply to editorial. *Journal of Volcanology and Geothermal Research,* 4, I-III.

Simkin, T. and Fiske, R.S. 1977, Volcanology. *Geotimes,* January, 42-43.

Simkin, T. and Fiske, R.S. 1983, *Krakatau: The volcanic eruption and its effects,* Smithsonian Institution Press, Washington, D.C.

Simkin, T. and Howard, K.A. 1970, Caldera collapse in the Gálapagos Islands, 1968, *Science,* 169, 429-437.

Simkin, T., Siebert, L., McClelland, L., Bridge, D., Newhall, C., and Latter, J.H. 1981, *Volcanoes of the world,* Hutchinson Ross, Stroudsburg, PA.

Simkin, T. and Siebert, L., 1981, Arc volcanism in space and time: A look at the global historical record, *International Association for Volcanology and Chemistry of the Earth's Interior, Conference Abstracts,* Tokyo, 342.

Simkin, T. and Siebert, L. 1982, Large historic explosive eruptions: How much warning? *Eos,* 63(18), 459.

Simkin, T. and Siebert, L. 1983, Explosive eruptions in space and time: Durations, intervals and a comparison of the world's active volcanic belts, in *Explosive volcanism: Inception, evolution and hazards,* National Academy Press, Washington DC, 110-121.

Smart, G.M. 1981, Volcanic debris control, Gunung Kelud, East Java, *International Association of Hydrological Sciences Publication,* 132, 604-623.

Smathers, G.A. and Mueller-Dombois, D. 1974, *Invasion and recovery of vegetation after a volcanic eruption in Hawaii,* National Park Service Scientific Monograph Series No. 5.

Snailham, R., 1978, *Sangay survived, the story of the Ecuador volcano disaster,* Hutchinson, London.

Sorensen, J.H. and Gersmehl, P.J. 1980, Volcanic hazard warning system: Persistence and transferability. *Environmental Management,* 4(2), 125-136.

Sorensen, J.H. 1981, Emergency response to Mount St. Helens' eruption: March 20 to April 10, 1980, University of Colorado (Boulder), Institute of Behavioral Science, *National Hazard Research Working Paper,* 43.

Sparks, R.S.J. 1976, Grain size variations in ignimbrites and implications for the transport of pyroclastic flows, *Sedimentology,* 23, 147-188.

Sparks, R.S.J., Pinkerton, H., and Hulme, G. 1976, Classification and formation of lava levees on Mount Etna, Sicily, *Geology,* 4, 269-271.

Sparks, R.S.J. and Wilson, L. 1976, A model for the formation of ignimbrite by gravitational column collapse. *Journal of the Geological Society of London,* 132, 441-451.

Sparks, R.S.J., Wilson, L. and Hulme, G. 1978, Theoretical modeling of the generation, movement, and emplacement of pyroclastic flows by column collapse. *Journal of Geophysical Research,* 83(B4), 1727-1739.

Sperry, F.L. 1903, The eruption of Colima. *American Journal of Science,* 15, 487-488.

Spinden, H.J. 1919, Shattered capitals of Central America. *National Geographic,* 35(3), 185-212.

Starbird, T.A. 1979, Taking it as it comes. *National Geographic,* 156(3), 398-425.

Stearns, H.T. 1925, The explosive phase of Kilauea Volcano, Hawaii, in 1924. *Bulletin Volcanologique,* 5 & 6 (3 & 4), 1-16.

Stefásson, K. and Sigurjónsson, J. 1957, The eruption of Hekla 1947-1948 III,2. Temporary increase in flourine content of water following the eruption. *Societes Scientiarum Islandica,* Reykjavik, 1-13.

Stehn, Ch. E. 1929, The geology and volcanism of the Krakatau Group. *4th Pacific Science Congress,* Pacific Science Association, Batavia-Bandoeng, 3-55.

Steinhauser, R. 1893, The tidal wave and the relief expedition from Finschhafen to the south-west coast of New Britain. *Westermann's Illustrierte Deutsche Manutshepte,* 71 (in translation).

Stephensen, M. 1785, Account of the volcanic eruption in Skaptefield's Syssel, in Hooker, W.J. 1813. *Journal of a tour in Iceland in the summer of 1809,* 2nd edition, Longman, Hurst, Rees, Orme, and Brown, and John Murray, London, 124-259.

Steward, J.E. 1982, Spirit Lake flood hazard task-force: situation assessment and recommendation, US Forest Service report.

Stoiber, R.E. and Rose, W.I. 1969, Recent volcanic and fumarolic activity at Santiaguito volcano, Guatemala, *Bulletin Volcanologique,* 33(2), 475-502.

Stommel, H. and Stommel, E. 1979, The year without a summer, *Scientific American,* June, 134-140.

Stothers, R.B. and Rampino, M.R. 1983, Volcanic eruptions in the Mediterranean before A.D. 630 from written and archaeological sources, *Journal of Geophysical Research,* 88(B8), 6357-6371.

Surjo, I. 1965, Casualties of the latest activity of the Agung volcano. *Bulletin of the Geological Survey of Indonesia,* 2(1), 22-26.

Sutherland, F.L. 1965, Dispersal of pumice, supposedly from the 1962 South Sandwich islands eruption, on southern Australian shores. *Nature,* 207, 1332-1335.

Suzuki, A. 1978, Effect of eruption on Mt. Usu on coastal fishery industry. *Usu Eruption and its Impact on Environment,* Hokkaido University, 251-256 (trans. US Geological Survey).

Swanson, D.A. and Christiansen, R.L., 1973, Tragic base surge in 1790 at Kilauea Volcano. *Geology,* 83-86.

Sylvester, A.G., Siggerud, T., Navrestad, T., Mathisen, J. and Hagen, O. 1973, *Surveillance of volcanic activity in Iceland as illustrated by Heimaey eruption.* University of Bergen Seismological Observatory.

Symons, G. (ed.), 1888, The eruption of Krakatoa and subsequent phenomena. *Royal Society of London, Krakatoa Committee Report.*

Tachibana, H. and Nasu, Y. 1978, Studies on the water quality of Lake Toya. *Usu eruption and its impact on environment,* Hokkaido University, 275-288, (trans. US Geological Survey).

Tanaka, J.C. 1983, *Volcanic hazards assessment for Mt. Rainier, Washington,* Washington State Department of Emergency Services, Olympia.

Tanakadate, H. 1926, The eruptive explosion of Tokati-dake, Hokkaido, in the year 1926. *Proceedings 3rd Pan-pacific Science Congress,* Tokyo, 784-786.

Tanakadate, H. 1927, Explosive eruption of Tokati-dake, Hokkaido, Japan. *Bulletin Volcanologique,* 13 + 14 (3 + 4),3-14.

Tanakadate, H. 1931, Volcanic activity in Japan and vicinity during the period between 1924 and 1931. *Japan Journal of Astronomy and Geophysics,* 9(1), 47-64.

Taniguchi, K., Tamaki, T. and Kudo, M. 1978, Temperature dependency of Na +, K +, –ATPase activity of mantle cord microsomes obtained from scallops grown at Mori and Toyura. *Usu eruption and its impact on the environment,* Hokkaido University, 239-241 (trans. US Geological Survey).

Taylor, G.A.M. 1958, The 1951 eruption of Mount Lamington, Papua. *Bureau of Mineral Resources of Australia,* Bulletin 38.

Taylor, G.A.M., Best, J.G. and Reynolds, M.A., 1957, Eruptive activity and associated phenomena, Langila volcano, New Britain. *Bureau of Mineral Resources of Australia,* Report 26.

Tazieff, H. 1952, *Craters of fire,* Hamish Hamilton, London (trans. E. Wilkins).

Tazieff, H. 1965, Confronting the irascible Irazú. *UNESCO Courier,* 18, Nov. 1965, 18-23, 30.

Tazieff, H. 1977, An exceptional eruption: Mt. Nyiragongo, Jan 10th, 1977. *Bulletin Volcanologique,* 49(3), 189-200.

Tazieff, H. 1979a, What is to be forecast: Outbreak of eruption or possible paroxysm? The example of the Guadeloupe Soufrière. *Journal of the Geological Society of London,* 136, 327-329.

Tazieff, H. 1979b, *Nyiragongo: The forbidden volcano,* Barron's, Woodburg N.Y. (trans. J.F. Bernard).

Tenison-Woods, J.E. 1884, The earthquake in the Straits of Sunda. *Sydney Morning Herald,* 16-18 January.

Thorarinsson, S. 1954, The eruption of Hekla, 1947-1948, II. 3, The tephra fall from Hekla on March 29th 1947. *Societas Scientiarum Islandica,* Reykjavik.

Thorarinsson, S. 1969, The Lakagígar eruption of 1783, *Bulletin Volcanologique,* 33(3), 910-929.

Thorarinsson, S. 1970, *Hekla, a notorious volcano,* Almenna Bókafélagid, Reykjavik (trans. J. Hannesson).

Thorarinsson, S. 1971, Damage caused by tephra fall in some big Icelandic eruptions and its relation to the thickness of the tephra layers. *Acta 1st International Scientific Congress on the Volcano of Thera,* Greece, 15-23 September 1969. Archaeological Services of Greece, Athens, 213-236.

Thorarinsson, S. 1974, The terms *tephra* and *tephrochronology,* in *Westgate, J.A. and Gold, C.M. (eds) World Bibliography and Index of Quaternary Tephrochronology,* INQUA and UNESCO, xvii-xviii.

Thorarinsson, S., 1979, On the damage caused by volcanic eruptions with special reference to tephra and gases, in Sheets, P.D. and Grayson, D.K. (eds) *Volcanic activity and human ecology,* Academic Press, New York, 125-159.

Thorarinsson, S. 1957, The jokulhlaup from the Katla area in 1955 compared with other jokulhlaups in Iceland. *Rejkjavik Museum of Natural History,* Miscellaneous Paper 16.
Thorarinsson, S. 1958, The Öraefajokull eruption of 1362. *Acta Naturalia Islandica,* 2(2), 1-99.
Thorarinsson, S. and Sigvaldasson, G.E. 1972, The Hekla eruption of 1970. *Bulletin Volcanologique,* 36, 269-288.
Thorarinsson, S., Steinthórsson, S., Einarsson, Th., Kristmannsdóttir, H. and Óskarsson, N. 1973, The eruption on Heimaey, Iceland. *Nature,* 241, 372-375.
Tilling, R.I., Koyanagi, R.Y., Lipman, P.W., Lockwood, J.P., Moore, J.G. and Swanson, D.A. 1976, Earthquake and related catastrophic events, Island of Hawaii, November 29, 1975: A preliminary report. *US Geological Survey Circular* 740.
Tomblin, J. 1979, Deontological code, probabilistic hazard assessment or Russian roulette? *Journal of Volcanology and Geothermal Research,* 5, 213-215.
Tolentino, B.S. 1973, *Philippine volcanoes and solfataric areas,* 2nd rev. ed., Comvol, Quezon City.
Tonkinson, R. 1979, The paradox of permanency in a resettled New Hebridean community. *Mass Emergencies,* 4, 105-116.
Trask, P.D. 1943, The Mexican volcano Paricutin. *Science,* 98 (2554), 501-505.
Trechmann, C.T. 1938, Relics of the Mt. Pelée eruption of May 8, 1902. *Nature,* 141, 435-437.
Trowbridge, T. 1976, Aniakchak Crater, in Alaska's volcanoes northern link in the ring of fire. *Alaska Geographic,* 4(1), 70-73.
Tsuya, H. 1930, The volcano Komagatake, Hokkaido, its geology, activity and petrography. *Bulletin of the Earthquake Research Institute,* 8(2), 237-270.
Tsuya, H. 1955, Geological and petrological studies of Mt. Fuji, V, 5, On the 1707 eruption of volcano Fuji. *Bulletin of the Earthquake Research Institute,* 33(3), 341-382.
Tubessing, S.K. (ed.) 1979, *Natural hazards data resources: Uses and needs.* Institute of Behavioral Sciences, University of Colorado, Program on Technology, Environment and Man #27.
Tylor, J.N. 1981, Rabaul area; communications network survivability report and disaster plan, Papua New Guinea Post and Telecommunication Corporation, Port Moresby.
Uman, M.A. 1971, *Understanding lightning,* Bek Technical Publication, Philadelphia PA.
UNDRO, 1976, Office of the UN Disaster Relief Co-odinator. *Disaster prevention and mitigation, a compendium of current knowledge vol. 1, volcanological aspects,* United Nations, Geneva.
UNESCO, 1979, Final report of the meeting of the experts on volcanic hazard zoning, Bandung 26-28 November, 1979.
US Army Corps of Engineers, 1980, *Lava flow control, Island of Hawaii, Review report and environmental impact statement,* US Army Engineer District.
Van Bemmelen, R.W. 1949, Report on the volcanic activity and volcanological research in Indonesia during the period 1936-1948. *Bulletin Volcanologique,* 9, 3-30.
Van Bemmelen, R.W. 1956, The influence of geologic events on human history (an example from Central Java). *Verhandelingen Geologisch Mijnbowkundig Genootschap voor Nederland en Kolonien Geologisch* 16, Brouwer Vol, 20-36.
Van Bemmelen, R.W., 1971, Four volcanic outbursts that influenced human history: Toba, Sunda, Merapi, and Thera. *Acta 1st International Scientific Congress on the Volcano of Thera,* Greece, 15-23 September 1969, Archaeological Services of Greece, Athens, 5-50.
Varenius, B. 1683, *Cosmography and geography,* edited and translated by Richard Blome, S. Roycroft, London.
Verbeek, R.D.M., 1886, *Krakatau,* Imprimerie de l'Etat, Batavia.
Vitaliano D.B. 1973, *Legends of the earth; their geologic origins,* Indiana University Press, Bloomington.
Vitaliano, D.B. and Vitaliano, C.J. 1971, Plinian eruptions, earthquakes, and Santorin: A review. *Acta 1st International Scientific Congress on the Volcano of Thera,* Greece 15-23 September 1969, Archaeological Services of Greece, Athens, 88-108.
Voight, B., Glicken, H., Janda, R.J. and Douglass, P.M. 1981, Catastrophic rockslide avalanche of May 18, in Lipman, P.W. and Mullineaux, D.R. (eds), The 1980 eruptions of Mount St. Helens, Washington, *US Geological Survey Professional Paper* 1250, 347-377.
Volcano, 1980, *Volcano – the eruption of Mount St. Helens,* staff of *The Daily News,* Longview and *The Journal-American,* Bellevue (eds), Longview Publishing Coy, Longview, WA.
Vlodavetz, V.I. 1959, The passive and active defense from the catastrophic eruption of volcanoes. *Bulletin Volcanologique,* 28, 79-91 (in Russian).

Wadge, G. 1978, Effusion rate and the shape of aa lava flow-fields on Mount Etna, *Geology,* 6, 503-506.
Waitt, R.B. 1981, Devastating pyroclastic density flow and attendant air fall of May 18 – stratigraphy and sedimentology of deposits, in Lipman, P.W. and Mullineaux, D.R. (eds), The 1980 eruptions of Mount St. Helens, Washington, *US Geological Survey Professional Paper* 1250, 439-458.
Waldron, H.H. 1967, Debris flow and erosion control problems caused by the ash eruptions of Irazú volcano, Costa Rica. *US Geological Survey Bulletin,* 1241-I, 1-37.
Walker, G.P.L. 1967, Thickness and viscosity of Etnean lavas. *Nature,* 213, 484-485.
Walker, G.P.L. 1973, Lengths of lava flows. *Philosophical Transactions of the Royal Society of London,* A, 274, 107-118.
Walker, G.P.L. 1974, Volcanic hazards and the prediction of volcanic eruptions. *Geological Society of London Miscellaneous Publication,* 3, 23-41.
Walker, G.P.L. 1980, The Taupo pumice: Product of the most powerful known (ultraplinian) eruption? *Journal of Volcanology and Geothermal Research,* 8, 69-94.
Walker, G.P.L. 1981, New Zealand case histories of pyroclastic studies, in Self, S. and Sparks, R.S.J. (eds) *Tephra studies,* Proceedings NATO Advanced Studies Institute, Laugarvatn and Reykjavik, June 18-19, 1980, Reidel, Dordrecht, Series C, 75, 317-330.
Walker, G.P.L., Heming, R.F. and Wilson, C.J.N. 1980, Low-aspect ratio ignimbrites, *Nature,* 283, 286-287.
Warden, A.J., 1970, Evolution of Aoba Caldera volcano, New Hebrides. *Bulletin Volcanologique,* 34, 107-140.
Warrick, R.A., Anderson, J., Downing, T., Lyons, J., Ressler, J., Warrick, M. and Warrick, T. 1981, *Four communities under ash – after Mount St. Helens,* Program on Technology, Environment and Man, Monograph 34, Institute of Behavioral Sciences, University of Colorado.
Washington, H.S. 1926, Santorini eruption of 1925. *Geological Society of America Bulletin,* 37, 349-84.
Weir, G.J. 1982, Kinematic wave theory for Ruapehu lahars, *New Zealand Journal of Science,* 25, 197-203.
Wenger, D.E., Dykes, J.D., Sebok, T.D. and Neff, J.L. 1975, It's a matter of myths: An empirical examination of individual insight into disaster response, *Mass Emergencies,* 1, 33-46.
West, S. 1980, The right to research. *Science News* July 26, 1980, 61-62.
Wharton, W.J.L. 1888, On the seismic sea waves caused by the eruption of Krakatoa, August 26th and 27th, 1883 in Symons, G. (ed.) *The eruption of Krakatoa and subsequent phenomena,* 89-107.
White, G.F. and Haas, J.E. 1975, *Assessment of research on natural hazards,* MIT Press, Cambridge, Mass.
White, W.T., Day, T.G., Bertek, D. and Fisher, J. 1980, Mt. St. Helens volcanic ash vs. Yakima wastewater treatment facility. *Institute for water resources newsletter,* 6(2), 4-8.
Wilcox, R.E. 1959, Some effects of recent volcanic ash falls with especial reference to Alaska. *US Geological Survey Bulletin,* 1028-N, 409-476.
Wilcoxson, K. 1966, *Volcanoes,* London, Cassel.
Will, J. 1902, Report on the medical relief expeditions to Martinique and St. Vincent in aid of the sufferers from the volcanic eruptions, May, 1902, in HMSO (1903), 8-15.
Williams, H. 1952, The great eruption of Coseguina, Nicaragua, in 1835. *University of California Publications, Bulletin of the Department of Geological Sciences,* 29(2), 21-46.
Williams, H. and McBirney, A.R. 1979, *Volcanology,* Freeman, San Francisco.
Williams, R.S. and Moore, J.G. 1973, Iceland chills a lava flow. *Geotimes,* 18(8), 14-17.
Williams, R.S. and Moore, J.G. 1980, *Man against volcano: The eruption on Heimaey, Vestmann Islands, Iceland,* US Geological Survey.
Williams, S.N. and Self, S. 1983, The October 1902 Plinian eruption of Santa Maria volcano, Guatemala. *Journal of Volcanology and Geothermal Research,* 16, 33-56.
Wilson, C.J.N. and Walker, G.P.L. 1981, Violence in pyroclastic flow eruptions, in Self, S. and Sparks, R.S.J. (eds) *Tephra Studies,* Proceedings NATO Advanced Studies Institute, Laugarvatn and Reykjavik, June 18-29, Reidel, Dordrecht, Series C, 75, 441-448.
Wilson, L. 1972, Explosive volcanic eruptions – II. The atmospheric trajectories of pyroclasts. *Geophysical Journal of the Royal Astronomical Society,* 30, 381-392.
Wilson, L. 1976, Explosive volcanic eruptions – III. Plinian eruption columns. *Geophysical Journal of the Royal Astronomical Society,* 45, 543-556.
Wilson, L. 1980, Relationships between pressure, volatile content and ejecta velocity in three types of volcanic explosion, *Journal of Volcanology and Geothermal Research,* 8, 297-313.

Wilson, R. 1962, The Los Angeles conflagration of 1961: The Devil Wind and wood shingles. *National Fire Protection Association Quarterly,* 55, 279.

Winchilsea, Earl of. 1669, *A true and exact relation of the late prodigious earthquake and eruption of Mount Aetna, or, Monte-Gibello,* T. Newcomb, Savoy.

Wolf, T. 1878, Der Cotopaxi und seine letzte eruption am 26 Juni 1877, *Neues Jahrbuch für Mineralogie, Geologie und Paleontologie,* 113-167 (in translation).

Wolfenstein, M. 1957, *Disaster: A psychological essay,* Routledge & Kegan Paul, London.

Wright, J.V., Smith, A.L. and Self, S. 1980, A working terminology of pyroclastic deposits, *Journal of Volcanology and Geothermal Research,* 8, 315-336.

Wright, J.V., Smith, A.L. and Self, S. 1981, A terminology for pyroclastic deposits, in Self, S. and Sparks, R.S.J. (eds) *Tephra studies,* Proceedings NATO Advanced Studies Institute, Laugarvatn and Reykjavik, June 18-29, Reidel, Dordrecht, Series C, 75, 457-463.

Yamada, M. 1978, On the influence of volcanic ejecta from Mt Usu on marine invertebrates, *Usu Eruption and its Impact on Environment,* Hokkaido University, December 1978, 227-232 (trans. US Geological Survey).

Yamada, T., Akitaya, E., Endo, Y., and Kuroiwa, D. 1978, On the stability of snow on the slope of Mt Usu covered with newly deposited volcanic products, *Usu Eruption and its Impact on Environment,* Hokkaido University, December 1978, 147-152 (trans. US Geological Survey).

Yamaoka, I., Fujita, M., Hasegawa, K. and Hoshi, K. 1978, A survey of volcanic ash-pumice flows in the Nishiyamagawa watershed (up to July, 1978), *Usu Eruption and its Impact on Environment,* Hokkaido University, December 1978, 101-120 (trans. US Geological Survey).

Yanev, P. 1974, *Peace of mind in earthquake country – how to save your home and life,* Chronicle, San Francisco.

Yokoyama, I. 1978, The tsunami caused by the prehistoric eruption of Thera, in Doumas C. (ed) *Thera and the Aegean World,* I, Thera and the Aegean World, London, 277-283.

Yokoyama, S. 1974, Mode of movement and emplacement of Ito pyroclastic flow from Aira caldera, Japan, *Science Reports, Tokyo Kyoiku Daigaku,* C, 12(114), 17-62.

Youd. T.L. and Wilson, R.C. 1980, Stability of Toutle River blockage: Mt. St. Helens hazards investigations, *US Geological Survey Open File Report,* 80-898.

Yoshioka, K. 1966, Development and recovery of vegetation since the 1929 eruption of Mt. Komagatake, Hokkaido. *Ecological Review* 16(4), 271-292

Zen, M.T. and Hadikusumo, D., 1965, The future danger of Mt. Kelut (Eastern Java-Indonesia). *Bulletin Volcanologique,* 28, 275-282.

Zollinger, H. 1885, *Besteigung des Vulkanes Tambora auf der Insel Sumbawa und Shilderung der Eruption desselben im Jahr 1815,* Winterthur, Wurster.

INDEX

In this index, volcanoes are identified with their location given in parentheses. Volcanoes are named following the system established in the Smithsonian Institution Volcano Reference File. Towns are identified as such where necessary.

B

C

F

G

T

Y

4 5 6 7 8 9 0 1 2 3
A B C D E F G H I J